FOOD from DRYLAND GARDENS

*An Ecological,
Nutritional, and
Social Approach to
Small-Scale Household
Food Production*

◆

by
David A. Cleveland and Daniela Soleri

illustrated by
Daniela Soleri

CENTER FOR PEOPLE, FOOD AND ENVIRONMENT
Tucson, Arizona USA
CPFE

with support from the

United Nations Children's Fund
New York, New York USA

FOOD from DRYLAND GARDENS
An Ecological, Nutritional, and Social Approach to
Small-Scale Household Food Production

Copyright © 1991 by
David Arthur Cleveland and Daniela Soleri

Published by the
Center for People, Food and Environment (CPFE)
344 South Third Avenue
Tucson, Arizona 85701, USA

with support from the
United Nations Children's Fund (UNICEF)
New York, New York, USA

Printed in the United States of America

ISBN: 0-9627997-0-X

Portions of this book may be copied, reproduced, or adapted to meet local needs, only if the portions reproduced or adapted are provided free of charge or at cost. Please let CPFE know of any such use. **Any copying, reproduction, adaptation, or translation done for commercial purposes, i.e. for profit, is not permitted without permission from the authors or CPFE.**

To order additional copies contact CPFE. This book is sold at a lower price to those living in poor countries and communities than to those in rich countries and communities. Discounts are also available for large orders.

To keep costs low, and make this book available to as many people as possible, we are not using a commercial publisher. This means we need help with distribution. If you have ideas about how to get this book to those who could use it, especially those working in dryland communities, please let us know.

Cover design: Paul Mirocha.

Cover photographs: front, harvesting leaves for soup in Zorse, northeast Ghana (David A. Cleveland); back cover, center, preparing leaves and okra for drying in Zorse, northeast Ghana (David A. Cleveland); left, Hopi-Tewa Norman Honnie in his garden in Arizona, USA (Daniela Soleri); right, marketing squash vine tips in northern Pakistan (Daniela Soleri); front piece, dates from the garden and kishk (fermented, dried wheat balls) are offered in hospitality, Upper Egypt (Daniela Soleri).

Preface

Household gardens are an important part of the indigenous agricultural production system in almost all developing countries. For poorer families, they typically make an important contribution to total household food intake, providing vital nutrients. Gardens also play a vital role in improving household well-being by providing income and savings, thus contributing to increased food security at the household level. All this has often been neglected or misunderstood by development planners. *FOOD FROM DRYLAND GARDENS* will help to change this misunderstanding.

The publication of *FOOD FROM DRYLAND GARDENS* is an important milestone in strengthening our understanding of and support for households in the Developing World. We, at UNICEF, are convinced that it will prove to be a key reference document for those individuals working in the fields of agriculture and nutrition. We are pleased to support its publication.

James P. Grant
Executive Director
UNICEF

Let Us Hear from You

The Center for People, Food and Environment (CPFE) is a non-profit organization devoted to research, education, and action for sustainable food systems. We believe that local control and self-reliance, social equity, cultural and biological diversity, and environmental conservation are essential ingredients of sustainable food systems.

FOOD FROM DRYLAND GARDENS is the first major project of CPFE. The authors of this book are co-directors of CPFE. Organizations or individuals interested in knowing more about CPFE can write to us.

We hope that those of you who use this book will take the time to tell us what you find most helpful, and will give us your suggestions for improvements.

If you have information that is relevant, including your own reports and observations, we would appreciate hearing from you and receiving copies of written works. We have found four types of information most useful in writing this book: a) reports of basic or applied research (from anthropology to hydrology) relevant to dryland household gardens, b) reports of how people in drylands are gardening, and the impact of gardens on nutrition, income and community development, c) manuals for field workers in related areas such as irrigation, d) reports of garden projects, including brief personal narratives, giving some detailed information on assessment, implementation, and evaluation. We would be grateful for any information of this sort you could share with us. We will fully acknowledge the source if any of the information you send us is used in subsequent editions or other publications. So let us hear from you. Write to:

CPFE
344 South Third Avenue
Tucson, Arizona 85701, USA

Thanks

Special thanks go to the gardeners and farmers who, with patience and good humor, taught us many things that have contributed to this book.

We thank the following people for comments on drafts of various chapters in this book: Stanely M. Alcorn, Department of Plant Pathology, University of Arizona; Lauren Blum, Helen Keller International; Thierry Brun, Cornell-China-Oxford Project on Nutrition, Environment, and Health, Cornell University; Roland Bunch, World Neighbors; Frank Carroll, Energy and Resources Group, University of California, Berkeley; Margaret A. Chapman, Department of Plant Pathology, University of Missouri; Mari Clark, Women In Development, US Agency for International Development; Arthur B. Cleveland; Joe Collins, Food First; Mahina Drees, Native Seeds/SEARCH; Steven R. Evett, US Department of Agriculture, Agricultural Research Service; Richard B. Hine, Department of Plant Pathology, University of Arizona; Alfredo Huete, Department of Soil Science, University of Arizona; Thomas H. Kruk, Department of Plant Pathology, University of Arizona; Susan Kunz; O.J. Lougheed, High Desert Research Farm, Ghost Ranch, New Mexico; Maura Mack, Office of Arid Lands Studies, University of Arizona; Steven P. McLaughlin, Office of Arid Lands Studies, University of Arizona; Gary P. Nabhan, Native Seeds/SEARCH; Robert McC. Netting, Department of Anthropology, University of Arizona; Mary W. Olsen, Environmental Research Laboratory, University of Arizona; Thomas V. Orum, Department of Plant Pathology, University of Arizona; Linda Parker, Native Seeds/SEARCH; Thomas Painter; Jennifer Pittet, Developing Countries Farm Radio Network; Lynda Prim, High Desert Research Farm, Ghost Ranch, New Mexico; Ian Robinson, Centre for Arid Zone Studies, University College of Northern Wales; Maria Carla Roncoli, Department of Anthropology, State University of New York, Binghamton; Cindy Salo; Judy Seoldo; Hope Shand, Rural Advancement Fund International; Daniel Toole, UNICEF; David Werner, Hesperian Foundation; Punch Woods, Community Food Bank, Tucson; Deborah Young, Yavapai County Extension Service, Arizona.

For their extra effort we thank Thomas V. Orum for his work on Chapter 13, and Steven R. Evett for his work on Chapters 9 to 12.

Ezzedine Boutrif, Food and Agriculture Organization; Thierry Brun, Nancy Ferguson, Department of Ecology and Evolutionary Biology, University of Arizona; Ian Martin, Thusano Lefatsheng, Botswana; and Thomas V. Orum shared useful information.

For assistance during the preparation of the final manuscript we thank Native Seeds/SEARCH for the use of its computer and printer; Kevin Dahl, Native Seeds/SEARCH; Paul Mirocha, Office of Arid Lands Studies, University of Arizona; and especially Emily Whitehead, Office of Arid Lands Studies, University of Arizona.

For support for printing and distribution we thank UNICEF, and Daniel Toole in particular.

How to Use This Book

This book is organized into four parts: Part I (Chapters 2-4) gives an overview of gardens in development and background information important for supporting gardens; Part II (Chapters 5-13) is about growing gardens; Part III (Chapters 14-16) discusses using the harvest from gardens; and Part IV (Chapters 17-20, plus an index) contains information on measurements, resources, and the complete list of references. some of which are annotated. It is not necessary to read the entire book, or to read it in the order in which it is laid out. Some readers will have very specific interests and needs and so may only use selected chapters. However, we encourage reading as much of this book as possible, and especially reading Part I before starting a garden project.

CHAPTER DIVISIONS Each chapter, except those in Part IV, starts with a brief introduction followed by a summary of the information and concepts discussed in the chapter. Within each chapter there are three levels of headings: the chapter number and title (e.g., 9 Soils in the Garden), the primary chapter divisions (e.g., 9.5 Soils and Plant Nutrients), and the secondary chapter divisions (e.g., 9.5.2 Nitrogen). There are also boxes and tables in some chapters.

BOXES Boxes are surrounded by a border and contain detailed information or techniques. The boxes are not necessary for understanding the discussion in the text. Boxes may occur anywhere in the text and are numbered sequentially within chapters. For example, Box 4.4 is the fourth box in Chapter 4.

TABLES Tables are summaries of information useful for quick reference. They may occur anywhere in the text and are numbered according to the chapter they are in. For example, Table 9.2 is the second table in Chapter 9.

CROSS REFERENCES Often in this book discussion of one topic refers to or builds on another that is discussed or illustrated in a different section. When this occurs we give the other section or figure numbers in parentheses in the text.

PRONOUNS This book reflects the fact that both women and men garden by using female pronouns (she, her, hers) to refer to the gardener in even-numbered chapters, and male pronouns (he, him, his) in odd-numbered chapters. Specific examples are sometimes exceptions to this.

NONENGLISH WORDS In the text, *italics* are used for all words in languages other than English, including the scientific names of plants, which are in Latin.

DEFINITIONS When a word or phrase is being defined in the text, it is ***italicized*** and printed in **bold**. In the index, page numbers in ***bold italics*** indicate where a word or phrase is defined in the text.

GARDEN CROPS In the text only the English common names of garden crops are given. In Chapter 18 the common names of all these crops are listed alphabetically, accompanied by their scientific names.

OTHER PLANTS When plants that are not garden crops are mentioned, their common and scientific names are given in the text.

REFERENCES For each chapter, references are indicated with superscript footnotes, referring to a list of abbreviated references at the end of that chapter. The complete references for all the material cited in the text are listed alphabetically in Chapter 20. Some of the most important references in Chapter 20 are followed by a brief comment.

MEASUREMENTS In the text all measurements are given in metric units and most are followed by the rounded-off English equivalents in parentheses. In examples of how to use a formula or equation, the measurements are only given in metric units. The glossary (Chapter 17) gives conversions and abbreviations for all measurements used in this book.

Table of Contents

Preface..i
Let Us Hear from You..i
Thanks...ii
How to Use this Book..iii
Table of Contents..iv
Detailed Table of Contents..v

1 Introduction...1

PART I
Gardens as a Development Strategy

2 Gardens and Nutrition in Drylands..10
3 Gardens, Economics, and Marketing...31
4 Assessment Techniques..47

PART II
Garden Management

5 How Plants Live and Grow..66
6 Growing Plants from Seeds...79
7 Vegetative Propagation...99
8 Plant Management..123
9 Soils in the Garden...153
10 Water, Soils, and Plants..188
11 Sources of Water for the Garden...207
12 Irrigation and Water-Lifting..227
13 Pest and Disease Management...241

PART III
Garden Harvest

14 Saving Seeds for Planting...285
15 Processing, Storing, and Marketing Food from the Garden..307
16 Weaning Foods from the Garden..327

PART IV
Resources

17 Glossary...339
18 Some Crops for Dryland Gardens..343
19 Resource Organizations..349
20 References...353
Index..373

Detailed Table of Contents

1 Introduction..1
   1.1 Some Definitions...2
   1.2 The Purpose of this Book..3
   1.3 The Organization of this Book..4

PART I
Household Gardens as a Development Strategy

2 Gardens and Nutrition in Drylands...11
   2.1 Summary...11
   2.2 Recommended Dietary Allowances and Nutrient Content of Foods....................12
       Table 2.1 Recommended Daily Dietary Allowances of Major Garden Nutrients........13
       Table 2.2. Rich Garden Sources of Some Nutrients...14
   2.3 Special Nutritional Needs in Drylands..14
       Table 2.3 Some Indicators of Nutritional Deficiencies in Drylands............................15
       2.3.1 Children's Special Needs...14
       2.3.2 Women's Special Needs..17
       2.3.3 Work..17
       2.3.4 Illness..18
   2.4 Energy...19
   2.5 Protein...19
       Box 2.1 Net Protein Utilization and Protein Complementarity.................................20
   2.6 Vitamins..21
       2.6.1 Vitamin A..21
       2.6.2 Vitamin D..22
       2.6.3 Vitamin C..23
       2.6.4 Folacin..23
       2.6.5 Thiamin (B_1)...23
       2.6.6 Riboflavin (B_2)...23
       2.6.7 Niacin..24
   2.7 Minerals..25
       2.7.1 Iron (Fe)..25
       2.7.2 Zinc (Zn)...25
       2.7.3 Calcium (Ca)...26
   2.8 Fats..26
   2.9 Fiber..26
   2.10 Anti-Nutrients...26
       Table 2.4 Anti-Nutrients Common in Dryland Diets..27
   2.11 The Effects of Gardens on Nutrition..28
       2.11.1 Nutrient Yields from Gardens..28
       2.11.2 Effects on Nutrition...28
   2.12 Resources...29

3 Gardens, Economics, and Marketing..31
   3.1 Summary...31
   3.2 People, Households, and Economics..31
       3.2.1 Production Efficiency..32
       3.2.2 Economic Rationality and Risk...33
       3.2.3 Control of Resources: Individual or the Group?...34

3.2.4 Economic Development and Well-Being..35
3.3 Garden Economics..36
    3.3.1 Garden Yields..36
    3.3.2 Income and Savings From Gardens..37
    3.3.3 Household Well-Being...39
3.4 Marketing Garden Produce...39
    3.4.1 Women and Marketing...40
    3.4.2 Risk, Investment, and Return...40
    3.4.3 Cooperation..42
    3.4.4 Garden Income and the Household...43
3.5 Resources...44

4 Assessment Techniques...47
4.1 Summary..47
4.2 Assessment, Monitoring, and Evaluation...47
4.3 From Whose Point of View?...49
    4.3.1 Assessment and Collaboration..49
    4.3.2 Representativeness...49
    4.3.3 Insiders and Outsiders...50
    4.3.4 Participant Observation..50
    4.3.5 Gardens for Whom?..50
4.4 What do Existing Gardens Tell Us?...52
    Box 4.1 Useful Information from Existing Gardens...52
4.5 Interviews..52
    4.5.1 Composing Questions...53
    4.5.2 Translating and Back-Translating..53
    4.5.3 Choosing a Sample...54
    Box 4.2 Statistics and Probability..54
    4.5.4 Pretesting..55
    4.5.5 Conducting the Interview..55
    4.5.6 Coding, Checking, and Analyzing..56
4.6 Seasonality...56
    Box 4.3 Possible Topics for an Annual Calendar..58
4.7 Food Distribution and Consumption...58
4.8 Maps...60
    Box 4.4 Useful Information That can be Shown on Maps..60
4.9 Long-Term Trends...60
4.10 Outside Sources...60
    Box 4.5 Outside Information...61
4.11 Resources..60

PART II
Garden Management

5 How Plants Live and Grow...67
5.1 Summary..67
5.2 The Vascular System in Plants..67
    5.2.1 Roots..68
5.3 Photosynthesis...70
5.4 Transpiration..71
5.5 Coping with Heat and Drought..71

5.6 Salt Tolerance..75
5.7 Seasonal Constraints to Plant Growth..75
    5.7.1 Daylength Requirements...75
    5.7.2 Temperature Requirements...76
5.8 Resources...78

6 Growing Plants from Seeds..79
6.1 Summary..79
6.2 Sexual Reproduction in Plants..79
    6.2.1 Life Cycles..79
    6.2.2 Flowering..80
    6.2.3 Pollination...82
        Box 6.1 Controlling Pollination..83
    6.2.4 Fertilization...85
6.3 Seed Germination and Dormancy..86
6.4 Suggestions for Planting Seeds under Dryland Conditions...87
    6.4.1 Preparing the Seeds...87
    6.4.2 Preparing the Planting Site..87
    6.4.3 Planting the Seeds...89
    6.4.4 Planting Density..89
    6.4.5 Covering the Seeds..91
6.5 Caring for Newly Planted Seeds and Young Seedlings..91
    6.5.1 Watering..93
    6.5.2 Mulching and Shading...94
6.6 Diagnosing Seed Planting Problems...95
    Table 6.1 Diagnosing and Treating Seed Planting and Seedling Problems...............96
    6.6.1 Testing Seed Germination..96
6.7 Thinning...98
6.8 Resources...98

7 Vegetative Propagation..99
7.1 Summary..99
7.2 Cuttings..99
    7.2.1 Trees...99
    7.2.2 Perennial Herbs...102
    7.2.3 Cassava..102
    7.2.4 Sweet Potatoes...103
7.3 Tubers, Tuberous Roots, and Bulbs...104
7.4 Offsets...105
    Box 7.1 Propagation by Offset—the Date Palm...106
7.5 Suckers...108
7.6 Grafting..108
    7.6.1 Compatibility for Grafting...109
    7.6.2 Effects of Stock and Scion on the Grafted Tree..111
    7.6.3 Approach or Attached Scion..111
    7.6.4 Budding...111
    7.6.5 Apical Grafting..114
    7.6.6 Topworking...114
        Box 7.2 Guidelines for Grafting...116
7.7 Layering...114
    7.7.7 Simple Layering..117

  Box 7.3 Guidelines for Simple Layering..118
 7.7.2 Air Layering...117
7.8 Resources..118

8 Plant Management..123
8.1 Summary...123
8.2 Nursery Beds and Container Planting...123
 8.2.1 Nursery Beds..123
 8.2.2 Container Planting..125
 8.2.3 When Direct Planting is Better...129
8.3 Planting Sites and the Sun...129
8.4 Transplanting...129
 8.4.1 Timing...129
 8.4.2 The Site..129
 8.4.3 Water..133
 8.4.4 The Transplant..133
  Box 8.1 Steps in Transplanting..136
8.5 Plant Interactions..137
 8.5.1 Mixed Planting..137
 8.5.2 Allelopathic Plants...138
 8.5.3 Crop Rotation..139
8.6 Weed Management..139
 8.6.1 Resource Use...140
 8.6.2 Effects on Pest Populations...140
 8.6.3 Timing...140
 8.6.4 Methods of Weed Control...141
  Table 8.1 Controlling Weeds in Dryland Gardens...141
8.7 Pruning...142
 8.7.1 Reasons to Prune...142
 8.7.2 Guidelines for Pruning Trees..143
  Table 8.2 Pruning Suggestions for Some Dryland Garden Trees...144
8.8 Trellising..149
8.9 Resources..151

9 Soils in the Garden..153
9.1 Summary...153
9.2 Soil and Land-Use Classification..153
 9.2.1 Indigenous Classification Systems...154
 9.2.2 The USDA Classification of Soils in Drylands...154
9.3 Physical Properties of Soils...155
 9.3.1 Soil Texture and Structure..155
  Box 9.1 Clay...155
  Box 9.2 Tests for Soil Texture..157
 9.3.2 Soil Porosity and Permeability...156
 9.3.3 Soil Color..156
 9.3.4 Soil Temperature...158
9.4 Soil Profile and Depth..158
  Box 9.3 Caliche..160
  Box 9.4 Plinthite and Ironstone...160
9.5 Soils and Plant Nutrients...158
  Box 9.5 Nutrient Uptake by Plant Roots..161

9.5.1 Soil pH and Plant Nutrition...161
  Box 9.6 pH..163
9.5.2 Nitrogen..163
  Box 9.7 Nitrogen Fixation..164
  Box 9.8 Commercial Chemical Fertilizers..166
9.5.3 Phosphorus and Potassium..166
9.5.4 Other Nutrients..167
9.6 Organic Matter...168
  Table 9.1 Approximate Nutrient Content of Some Dryland Organic Matter..........169
  Box 9.9 Advantages of Organic Matter..170
9.6.1 Animal Manures..169
9.6.2 Composting..171
  Table 9.2 Examples of Approximate C:N Ratios of Dryland Organic Matter........173
  Box 9.10 For a Fast Compost Pile..174
9.7 Preventing Soil Erosion...172
9.7.1 Decreasing Runoff..175
  Box 9.11 Determining Contours..178
9.7.2 Decreasing Raindrop Impact..181
9.7.3 Increasing Soil Resistance to Erosion..181
9.7.4 Reducing Wind Erosion...182
9.8 Building Garden Beds..183
9.8.1 Sunken Beds..183
9.8.2 Raised Beds...183
9.9 Resources...185

10 Water, Soils, and Plants...189
10.1 Summary..189
10.2 Dryland Garden Water Management...189
10.3 Water, Soils, and Plants..190
10.3.1 Water Storage in the Soil...190
  Box 10.1 How Water is Held in the Soil...192
10.3.2 Water Movement in the Soil..192
10.3.3 Evaporation..193
10.3.4 Water Uptake and Transport by Plants..193
10.4 Soil Water and Garden Yield...195
10.5 How Much Water?..196
  Box 10.2 Estimating Water Requirements for Large Garden Projects...........198
10.6 Measuring Water Applied to the Garden..198
10.7 When to Water..200
  Table 10.1 Soil Texture and Water Deficit..200
  Box 10.3 Calculating Irrigation Frequency for Large Garden Projects..........202
10.8 Mulches, Shades, and Windbreaks...202
10.8.1 Surface Mulches..203
10.8.2 Vertical Mulches..203
10.8.3 Windbreaks, Shades, and Cropping Patterns..204
10.9 Resources...206

11 Sources of Water for the Garden..207
11.1 Summary..207
11.2 Water Quality for Plants..208
  Box 11.1 Measuring Water Salinity..208

11.3 Water Quality for People..208
    Box 11.2 Human Disease and Water Quality..209
11.4 Rain...210
    11.4.1 Rainfall Records..210
        Box 11.3 Calculating Rainfall Probabilities from Records..............................211
        Table 11.1 Calculating Rainfall Probabilities...211
    11.4.2 Measuring Rainfall..210
11.5 Harvesting Rainwater for Dryland Gardens...212
    11.5.1 Patterns of Water Harvesting...213
    11.5.2 Building on Local Knowledge..213
    11.5.3 Catchments and Runoff..214
        Table 11.2 Comparison of Desirable Characteristics for Catchment versus Growing Areas for Rainwater Harvesting..215
        Box 11.4 Measuring a Slope...217
    11.5.4 Estimating the Catchment to Garden Area Ratio...................................217
        Box 11.5 Designing a Microcatchment Garden Plot.......................................220
11.6 Harvesting Stream Flow and Floodwater...219
    11.6.1 Water Spreading..220
    11.6.2 Flood Recession Gardening..220
11.7 Groundwater and Wells...221
    11.7.1 Groundwater..221
    11.7.2 Locating a Well..221
    11.7.3 Hand-Dug Wells..222
    11.7.4 Small-Diameter Wells..223
11.8 Water Storage..223
11.9 Resources...224

12 Irrigation and Water-Lifting...227
12.1 Summary..227
12.2 Irrigation Efficiency..227
12.3 Surface Irrigation..228
    12.3.1 Transporting Water to the Garden...228
    12.3.2 Basin Irrigation..229
    12.3.3 Furrow Irrigation...230
    12.3.4 Trickle Irrigation..231
12.4 Root Zone Irrigation...233
    12.4.1 Pitcher Irrigation..233
    12.4.2 Water Table Irrigation...233
12.5 Sprinkler Irrigation...233
12.6 Irrigation Problems...234
    12.6.1 Waterlogging...234
    12.6.2 Salinity..235
12.7 Water-Lifting...236
    12.7.1 Lifting with Human and Animal Power...237
    12.7.2 Lifting with Other Power Sources...239
12.8 Resources...239

13 Pest and Disease Management...241
13.1 Summary..241
13.2 An Ecological Approach..241
    13.2.1 Pest and Disease Management by the Crop Plant.................................242

13.2.2 Environmental and Mechanical Management of Pests and Disease..................................243
13.2.3 Pest and Disease Management Using Other Organisms...244
    Box 13.1 Biological Control Using Insect Pathogens...246
13.2.4 Pest and Disease Management with Chemicals...245
    Box 13.2 Safe Homemade Pesticides..247
    Box 13.3 Synthetic Pesticides...248
    Box 13.4 Emergency First Aid for Pesticide Poisoning..250
13.3 Examples of Pest and Disease Management...250
  13.3.1 Insects..250
  13.3.2 Nematodes...255
  13.3.3 Large Animals as Pests..259
  13.3.4 Diseases...261
13.4 Diagnosing Pest and Disease Problems...265
  13.4.1 Wilts...273
    Table 13.1 Wilts...274
  13.4.2 Leaf Problems..273
    Table 13.2 Leaf Problems...276
  13.4.3 Abnormal Growth..273
    Table 13.3 Abnormal Growth..278
  13.4.4 Fruit Problems...273
    Table 13.4 Fruit Problems..280
13.5 Resources..273

PART III
Garden Harvest

14 Saving Seeds for Planting..285
14.1 Summary..285
14.2 Seeds, Gardens, and Diversity...286
    Table 14.1 The Diversity Continuum at Different Levels in Agriculture....................287
  14.2.1 Diversity in the Seed..287
    Box 14.1 Genetic Diversity in the Seed...288
    Box 14.2 Hybrid Seeds...290
    Box 14.3 Southern Corn Leaf Blight..292
  14.2.2 Diversity in the Garden...292
  14.2.3 Conserving and Using Genetic Diversity: How and for Whom?........................293
14.3 Seed Saving..295
  14.3.1 Seed Harvest and Processing..295
    Table 14.2 Seed Harvest and Processing..296
  14.3.2 Seed Drying...299
14.4 Saving Seed from Trees..300
  14.4.1 Cold Stratification..300
14.5 Seed Storage...301
    Table 14.3 Summary of Dryland Seed Storage Problems and Responses..................301
  14.5.1 Moisture and Temperature...301
  14.5.2 Pest Control...302
14.6 Resources..305

15 Processing, Storing, and Marketing Food from the Garden...307
15.1 Summary..307
15.2 Harvesting Garden Foods..307

15.3 Cooking and Using Garden Foods..........310
    Table 15.1 Cooking and Nutrient Content of Foods..........310
    15.3.1 Fresh Foods..........310
    15.3.2 Dried Foods..........311
15.4 Food Drying..........311
    15.4.1 Materials for Drying..........312
    15.4.2 Preventing Contamination..........312
    15.4.3 Selecting and Preparing Produce for Drying..........312
        Table 15.2 Summary of Drying Methods..........314
        Box 15.1 Blanching..........316
15.5 Sprouting and Malting..........315
    15.5.1 Sprouting..........315
    15.5.2 Malting..........317
15.6 Fermentation..........318
    15.6.1 Pickling..........319
15.7 Storing Garden Foods..........319
    15.7.1 PreHarvest Storage..........320
    15.7.2 Postharvest Storage of Fresh Produce..........320
    15.7.3 Storing Dried Produce..........321
    15.7.4 Storing Other Processed Garden Foods..........322
15.8 Marketing Garden Produce..........322
    15.8.1 Harvesting for Market..........322
    15.8.2 Transport from Garden to Market..........322
    15.8.3 Protecting Produce Quality at the Market..........324
15.9 Resources..........325

16 Weaning Foods from the Garden..........327
16.1 Summary..........327
16.2 The Role of Weaning Foods..........327
16.3 Nutrient Density..........329
    16.3.1 Energy..........329
    16.3.2 Protein..........330
    16.3.3 Vitamins and Minerals..........331
    16.3.4 Weaning Food Consistency..........332
16.4 Hygiene..........334
    Box 16.1 When the Weaning Child Becomes Sick..........335
16.5 Weaning as a Part of Daily Life..........335
16.6 Resources..........336

Detailed Table of Contents

PART IV
Resources

17 Glossary..339
   17.1 Abbreviations Used in Measurements...339
   17.2 Equivalencies in Units of Measurement..339
   17.3 Atomic Symbols and Molecular Formulas...340
   17.4 Other Abbreviations and Acronyms..340

18 Some Crops for Dryland Gardens...343
   18.1 Common English and Scientific Names for Some Crops.....................343
   18.2 Important Dryland Garden Plant Families..346

19 Resource Organizations..349

20 References...353

Index..373

FOOD FROM DRYLAND GARDENS

1
Introduction

On the outskirts of a city in northern Egypt a man raises water from the wide, brown, Nile River with a bucket hanging from the end of a long wooden pole that seesaws on a support near the river's edge. He swings the full bucket over the river bank, emptying it into a small canal that carries the water to tomatoes and eggplants growing in a narrow plot. In rural, northern Mexico a woman picks ripe pomegranates from a tree growing by her house. The tree is surrounded by a tangle of squash vines, maize plants, and herbs, and chickens run between the plants to catch and eat insects.

In northern Arizona, USA, Hopi women leave their stone houses and descend a steep path down the side of the mesa to a cluster of over 100 terraced garden plots. They water the plots through a network of small canals fed by a spring. The women talk and laugh as each one harvests chilis in her own plot. Although very different, these are all examples of people gardening in drylands around the world.

Inside a home compound in northern Pakistan a hand-formed watering basin topped with thorn branches protects a young jujube tree from animals. Children love to eat the sweet jujube fruits. In the irrigated fields nearby a man has planted a patch of squash, eggplants, and chilis along a small irrigation ditch (Figure 1.1). The squash vines sprawl out along the canal and between the other plants, their clipped ends showing where the vine tips have been harvested to add to soups and sauces for the family meals.

Figure 1.1 A Garden in Northern Pakistan

In a Mexico City slum a woman has cut a hole in the side of an empty shampoo bottle, filled it with good soil she brought from another area, and planted mint in it. In front of her neighbors' shack, chili plants are growing in a stack of old tires, and a young fig tree has been planted in a large tin can found at the nearby garbage dump.

In the savanna of northern Ghana a woman empties a clay bowl of water from washing onto a patch of okra growing outside the gate of her mud-walled compound. At the end of the rainy season she will dry the okra and store it for later use. In the dry season her husband will clean out a shallow well in the bed of a seasonally flowing stream, repair the thorn branch fence, and plant tomatoes and sweet potatoes, some of which he will sell in the market along with mangoes from two trees which are also growing there. At the beginning of the rainy season when food supplies are low, their children will gather leaves of weeds growing in the fields and will climb the giant baobab tree near their house to pick its young leaves for soup.

These people are all gardening—in the wet season, in the dry season; in cities and in rural areas; near their houses, in fields, and alongside roads, canals and rivers; in separate plots, and on individual plots in communal gardening areas; on land that they have a right to cultivate because of the family they belong to, on land that they have borrowed or rented, and without permission on land owned by the government or a railroad company. The crops and varieties they grow are chosen primarily from among those that have been handed down from parents and grandparents. They are adapted to the climate and soils, resistant to local insects and diseases, and are easy-to-cook, good-tasting ingredients of the meals that are part of their cultural identity. At the same time, other crops and varieties are new to the gardeners and are being grown as experiments.

Crops harvested from the garden are sometimes sold in local markets, bartered, or given as gifts, but some are always eaten. These garden foods can provide many nutrients but are especially important because of their contributions of vitamins, minerals, and special foods such as those used to wean children.

1.1 Some Definitions

We have written this book to encourage gardens to improve nutrition, income, and self-sufficiency in rural and urban communities in the drylands of the Third World.

In the *drylands,* lack of water limits plant growth for at least several months of the year (Figure 1.2). Drylands include the deserts and savannas as well as the subhumid regions where there is a long dry season, including west, east and southern Africa, southern Europe, north Africa and southwest Asia, south Asia, southwestern North America, northeastern Brazil and western South America, and most of Australia.

In this book *Third World* does not refer to a geographic region but to a situation where communities are not in control of their own resources, and are often exploited by outside markets, organizations, or governments on which they are dependent. Third World people are relatively poor, unhealthy, and malnourished compared with most people in the wealthier industrial world, and the majority of people in drylands live in the Third World. The Third World includes not only the majority of the population in Africa, Asia, and Latin America, but also many communities within the rich industrialized nations of Europe, the USA, Canada, Japan, and Australia. We use the terms "Third World" and "industrial world" instead of "developing" and "developed" because these last two imply a single cultural and socioeconomic path toward a goal the industrialized countries have attained and to which the Third World must aspire.

A *household* is a group of people who regularly work and eat together. Gardens, like the ones just described, have been a part of household food production systems around the world for hundreds and thousands of years. They continue to be an important part of households' production and consumption strategies into the 1990s. Gardens can be identified primarily by their function, rather than their form, location, size or the types of crops grown. Whether controlled by the household or by an individual in the household, *household gardens* are secondary sources of food and income, while field production, animal husbandry, wage labor, professional services, or trading are the major sources of support.

The value of household gardens lies not only in what they can do but in how they can do it. Improving nutritional and economic conditions is the goal of many "development" efforts. However, it is rare to find those goals pursued in ways that support local participation and control and equity, while striving for sustainable use of resources. Garden projects are no exception. Many garden projects are based on the promotion of an industrial garden model, rather than on the indigenous gardens which people in local communities are already growing (Part I). We use the term *indigenous* to describe locally developed knowledge, practices, and resources including crop varieties and gardens.

1.2 The Purpose of This Book

FOOD FROM DRYLAND GARDENS was written to encourage gardens that serve local needs, that are based on local knowledge and resources, and that conserve natural resources and the biological diversity of traditional crops. It was written for field workers, extension agents, students, project workers, and program planners. Both a beginner's guide as well as a reference for those with more experience, this book helps the reader observe and work with local people to ask appropriate questions about the community, the environment, and the potential for gardens to improve nutritional, economic, and social well-being.

Because every location, household, and community is different, the solutions to their gardening problems must be unique. Finding locally appropriate solutions is best accomplished through an appreciation of the adaptedness of indigenous gardens, and of the fact that gardeners manage these gardens according to the same principles on which Western, formalized science is based. While the application of horticultural science may lead to improvements, the foundation from which to begin any garden project is existing local knowledge because it supports equity through self-sufficiency and local participation and control. This is true even where great change has occurred, as in refugee camps, crowded urban areas, or environmentally degraded rural areas. Therefore, although many specific techniques are included, this is not a how-to cookbook. The basic principles of nutrition, agriculture, ecology, and social science are described, along with examples of indigenous gardening, with the goal of encouraging experimentation and adaptation to each situation.

We emphasize long-term environmental and social sustainability throughout FOOD FROM DRYLAND GARDENS. *Environmental sustainability* means the management of soil, water, and biological resources so that all future generations can also use them. To be *socially sustainable* gardens must improve nutrition and income in ways that are cost effective and promote local self-reliance and a just distribution of resources. This means gardens that use local resources available to all households; these resources include indigenous

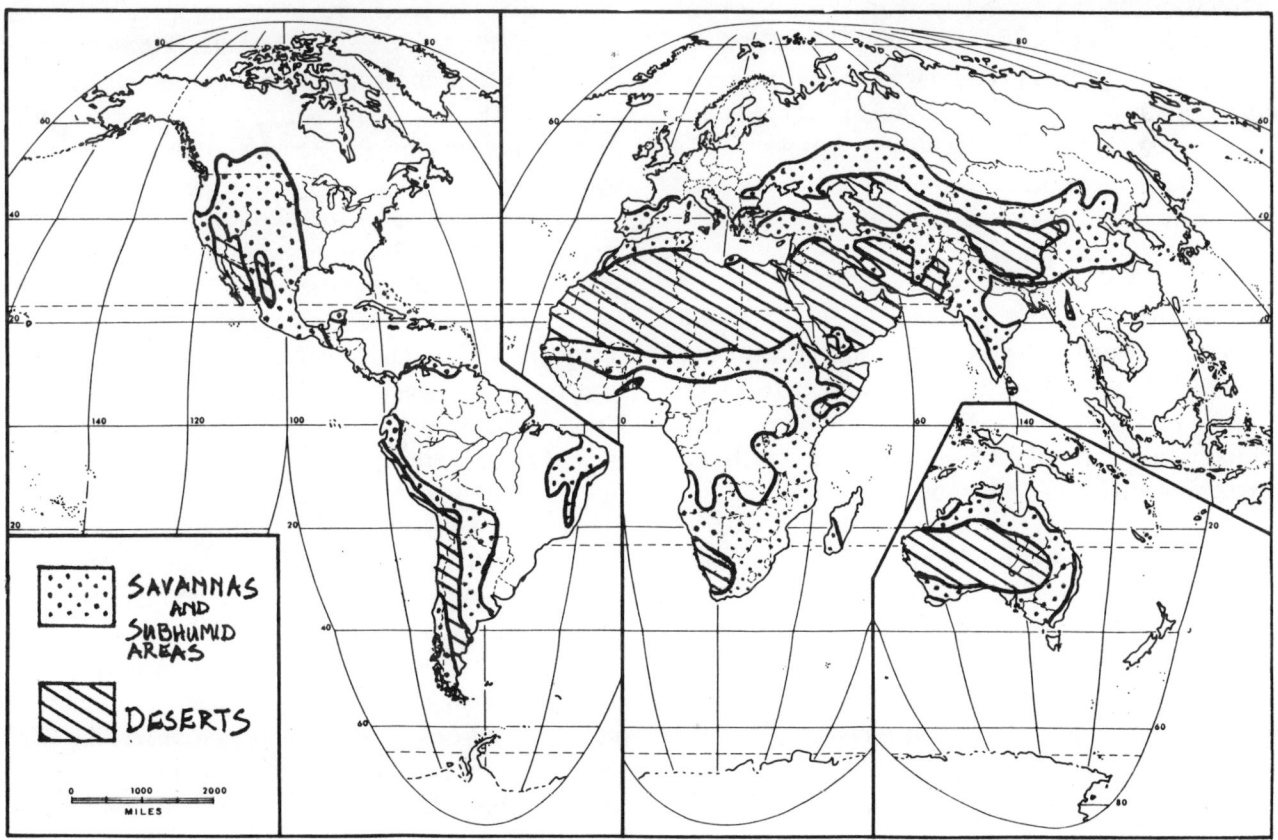

Figure 1.2 The World's Drylands (After UNESCO 1977)

gardening techniques, and indigenous trees and other garden crops.

Promoting social sustainability requires an understanding of relationships within the household, and between the household, the community, the nation, and the rest of the world. It means ensuring participation and equity for minorities, including many indigenous peoples, and for women, children, and the handicapped. To improve the well-being of the poor and hungry of the drylands their needs, desires, resources, and skills must be kept at the center of the project and they must have control over changes affecting them.

When garden projects from the largest to the smallest fail, it often seems obvious in retrospect that a major cause was the lack of understanding by project workers of the human side of the food system. Trying to understand whole systems can be frustrating and time consuming at first and mistakes will be made. This book will increase awareness of gardens as part of ecological and social systems, and garden projects based on this awareness will have a greater chance of success. However, this book alone is not enough: gaining a firsthand understanding of the social, environmental, and economic ways in which the local system works is most important, and should be a prerequisite for all project workers. Like other improvements, gardens promoted in this way may not produce showy and spectacular results at the beginning, but are more likely to respond to real needs, and persist and grow beyond the life of the project.

1.3 The Organization of This Book

Part I, Gardens as a Development Strategy, summarizes the basic principles of nutrition and economics as they apply to gardens in the Third World, the evidence that household gardens are a viable development strategy, and ways of assessing whether gardens are appropriate in a given situation.

Chapter 2 reviews the special nutritional needs of women and children, and the effects of work, illness, and seasonality on nutrition in drylands. The function, requirements, and dietary sources of specific nutrients are discussed, but the emphasis is on the important effects that the combination of foods in household meals can have on the total nutritional value of the diet, and on the primary goal of assuring adequate energy intakes. Gardens have the potential to improve overall dietary diversity and contribute critical nutrients such as viamin A, iron, and energy, often when other sources are not available.

Gardens can improve household well-being by providing income and savings. Chapter 3 discusses the need to understand gardeners' economic decision making, including concepts of production efficiency, economic rationality, and control over resources. Storage and processing techniques and organizing into cooperatives are discussed as ways of reducing the risks involved in marketing garden produce. Women's roles in production and marketing must be explicitly considered so that they are not excluded from the economic benefits of gardens.

In Chapter 4 the essential role of assessment, monitoring, and evaluation in garden projects is reviewed. Community control of the assessment process, representativeness, understanding existing gardens, and some specific techniques, such as interviewing, are discussed.

Part II, Garden Management, covers the basic principles, indigenous practices, and specific suggestions for managing plants, soils, water, pests, and diseases in dryland gardens. The emphasis is on managing the whole garden as an ecological system, and on the garden as only one of many household activities. This means that the use of resources in specific garden management strategies must be balanced against the potential use of those resources in other garden and household activities. The goal is not maximizing production, but maximizing household and community well-being.

Chapter 5 discusses the basic principles of plant biology in relationship to heat, drought, salinity, and seasonality. Sexual reproduction and growing plants from seeds is the topic of Chapter 6, which includes many specific suggestions for planting seeds under dryland conditions and diagnosing planting problems. Vegetative propagation by cuttings, tubers, bulbs, offsets, suckers, grafting, and layering are discussed in Chapter 7. Chapter 8 covers a wide range of practices for maintaining healthy and productive dryland gardens: nursery bed and container planting, transplanting, plant interactions, weed management, pruning, and trellising.

Chapter 9 on soils emphasizes the importance of reducing wind and water erosion, and of maintaining adequate soil organic matter to ensure fertility and water-holding capacity. The movement of water in soils, and the relationship of soil, water, and garden yield are discussed in Chapter 10 as the basis for specific techniques to improve water management. Chapter 11 describes sources of water for the garden and various indigenous and other techniques for capturing this water through rainwater harvesting, floodwater gardening, and hand-dug wells. Chapter 12 discusses water-lifting and the application of water to the garden through surface irrigation, root zone irrigation, and sprinkler irrigation. It also addresses ways to avoid salinity and waterlogging.

Chapter 13 advocates an ecological approach for dealing with pests and diseases in which total garden management rather than the use of toxic pesticides is the most efficient, self-sufficient, and ecologically sustainable strategy. It includes four tables with accompanying figures for diagnosing and managing garden problems.

Part III, Garden Harvest, covers harvesting and using garden produce, including seed saving for future planting. Chapter 14 discusses the value of local control of folk crop varieties for genetic diversity, sustainability, and self-sufficiency, and methods for seed saving and storage. Indigenous and other techniques for harvesting, cooking, drying, sprouting, malting, fermenting, and storing garden produce to increase its contribution to diets throughout the year are the topic of Chapter 15. Weaning foods are one of the most important dietary contributions gardens can make. Chapter 16 describes how many garden foods can be processed to provide nutrient- and energy-rich weaning foods, often in quantities and at times when other food sources are not sufficient.

Part IV, Resources, contains a glossary (Chapter 17), a list of all garden crop species mentioned in the text with their scientific names and a list of important garden crop families (Chapter 18), an annotated list of resource organizations (Chapter 19), and a list of references cited in the text (Chapter 20), the most useful of which are annotated. An index is also included.

Part I
Gardens as a Development Strategy

A vast part of the world's population suffers from poverty, malnutrition, and environmental degradation,[1] and gardens are often a part of peoples' struggle to cope with these problems. Gardens contribute a great deal to the nutritional, economic, and social well-being of dryland households, and they have the potential to contribute much more. Why then do so many garden projects fail? In many cases the answer is because they start out by establishing a model garden and trying to convince local people to adopt the model without first understanding existing local gardens, resources, or knowledge.[2] Whether intentional or not, this reflects an assumption that the project workers know more than people in the community, and that learning will be a one way process with the project providing the answers. This is an example of the "top-down" approach to development.

We believe that to be sustainable, development that involves outsiders must be a cooperative venture. Local people guide this process and project workers are resources for them. Community members must take pride in themselves, demand control of the changes affecting them, and work with and learn from project workers as equals. Project workers, especially those from outside, must also learn to work with community members as equals, while recognizing that the local people must guide the project's direction. Project workers must respect and support local skills and

The Top-Down Approach to Development

Development is Cooperation Between Equals

knowledge, and always keep in mind the ultimate goal of improving people's well-being in a way that is both socially and environmentally sustainable. Gardens that support self-sufficiency by using local resources, improving nutritional status and incomes, and protecting the environment, can make an important contribution to finding solutions.

Model gardens are gardens that are developed without regard to the local circumstances where they are to be promoted. They are what someone from outside the community believes gardens should be. Model gardens are often inappropriate in many ways. They may require more time, water, or land than local people can afford, use seeds and techniques that are not locally adapted, or produce foods that people do not like. Promoting model gardens also ignores both local gardening skills and local gardening problems.

There are some distinct approaches to household gardens for improved well-being that reflect different values in the field of development. These approaches can be most usefully distinguished according to whether they are based on models brought in from the outside, or are built on local, *indigenous knowledge*. Today agriculture, nutrition, health, and rural development projects often promote gardens in recognition of their potential contribution to household well-being, but frequently these projects promote an industrial garden model which is very different from the gardens already existing in the area. *Industrial gardens* are based on agriculture in industrial countries and include crops, tools, inputs, production techniques, marketing organization and nutrition education which are usually inappropriate for the local situation in the Third World, and are not sustainable.

A much less common development approach is to support *indigenous gardens*, those that are developed by the gardeners themselves, based on local knowledge and resources, and adapted to local needs.[3] Indigenous household gardens are valuable because they adapt to so many different human needs and physical environments in such a great diversity of ways, and persist even after, or sometimes in spite of, the introduction of "modern" agriculture and gardens. In fact, indigenous gardens are not only widespread in the Third World, but are also popular in urban and rural areas of industrialized countries like the United States, Canada, Great Britain, and Poland where they have important economic, nutritional, and social functions.[4]

An approach to gardens in development based on indigenous gardens cannot use models because indigenous gardens are often unique to specific locations. Using an indigenous garden from one area as a model for gardens in another area can be as inappropriate as using any other garden model. New ideas are valuable and needed, but their appropriateness should never be assumed until tested and evaluated by gardeners themselves.

One reason for the lack of attention to indigenous gardens in development projects is that they are not well documented or understood in the horticulture, economic, nutrition, or social science literature that is the source of information for most project planners and field workers. European colonialism in the Third World did much to establish this bias against indigenous food production.[5] Colonialism contributed to the belief still held by many today that indigenous food production expertise in the Third World is inferior and not suited to the modern world, and industrial, large-scale, capital- and resource-intensive agriculture is the only way to improve the situation.[6]

However, while development strategies like the "green revolution," which are based on an industrial agriculture model, sometimes result in increased production, they have often led to increased inequities in the Third World countries where they have been applied. These strategies have frequently perpetuated dependence on the industrialized nations and the international markets they control.[7] Meanwhile, malnutrition and poverty persist as major problems.

Indigenous gardens appear to have suffered from both the bias against indigenous agriculture, as well as from neglect because gardens were not considered to be a significant part of the food system. As a result, most of what has been written about indigenous gardens is brief and descriptive, and does not analyze the production techniques or the effects of gardens on income or nutrition. The assumption often follows that indigenous gardens are not based on scientific principles. Yet nothing could be further from the truth. In fact, the more that is learned about indigenous food production, the more obvious it is that it is based on the same principles as Western science. It is also obvious that both are influenced by the experiences and values of the people who practice them.

For example, Western agricultural science today is very much under the influence of a world economic system that emphasizes maximizing production and profits (section 3.2). The majority of research carried out is on strategies that increase farmer dependence on the market. Relatively little research is done on strategies that increase small farmer and gardener self-reliance or on minimizing destruction of the environment. The strong influence of values on the direction of research has led to a vicious cycle; because alternatives are not documented they are not believed to be valid, and those who might be interested are discouraged from researching and documenting them.[8] Those who are practicing these alternatives, such as indigenous gardeners and farmers, are told that their skills and knowledge must be abandoned in favor of a system over which they have no control. However, the need for such alternatives is increasing, and nowhere is this more obvious than in the world's drylands.

People in drylands—such as migrants to cities and to marginal rural areas, participants in large-scale irrigation schemes, and refugees fleeing across borders to temporary camps—are increasingly faced with new situations. Rising population densities, environmental degradation, water scarcity, and rapid social and cultural change mean new conditions for everyone. Without any outside encouragement, many of these people are growing gardens as part of their survival strategy. But the gardens they are familiar with may not be the most appropriate for their new, difficult conditions. These people do not need to be told how to garden, but they do need assistance as they work to develop gardens appropriate for these new circumstances.

While the problems of poverty and powerlessness facing the poor in drylands can only be eliminated by addressing their social and economic roots in colonialism, global inequity, and dependency, gardens can provide immediate benefits, and most importantly, can provide those benefits in a way that contributes to the solution of the larger problems.

References

[1] Durning 1990.

[2] Bittenbender 1985; Brownrigg 1985:100-112; Cleveland 1986; Cleveland and Soleri 1987; Niñez 1987; Pacey 1978:23-24.

[3] Dupriez and De Leener 1987; Sommers 1984; UNICEF 1985.

[4] Crouch and Ward 1988; Gladwin and Butler 1984; Kleer and Wos 1988; Omohundro 1985.

[5] Bodley 1990:13-14, Richards 1986:138-140.

[6] For example, see Todaro 1985:285-310.

[7] Latham 1990; for Mexico see DeWalt 1985.

[8] Warren, et al. 1989.

Figure 2.1 Good Nutrition is Essential for Good Health

2
Gardens and Nutrition in Drylands

Nutrients are the chemical compounds that living plants and animals need for growth, physical maintenance, work, reproduction, and combating disease and other stresses. *Nutrition* is the study of the intake of food and the use of food nutrients by living organisms. Good nutrition is essential for good health (Figure 2.1). *Malnutrition*, the lack of required nutrients, is a major problem in drylands, especially for poor households. Malnourished people are more vulnerable to disease; at the same time, disease often contributes to malnutrition. People suffering from disease and malnutrition are not able to work as productively as healthy, well-nourished people. This increases dependence on outside help and reduces the quality of life (Figure 2.2). This can become a self-perpetuating cycle that affects a household, a community, and even a nation.

In this book we apply the concepts of nutrition to both plants and people. In Part II we discuss the nutritional needs of plants, sources of these nutrients, and the effects of nutrient deficiencies. In this chapter we consider the same issues for human nutrition in drylands. We include this chapter in a book on gardens because malnutrition is such a serious problem in drylands and because gardens can contribute a great deal to solving this problem. Gardens can do this directly by providing nutrients, or indirectly by increasing household income or savings which may be used to improve nutrition (Chapter 3). The nutritional effects of processing and preserving garden produce are discussed in Chapter 15 and the preparation of weaning foods from the garden in Chapter 16.

The relationship between food production and nutrition is very complex and improving human nutrition can be difficult. Human values and the assumptions and policies that are based on them have a tremendous effect on nutrition. Today there is growing recognition that assumptions about "development" and "progress" in diets, lifestyle, and agriculture that are modeled on Western, industrialized countries must be carefully reexamined.[1] Clearly, malnutrition is not simply a problem of inappropriate or inadequate food. It is also a consequence of policies and assumptions at the local, regional, national, and global levels that affect the production, processing, promotion, and distribution of food. Increasingly, malnutrition is a visible symptom of inequity, powerlessness, and greed. Both over-consumption by the world's rich minority (Figure 2.3), and rapid population growth among the poor majority (frequently an individual or household response to these same problems), intensify the stress put on the finite natural resources on which we all rely.

2.1 Summary

Children, sick people, and pregnant and breast-feeding women need extra nutrients to avoid becoming malnourished. Those working hard in the home, field, or factory also need extra food. Food consumption and distribution is determined not only by need, but by beliefs and traditional dietary patterns, and by patterns of control over resources in the household, the community, the nation, and the world.

Energy, protein, vitamins A, C, D, folacin, thiamin, riboflavin, and niacin, and minerals such as iron, zinc, and calcium are nutrients essential for good health—however, they are often inadequate in dryland diets. Gardens are good sources for many of these nutrients.

Anti-nutrients are substances in food that are poisonous or that reduce nutritional value, and are present in all diets. Traditional processing techniques often help eliminate these.

Figure 2.2 The Cycle of Malnutrition, Disease, and Lower Food Production

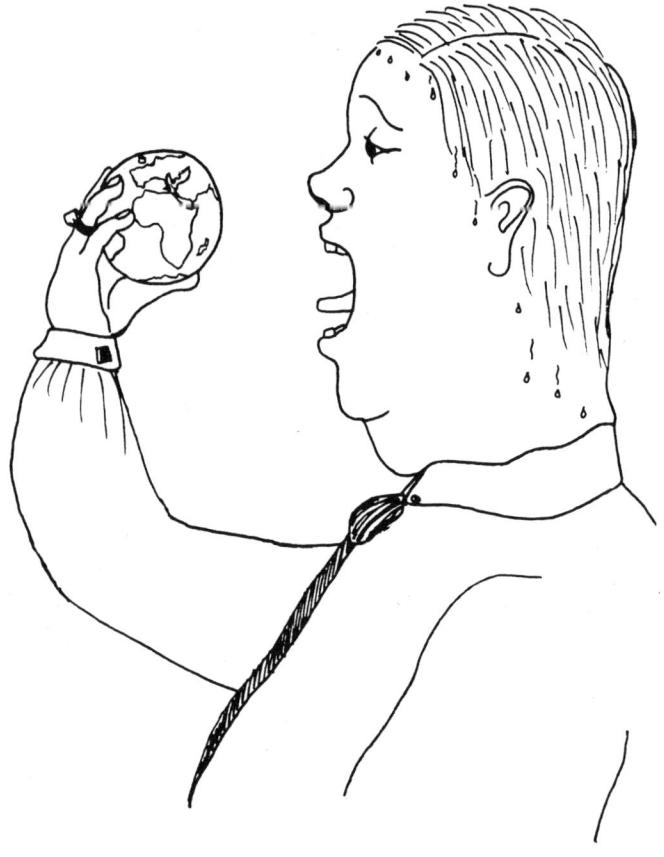

Figure 2.3 Overconsumption by the Rich Threatens the Earth and the Future.

2.2 Recommended Dietary Allowances and the Nutrient Content of Foods

Recommended Dietary Allowances (RDAs) are nutritionists' estimates of the amounts of nutrients required each day for the majority of healthy people in a given population to remain in good health.[2] These amounts include an increase over the average person's needs to include healthy people whose requirements are greater than average.

The RDAs and tables listing the nutrient content of foods are often used together to estimate the nutritional adequacy of foods and diets, and to plan projects to increase the nutrient content of deficient diets. However, several points must be kept in mind when using RDAs:

- RDAs should be used for evaluation of populations, not individuals.
- Daily nutrient intakes can be averaged over a week.
- RDAs do not take into consideration unusual stress due to disease, strenuous work, climate, and previous malnutrition.
- There is much disagreement among the "experts" over the amount of some nutrients that is adequate.

In most of the world's drylands there has been very little work done to establish appropriate RDAs for local populations. The United States' RDAs[3] are often used, but they were calculated for an unusually affluent, inactive, and well-nourished population. RDAs

Table 2.1 Recommended Daily Dietary Allowances of Major Garden Nutrients [a]

	Age (yrs)	Weight (kg)	Energy[b] (kcal)	Protein[c] (gm)	Vit. A (RE)[d]	Vit. C[e] (mg)	Iron (mg)
Children							
	<1	7.3	820	14	300	30	5-10
	1-3	13.4	1,360	16	250	30-60	5-10
	4-6	20.2	1,830	20	300	30-60	5-10
	7-9	28.1	2,190	25	400	30-60	5-10
Males							
	10-12	36.9	2,600	30	575	30-60	5-10
	13-15	51.3	2,900	37	725	30-60	9-18
	16-19	62.9	3,070	38	750	60	5-9
Females							
	10-12	38.0	2,350	29	575	50	5-10
	13-15	49.9	2,490	31	725	50	12-24
	16-19	54.4	2,310	30	750	60	14-28
Adult man (moderately active: using 2,600-3,400 kcal/day)							
		65.0	3,000	37	750	60	5-9
Adult woman (moderately active: using 2,000-2,400 kcal/day)							
		55.0	2,200	29	750	60	14-28
Pregnant			+350	38	750	+10	28[f]
Breast-feeding			+550	46	1,200	+30-35	28[f]

[a] Based on Passmore, et al. 1974.
[b] Strenuous activity will increase requirements for energy.
[c] Diets with plant foods as the main source of protein require higher intakes of protein.
[d] RE = retinol equivalent; see section 2.6.1.
[e] From NRC 1989:117-120.
[f] From Latham 1979; Cameron and Hofvander 1983.

established by the World Health Organization (WHO) may be more appropriate for most dryland areas.[4] Using RDAs to estimate nutritional needs provides an overview of the nutritional situation and suggests areas that need further investigation. However, they should never be used to give quick and definitive answers about nutrition problems. In Table 2.1 we provide RDAs for some nutrients covered in this chapter.

Tables listing the nutrient content of foods are estimates. As with the RDAs, they should be used with caution since the content of foods listed in these tables is based on samples and is affected by the following factors:
- The particular variety of the crop tested.
- Where and how the crops were grown.
- At what point in maturity the food was harvested.
- The time between harvesting and the time the analyses were conducted.
- Type of processing, if any.
- The combinations of foods analyzed.
- The technique used for the analyses.

Tables based on foods eaten locally are best, but these are not always available. The Food and Agriculture Organization (FAO) has published regional tables for Africa, the Near East, Latin America, and East Asia.[5] The U.S. Department of Agriculture publishes a series of detailed volumes that are often used outside the United States.[6] Table 2.2 gives examples of some garden foods rich in the nutrients we will discuss.

Table 2.2 Rich Garden Sources of Some Nutrients [a]

Nutrient	Garden sources	Example of nutrient/100 gm of edible portion
Energy	Sweet tree fruits, dried fruits and vegetables, nuts, seeds	Fresh tiger nut[b] 450 kcal/100 gm
Protein	Pulses, seeds, dried DGLVs*, especially for protein complementarity with staple	Dried sunflower seeds 23 gm/100 gm
Vitamin A	DGLVs, dark orange and yellow fruits and vegetables	Cooked amaranth leaves 277 RE[c]/100 gm
Vitamin D	Exposure to sunlight	Depends on altitude, latitude, season, and skin pigmentation
Vitamin C	DGLVs, fruits	Fresh guava 183 mg/100 gm
Folacin	Fruit, DGLVs, pulses	Avocado 66 mcg/100 gm
Thiamin (B_1)	Vegetables, fruits, seeds, nuts	Cooked cowpea leaves 0.3 mg/100 gm
Riboflavin (B_2)	DGLVs	Cooked amaranth leaves 0.1 mg/100 gm
Niacin	Fruit, nuts, seeds	Banana 0.5 mg/100 gm
Iron	Dates, figs, nuts, seeds	Dates 1.2 mg/100 gm
Zinc	Seeds, DGLVs	Squash seeds 7.5 mg/100 gm
Calcium	DGLVs, dried fruits, pulses, sesame seeds	Dried cassava leaves[d] 313 mg/100 gm

[a] Based on USDA 1982, 1984a, and 1984b, unless otherwise noted.
[b] Irvine 1969.
[c] RE = retinol equivalents; see section 2.6.1.
[d] Leung, et al. 1968.
* DGLVs = dark green leafy vegetables.

2.3 Special Nutritional Needs in Drylands

Table 2.3 summarizes the diet and health indicators of nutritional deficiencies in drylands. We will discuss them further in the sections on individual nutrients. In this section we cover the special needs of children, women, those doing heavy work, and sick people.

2.3.1 Children's Special Needs

Because of the nutritional demands of growth and development, especially in younger children, adequate nutrition is essential to produce healthy and productive adults (Figure 2.4). For example, girls 1 to 3 years old need twice as much vitamin A and C per kg of body weight as adult women who are not pregnant or lactating. Children with protein-energy malnutrition and vitamin A deficiency are at great risk of becoming blind if they get measles, a common disease in much of the Third World (section 2.6.1). The daily requirement for energy in infants under 3 months is 120 kcal/kg of body weight, but drops steadily to 40-45 kcal/kg by age 20.[7] The weaning period beginning at 6 months is critical; breast milk no longer provides enough nutrients, and nutritious weaning foods are often lacking (Chapter 16).

Child growth is an important sign of nutritional

Table 2.3 Some Indicators of Nutritional Deficiencies in Drylands

Nutrient	Function	Dietary indicators	Health indicators
Calories	Energy for work, to maintain body, to fight sickness	Not enough to eat; lack of concentrated energy sources, especially for children	Children underweight for age; marasmus; adults unable to work long or hard
Protein	Growth and repair of body	Not enough to eat; lack of high protein foods like nuts, seeds, legumes	Children short for age, hair color lighter than usual; kwashiorkor
Vitamin A	Vision, bone growth, healthy skin, fighting infectious disease	Lack of DGLVs*, orange fruits or vegetables, and fat	Loss of night vision, can lead to blindness, xerophthalmia; increased infectious disease, malnutrition, death
Vitamin C	Iron absorption, development of skin, bones, and teeth	Lack of fresh fruits and vegetables or overcooking them	Scurvy: bleeding, slow wound healing, poor development of bones and teeth; anemia (see iron)
Iron (Fe)	Formation of blood	Lack of nuts, seeds, legumes, fresh fruits and vegetables	Anemia: tiredness, breathlessness, pale under eyelids and fingernails
Vitamin D and Calcium (Ca)	Development of bones and teeth especially in children	Lack of exposure to sunshine, diet high in phytic and oxalic acids	Rickets: softening and malformation of bones and teeth
Thiamin (B_1)	Function of central nervous system	Diet of milled or polished grains, pulses, some root crops	Beriberi: weakness in legs, weight loss, marasmus in infants
Riboflavin (B_2)	Use of energy and protein, tissue growth and repair	Lack of vegetables	Cracked, dry, itching skin, especially on face and genitals, failure to grow in children
Niacin	Metabolism	Diets based on maize or sorghum	Pellagra: dementia, dermatitis, diarrhea
Folacin	Production of red blood cells	Lack of DGLVs	Anemia (see iron)
Zinc (Zn)	Growth and repair of tissue	Lack of whole grains, pulses or animal products	Failure to grow, abnormal development of sexual organs, wounds slow to heal

*DGLVs = dark green leafy vegetables.

status. If children in the community go regularly to a health clinic they may have clinic cards with their growth plotted against standard growth curves. As with RDAs, most growth standards in use are based on better-nourished populations in the United States or Europe, and are inappropriate for judging individual children's height or weight in the Third World. They are, however, a valuable tool in judging the growth pattern of a child or group of children. The pattern of growth for well-nourished children from all parts of the world is essentially the same, and a deviation from that pattern is a sign of malnutrition.

Figure 2.5 is a sample child growth chart similar to those used by child health clinics all over the world. The upper line on the chart shows the 50th percentile growth pattern in kilograms for boys based on WHO reference weights.[8] That is, in the WHO sample, 50% of boys weigh more and 50% weigh less at those ages. The

Figure 2.4 Children Need Good Nutrition to Grow, Work, and Play

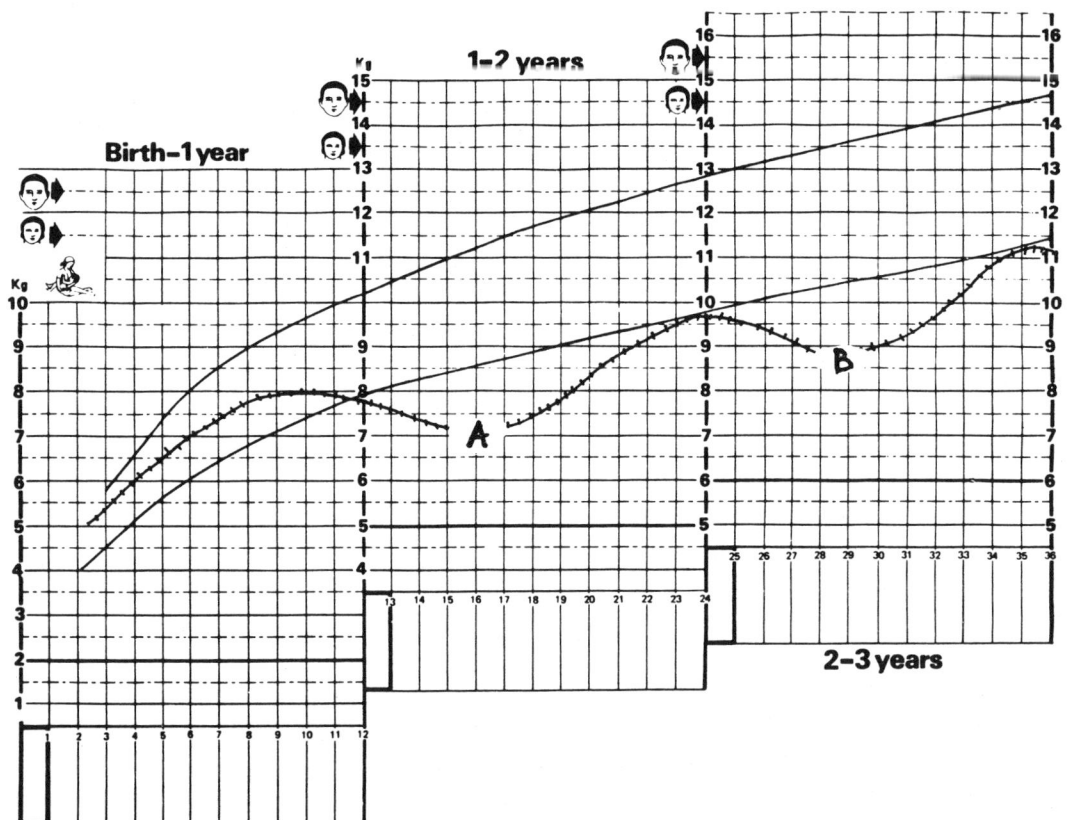

Figure 2.5 A Sample Child Growth Chart (from Werner 1977)

lower line is the third percentile for girls; 97% of girls weigh more, only 3% weigh less. The space between these two lines is called the "Road to Health" and it includes a wide range of possible weights for healthy children at different ages. But most importantly, the steady upward progress of the road to health outlines a healthy growth pattern. If a child's growth fluctuates up and down within the road to health or drops below it, she is much more likely to become seriously ill, and her diet needs immmediate improvement. The pattern in Figure 2.4 is very common in drylands and shows an overall lack of nutrients after weaning begins at approximately 6 months (A), and a seasonal drop in weight each year (B).

2.3.2 Women's Special Needs

Dryland diets often lack the extra nutrients needed by women during pregnancy, breast-feeding (lactation), and menstruation. Maternal malnutrition during pregnancy and breast-feeding can be harmful for the woman and may cause permanent physical and mental damage to the baby. Women who are malnourished during pregnancy are more likely to get sick and have low-birth-weight babies weighing less than 2.5 kg.[9] These babies often grow slowly, have decreased mental ability, and are more likely to get sick and die than babies with higher birth weights.

In general, the amount of all nutrients in the diet should be increased during pregnancy and lactation because, in addition to meeting her own needs, the woman is contributing to the nutritional requirements of her baby. For example, vitamin A requirements increase 50-60% during breast-feeding,[10] and requirements for vitamin C increase 15% in pregnancy and 50-60% during breast-feeding.[11]

Women may need as much as 150 kcal/day extra energy during the first 3 months of pregnancy, and 350 kcal/day extra during the last 6 months.[12] This is in addition to the 2,200 kcal/day required by a 55-kg woman who is moderately active. If the woman is doing heavy work during pregnancy, as many women in the Third World do (Figure 2.6), then her need for energy will increase even more. For example, hoeing requires about 3 kcal/min more energy than that needed for moderate activity such as light work or walking. If the woman was hoeing for 4 full hours (4 x 60 min = 240 minutes) each day, she would need 720 kcal/day more (240 minutes x 3 kcal/min). Therefore, a woman who is 6 months pregnant and hoeing weeds for 4 hours a day would need about 3,270 kcal/day (2,200 + 720 + 350), or 50% more energy than usual. Breast-feeding also requires added energy, estimated to be at least an additional 550 kcal/day, especially during the first 6 months.[13]

Another nutrient important during pregnancy and lactation is calcium. In pregnancy extra calcium is needed for growth of the baby's skeleton. Breast-feeding women lose approximately 245 to 280 mg of calcium each day.[14] If these needs are not met in the diet, calcium is obtained from the mother's bones in a process called *demineralization*, making her bones weaker and more susceptible to breaking. For this reason calcium intake should more than double, going from 400-500 mg/day for adult women to 1,000-1,200 mg/day during pregnancy and lactation. Because of the loss of blood during menstruation and childbirth, the iron RDA for women in their reproductive years (14-28 mg/day) is three times higher than that for men (5-9 mg/day).

2.3.3 Work

In farming communities the labor demands of food production take priority over all others. However, many other important activities make demands on labor, for example, collecting water and fuel, child care, animal husbandry, gathering wild foods, craft production, house building, and repair. In urban areas wage labor and marketing may be the priorities.

The greatest demand for labor in many farming communities comes when staple food supplies from the previous harvest are low. This period is called the *hungry season*, and is common during the early rainy season in semiarid and subhumid drylands where the staple crops are rain-fed. This is the time when crops have to be planted and fields weeded. Without enough to eat people may be unable to do the necessary work, and so yields are low.[15] People seen resting when there is work to be done may literally not have the energy to do it.

The time which Mossi men in Burkina Faso spend resting during the dry season may be necessary for them to recuperate from their strenuous agricultural work during the rainy season.[16] Mossi women also work hard at farming during the wet season, but they do not have the benefit of a dry season rest period, since they must maintain the household and continue marketing and craft activities during that season.[17] Although men expend more energy during the peak

Figure 2.6 In Many Places Pregnant and Lactating Women do Hard Physical Work

labor period, the fact that they may rest while women must continue working means that women may actually be more overworked.[18] Although these people do not have dry season gardens, some nearby villages do. Neither women nor men would probably be interested in dry season gardens unless these gardens could eliminate the need for some current work by providing substantially more food or income.

2.3.4 Illness

Illness increases the need for energy and other nutrients. For example, diarrhea reduces the absorption of nutrients and fever uses up extra energy. Any disease causing loss of blood such as schistosomiasis or hookworm increases the body's need for iron. Therefore a sick person often needs special foods, and those recovering from an illness may also need extra food even after the symptoms are over. As already mentioned, it is important to care for these needs because malnutrition increases the effects of the illness.

The occurrence of many diseases in drylands varies according to the season. For example, the incidence of malaria increases dramatically in the rainy season when the number of mosquitoes increases. Gastrointestinal diseases may occur more frequently during the harvest or planting seasons because long work days often mean that food may be prepared early in the morning, long before being eaten later in the day. This delay allows disease-causing microorganisms to contaminate foods, making those who eat them sick.

2.4 Energy

Energy is needed for all of the body's functions. People who are working hard, who are sick, or who are exposed to cold, need extra energy, as do growing children, and pregnant and lactating women. The prevalence of malnutrition among the poor in the Third World is due not only to hunger and poverty, but also to the fact that in comparison to the wealthy in the industrialized world they do more physical work, are sick more often and more severely, and frequently have inadequate clothing and housing.

In many dryland areas energy requirements vary seasonally with changes in the type and amount of work being done. For example, in central Burkina Faso Mossi farmers use the greatest amount of energy in the rainy season when most farm work is done.[19] Men use approximately 2,410 kcal/day and women 2,320 kcal/day in the dry season. However, in the rainy season this increases to 3,460 kcal/day and 2,890 kcal/day, respectively.

Carbohydrates (CBHs) are the sugars and starches burned for energy in the body. Carbohydrates in the form of staple cereals or roots are the major source of energy in diets of the poor in drylands. Fats, from animal or plant products, are a very concentrated source of energy but are usually more expensive than carbohydrates and make up a relatively small proportion of most Third World diets. Protein will be used for energy only if the body does not have adequate amounts of other energy sources.

While dryland gardens will not be the main source of energy, they may supply energy in convenient forms and at times of the year when major sources are unavailable. This is especially important for weaning children, the group in the population most likely to become malnourished. For example, the caloric density of most fresh fruits and vegetables is only about one-sixth or less than that of grain, but this increases several times when they are dried. The energy content of some vegetables grown in household gardens such as tiger nut, and dried sweet potato, yam and cassava, and of many dried fruits, approaches or exceeds that of grains, which is about 350 kcal/100 gm.

2.5 Protein

Proteins are composed of *amino acids*, organic compounds of carbon, hydrogen, oxygen, and nitrogen atoms. The protein in food is broken down in the body into amino acids. The amino acids are then used to make the kind of proteins humans need to build skin, liver, brain, and all of the other tissues and organs, and to replace daily losses in sweat, feces, urine, skin, hair, and nails. *Enzymes* are made of proteins and play a vital role in chemical reactions like digestion, vision, movement, reproduction, and thinking.

The protein in the human body is made up of 22 different amino acids. If not present in the diet, most of these can be manufactured in the body from other amino acids. However, eight of them cannot, and must be contained in the diet. These are called the *essential amino acids* (EAAs) (Box 2.1). Infants require a ninth amino acid, histidine, which they receive in breast milk. It is also present in other animal milks and eggs.

In many Third World drylands, *cereals* (grains) are the staple foods, providing not only most of the calories, but most of the dietary protein as well. Therefore, diets that lack an adequate quantity of the staple food lead to *protein-energy malnutrition* (PEM, also referred to as PCM, protein-calorie malnutrition), a significant problem in many drylands.[20] A severe deficiency of food causes *marasmus*. People suffering from marasmus are extremely underweight; children with marasmus fail to grow because they are literally starving. Marasmus is common in weaned children, and also occurs in breast-feeding children who are not receiving enough breast milk or supplemental foods.[21]

Kwashiorkor is a form of PEM caused by a relatively greater lack of dietary protein than calories, and is especially common in weaning-age children.[22] The growth rate of these children is slow and they have a

> **Box 2.1**
> *Net Protein Utilization and Protein Complementarity*
>
> Amino acids are not stored in the body, but are used soon after digestion for *protein synthesis*, that is, they are assembled in different ways to form various types of protein molecules. Amino acids not used for protein synthesis are excreted. Therefore, the most useful dietary protein is that which contains the EAAs in the same proportions as those required by human beings for protein synthesis. The use of protein in the diet is limited by that EAA which is most deficient compared with how much of it is required for protein synthesis. The value of all other EAAs present will be proportionally reduced (Figure 2.7). *Net protein utilization* (NPU) is the percentage of dietary protein in a food or a meal that can be absorbed and used by the body. For example, sesame seeds contain about 18% protein and have an NPU of 53%.[23] Thus 100 gm of sesame seeds would provide a person with 9.5 gm of usable protein (100 gm seeds x 0.18 = 18 gm protein; 18 gm protein x 0.53 = 9.5 gm).
>
> Animals are more closely related to humans evolutionarily than plants are. Therefore animal proteins in eggs, meat, and dairy products have a higher NPU than most plant proteins. However, when plant proteins are eaten in the right combination the resulting NPU can easily equal that of animal protein[24] because of protein complementarity. *Protein complementarity* means that the combination of foods in a meal compensates for shortages of specific EAAs in individual foods, making more protein available to the body. Through protein complementarity, fruits, vegetables, nuts, and seeds from the garden can increase the amount of protein available in the diet. For example, many dryland garden vegetables are high in the amino acid lysine but low in tryptophan, so combining them in meals with most cereals, which are low in lysine but high in tryptophan, increases the protein contribution of the garden.

swelling called *oedema*, which begins in the feet and legs, later spreading to the rest of the body. When the depression caused by pressing a finger on the flesh above the ankle is slow to disappear it is a sign of oedema.

Although the symptoms are not as obvious, *mild* or *moderate PEM* is far more common than the severe conditions of kwashiorkor or marasmus. For example, it is estimated that in the African Sahel 30% of the children between 1 and 2 years old suffer from moderate PEM.[25] The main signs of this condition are slow or no growth, loss of muscle tissue, a protruding belly, thin arms and legs, and sometimes a lighter than normal hair color. Children suffering from PEM are listless, weak, and unhappy. Protein-energy malnutrition also makes children more vulnerable to disease which, in turn, worsens their nutritional condition even further.

As with energy, gardens are not the main source of protein. However, they may supply protein in convenient forms and at times of the year when major sources are unavailable. Again this is especially important for weaning children. Dried seeds and pulses are often recognized as good protein sources, but other garden produce can also provide concentrated protein, especially when dried. For example, the dark green leaves of jute, cowpea, and pumpkin, widely eaten in Third World drylands, are only 4% or more protein by weight when fresh, but are 20% to 35% protein when dried.[26]

While many nutrition programs have emphasized increased consumption of protein in the form of animal products, this strategy is not necessary for improved nutrition and can contribute to other health problems, as well as having serious negative social and environmental consequences.[27] Increased consumption of animal products is often associated with increased cardiovascular disease, obesity, and cancer. The social and environmental costs of increased consumption of animal products occur because the animals are fed large quantities of grains and pulses. For example, it has been estimated that 16 lbs of grain and soybeans are required to produce 1 lb of beef.[28] As this food is diverted from feeding people directly to producing meat and milk, less food is available to people, and it is produced at a higher cost. In many areas increased consumption of animal products is accompanied by an accentuated division between the small, affluent minority who are consuming those products and the vast majority whose diets are wors-

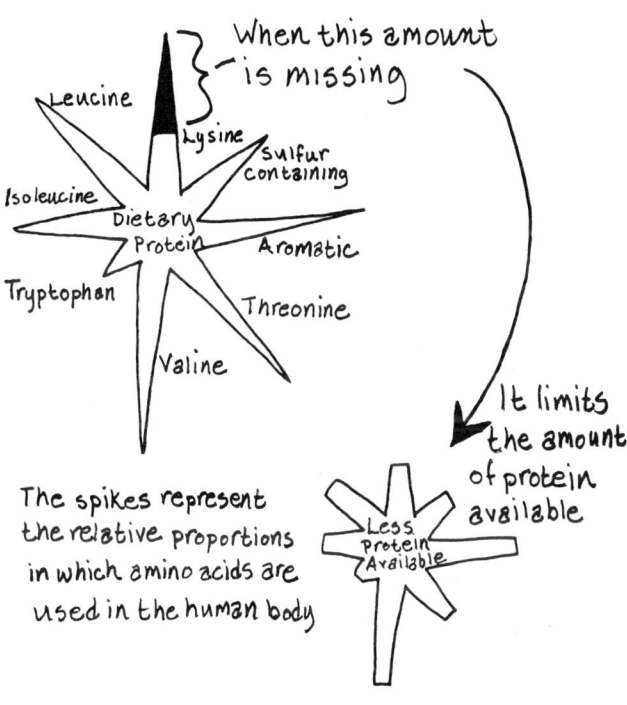

Figure 2.7 *Limiting Amino Acids* (from Lappé 1982:175)

ening, as in Mexico and Africa.[29] Not only are grains diverted from feeding people to feeding animals, but valuable land, soil, and water are invested in feed production as well. When grain production is not enough to meet the demands of the rich for animal products, grain is imported, using scarce foreign exchange.

2.6 Vitamins

Many fruits and vegetables from the garden are rich in vitamins, especially A and C. *Vitamins* are organic substances that occur naturally in plant and animal tissue. They are essential for the functioning of the body and must be contained in the food we eat because the human body cannot make them. An exception is vitamin D which is produced in the skin when it is exposed to sunlight. *Fat-soluble vitamins* (A, D, E, K) require fat in the diet to be absorbed by the body, and they can be stored in the fatty tissues of the body. *Water-soluble vitamins* (including vitamin C, folacin, thiamin, riboflavin, and niacin) are those that dissolve in water, so that any excess is excreted; foods containing them must be eaten more frequently. Here we discuss seven of the most important vitamins for dryland diets: A, C, D, folacin, thiamin, riboflavin and niacin. See Table 2.2 for examples of vitamin content of foods grown in dryland gardens.

2.6.1 Vitamin A

FUNCTION One of the most important functions of vitamin A, also called *retinol*, is in vision. A deficiency of this vitamin leads to a deterioration of the eyes, a disease known as *xerophthalmia*. Early signs include loss of vision in dim light (night blindness), while continued deficiency leads to blindness, especially in young children. Vitamin A is also necessary for bone growth and healthy skin, and vitamin A deficiency is associated with increased risk of sickness and death. Evidence from recent research strongly suggests that vitamin A deficiency increases the risk of death due to infectious diseases (such as measles, pneumonia, tuberculosis, dysentery, and gastrointestinal infections).[30] In turn, infectious disease and PEM may increase the risk of vitamin A deficiency.[31] Vitamin A appears to be important for maintaining the tissues that protect the respiratory, stomach, and gastrointestinal systems, and the genital organs. A deficiency results in a breakdown of this protective barrier, leaving the body vulnerable to infection. This vitamin may also play a role in enhancing the body's immune system.[32] For these reasons vitamin A deficiency is seen as a serious threat to child health and survival and is becoming an important focus for efforts to improve children's nutrition worldwide.

PROBLEMS IN DRYLANDS Vitamin A deficiency is one of the major nutritional deficiencies in the world. It often fluctuates seasonally, increasing when green vegetables and fresh fruits are not available. In India, for example, it is estimated that vitamin A deficiency contributes to 52,000 children becoming blind and 110,000 to 132,000 partially blind every year.[33] Vitamin A deficiency is probably high in most drylands, though there is not much detailed information for Africa or the Middle East. Some food surveys in the African Sahel suggest that in the dry season only 50% of vitamin A needs are met by the diet.[34] Deficiency can result not only from a lack of vitamin A in the diet, but from a low-fat diet, since vitamin A is a fat-soluble vitamin. Deficiency may also occur due to overall poor nutritional status, especially PEM, and due to infectious diseases, all of which reduce the body's ability to absorb vitamin A.

REQUIREMENTS A daily intake of 750 mcg of retinol is suggested for adults and at least 300 mcg for young children. This means that just 30-100 gm (1-3 oz) of vitamin A-rich food can meet the RDAs for infants and children. Breast-feeding women who provide their infants with retinol in breast milk need approximately 1,200 mcg daily. Several months' supply of vitamin A can be stored in the liver, so it is a good idea for people to eat lots of vitamin A-rich food when available, to make up for a shortage later on in the year. In savanna West Africa, for example, mangoes are rich in vitamin A and are often abundant at the end of the dry season. Children love mangoes and will devour lots of them if they get the chance.

SOURCES Fruits and vegetables contain *provitamin A* (carotenes, the precursors of the vitamin) primarily in the form of *beta-carotene*, which is transformed in the body into the active form of vitamin A, retinol (also called *preformed vitamin A*). Carotenes are substances with yellow/orange pigment which gives carrots, mangoes, sweet potatoes, loquats, and papayas their typical color. Dark green leafy vegetables, especially young leaves, are also rich in carotene, but the color is masked by the green of the chlorophyll. In general the darker orange or green the vegetable, the higher its carotene content. For example, there is much more carotene in dark green leaf amaranth than in light green head cabbage or lettuce. Nonindigenous fruits and vegetables with high prestige value often have lower vitamin A content than indigenous fruits and vegetables, leading to poor vitamin A nutrition as the former replace the latter. This is the case in northeast Brazil where apples, pears, oranges, and lemons are replacing local palm fruits, mangoes, passion fruit, and papayas in the local diet, contributing to vitamin A-deficiency in the region.[35] Some animal products such as eggs, butter, milk fat, and liver contain retinol, but these foods are often not available or are too expensive, except, perhaps, for people who herd their own animals.

Red palm oil is sometimes advocated as a source of vitamin A in drylands. Oil palms grow in the humid tropics, however, and buying imported oil is likely to be too expensive for poor dryland households. Sometimes vitamin A is added artificially to foods, such as biscuits, sugar, or monosodium glutamate (MSG, a traditional seasoning in Asia). However, their purchase may not only limit more important uses of household income, but these items may themselves be associated with increased health problems. Fortunately there are many rich sources of vitamin A that can be easily grown in household gardens.

Currently the standard practice is to give the vitamin A content of foods in retinol equivalents. However, some food composition tables still use international units (IU) of beta-carotene. To convert these to the retinol equivalents used in this book, the following equivalencies can be used:[36]

1 retinol equivalent (RE) equals
= 1 mcg retinol
= 6 mcg beta-carotene
= 12 mcg other provitamin A carotenoids
= 3.33 IU vitamin A activity from retinol
= 10 IU vitamin A activity from beta-carotene.

2.6.2 Vitamin D

FUNCTION This fat-soluble vitamin is important for the regulation of calcium and phosphorus in the body to make strong bones and teeth, especially in infants and children (section 2.7.3). The ultraviolet rays in sunlight cause vitamin D to be formed from a naturally occurring substance in the skin. It is also present in certain foods. *Rickets* is a disease resulting from vitamin D deficiency and causes softening and malformation of the bones. PROBLEMS IN DRYLANDS In most dryland tropical and subtropical countries there is enough sunlight to supply people's vitamin D requirements. However, there are vitamin D deficiencies in Ethiopia; in the larger, crowded cities of drylands in the Near East and Asia;[37] and where cultural or religious values stress covering most of the body, such as among some Moslems and Hindus.[38] Rickets may become a problem during and after weaning if the vitamin D present in breast milk is not replaced by exposure to the sun or supplied by another dietary source.

REQUIREMENTS The amount of sunlight needed to provide someone with their vitamin D requirement depends on the *altitude* (height above sea level) or *latitude* (how close to the equator) of where they live, the season, the frequency and length of exposure, and how dark their skin is.[39] People with darker skin colors must be exposed longer than those with lighter skin colors. The higher the altitude or the closer to the equator, the less exposure required. The elderly can only synthesize half as much vitamin D from sun exposure as a similar younger person can. It is difficult to set a dietary requirement since diet is not the only source of this nutrient. SOURCES Dryland gardens may contribute in two ways to meeting vitamin D requirements. They can provide eggs whose yolks are a good source. They can also provide a place of seclusion where, while doing other work, mothers can expose a portion of both their own and their baby's skin to sunlight for a brief period every day. Unless fortified, cow's milk is not a good source of this vitamin.

2.6.3 Vitamin C

FUNCTION Vitamin C (*ascorbic acid*) is a water-soluble vitamin needed for absorption of iron in the diet. This vitamin is also important for the formation and maintenance of the skin, bones, and teeth. **PROBLEMS IN DRYLANDS** Vitamin C deficiency is often associated with the iron deficiency anemia so prevalent in drylands (section 2.7.1). Lack of vitamin C also causes *scurvy*, characterized by bleeding, slow healing of wounds, and poor development of bones and teeth in children. Severe cases can cause death.

REQUIREMENTS Since vitamin C is constantly being lost from the body, a continual supply is needed in the diet. The following RDAs are for the United States and the requirements of Third World dryland populations may be greater. A daily intake of 60 mg/day for adults and 30 mg/day for children maintains an abundant quantity in the body,[40] but requirements for vitamin C increase during the infections and fevers common among children and adults of the drylands. Between infancy and adulthood the requirement increases from 30 to 60 mg/day. Pregnant women transfer vitamin C to their babies so they need about 10 mg/day more than the adult requirement. Breast milk is high in the vitamin and so lactating women need 30-35 mg/day more to compensate for this.

SOURCES Many dryland fruits and vegetables are excellent sources of vitamin C. Some examples are guavas (one of the richest sources with 326 mg/100 gm), tomatoes, citrus, papayas, chilis, sweet peppers, mangoes, and DGLVs like amaranth or baobab leaves (Figure 2.8).

2.6.4 Folacin

FUNCTION Folacin (also called *folate* and *folic acid*) is a water-soluble vitamin whose deficiency is the cause of the second most common type of nutritional anemia after iron deficiency anemia.[41] **PROBLEMS IN DRYLANDS** Anemia due to folacin deficiency is especially common in pregnant women. **REQUIREMENTS** Daily intakes of 100 mcg for children, 200 for adults, 400 for pregnant women, and 300 for breast-feeding women have been suggested.[42] **SOURCES** Dark green leafy vegetables, fruit, yeast, and *pulses* (dried legume seeds, such as beans, which are cooked for eating) are good sources of folacin, as are kidney, liver, and cow's milk.

2.6.5 Thiamin (B_1)

FUNCTION Thiamin is necessary for functioning of the central nervous system. A deficiency in this water-soluble vitamin results in loss of reflexes and muscle control, especially in the legs. *Beriberi* is a disease of the nervous system which results from thiamin deficiency.

PROBLEMS IN DRYLANDS In cereals thiamine is found primarily in the germ and outer seed coat and so deficiencies of this vitamin occur where polished rice or other refined grain is the staple. Starchy roots (such as cassava) are also very low in thiamine and diets based on these roots can lead to deficiency.

REQUIREMENTS Only about six weeks supply of thiamin is maintained in the body. The daily adult intake suggested for thiamin is approximately 0.4 mg/1,000 kcal with increases of 0.1-0.2 mg/day more for pregnant and lactating women.[43] Because thiamin is essential for the metabolism of carbohydrates, diets that rely heavily on carbohydrates for energy require more of this nutrient; high-fat diets need less. **SOURCES** Thiamin is present in green vegetables and fruit, seeds, nuts, yeasts, pulses, and unrefined cereals. Fish, meat, and milk, including breast milk, are also sources of thiamin. Especially good dryland garden sources (per 100 gm edible portion) include cashew nuts (0.2 mg) and cooked, leafy cowpea vine tips (0.3 mg).

2.6.6 Riboflavin (B_2)

FUNCTION Water-soluble riboflavin is necessary for chemical reactions in the body which synthesize protein, release energy from foods, and grow and repair body tissues. **PROBLEMS IN DRYLANDS** A deficiency of riboflavin causes lesions of the skin, especially the lips and corners of the mouth, and the tongue becomes a purplish red color. **REQUIREMENTS** Requirements for riboflavin are currently being reexamined. Based on extensive studies in China, some researchers are suggesting an RDA of 0.6 mg/day for adults, half the previously accepted RDA.[44] The research in China has not yet led to new RDAs for riboflavin during pregnancy and lactation. The current official RDA for the United States is an additional 0.3 mg/day during pregnancy and 0.5 mg/day during lactation.[45]

SOURCES Riboflavin is found in dark green vegetables, yeast, milk and milk products, insects, eggs, fish, and meat. Riboflavin in pulses such as lentils and fava beans is increased by sprouting (section 15.5.1). Fermentation increases the riboflavin content of milk.

Figure 2.8 Green Leaves are a Good Source of Vitamin C

2.6.7 Niacin

Function Niacin is a water-soluble vitamin required by the body to metabolize carbohydrates, fats, and proteins. Without it many essential chemical reactions cannot occur. The body also creates niacin from the EAA tryptophan at the rate of 1 mg of niacin from 60 mg of dietary tryptophan. **Problems in Drylands** The disease resulting from a deficiency of niacin is called *pellagra*. Symptoms include skin lesions, changes in tongue color, diarrhea, and mental disturbances (the three Ds: dermatitis, diarrhea, dementia). Niacin deficiency is common in Africa,[46] especially where maize is the staple because the niacin in maize is unavailable to humans, and that grain is also low in tryptophan. Niacin deficiency can be avoided by supplementing maize-based diets with beans and other sources of tryptophan, as is often done traditionally. Niacin deficiency is also found in diets based on cassava tubers and may be associated with a staple diet of sorghum as this cereal is high in the amino acid leucine which can interfere with tryptophan and niacin metabolism.[47]

REQUIREMENTS Approximately 5-15 mg of niacin/day is recommended for small children, 15-20 mg/day for adolescents and adults, and increases of 2 and 5 mg/day for pregnant and lactating women, respectively.[48] SOURCES Bananas, groundnuts, cowpeas, dried chilis, and sesame and sunflower seeds are all excellent sources of niacin. Another good source is yeast from beer brewing; in some countries of savanna West Africa yeast is often added to soups. Most pulses and cereals do not have a high niacin content, but where they are the staple they may supply most of the niacin in the diet. Insects and meat are also good sources of niacin.

2.7 Minerals

Like plants, people need mineral elements to stay alive. Seven elements account for 60-80% of the minerals in the body: sodium (Na), potassium (K), chlorine (Cl), sulfur (S), calcium (Ca), phosphorus (P), and magnesium (Mg).

In addition, eight other minerals, referred to as *trace elements*, are needed in very small (trace) amounts: iron (Fe), zinc (Zn), iodine (I), copper (Cu), manganese (Mn), chromium (Cr), selenium (Se), and molybdenum (Mb). We discuss only iron, zinc, and calcium because deficiencies of these minerals are the only ones common in drylands that can be easily addressed by household gardens.

2.7.1 Iron (Fe)

FUNCTION Iron is necessary for the formation of *hemoglobin*, a protein in red blood cells that carries oxygen from the lungs to all cells of the body. PROBLEMS IN DRYLANDS Iron deficiency anemia is a major health problem in drylands. This nutritional anemia is caused by dietary lack of available iron or of vitamin C. Vitamin C aids in iron absorption if eaten at the same time as foods containing iron. Signs of this anemia include tiredness, breathlessness, and pale skin (caused by lack of red blood) under the eyelids, inside of cheeks, and beneath fingernails. A spoon-shaped deformity of the nails on both hands may also be present. Vitamin C in the diet should be increased after any loss of blood, for example from hookworm, menstruation, and childbirth.

REQUIREMENTS[49] Depending upon diet and work, the RDAs for men and postmenopausal women are 5-9 mg/day and for young children, 5-10 mg/day. The higher requirements are for populations whose diets are primarily based on plant foods, as is common in the Third World. For women in their reproductive years, 14-28 mg of iron/day is suggested, the most being required during pregnancy and breast-feeding.

SOURCES There is a high iron content in many dryland garden crops including dates, figs, nuts, seeds, beans, and asparagus, as well as most DGLVs. However, only 5-10% of this iron is in a form that can be absorbed by humans. The iron in animal products is several times more absorbable than that in plant foods. Also, the iron content of plants depends on the iron content of the soil. Millet, wheat, yeast, and organ meats (e.g., liver, kidney) are good sources of this nutrient, although phytates in some cereals, nuts, and pulses lower iron absorption (section 2.10). Using iron utensils and cooking pots increases the amount of iron contained in the food.

2.7.2 Zinc (Zn)

FUNCTION Zinc is necessary for normal growth, sexual development, and reproduction. A deficiency leads to loss of appetite, failure to grow and to develop sexually, slow wound healing, decreased sense of taste, and changes in skin texture. PROBLEMS IN DRYLANDS Zinc deficiency is found in the Near East, especially Iran and Egypt. The availability of dietary zinc is reduced by binding with phytates and fiber,[50] which may be a problem with some dryland diets. REQUIREMENTS The RDA for the United States of 15 mg/day for men and 12 mg/day for women[51] assumes regular consumption of animal products. Absorption of zinc in persons on largely vegetarian diets may be limited, due in part to phytates and fiber in the diet, and so the RDA may be higher. SOURCES Most seeds are high in zinc; for example, pumpkin seeds have 7 mg/100 gm. Zinc is also found in DGLVs, pulses, eggs, milk, seafood, and meat. The zinc content of plants (and of animal products formed from them) depends upon the zinc content of the soil.

2.7.3 Calcium (Ca)

FUNCTION Calcium is essential for the formation and repair of bones and teeth which contain 99% of the body's calcium. In the blood stream this mineral is needed for the muscles and nerves and for coagulation of blood.[52] **PROBLEMS IN DRYLANDS** Dryland diets high in phytates and oxalic acid may lead to calcium deficiencies (section 2.10). **REQUIREMENTS** Within limits, the body appears to adapt to varying intakes of calcium by adjusting the amount of the mineral excreted.[53] In addition, the more protein consumed the less efficient the body's use of calcium, with large amounts excreted by individuals on high-protein diets. For this reason the RDA for calcium is higher for populations with high-protein diets, for example, children and adults in the United States (800 mg/day), as compared with the RDAs for populations whose diets contain less protein, for example, children (400-700 mg/day) and adults (400-500 mg/day) in Africa.[54] **SOURCES** DGLVs, some pulses, hulled sesame seeds, yeast, and dairy products are all sources of calcium.

2.8 Fats

Dietary fats can be either solid like butter or lard or liquid like groundnut oil. They provide twice as much energy per volume as protein or carbohydrates, and are therefore good additions to weaning foods (section 16.3.1). Fat in the diet is needed to digest the fat-soluble vitamins A, D, E, and K. Fats contain fatty acids, some of which are essential for the human body.[55] Plant oils like those found in avocados, olives, groundnuts, and seeds such as sesame and melon are a good source of fatty acids. In the body, fat is the storage tissue formed when more energy is consumed than is immediately needed.

2.9 Fiber

Dietary fiber such as *cellulose* is an indigestible carbohydrate, or carbohydrate-like substance in plant food. Fruits, vegetables, and pulses from the garden are sources of fiber, as are unrefined grains. Fiber is important for the movement of food through the intestine. It encourages the growth of beneficial bacteria there and helps prevent some diseases in the digestive and excretory systems. The kind of fiber in fruit, vegetables, and pulses also helps reduce cholesterol levels.

Cholesterol is a waxy substance in the blood essential for body functioning—but too much of it clogs the arteries causing heart disease, a major problem in the industrial world and some areas of the Third World.

Too much fiber in the diet, however, can reduce absorption of many minerals because the food is moving too quickly through the body. This is especially important in the diets of young children. Some of the processing and preparation methods discussed in section 16.3, such as removing bean "skins," help reduce fiber content of garden produce used in weaning foods.

2.10 Anti-Nutrients

Anti-nutritional factors, or *anti-nutrients*, are substances found in most foods, and that are poisonous or in some way limit the nutrients available to the body. Plants have evolved these chemicals to protect themselves from being eaten.[56] Frequently the anti-nutrients occur in such small quantities that they cause no harm. But if the diet is not varied some of these toxins can build up in the body to harmful levels. In many areas traditional food processing techniques effectively eliminate any harmful effects of anti-nutrients occurring in the diet. The anti-nutrients discussed below are summarized in Table 2.4.

Some of the most common anti-nutrients found in dryland garden foods are those that form insoluble salts with minerals like calcium and iron, reducing absorption of these minerals by the body. *Phytates* occur in cereals, pulses, and nuts. People consuming foods high in phytates may develop symptoms of calcium deficiency, especially in areas or seasons where calcium intake is low. Phytates also appear to decrease iron and zinc absorption.[57] To reduce phytate content one may soak pulses and discard the water or sprout them.[58] The slow fermentation process used to leaven many traditional grain breads also reduces phytates. The longer the fermentation the fewer the anti-nutrients that remain in the bread.

Relatively large amounts of *oxalates* are found in some DGLVs like leaf amaranths and onion greens, as well as in purslane and in members of the chenopod family such as spinach and chard. A diet high in oxalates can cause calcium deficiency and could eventually lead to kidney damage.[59] Boiling or steaming these vegetables and then rinsing them and discarding the water reduces their oxalate content. Sesame seed hulls contain oxalates and should be removed to make the calcium in the seed available.

Table 2.4 Anti-Nutrients Common in Dryland Diets

Anti-nutrient and source	Effect	Ways to minimize
Phytates: cereals, pulses, nuts	Reduce Ca and Fe absorption	Soak (discard water), and sprout cereal before cooking, slow fermentation of bread doughs
Oxalates: DGLVs, sesame seed coats, fruit	Reduce Ca absorption, encourage formation of kidney stones	Steam or boil greens, rinse and discard the water, hull sesame seeds, eat fruit soon after harvest
Tannins: dark-colored pulses, sorghum	Reduce digestion of proteins and CBHs	Soak (discard water), sprout and cook, remove dark seed coats
Oligosaccharides: pulses	Intestinal gas, discomfort, loss of appetite	Sprout, ferment, cook thoroughly
Hydrogen cyanide (HCN): crucifers, sorghum sprouts, cassava, cashews, groundnuts	Contribute to goiter if diet is low in iodine	Varied diet, more iodine consumption, cook sorghum sprouts, ferment cassava, eat crucifers that are young because age increases HCN content
Aflatoxin: many pulses, nuts, and seeds	Liver damage, cancer	Dry, cool storage conditions, do not store damaged pulses, nuts, or seeds

Some vitamins in food may be destroyed by anti-nutritional substances. *Ascorbase* is an enzyme released by plant cells in response to damage such as harvesting.[60] This enzyme starts a reaction eventually changing ascorbic acid to oxalic acid, an oxalate. The ascorbase content of fruits and vegetables varies both between species and varieties. Because this reaction increases during storage, especially in warm conditions, it is best to harvest most garden produce as close as possible to the time when it will be eaten (section 15.2).

Tannins are substances in food that reduce digestion of proteins and carbohydrates. The foods from dryland gardens highest in tannins are pulses, especially their seed coats. Tannins have a dark color; pulses with light-colored seed coats contain a negligible amount of tannins. Removing the seed coat, as is traditionally done when making *dhal* in India, removes 83-97% of the tannins in pulses.[61] Simple practices, already used in the preparation of pulses in many areas, such as soaking and discarding the soaking water, germination, and cooking are all ways to remove tannins. Sorghum can also contain significant amounts of tannins. In southern Africa where a variety of drought-resistant sorghum with a high tannin content is eaten it is associated with an unusually high incidence of cancer of the esophagus.[62]

Other anti-nutritional factors found most often in pulses are *oligosaccharides* which increase the formation of intestinal gas, often causing discomfort and loss of appetite. Sprouting, fermentation, and thorough cooking reduce oligosaccharides.[63]

Some foods contain substances that react to form the poison hydrogen cyanide (HCN) either during processing or digestion. Where diets are low in iodine, HCN is a cause of a disease of the thyroid gland called *goiter*. Sources of HCN in dryland diets are cashew nuts, groundnuts, and many vegetables in the crucifer family. For people not suffering from goiter, consuming these foods as part of a varied diet is not a problem. Sorghum sprouts and cassava are high in HCN. However, if the sprouts are cooked or malted and the cassava is fermented, both traditional practices in West Africa, the HCN content is reduced to safe levels.[64] Lima beans are also high in cyanides and should not be eaten raw. Lima beans should be presoaked, drained, boiled, and drained again to rid them of most of this toxin.

Some molds growing on crops (section 13.3.4) and foods produce poisons called *mycotoxins*. The mycotoxin *aflatoxin* is found in molding dryland garden foods such as groundnuts and other pulses, nuts, and seeds. If eaten, it can cause severe and sometimes deadly liver damage. Aflatoxin-producing molds grow under warm, moist conditions, and so care must be taken when storing these foods during the rainy season. Damaged and broken pulses are especially vulnerable, and any that look discolored, are moldy, or have a bad smell should be composted, not eaten. They should not be fed to animals as they too can be poisoned and the toxin can be passed on to people through meat and milk.[65]

2.11 The Effects of Gardens on Nutrition

The nutritional effect of gardens has seldom been measured, partly because it is so difficult to do. However, gardens can make a significant contribution to solving three of the most important dryland nutritional problems: PEM of infants and children, vitamin A deficiency, and anemia resulting from lack of iron and vitamin C.[66]

Even when working with gardens on the community and household level, the path between food production and nutrition is complex. Availability and access to land, good soil, water, and seeds or cuttings determine whether food can be produced. When gardens can be grown their influence on the nutritional status of household members depends on many factors including:
- The amount of different foods harvested from the garden during each season of the year.
- The quantity and quality of the nutrients in garden produce.
- The availability of nutrients in garden produce.
- Methods of storage and processing.
- Distribution of garden produce to different members of the household.
- The amount of produce not consumed by the household, e.g., because it is sold.
- The way that food from the garden is combined in meals with other foods.
- The health and activities of household members.

For example, an assessment may find that children in households with gardens are just as deficient in vitamin A as children in households without gardens. The gardens are not providing the children with the benefits they need. In surveys in Indonesia, for example, it was found that 80% of families with children who had a vitamin A-deficiency-caused eye disease consumed DGLVs once a day, and 99% once a week, the same frequency as families of children without the vitamin A deficiency disease.[67] To understand the potential contribution of gardens to improving vitamin A nutrition in this case, we would need to know what causes the difference in the vitamin A nutrition of children in families who consume similar amounts of DGLVs. Is there a difference in the distribution of DGLVs in these families? Are foods prepared differently? Are children's eating habits different?

Many of the dryland garden crops that are excellent sources of energy, protein, vitamins, minerals, fats, and fiber have been presented in sections 2.4 to 2.9. In the following sections we discuss the nutrient yields of gardens and the contribution of gardens to nutrition.

2.11.1 Nutrient Yields from Gardens

The nutrients produced in indigenous Third World gardens have rarely been studied, but data from other types of gardens is available. Research on two urban desert gardens (77.4 and 58.3 m^2, 833 and 627 ft^2) in Arizona, USA, recorded a year-round harvest that provided the gardeners with significant proportions of the RDAs for 10 nutrients, including over 50% of the RDA for vitamins A and C for more than half the months of the year. Only 2 to 3 hours per week were spent gardening.[68] Perhaps the most ambitious study to date was carried out in experimental gardens in the humid tropics by the Asian Vegetable Research and Development Center (AVRDC) in Taiwan.[69] Results from the third year of the study (1983-84) showed yearly production of RDAs for a family of five, determined quarterly on samples from the gardens, as follows: 13-18% protein, 33-42% calcium, 56-82% iron, 82-125% vitamin A, and 336-374% vitamin C.

2.11.2 Effects on Nutrition

The ways in which the abundant nutrients potentially available from household gardens are actually translated into improved nutritional status depend not only on the place of garden foods in the diet, but on the complex social dynamics of households, which determine who receives which foods at what times. Working males, for example, may be the first to eat, then older women, and finally younger women and their children. This is sometimes due to women's weaker bargaining power in the household, or to their low social, economic, and cultural status outside of the household. One study in the Philippines showed that this can result in inadequate consumption of calories by women as compared with men.[70]

Having more household food may result in more reaching the last in line. When women control the garden and its produce, household food distribution may also change for the better nutrition of children, especially during weaning.[71] Yet it can never be assumed that women's gardens will automatically have this benefit, since there is very little evidence on how frequently it occurs, or what conditions favor it. Therefore, garden projects promoting women's gardens for improving nutrition of women and children must be based on an understanding of how household dynamics affect food distribution and use of income in particular communities.

The combination of foods in various household meals can have a great effect on the total nutritional value of the diet. In theory, gardens facilitate the continual consumption of small amounts of a variety of nutrients which complement the rest of the diet.[72] For example, the essential amino acid patterns in the protein of vegetables complement those of many grains, seeds, and nuts (section 2.5). In addition, eating fresh produce from the garden soon after harvesting avoids the postharvest nutrient losses that occur due to storage, handling, exposure, and processing.

Most of the research on gardens, food consumption, and household nutritional status is from humid areas. For example, in Tabasco, Mexico, it was found that fruits and vegetables were not eaten unless they were grown in the family garden because they were otherwise too expensive to purchase.[73] A study in Java found that low-income households consumed the least amount of rice, but the greatest amount of leafy garden vegetables high in vitamin A, and that gardens provided up to 40% of the household requirement for energy.[74] In Puerto Rico, "homemakers" were more likely to have diets adequate in vitamins A and C, and preschoolers in vitamins A, C, and riboflavin, and calcium, energy, and protein, when their households produced fruit and vegetables for consumption in home gardens.[75]

In southern India it was found that garden production was positively correlated with the nutritional status of weaning age children.[76] This was especially true in the slack season for off-farm employment when garden produce, or the income from selling it, kept child nutritional status from falling. A garden project in the Philippines found gardens were positively associated with higher levels of vitamin A in children who had the lowest levels before the project, and with a significant increase in weight for height of children.[77]

Nutrition education can be a vital part of garden programs aimed at improving nutrition, especially when gardens are being introduced for the first time, or when new types of garden crops become available. A garden project in Ilesha State, Nigeria, in the late 1960s emphasized traditional crops and gardens and included a strong nutritional education component directed at local women.[78] This project is said to have reduced child death due to malnutrition among gardening households from 10% to 6% in three years. However, research on the nutritional impact of a garden project 20 years after its initiation in Senegal shows no improvements among participating households.[79] The researchers believe this is due to lack of nutrition education and because most of the produce was being sold with only a small percentage of that income used directly for food purchases.

Nutrition education is also vital to counter the increasing consumption of prestige foods associated with a "modern" way of life, many of which are not as nutritious as indigenous local foods. For example, in northeast Brazil where there is vitamin A malnutrition, nutrition education may be necessary to counter the replacement of local fruits and vegetables high in vitamin A with fruits and vegetables imported from the south. These imports are more popular because of their association with the more "modern," affluent section of the country.[80]

It is obvious that there are many links between the garden and improved nutritional status. Problems can occur with any one of the links which will decrease the nutritional contribution of the garden. However, some ways to improve the chances that gardens will succeed in having a long-lasting, positive effect on people's overall nutritional status are to:

- Base new or improved gardens on indigenous gardens and indigenous crops.
- Encourage gardeners to grow and eat a wide variety of foods.
- Support or introduce simple harvesting and processing techniques that preserve the nutritional quality of garden foods.
- Recognize and address intra-household differences in consumption.
- Include participatory nutrition education at all stages of the work.

2.12 Resources

The best source of information for those working to improve nutrition is the diet of healthy people in the community. Secondary school or college nutrition textbooks are good references for learning the physiology of human nutrition. Detailed information on the RDAs can be found in NRC (1989), FAO (1973), and Passmore, Nicol, and Rao (1974).

Diet for a Small Planet (Lappé 1982) is a good exploration of the social, economic, environmental, and nutritional impacts of protein in the diet.

One of the best practical books on nutrition for field workers is *Human Nutrition in Tropical Africa* by Latham (1979). Although written from the author's experience in East Africa, the combination of simple theory and practical information give this book a wider audience.

"What to Eat to be Healthy," Chapter 11 in *Where There is No Doctor* (Werner 1977), discusses nutrition and health. *The Manual on Feeding Infants and Young Children* (Cameron and Hofvander 1983) focuses on infants' and children's special nutritional needs. Brownrigg (1985) includes some case histories and discussion about garden projects and nutrition.

References

[1] Latham 1990.
[2] NRC 1989:10-12.
[3] NRC 1989.
[4] Passmore, et al. 1974.
[5] Leung, et al. 1968; FAO 1982a; Leung and Flores 1961; Leung, et al. 1972, respectively.
[6] For gardens see USDA 1982, 1984a, 1984b, 1989.
[7] Cameron and Hofvander 1983:38-39.
[8] Cameron and Hofvander 1983:11.
[9] Cameron and Hofvander 1983:3.
[10] Calculated from Passmore, et al. 1974.
[11] Calculated from NRC 1989.
[12] FAO 1973:36.
[13] FAO 1973:36.
[14] 700-800 ml breast milk produced/day x 35 mg Ca/100 ml breast milk, calculated from Cameron and Hofvander 1983:84-86.
[15] Levi and Havinden 1982:55, 62.
[16] Brun, et al. 1981.
[17] Bleiberg, et al. 1980.
[18] Brun, et al. 1981.
[19] Bleiberg, et al. 1980; Brun, et al. 1981.
[20] IDRC 1980:15.
[21] Latham 1979:123.
[22] Latham 1979:113.
[23] Lappé 1982:420.
[24] Lappé 1982:172-182.
[25] IDRC 1980:11.
[26] Calculated from Leung, et al. 1968.
[27] Campbell, et al. 1990.
[28] Lappé 1982:70.
[29] DeWalt 1985; Youtopoulos 1985.
[30] Sommer, et al. 1983, 1986.
[31] Charoenkiatkul, et al. 1985; Sommer, et al. 1987.
[32] Eastman 1988:23-37.
[33] WHO 1982.
[34] IDRC 1980:15.
[35] Shrimpton 1989.
[36] NRC 1989:80-81.
[37] Latham 1979:107.
[38] Waterlow 1982.
[39] NRC 1989:93.
[40] NRC 1989:117-120.
[41] Latham 1979:85.
[42] Passmore, et al. 1974:68-69.
[43] Latham 1979:81.
[44] Campbell, et al. 1990, n.d.
[45] NRC 1989:135.
[46] Latham 1979:83.
[47] Latham 1979:83.
[48] Passmore, et al. 1974.
[49] Passmore, et al. 1974:68-69.
[50] NRC 1989:207.
[51] NRC 1989:209.
[52] Latham 1979:66-69.
[53] NRC 1989:175,178.
[54] Latham 1979:69.
[55] NRC 1989: 46-49.
[56] Ames 1983; NAS 1973.
[57] Oberleas 1973:367-368.
[58] Akpapunam and Achinewhu 1985; Khokhar and Chauhaw 1986.
[59] Ferrando 1981:32.
[60] Ferrando 1981:39.
[61] Rao and Deosthale 1982.
[62] Singleton and Kratzer 1973:329.
[63] Achinewhu 1986; Akpapunam and Achinewhu 1985.
[64] Dada and Dendy 1987; Odunfa 1985.
[65] Ferrando 1981:66-68.
[66] Latham 1984.
[67] Tarwotjo, et al. 1982.
[68] Cleveland 1982.
[69] Gershon, et al. 1985
[70] Folbre 1984.
[71] Carloni 1981.
[72] Grivetti 1978; Longhurst 1983.
[73] Dewey 1981.
[74] Stoler 1979.
[75] Immink, et al. 1981.
[76] Kumar 1978.
[77] Solon, et al. 1979.
[78] Brownrigg 1985:68-76.
[79] Brun, et al. 1989.
[80] Shrimpton 1989.

3
Gardens, Economics, and Marketing

One reason why many people garden is because of the income they can earn by marketing garden products. More and more households need this income to pay for medicine, school fees, clothing, and food.

Economic data on dryland gardens is very rare. There is little doubt, however, that gardens make economic sense to the women, men, and households who have them. Even in the midst of large-scale development projects, as along side irrigation canals in savanna West Africa, Pakistan, or Egypt, gardens tend to emerge spontaneously. The questions that are important for the readers of this book are:
- What are the economic contributions of existing gardens?
- Can they be improved, and, if so, how?
- Under what conditions will gardens make economic sense for households that do not have them?

The answers to these questions must be sought in each local situation. The goal of our discussion of garden economics and marketing in this chapter is to help readers ask these questions in ways that will lead to answers.

Many of the assumptions of conventional economic theory ignore social and ecological reality.[1] The primary goal of development programs based on this theory is economic growth. These programs are a major cause of poverty and environmental destruction in both industrial and Third World countries. We believe that economic development should be:
- Environmentally sustainable by not destroying resources that will be needed by future generations.
- Socially sustainable by providing benefits equitably, which may include redistribution of control over resources.

To meet these goals development projects should:
- Encourage local self-reliance by building on local knowledge and resources.
- Encourage both biological and cultural diversity.
- Encourage community organizations that place a priority on social well-being.

The kind of gardens promoted in this book are an important part of this type of economic development.

3.1 Summary

Gardens provide both income and savings, but their effects on the whole household or individual household members depend on many factors both within and beyond the household. To contribute to sustainable development, each garden or garden project needs to be adapted to the local social system and environment, and not based on the faulty assumptions of conventional economics. This demands an understanding of gardeners' economic decision making and the forces that affect it. For example, while women are often the gardeners in the household, they may not have control over productive resources like land, or over income from marketing garden produce. Marketing and processing techniques can help to reduce gardeners' risks and to increase benefits to the household. Forming cooperatives can spread risks and is often an appropriate way to organize market gardening. In many places indigenous or spontaneous social groups become the basis for successful marketing cooperatives.

3.2 People, Households, and Economics

Economic development is dominated by the economic theories of Western "capitalist" nations, increasingly so as "socialist" states like the USSR and those of

Eastern Europe adopt the Western model. However, it is becoming more apparent that many of the common economic assumptions are ethnocentric, and are not compatible with sustainable development, especially in the Third World.[2] Some of the key assumptions on which conventional economic theory is based include:

- Economic development means economic growth.
- Resources for continued growth will be available.
- Any destructive effects of growth on society and the environment are only important if they have an effect in the market, and if they do, they can be reduced by better technology and more economic growth.
- Individual behavior is motivated only by self-interest.
- Resources are best allocated in markets where everyone is trying to maximize their individual profit.

New information from ecological and social research is showing the negative effects of this kind of economic development on the environment, indigenous cultures, and the poor. In the industrial capitalist countries (the United States, Western Europe, Australia, and Japan), industrial socialist countries (the USSR and Eastern Europe), and in Third World countries, there is a small but growing movement toward alternatives to conventional economics, including agricultural systems that are less environmentally and socially destructive.[3] Part of this movement is a recognition that economic development does not necessarily mean economic growth, especially at the global level.[4]

In the sections that follow, we discuss four aspects of economic development that are important for understanding the role of gardens. Sometimes the lack of information about small-scale food production and the unfounded assumptions of conventional Western economics make gardens, especially indigenous gardens, seem an unwise investment of resources. In fact, indigenous gardens often make economic sense as well as social and environmental sense—they have the potential to be a viable development strategy.

3.2.1 Production Efficiency

An important question concerning garden economics is whether gardens are worth the investments they require. *Efficiency* is a measure of the ratio of investments to returns, or output to input (output/input). Irrigation efficiency, for example, is the ratio of irrigation water delivered to the root zone of the plants in the garden, to the amount of water extracted from a river, well, or other source (section 12.2). Irrigation efficiency increases as the water in the root zone increases as a proportion of the total amount of water extracted.

The same operation can have high or low efficiency depending on how investments and returns are measured and what factors they include. For example, there is a tendency to think that large-scale, industrial food production is more efficient than small-scale, non-industrial production.[5] This is because industrialized nations emphasize labor efficiency, and are able to increase this through greater inputs of machinery, fossil fuels, and commercial chemical pesticides and fertilizers. This greatly increases the amount of harvest per worker, but greatly decreases the amount of harvest per unit of energy and other resources invested.[6] Simply because of the shortage of capital at the household and national level for buying machinery and other imported inputs, low-input gardens are an attractive economic alternative to large-scale or capital-intensive production. In addition, it is also widely agreed that in terms of efficiency of land use, the yield per unit of land and other resources tends to be greater on smaller production units[7] (section 3.3.1).

Low-income households in drylands often have more labor than they have capital to purchase labor-saving inputs like fuel and machinery. Productive resources that are often scarce for these households are land, organic matter, and especially water. Under such conditions it makes sense to get the most from scarce resources by labor-intensive, small-scale production, like household gardens. However, not all Third World households in rural or urban areas have an abundance of labor. For these households, like the Mossi people of Burkina Faso described in section 2.3.3, increasing the efficiency of (i.e., returns to) labor might be possible. However, increasing the amount of labor invested, even with higher rates of return, is not possible unless the households have more food to eat, or more time.

We need to go beyond economic measures in evaluating the potential of gardens. Conventional measures of efficiency calculate output in terms of harvest size, or monetary profit. Yet, even though harvest size and profits are important, they are often not very good measures of people's well-being, especially over the long term (section 3.2.4). We should instead be more interested in output as measured by diet, nutritional status, health, equity of distribution, environmental degradation, local self-sufficiency, and self-esteem.

Using local needs as the criteria, efficiency and the effects of gardens or garden projects on efficiency could be more meaningfully measured by such out-

put/input ratios as:
- Vitamin A nutritional status/investment of time and resources in household gardens.
- Number of mothers feeding nutritious weaning foods from the garden/number of gardens established in a community.
- Number of poor households using garden plots/total number of plots at a community garden site.

Another problem with conventional measures of efficiency is that measures of inputs seldom include the costs of long-term environmental or social degradation.[8] For example, the cost of irrigation water does not include the losses in production due to soil salinization or waterlogging which irrigation usually causes, or the cost of expensive methods to reduce these problems. The cost of insecticides does not include the costs of damage to human health, the loss of bees and other beneficial insects, or water pollution. As this damage occurs, the costs are borne by the society at large, and by future generations. If these costs were to be included in calculations of efficiency, many small-scale, low-input systems would be recognized as much more efficient than the large-scale, industrial systems that are promoted by so many as a model for the Third World.

3.2.2 Economic Rationality and Risk

Until recently, a common misunderstanding was that small-scale producers in the Third World were economically irrational. It was thought that they were not able to evaluate possible economic returns to alternative investments of their resources as a basis for decision making. It is now accepted by many, though not all, development experts that Third World gardeners and farmers understand these concepts very well, and only appear to be acting "irrationally" because the outsider does not understand the local conditions.[9] However, it is still assumed that the main criterion gardeners and farmers use in making decisions is how to optimize their economic return.

Figure 3.1 shows some of the relationships that may determine how much investment of time, including labor and management, as well as other resources such as water, organic matter, or land a gardener will be willing to invest to increase the amount he is able to harvest from the garden. This *production possibility curve* illustrates the general rule that as inputs (such as hours of labor) increase, the return (garden harvest) increases at a slower rate. Economists refer to this as *diminishing marginal returns*.

For example, a gardener is working in his garden 1

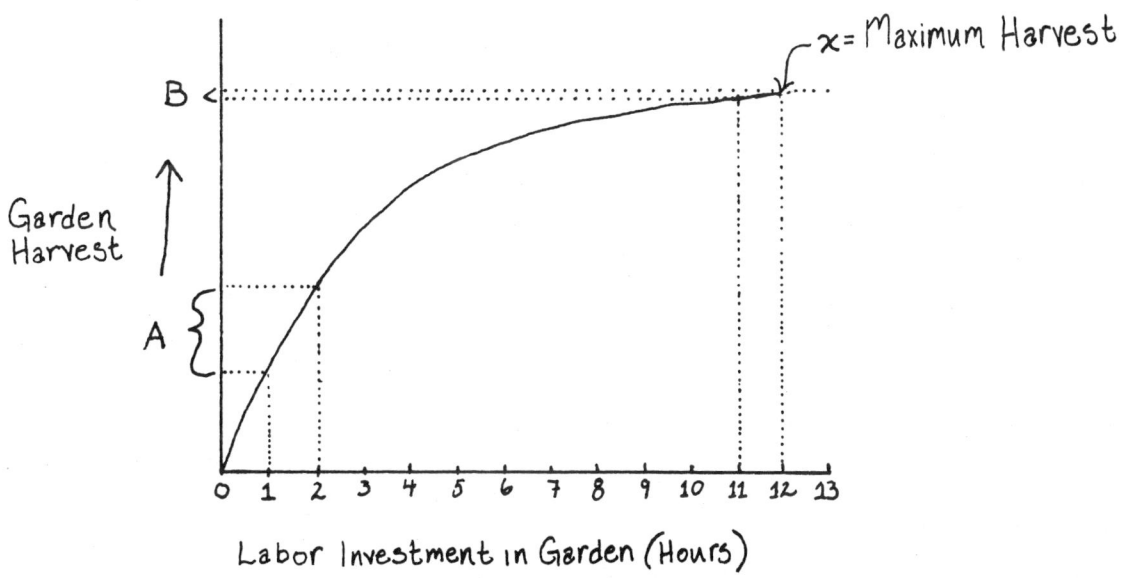

Figure 3.1 Production Possibility Curve: The Relationship Between Investments and Returns

hour a week, and decides to work 1 additional hour (see Figure 3.1). The increase in his returns (A) for that additional hour will be greater than the increase in his returns from an additional 1 hour a week if he had already been working in the garden 9 hours a week (B). This is because the more a person works the more tired he becomes, and because the first jobs done in the garden are those with the highest output per unit input, for example watering, pulling the biggest weeds, transplanting, and managing the most harmful pests. Minor tasks that are done later will therefore have a lower return, for example, pulling smaller weeds, and managing the less damaging pests. If the gardener is sick, or lacks energy because of food shortages, his interest and ability to work in the garden will also be affected (sections 2.3.3 and 2.3.4). Adopting different techniques can sometimes increase the productivity of labor. For example, a person who keeps his garden well mulched in the hot, dry season, or who uses a hoe, may be able to produce more per hour of labor than someone who does not mulch, or who does not use hand tools.

Another reason why gardeners and farmers do not often invest the time and resources to produce the maximum possible yields or profits (point X in Figure 3.1) from their plots is because of their perception of risk. *Risk* is the gardener's judgment about the probability of failure due to his inability to predict uncertain events. For example, a gardener may try to plant as early as possible in the rainy season to have harvests sooner. However, the earlier he plants, the greater his risk of seedlings dying for lack of rain, since the timing and amount of rains at the beginning of the rainy season are very uncertain. Information on past events, like records of total monthly rainfall, or years or even generations of past experience, can decrease uncertainty about future events and the risks involved in basing decisions on these events. But this information does not eliminate risk (Box 11.3 in section 11.4.1).

Risks often increase as production approaches the maximum, in part because increasing production decreases diversity on many levels (section 14.2). Modern crop varieties, for example, may produce larger harvests than folk varieties if they are supplied with optimal water, nutrients, and protection from pests. However, when a drought, pest invasion, or other unforeseen event occurs, the harvest will be lower than with folk varieties (section 10.4). The gardener's losses will be even greater than the loss of harvest if he invested in inputs such as chemical fertilizer. Thus with modern varieties the gardener's risk is often greater.

Again, it is necessary for us to go beyond conventional economics. The goal of gardeners and farmers is not to produce as much as possible, but to use resources for a number of different purposes to achieve the greatest overall benefit. Therefore, deciding how much time and how many other resources to invest in the garden (Figure 3.2) will also depend on the returns to alternative uses of the same resources, including investing in field crop production, wage labor, crafts, school, religious activity, or building and maintaining social relationships.[10] People will stop gardening when they believe that the returns are not worth the effort.[11]

Figure 3.2 Work is an Investment in the Garden

3.2.3 Control of Resources: Individual or the Group?

Conventional thinking in the Western industrialized countries is that maximizing profit on privately owned resources in the marketplace is the best way to ensure productive, sustainable use of those resources. However, this has not been the case in these countries. This is because markets have no way of sensing justice or sustainability, and in fact without outside social control, markets tend to result in inequitable distribution of economic benefits and to destroy natural resources.[12]

In many nonindustrial societies resources have traditionally been controlled by local communities as a group, and even resources controlled by an individual are often redistributed to other community members periodically.[13] This is in contrast to capitalist market economies where most resources are considered to be private, and even public resources are often open to exploitation by influential individuals. In socialist industrial societies, many resources are controlled by the state, rather than local communities, and once again, influential people may be able to exploit those resources. While exploitation and concentration of power may also occur in societies with community control of resources, participation and equity are much more feasible because the group is joined by common interests and personal contact.

Resources controlled by the local community, called *common property resources* by researchers, are more responsive to changing community needs than resources controlled privately or by the government.[14] This responsiveness is especially important for poor households who are most vulnerable to changes such as drought and food shortages; common property resources are often the only ones that these households have access to. Community control in cooperation with larger regional or national management organizations can be an effective way of organizing management of resources such as rivers or forests that are shared by more than one community.[15] For example, nationalization of forests in Nepal, as in many other countries, undermined local management groups and resulted in widespread destruction because local people had lost control of the resource and the national government could not afford to protect the forests. However, local communal management groups are again being allowed to operate. Cooperation between the local and national levels has ended uncontrolled destruction of those forests, an essential resource for the communities that manage them.

It is important to distinguish common property management from no management at all, although some researchers and policymakers have failed to make this distinction. A resource for which there is no form of participatory community control is often rapidly exploited by individuals acting in self-interest. This has been called the "tragedy of the commons," but this is incorrect because such resources are not truly "commons." These resources are not seen as belonging to everyone, but rather to no one, as was the case with state control of the forests in Nepal. The advantages of common property and cooperation in marketing garden produce are discussed in section 3.4.3.

In communities where resources are treated as common property, there is a tendency for more intensively used resources to be under more individual control. This is because of the greater long-term investment, for example, of trees, manure, and irrigation canals, in intensively cultivated land. Household gardens are often the most intensive part of the household food production system and therefore the land, water, and other resources used tend to be under individual control. However, this is not the same as individual property rights as defined in Western market economies. The individual plots may often remain within a system of local community control that determines the amount of land an individual or household can use, and how that land may be used, inherited, and transferred.

In considering control over resources, it is important to examine how household resources are managed within the household. A common false assumption of conventional economics is that a household acts as a unified economic unit in deciding how to use resources. Evidence shows that inequities in the legal, social, and economic positions within the household, for example, between men and women, lead to patterns of control over resources that detract from the optimal well-being of the household. These patterns can result in the poor nutritional status of women or young children (section 2.11).[16] Garden projects can make the distribution of household resources more equitable by providing new sources of income. This is especially important for those who may not be able to go far from the home because of other work or cultural obligations, as in the case of women, or because of physical difficulty, as with the elderly or disabled.

3.2.4 Economic Development and Well-Being

A major aim of most development programs is to increase national economic growth, commonly measured by gross national product (GNP), which is assumed to reflect overall national well-being.[17] If better-off groups benefit most from this growth, it is often justified by saying that the benefits will "trickle down" to the poor. Often, however, such programs further increase inequality, and may even lead to greater poverty and malnutrition. This has been a consequence of the green revolution in countries like Mexico.[18]

The growth of poverty and malnutrition in the Third World, in spite of billions of dollars spent on development in the last 40 years, has led to a realization that if alleviating poverty and malnutrition are goals of development, then development programs must be specifically designed to do this. In many areas household gardens are one way of improving the nutrition and income of those most in need without excessive reliance on already overtaxed national bureaucracies and budgets.

Another major false assumption is that greater income and consumption—at the national, household or individual level—always results in increased well-being.[19] Although increased household income can improve well-being, it should not be assumed that it will, especially as more and more "consumer goods" become available all over the world, even in remote areas. In some cases the desirability of these goods and their association with social status is persuasively promoted through aggressive advertising. These advertisements are successful both in industrialized countries and the Third World.

For example, if commercially manufactured "junk" foods such as carbonated drinks, candies, and other snack foods are easily available, low-income households frequently tend to purchase these rather than fruits or vegetables.[20] Junk food ingredients such as sugar, water, salt, and flour, are usually so inexpensive that even at a relatively low selling price the manufacturer is able to make a sizable profit. These foods taste good, and may give the consumer a temporary surge of energy because of their high sugar and fat content. However, these are empty calories in that they do not provide significant amounts of vitamins, minerals, or protein. **Commerciogenic malnutrition** is the result of diets high in these foods and lacking in many necessary nutrients;[21] it is found in both industrial countries and the Third World.

Additional household income is often spent on cigarettes and alcohol. These widely advertised, popular consumer goods harm people's health in every population where they are consumed. They can cause cancer and heart disease, and can permanently damage unborn children.

The tremendous pressure of advertising is one reason why nutrition and health education is often essential for ensuring that garden income contributes to improved well-being. Community control over advertising and sales of harmful items like junk food, cigarettes, and alcohol may also be necessary.

On the other hand, some purchased goods that seem frivolous or a sign of poor judgment to outsiders do provide benefits. For example, the jewelry purchased by Peulh and Toucouler women in Mauritania (section 3.4.4) not only indicates their social standing but provides security for them in times of crisis and old age.

In many cultures hospitality and gift giving are essential social gestures that establish and reinforce relationships of mutual support. Income or savings may be used to purchase the specific goods considered appropriate as gifts. In the West African Sahel, purchased kola nuts (*Cola* spp.), imported from the more humid regions in the south, are one of the most desirable offerings. In North Africa, tea made with purchased sugar and imported green or black dry tea leaves is a regular part of the daily diet. However, it is also essential to serve generous quantities of this heavily sweetened tea as a gesture of hospitality to visitors (Figure 3.3). In commercialized market economies hospitality and gift giving are no longer a large part of everyday social life. As a result the relationships they reinforced and the services provided by those relationships like care of children and the elderly must be purchased.

3.3 Garden Economics

The term *economy* is based on a Greek word meaning the skillful management of household resources for the benefit of the household. However, in conventional economic theory and in popular use, economy has become strongly identified with money, giving rise to a very narrow interpretation of the economics of many activities,[22] including gardening (Figure 3.4). Garden economics considers the effects that income or savings from gardens have on the well-being of those who are gardening, and the people they take care of. Since household incomes in Third World drylands are so low, frequently the equivalent of only a few hundred US dollars a year, even small amounts of savings or income from selling garden produce can make a big improvement in household well-being.

3.3.1 Garden Yields

Quantitative studies of household gardens are rare, but studies of small-scale agriculture suggest that we should expect yields to increase as the size of the farm decreases.[23] Household garden yields can be high. A study in eastern Nigeria shows that dry weight yields per area from "compound" gardens are twice as large as those from more extensively cultivated outer fields.[24]

As intensity in small-scale agriculture increases, that is as the number of times crops are planted and harvested per year increases, we should also expect annual yields to increase, although returns to labor generally decrease. Unlike field crop production, intensive garden production may not mean lower labor productivity. In indigenous mixed gardens returns to labor may actually increase because of greater biological diversity, continuous harvesting, and a large pro-

Figure 3.3 In Some Areas Purchased Goods are Important for Cultural and Social Life

portion of perennials. Continual harvesting may raise annual yields and encourage fine tuning of management strategies. The mixed compound gardens in eastern Nigeria mentioned earlier yield returns to labor that are four to eight times greater than those in outer fields. A study of two household gardens in an urban desert environment in Arizona, USA, showed yields between 1.2 and 6.5 kg/m^2 (0.25 and 1.33 lb/ft^2).[25] Results from a number of studies of experimental and demonstration gardens show yields between 2.5 and 15.5 kg/m^2 (0.51 and 3.18 lb/ft^2).[26] These results can be compared with commercial vegetable production in the United States, yielding on average 1.7 kg/m^2 (0.35 lb/ft^2). Garden returns to labor from less than 1 to almost 16 kg/hr (2.2 - 35.28 lb/hr), are much lower than in large-scale, commercial agriculture, but this comparison is deceptive. The high labor productivity in large, mechanized fields is made possible with large amounts of expensive, nonrenewable energy in the form of pesticides, fertilizers, electricity, and fuel. For this reason their production efficiency is debatable, especially now, with growing recognition of the serious environmental problems their use can cause.

3.3.2 Income and Savings from Gardens

Gardens can make economic contributions to household well-being in two ways: a) gardens can help save money by providing food as well as medicines, fodder, fiber, building and craft supplies, or other materials

Figure 3.4 Garden Economics?

that would otherwise have to be purchased; and b) income from the sales of garden produce can be used to buy food and other items.

Household market gardens may look quite different than those used primarily for consumption. Rather than a continuous harvest of many different fruits and vegetables throughout the year, market gardens may concentrate on a limited number of crops that can be harvested in quantities large enough to make marketing them worthwhile. Frequently, however, household gardens serve both functions, with the same products being used by the household and marketed. If markets are close by, and there are few garden expenses, then proceeds from even small amounts of marketed produce can frequently be profitable for gardeners. This seems to be the case, for example, in gardens surveyed along the Senegal River in Mauritania.[27] If marketing opportunities exist, development project gardens started for household consumption usually end up producing something for the market.[28]

In a market gardening project in Senegal, women's gardens responded to the demand for fresh produce from hotels serving a booming tourist trade.[29] While eggplant, tomatoes, onions, and peppers were both sold and consumed by gardeners' households, salad greens were sold mostly to hotels. Home consumption accounted for 30% to 50% of the harvest from these gardens. In Java, lower-income households with smaller gardens were more likely to sell some garden produce, with gardens providing up to 20% of household income.[30]

Household gardens can also provide savings. For example, a survey of 250 households in low-income areas of Lusaka, Zambia, found that 57% of the households cultivated either a rainy season garden on the outskirts of the city, a plot garden near the house, or both.[31] Of those with gardens, 77% of the plot gardens and 85% of the rainy season gardens were started to save money because of the rising price of vegetables in the market, and the need to supply the household with relish.

Similarly, in a survey of urban gardeners in Ibadan, Nigeria (40 people), and Freetown, Sierra Leone (60 people), most of the gardeners were low-income residents of urban shanty towns and poor neighborhoods.[32] All of the low-income gardeners said that they gardened to feed their family and/or to supplement their low incomes.

In the industrial world, household gardens also provide important savings. Among farmers in Florida, USA, for example, household gardens contribute up to 10% of net farm income.[33] A sample of more than 10,000 households in the United States showed that 27% used homegrown vegetables; that they saved 22% of the amount of money spent on vegetables by households without gardens; and that these savings were an important consideration in the decision to garden.[34] Savings on food purchases were also found to be an important incentive for gardeners in Poland[35] and Great Britain.[36]

3.3.3 Household Well-Being

The most important consideration about marketing garden produce is the ultimate effect it has on people. Complex relationships between a large number of variables determine how the produce from gardens affects household nutrition, income, and general well-being. Whether or not increased household income from the garden leads to improved household well-being (section 3.2.4) depends on how this additional income is spent. There is some evidence that household income does not correlate with improved nutrition,[37] and there is little evidence to support the assumption that women are more likely than men to spend income on food.[38] There simply are not many good studies of the effects of garden income or savings on well-being.

One study in humid southern Kerala state in India did show that increased income from women's household gardens contributed to improved child nutrition, especially during the season when wage labor was scarce (section 2.11.2).[39] A study of a 20-year-old garden project in Senegal found that the women gardeners spent only a small fraction of their net income from gardening on food.[40] However, the effects of increased income can be indirect. Even though garden income may not directly improve nutrition, it may increase individual or household well-being in other ways such as improved social status, or it may free up other sources of income that can be spent on food at other times of the year.

Finally, while gardens certainly have the potential to make an economic contribution to household well-being, their value and relevance must always be considered within the larger socioeconomic setting. As one researcher has pointed out, nutrition programs often avoid the question of improving incomes and access to resources.[41] Focusing on improved use by the poor of existing resources may avoid the real issue. For example, it is possible that even though gardens can provide immediate benefits to households, they can also support an overall structure that works against households trying to improve their situation. African slaves on Caribbean plantations were given plots of marginal land for gardens by plantation owners. Forcing the slaves to grow some of their own food relieved the slave masters of this expense, but did not improve the slaves' status.[42] Similary, in 19th century England, garden plots provided for low-income households by large landlords or the state served to subsidize low wages paid by the landlords or industry.[43] Thus, encouraging the poor to grow gardens may be a way of avoiding larger issues of inequity in the system that keeps them poor. Ultimately, improved well-being does not come from making an exploitative system more efficient, but from changing its structure. For example, a study of urban gardening in Buenos Aires, Argentina, found that programs that encouraged gardens as a way of increasing self-reliance overlooked the need for structural changes such as income redistribution and support services.[44] Local participation in and control over garden programs, and education, will help eliminate such possible negative effects of gardens.

3.4 Marketing Garden Produce

Trade between different areas of the world has gone on for much of human history, for example, in the drylands of the Middle East and the Saharan-Sahel region of Africa. This trade involved exchange of a good produced in one area that was not produced in another. Markets today offer many goods that cannot be produced locally, such as salt, radios, batteries, or some medicines. In some cases the same goods are both produced locally and imported, like fruits and vegetables, snack foods, tools, clothing, and shoes. When locally produced goods are sold at small, local markets the producer and consumer share the same resource base and similar living conditions, and their exchange remains in the community. Buying an import that sells for less than a local product makes sense in the short run for the individual or household, but can have negative consequences for the community. As in any situation where there is no local control over decisions that have local effects, difficulties may arise. Most money spent on imports leaves the community, except for a small portion if a local middleperson is involved. The effect is frequently felt most by the poor because it is often their production activities that are displaced.

The vulnerability of communities that are less and less involved in production for their own markets is evident not only in the Third World but also in industrialized countries. In the United States, for example, textile workers and others are losing jobs to factories in the Third World where production costs are much cheaper. These factories keep their costs low by paying their workers very low wages; these workers must also endure extremely poor living and working conditions.

3.4.1 Women and Marketing

Money necessary for household maintenance and child care must often be provided by women, and does not come from a common household fund. Where this is true, marketing garden produce can be important as a source of independent income for these women (Figure 3.5). In some areas where women's market gardening has become more profitable, men have started competing with them after seeing how much income can be earned by market gardens.

In a project in Botswana a cooperative of 29 women and 4 men from the poorest households in the community work 33 garden plots that are hand irrigated with water stored behind an adjacent dam (section 3.4.3).[43] Produce is eaten by the gardeners, shared with those who helped in the garden, or marketed. For 21 of the 33 gardeners, gardens are their major or only source of cash income. But as the commercial potential of such gardens is realized, better-off individuals have started claiming entire dam sites for themselves, and some have suggested that it may become necessary to reserve gardening sites for groups of the poor, most of whom are women.[46]

On the Tonga plateau of Zambia, a fruit and vegetable growing cooperative was begun by local villagers. Most of the women involved used land borrowed from their husbands, and as the project became more profitable the husbands took over. In response, 26 of the 33 women in the co-op now obtain their garden plots independently of their husbands and many of the women (44%) feel that their garden income has made them less dependent on their husband's income.[47]

Sometimes the opposite situation can occur. In an area of southern Senegal, extension agents first promoted gardens to men.[48] But the men soon stopped gardening and the women took over. The men said this was because gardening involved fetching water, a women's activity which the men felt was inappropriate for them.

Marketing garden produce provides women an opportunity to leave the house and socialize, especially with other women. In the market, relationships can be formed or reinforced, and information, goods, and services are exchanged. Market days may be one of the few times available for important activities such as these.

Where women's activities are restricted by religious customs, such as the seclusion of Moslem women, earning income must be done in or near the home compound. The possibilities include gardening, food processing, sewing, craft work, teaching and raising small animals.[49] Garden produce can be sold to customers who come to the women's homes, or sold outside the home by children.

3.4.2 Risk, Investment, and Return

Like gardening, marketing involves risk to the gardener because it is not possible for him to control or predict all the factors that influence its success. The gardener does not know exactly how much rain will fall and at what times, or which crops will be least affected by pests and diseases. Neither does he know exactly which fruits and vegetables will sell, how many people will want to buy them, how much they will be willing to pay, or how many others will be selling the same produce. The risks involved in making decisions based on these uncertain events must be carefully considered in any garden project promoting marketing. A market survey (section 4.7) can help assess the risk of market gardening.

To minimize the risks, marketing should start small, because the smaller the investments of time, labor, money, and water, the smaller the gardeners' losses if there are problems. Another way to reduce risks is to only grow market crops that the household can use if they cannot be sold.

Processing and storing garden produce can also reduce the risks of marketing. If it is not possible to get fresh produce to market, or if the market price is not sufficient to repay the gardener's investment, the goods may be processed, and stored, and consumed or sold later. Because processing adds value, the price of processed garden products such as dried amaranth leaves and tomatoes is almost always more than equivalent amounts of unprocessed ones. Such a reduction in risk and/or increase in income makes the extra work of processing worthwhile.

3 *Gardens, Economics, and Marketing* 41

Figure 3.5 Marketing Garden Produce can be an Important Source of Income for Women

For example, a Dogon gardener in Mali took his large onion harvest to market.[50] The fresh onions weighed 52 kg (115 lb), which, at the current market price of 75 Malian Franks (MF)/kg (34 MF/lb), would have given him (52 kg x 75 MF/kg) 3,900 MF. Because he had no immediate need for money he could afford to dry his onions and sell them later. When dried, his harvest weighed 8 kg (17.5 lb) and sold for 500 MF/kg (227MF/lb) earning a total of 4,000 MF. If this gardener had stored the dried onions for several more months until onions of any kind become scarce he could have sold them for over 600 MF/kg (272 MF/lb). At that price the gardener would have made at least 4,800 MF, approximately 900 MF more than what the fresh harvest would have earned. Of course this is assuming that the gardener did not need the cash immediately and that the onions could have been stored without damage from pests or weather.

The value of the time and labor spent marketing garden produce or processing it for sale depends on the other possible uses of this time and labor (section 3.2.2). If the garden produce must be harvested and dried during a period of heavy labor demand, such as the time for weeding the fields, the cost to the household of labor diverted to harvest and dry garden produce would be extremely high and the market selling price would need to be high enough to compensate for this. However, at a less busy time for the men, such as during the dry season, the cost to the household of investing their labor in gardening may be much lower. This may differ for women because the demands on their labor are less seasonal due to their year-round responsibility for maintaining the household (fetching water and fuel, child care, cooking), in addition to food production.

When more than one activity is carried out in a given time period, the amount of time invested in each activity is decreased. For example, if a gardener spends three hours selling her produce on each trip to the market, this time is considered part of the cost of marketing. However, during those three hours she may also make arrangements for religious ceremonies, buy food for her family, and visit with friends and relatives. This means that the time actually invested in marketing is reduced, making marketing less costly.

Just as investments or costs may have different values according to the situation, the same may be true of the value of returns, or cash income. For example, if money is needed to pay for an emergency such as medicine, that income will have far greater value and therefore be worth a greater investment than usual. Thus a man may be willing to spend however long it takes for him to sell US $5 worth of garden produce because he urgently needs that amount of cash. When the need is less urgent, his willingness to invest time and labor will also be reduced.

Gardeners sometimes choose not to sell their products directly to consumers but rather to a *middleperson* who then sells to the consumers. This saves the gardener time and sometimes expenses because middlepeople do the marketing and often transport the goods to market as well. Usually the gardener will not receive as high a price from a middleperson as he would have from consumers themselves. At the market the middleperson will have to sell the produce at a price that will earn him or her a profit. Sometimes this means that the gardener will receive a lower price from the middleperson than if selling directly to the consumer. However, if a middleperson is buying garden produce to sell in another area where market prices are higher, such as a large city, the gardener may be able to sell his produce to the middleperson for as much or more than the local market price.

Whether a gardener decides to sell his produce through a middleperson or directly depends upon market possibilities, transportation, the value of his time, and the return he hopes to receive. A fair and honest middleperson can work with a gardener or group of gardeners to the benefit of all. Middlepeople frequently have many contacts and access to resources like transportation, which are essential for marketing. But some middlepeople can be exploitative, seeing the relationship with the gardeners as an easy way for them to make money. Working with a middleperson can be especially risky when he is not well known in the local community. Problems can arise if the community is isolated and must rely on the middleperson's honesty for information about current prices in distant markets.

Timing also affects the returns gardeners can get for their produce. Processing garden produce, as done by the Dogon gardener in Mali described earlier, is one way to time produce marketing to obtain the best prices. Hausa men with market gardens in northern Nigeria have devised ways to time the production of their garden crops, such as onions, to take advantage of seasonal changes in market prices.[51] In this area onions are grown in the dry season and sell for the highest price during the rainy season, however, the gardeners need some income before that peak period. To meet that need some gardeners plant an early crop to sell locally and then plant again, later, to sell in the more profitable rainy season. Others plant only one crop of onions that are stored for sale during the rainy season. To meet their need for income before that time these gardeners sell fruits such as guavas, limes, and mangoes. Both these marketing strategies give the gardeners some early income, allowing them to wait and sell a crop of onions at the most profitable time.

3.4.3 Cooperation

Working together often reduces the risks of marketing, especially for poor gardeners. Cooperation does not necessarily eliminate competition in selling produce, but it can help prevent concentration of the benefits of marketing in the hands of a few. Projects to support market gardening frequently include the establishment of formal cooperatives with elected officials.

In Botswana, a small group of gardeners, most of them poor women (29 women and 4 men), joined together in 1981 to create a formal group under the Ministry of Agriculture called the Tshwaragano Vegetable Production Group (section 3.4.1).[52] Group members pay a small annual fee and elect a grower's committee, the group's governing body. While each gardener controls her or his own plot, the group establishes and implements rules for participation in the garden and it negotiates with individuals and organizations outside the group for such things as transport to market and technical assistance. In addition to the individual plots whose profits go entirely to the plot holder, some areas are worked collectively. The profits from the sale of this produce are added to the annual fees and used to purchase tools and cow and chicken manure for fertilizer.

Traditional community groups, such as those based on age, may provide the basis for either formal or informal garden cooperatives. Informal cooperation is common among friends, relatives, neighbors, and residents of a community. An example of informal cooperation can be seen among the Hausa market gardeners in northern Nigeria described in section 3.4.2.[53] To minimize the risks of local markets being glutted, they sell their goods in groups made up of close male friends who are often relatives. These men put their goods together to be sold as a package to a middleperson. Everyone in the group receives the same per unit price for their goods. With this arrangement they do not have to compete with each other, which would lower the returns many would receive, and might even prevent some gardeners from selling anything.

Groups of market gardeners from rural areas outside of Rio de Janeiro, Brazil, made contracts with groups of residents in the city to supply pesticide-free fresh produce.[54] The gardeners bypassed the middlepeople who had forced them to use chemical insecticides and fertilizers, and the city residents got healthier produce. Both groups realized the necessity for political organization to make such arrangements work.

Market gardeners can cooperate in producing or processing goods from the garden; watching over and selling the goods of another gardener who cannot come to the market; agreeing on a minimum price below which they will not sell a particular good; or uniting to negotiate with truck drivers, middlepeople, merchants, and others. An example of women gardeners in Senegal joining to form a cooperative for providing transport of their garden produce to market is described in section 15.8.2.

Whatever the form, cooperation can provide support for group members and a stronger voice for pursuing needs and interests that group members share.

3.4.4 Garden Income and the Household

Who in the household controls the income received from marketing garden produce depends upon who grew it, who sold it, and their sex, age, and relationship to other household members.

In the Yatenga area of Burkina Faso, Mossi men and women garden separately on land belonging to their clans. A portion of their produce must be contributed to the family, but the rest can be sold by the gardener. Both men and women prefer not to ask their spouses for help in the garden, since they would then have to share the income. Instead they pay younger siblings (brothers or sisters) or other children.[55]

Among the Dogon of Mali, market gardening of onions is done mainly by men.[56] Even 14-year-old boys are given garden space to grow a crop of onions for market. Women, however, are not included in the distribution of resources like land for onion gardening, and often they do not enjoy any benefits as a result of income earned by male members of their household from selling onions.

Peulh and Toucouler women in Mauritania must provide the sauces that are an essential part of the meal, as well as tea, sugar, housewares, soap, clothing, jewelry, and their daughter's dowries.[57] Clothing, jewelry, and dowries are indicators of these women's social status and thus affect their position and voice in the community. Since the drought in the early 1970s, irrigated gardens have become more and more important for these women as a source of income for purchasing such goods.[58]

In contrast, married women in lower Egypt live with their husband's family, and the mother-in-law has complete authority over her daughter-in-law.[59] Especially in the early years of marriage, the mother-in-law controls all income earned by her daughter-in-law, including that earned from selling garden produce, eggs, and dairy products. Whatever money the daughter-in-law receives, if any, comes from her husband who is allowed to keep only a portion of his own income. All other earnings are given to the husband's parents who run the household.

As the household changes through time, its need for income may also change. When the children grow up and are married, money will no longer be needed for school fees or dowries. Different needs will arise and the responsibility for providing the household with income may also shift, for example, from the older generation to the younger one.

3.5 Resources

The ability to read critically and recognize authors' assumptions is important when reading any book, including those on economics. Sometimes basic assumptions are not stated, even though the "logic" of the book's whole argument is based on them. Questioning those assumptions can improve the reader's understanding of the author's argument. Questioning assumptions and exploring more realistic or desirable alternatives is an important step in changing and improving our lives and the world.

Resources about economics can often be divided into two categories: those that look at macroeconomics on the level of regions, nations, and even the world; and those that focus on microeconomics at the level of the household or community. While both perspectives are useful, the microeconomic approach is most appropriate for the topics addressed by this book. However, we feel that it is essential to explore the relationship between these two levels of economic activity when trying to understand how local gardens and markets work.

A standard, well-written text for the macroeconomic perspective in Third World development that is firmly based on conventional economic assumptions is *Economic Development in the Third World* (Todaro 1985). Several books, for example, those by Wallerstein (1974) and Worsley (1984) focus on the effect of the conventional model on the Third World especially as a vehicle for furthering colonial and neocolonial agendas.

Representing the emerging discipline of "ecological economics," Daly and Cobb (1989) state that the high social and environmental costs of conventional economics, also referred to as neoclassical economics, make this perspective dangerously destructive and unsustainable. Instead the authors propose a new direction for economics based on the ecological reality of our finite resources and more humane priorities concerning social responsibility. Schumaker's *Small is Beautiful* (1973) is a classic work that advocates making the fulfillment of the basic needs of the majority the central focus of economic activities and policy. The small is beautiful approach is based on many of the same recommendations made by Mahatma Ghandi earlier this century in India. These include decentralization of decision making and production and the support of many small-scale, local industries.

One of the best attempts at understanding economics from the perspective of small-scale agriculturalists in the Third World is Levi and Havinden (1982). They demonstrate why many of the assumptions of Western economics are not appropriate for the situation of African farmers. Brownrigg (1985) contains a review of some economic analyses of home gardens.

References

[1] Daly and Cobb 1989.
[2] Cleveland 1990; Daly and Cobb 1989.
[3] NAS 1989a, for the USA, for example.
[4] Daly 1989.
[5] Todaro 1985:292.
[6] Pimentel and Pimentel 1979.
[7] Cleveland and Soleri 1987; Cornia 1985.
[8] Daly and Cobb 1989.
[9] Todaro 1985:305.
[10] Levi and Havinden 1982.
[11] Haswell 1975; Levi and Havinden 1982.
[12] Daly and Cobb 1989:49-58, 145-146.
[13] NAS 1986.
[14] FAO 1989:96ff.
[15] Berkes, et al. 1989.
[16] Dixon-Mueller 1985; Folbre 1984.
[17] Daly and Cobb 1989:62-94.
[18] DeWalt 1985.
[19] Daly and Cobb 1989:77.
[20] E.g., Dewey 1981.
[21] Jelliffe 1972.
[22] Daly and Cobb 1989:138-139.
[23] Cornia 1985.
[24] Lagemann 1977:55.
[25] Calculated from Cleveland, et al. 1985.
[26] Cleveland and Soleri 1987.
[27] Stone, et al. 1987.
[28] Brownrigg 1985:113-114.
[29] Yoon 1983.
[30] Stoler 1979.
[31] Sanyal 1986.
[32] Tricaud 1987.
[33] Gladwin and Butler 1984.
[34] Blaylock and Gallo 1983.
[35] Kleer and Wos 1988.
[36] Crouch and Ward 1988.
[37] Kennedy 1983.
[38] Piwoz and Viteri 1985.
[39] Kumar 1978:44-45.
[40] Brun, et al. 1989.
[41] Kumar 1978.
[42] Brierly 1976.
[43] Crouch and Ward 1988:67,103,216.

44 Gutman 1987.
45 Duggan 1985.
46 Duggan 1985.
47 Milimo 1985.
48 Brun, et al. 1989.
49 Barkow 1972.
50 Eskelinen 1977:40-41.
51 Scott 1976:123.
52 Duggan 1985.
53 Scott 1976:120.
54 La Rovere 1985.
55 Hammond 1966:82.
56 Eskelinen 1977:45.
57 Smale 1980:xviii,25,71.
58 Smale 1980:19.
59 Zimmerman 1982:47-53.

4
Assessment Techniques

Most poor households in drylands need clean water, land, medical care, improved sanitation and nutrition, increased income, and control over decisions affecting their lives. An assessment helps communities organize themselves to address their needs by identifying those needs and the community's ability to meet them. The potential for collaboration with outside organizations is also considered. Assessment of progress while a project is being implemented can help development programs reach community goals, and not be diverted to serve the special interests of a powerful, rich minority, outside development organizations, or governments.

An assessment also helps outside project workers learn enough about the local situation so that they can support community members in their own efforts. Project and field workers should understand what gardens are and what the potential contribution of gardens to improving well-being is, including the possibility that gardens may not be the best investment of time and resources. To ensure real benefits, it is important for an assessment to be as accurate as possible. So that project goals are not based on false preconceptions, the assessment has to be grounded in the values of the local people, as well as an accurate evaluation of objective factors.

4.1 Summary

An assessment gathers information about local conditions, needs, and resources and is a vital part of planning for any project. In addition, assessments are valuable for monitoring projects while in progress and for evaluating their impact after completion.

Different individuals and groups have different perspectives on local conditions and different ideas about how projects should be done. All perspectives are useful but the views of community members, especially those who will be directly affected, are most important.

Careful observation and participation in day-to-day activities give the field worker insights into local conditions and help to establish understanding and friendship. This includes gardening using only local resources. Existing gardens provide a wealth of information for any assessment.

Interviews are useful for assessment—careful design improves their accuracy and acceptability to those being interviewed. Reports, censuses, and other outside sources can also provide useful information. When collecting and analyzing information on climate, work, health, and availability of resources it is important to consider how these are affected by seasonality.

Patterns of food distribution and consumption, maps of community resources and landmarks, and long-term social and environmental trends are other types of information that can be useful when conducting an assessment and planning for a project.

4.2 Assessment, Monitoring, and Evaluation

We define a *project assessment* as the gathering and analysis of information for planning future activities and evaluating present and past ones. An assessment is a learning tool that helps a community organize itself to address its needs, and helps project workers understand a community. Assessment should be a part of all development projects, but collecting information takes time and money which are often scarce resources. When resources are limited, assessments must be short and narrowly focused. The larger the project area, the more money being spent, or the greater the project's impact, the more thorough the assessment should be. In this chapter we discuss useful assessment techniques with an emphasis on information relevant to projects including gardens.

In many ways the best people to conduct an assessment are the community members themselves. They speak the local language and have a strong personal interest in any project that will follow the assessment. Yet, for some of these same reasons, when local people conduct an assessment their membership in a particular faction or kin group and their personal interests can bias the design of the assessment, the community's response to it, and their own interpretation of it. A community has a great challenge in selecting open-minded, fair, and respected members to conduct an assessment. Where local elites have a long-established and powerful hold on the community, such as in irrigation districts in northern Pakistan, the only way to ensure an assessment reflecting the needs of the majority is to have outsiders conduct it. Perhaps the best assessment is one in which outsiders and locals are equal collaborators.

We have written this chapter about assessment as though the people doing the assessment were from outside the community, although many of the concepts and techniques are also appropriate for assessments by community members. Whoever conducts an assessment must try to be as open and objective as they can. For an outsider, the best way to begin is by observing and participating in daily activities like hauling water, preparing food, gardening, and weeding the fields. The project worker should listen, offer support, and develop trust and friendships. This makes it possible for the community to get to know the project worker's intentions and decide if they can trust her. Similarly, the project worker can learn more about the community, which helps her focus information gathering on the most important topics. After this, formal methods (section 4.5) can be used to gather information, if necessary.

Community members usually have many questions to ask and should be encouraged to voice opinions and ideas. One way to do this is through group discussions about needs and how to solve them.[1] This can begin with the group making a list of all the members' comments about local problems. Similar comments are grouped together into categories, the cause of each category is explored by the group, and the group then decides which category or issue requires response.

The first assessment conducted before starting the project itself is sometimes called a *feasibility study*. Its purpose is to see whether a project is needed in an area, and whether it is possible. A *baseline survey* may be done as part of the feasibility study, or after it. A baseline survey documents conditions in the community before the project begins. Data from the baseline survey are what future assessments will be compared with to see if any changes have occurred.

Assessment done during a project is often referred to as *monitoring*. Monitoring is useful for keeping projects responsive and flexible, ensuring that the participants do not lose sight of the project goals. This is important because unforeseen situations often arise during a project. For example, project plans can be upset by transportation difficulties, personality disputes, climatic change, a sudden drop in market prices, money being spent twice as fast as planned, or men expressing an interest in gardens meant for women. Adjustments are needed so that the project continues to work toward its original goals, or the goals need to be adjusted by the community.

Post-project assessment is called *evaluation*.[2] An evaluation determines whether a project has accomplished its stated goals by comparing conditions at the time of the evaluation to those before the project began. Evaluations can be done after the project is completed, or during the project when specific goals were scheduled to be completed. An evaluation asks the questions, "Have the goals been reached?" and "Are these long-lasting, sustainable changes?" Community opinion is a critical part of any evaluation.

Attempts have been made to establish formal guidelines for garden evaluations.[3] Their basic approach has been calculating efficiencies, that is output/input or benefit/cost ratios. In the most common benefit/cost analysis this means calculating market value of garden produce and dividing by project and production costs. Indicators of nutritional status can also serve as outputs when converted to economic terms. Output/input ratios are important considerations and these guidelines can be useful for stimulating thinking about garden evaluations especially for large-scale projects. However, these guidelines tend to be narrowly focused and do not take into account the social and ecological complexity of gardens or communities. In addition, the cost of collecting the detailed quantitative data they require would be far too expensive except for the largest projects. For most garden projects of the kind we recommend, based on indigenous knowledge and local resources with miminal external inputs and project costs, an alternative to formal benefit/cost analysis might be to let the gardeners themselves calculate benefit/cost ratios. They will do this very quickly, and the results will be obvious as evidenced by whether or not project participants continue to garden, or continue the changes recommended by the project. Evaluation efforts could then focus on whether changes promoted by the project meet overall goals of improving well-being.

Evaluation findings are compared with project goals and for change, in comparison with the baseline survey. For example, if a project goal was to improve the nutrition of weaning-age children by encouraging households to grow nutritious weaning-food ingredients in gardens, the evaluation should focus on those children, their diets and nutritional status, and the gardens.

Evaluations should be sensitive to unanticipated positive and negative effects of the project. For instance, was the garden produce sold at the market instead of being used for weaning foods? Another evaluation months or years later will help determine if the project created long-lasting changes.

Evaluations can shed light on the relationship between different project goals, and can provide guidelines for future projects. For example, we visited a one-and-one-half-year project in rural Egypt which was devoted to starting household gardens to improve nutrition. Twenty gardens were established during the project and one year later only a few of those were still in existence. Those gardens were in the households of wealthy community members, some of whom hired servants to do the gardening. The evaluation showed that the number of households who adopted the project's gardening advice was low, and even more importantly, showed which households continued to garden and why. The project defined household gardens as being near the house, and most poor households were not interested as they did not have spare land near their houses in the densely populated village.

4.3 From Whose Point of View?

It is important for anyone involved in an assessment to realize that there is no such thing as a completely "objective" or "impartial" assessment. No matter how objective or "scientific" an assessment is, it will always reflect the biases and values of those who conduct and interpret it. This is true because science itself is embedded within the cultural values of society. A "good" assessment is one that honestly acknowledges these biases and values, and strives to fairly reflect the interests of the community, while at the same time being as objective as possible.

4.3.1 Assessment and Collaboration

Meetings of existing community groups such as women's groups or village elders can be a good place to announce and discuss a survey. When community members and leaders have been involved in the assessment from the beginning and have helped design the survey, it will be easier to obtain the cooperation of all the households included.

Project field workers should be sensitive to the relationship between the community and those identified as leaders. Leaders are often respected and recognized authorities in their communities, especially if they hold traditional positions. But in some cases leaders may not represent their community and are disliked or mistrusted. Perhaps the most extreme examples of this are the "leaders" created by colonial powers to implement their policies among local populations. This has occurred all over the world. The goals of such figureheads do not reflect the best interests of the community and these "leaders" are not trusted. Field workers and projects that align themselves with such leaders will have a difficult time being accepted and working effectively with the community.

4.3.2 Representativeness

Individuals and groups within a community often have different values, needs, and interests. Assessments should be *representative*, that is, they should include households from all segments of the community, such as different religious and ethnic groups, castes, economic levels, occupations, and geographic locations. It is also important to make sure that the needs of different individuals within the household are represented in the assessment. The different members of a household, including men and women, adults and children, or in-laws and blood relatives, often have very different responsibilities and power. A representative assessment may require talking to household women, not just the male "head of household," and making special inquiries about children, the handicapped, and the elderly. Cooperatives, community elders, local clinics, traditional healers, and women's, farmers' and students' organizations are examples of specialized community groups whose ideas are valuable for a representative assessment.

Whenever possible people should speak for themselves. There are always some people who are easier to talk to than others and in some cases certain groups of people may be less accessible or less accustomed to having their viewpoint valued. However, it is still important for their voices to be heard, and not have them represented by the opinions of other people, no matter how good those other people's intentions are.

4.3.3 Insiders and Outsiders

There can also be significant differences, as well as agreement, between the perspectives of local people and outsiders, such as project field workers. Either way, discussion helps insiders and outsiders understand each other and come to an agreement on the community's most pressing needs and ways of addressing them. For example, local people may have many uses for the "weeds" growing in their gardens, while an outsider may only see these plants as a factor causing decreased garden production. On the other hand, an outsider may be able to see needs of which the local people are not aware, and help them find ways to meet those needs. For example, an assessment that includes testing of the local water supply may find that it contains a lot of disease-causing bacteria, very likely responsible for much illness and some deaths in the community. While local people will recognize the illness as a problem, they may attribute it to other causes. In this case there must be discussion with the community explaining the connection between their problem and its cause, and addressing their questions and concerns. A project should never proceed without community understanding, support, and participation.

4.3.4 Participant Observation

Living, eating, working, talking, and relaxing with people is the best way to gain some insight into their world as they perceive it. This *participant observation* is a vital part of any assessment and includes conversation and informal interviewing, as well as observations of physical surroundings, people, and activities.[4] Relaxed social settings encourage candid and open discussions. Feelings of goodwill, trust, and mutual respect, so essential in community development work, are established through this sort of interaction.

Good observation involves clearing the mind as much as possible of preconceptions and expectations, focusing on the environment, and asking questions about what is seen, heard, smelled, tasted, and felt. Making brief notes in the field and later expanding on them may help improve the powers of observation. However, it is very important not to let note-taking in the field come between the field worker and the people she is working with. If people seem uncomfortable or offended, it is wise to stop taking notes.

We feel very strongly that the first thing any garden project field worker should do is try to grow a garden just as the people in the community do. If there are no local gardens then the field worker's garden should use only resources that are readily available to local households. By doing this she gains an understanding of gardening conditions in the area, including local resources, skills, and problems. She also demonstrates that local people are the focus of the project and that she recognizes their skills and knowledge. Only then will she begin to appreciate local conditions and be able to work with gardeners or those interested in gardening to support and improve gardening in the community.

An excellent way to start understanding indigenous gardens is by making a list or catalog of local garden plants. Such a catalog should include a sample or drawing for identification, the local name, how the plant is grown, and how it is used. These catalogs are also very useful for learning about indigenous knowledge regarding other topics including soils (section 9.2.1), water, garden pests and diseases, and food.

4.3.5 Gardens for Whom?

Project objectives can be greatly influenced by the assumptions of the development organization.[5] Such assumptions may not be stated explicitly and project workers are often unaware of them. However, these assumptions affect how needs are identified and how they are addressed. For example, if a development agency assumes that the solution to vitamin A and C deficiencies is increased production of dark green leafy vegetables (DGLVs), they may overlook important factors such as food preparation, distribution, and consumption. Among poor households the extra produce may be sold to obtain money for debt payments, and so will not contribute to improving household nutrition.

Some agencies and people working for them may be unwilling to endorse an assessment that excludes them from future project activities. For example, an agency specializing in irrigation may be reluctant to recognize an assessment that indicates improved health care and nutrition education as the priorities. Agencies and project workers must thoughtfully examine their assumptions to make sure that these do not interfere with the goal of improving people's well-being.

When answering strangers' questions, many of us tend to give the answers we think the interviewer expects. This is also a common problem when outsiders talk to people about gardens. The local people being interviewed may assume that outsiders are only interested in industrial-style gardens, an accurate assumption based on the approach taken by many garden projects (section 1.2). We have visited villages where residents had cultivated fruits and vegetables for their households for many years. However, local extension agents believed there were no gardens and went to great efforts to persuade the residents to learn less appropriate, industrial gardening methods[6] (Figure 4.1).

A major reason for these problems is that the Western, industrial-style garden has come to dominate the definition of gardens in development. Emphasizing a functional definition of gardens as we describe in

Figure 4.1 Assumptions About What Gardens are Have a Big Effect on Garden Projects

section 1.1 can help overcome this stereotype. Defining gardens can be further complicated by problems of translation. In Egypt we found that some bureaucrats and nutritionists used an Arabic term for gardens which meant formal pleasure gardens. Obviously poor villagers did not have anything that would fit that description, and an interview using this word to ask if people had gardens would get "no" for an answer. However, many poor households do cultivate small quantities of fruits and vegetables for their own consumption, fitting our functional definition of gardens.

4.4 What Do Existing Gardens Tell Us?

The first and most important step in assessing the need for gardens is understanding how gardens function in the households that already have them, why those households have gardens, and why other households do not have them. Answers will come through keen observation, patient listening, and by asking relevant questions in conversation and formal interviews. Nothing should be assumed. How the gardens work, and how they contribute to the household and community should be investigated. If there are no gardens in the community the project worker must find out why. Local gardening, or lack of it, can be compared with gardening in nearby communities, or even in more distant ones, if they are similar to the one being assessed.

Not only is understanding existing gardens an important first step when considering a project in a community, it is also essential throughout the process of assessment and project implementation. At each step, and for every topic, project workers need to ask questions like, "How is this task accomplished in existing gardens?" and "Do existing gardens meet this need or address this problem?" (Box 4.1).

4.5 Interviews

A *survey* is a tool for assessment in which information on the same topics is recorded for each household, garden, or other unit in the sample (section 4.5.3 discusses samples). Information can be gathered by observation or an interview, which can be either formal or informal. When the questions are numerous or complicated, or a large number of people are to be interviewed, a formal survey is better. Formal interviews should only be done after participant observation has established a good relationship between field workers and the community, and informal surveys have identified topics on which more information is needed. In this section we focus on formal surveys.

The value of a formal interview is that it provides a structured, standard format for making observations and asking questions. Because of this the information gathered can be summarized, subgroups identified,

---

Box 4.1
Useful Information from Existing Gardens

The following questions about existing gardens can be answered by casual observations and conversations, as well as with formal surveys or interviews.

- Which households are gardening? Do they belong to a particular social, economic, ethnic or other group?
- Where are these households? Are their gardens next to their houses, in their fields, along canals, in a community garden area?
- Who controls access to land and water for gardening?
- What are the age and sex of household members who garden the most?
- How long have they been gardening?
- How much time do they spend gardening? Does this differ by age or sex?
- What is the daily and seasonal garden schedule?
- How large are the gardens? Is there a relationship between household size and garden size?
- What is grown and when? Are there differences between wet and dry season gardens?
- Where do the planting materials (seeds, cuttings, etc.) come from?
- What other resources are used? Where do they come from?
- What foods from the garden are eaten? By whom? How are the foods prepared?
- Is any produce sold, traded or given as gifts? By whom?
- Who controls the income? What do they use it for?

and comparisons made between them. For example, if information about individuals' land resources is gathered for a community, women's access to land can be compared with men's and the implications discussed. In some parts of Mali where men have easy access to land for onion gardening but women do not, the lucrative business of onion marketing is not available to most women.

Formal interviews should be accurate and representative of various groups in the community. It is better to collect a small amount of useful, good-quality information rather than a large amount of information that is difficult to analyze or use because of its size and inaccuracies. The interview should be kept short (no more than one page long, at least for the first survey) and should be designed to obtain practical information for the assessment and the project (Figure 4.2). Another reason to keep interviews short is because people are busy and long interviews are tiring and irritating. The longer the interview, the less likely the interviewee will be to give accurate, thoughtful answers. It also becomes less likely that they will want to talk with the interviewer again, let alone consent to another interview in the future.

The following steps for a survey using a formal interview will be briefly discussed: composing questions, translating and back-translating, sample size and selection, pretesting, administering the interview, and coding, checking, and analyzing.

Figure 4.2 A Formal Interview

4.5.1 Composing Questions

There are several types of questions that can be used for a formal interview:
- Yes/no questions, for example, "Do you have a garden?"
- Precoded selection of responses, for example, "During which month does the household have the least food?" 1=January, 2=February,....12=December.
- Open-ended questions which may prompt a long, unstructured response, such as, "Why do you think people in this village do not have gardens?"

Yes/no questions are easiest to obtain answers for, and open-ended questions the hardest. Answers to yes/no or precoded questions can be summarized or coded along one margin of the questionnaire form. Coding makes it easier and quicker to tabulate and analyze the information. All questions should be as clear as possible, and should not be offensive. For example, it is considered rude in northern Ghana to ask how much food a household has stored, or to ask too many details about a person's health. All cultures have such areas of sensitivity. In the United States it is considered rude to ask people how much money they make. Answers to such questions may not be accurate, and these types of questions may make the respondent reluctant to answer other questions.

Questions should be composed so that a bias or expectation on the part of the interviewer does not interfere with the response. For example, using male pronouns (he, his) to refer to gardeners may indicate to the person being interviewed that the interviewer is not interested in knowing about women gardeners, or vice versa. Similarly, expressing an interest in, and approval of, local crops encourages people to discuss them, and not just the new commercial varieties being sold in the area. Each question needs to be reviewed for clarity and relevance. Can the question be understood? Will the answers provide information that will help make decisions about household gardens?

4.5.2 Translating and Back-Translating

If the interview will be given in a language other than the one it was composed in, several people should translate it independently, and their results should be compared to eliminate effects of personal biases and limitations in language skill. A separate set of people should then translate the interview back into the original language to make sure that the original meaning has not changed. This is called *back-translating*. For example:

Translation: a) English —> b) Hausa
Back-translation: b) Hausa —> c) English

Do a and c match? The wording should be corrected and adjusted until they do. Failure to back-translate is the cause of much inaccurate information being gathered.

4.5.3 Choosing a Sample

Deciding on which people are to be interviewed depends on the specific purpose of the garden project being considered. A *population* is the whole group of people that the survey is about. For example, the population could be all households in a community or neighborhood, or a particular category of people such as all women of child-bearing age in the district of Kowanga. When it is not possible to interview everyone or every household in a population a *sample* or subgroup representative of that population is selected.

The size of a sample will depend on a number of factors. Limited project time and resources often have a big effect on the size of the sample taken. A useful principle to keep in mind is that the larger the total population, the smaller the proportion of the population included in the sample needs to be for obtaining accurate, representative information. For example, in a community of 3,000 households, a 10% sample would be 300 households, enough to obtain information that is not dominated by the extreme or unusual responses of just a few households. If 150 households in the sample (i.e., 50%) say they are interested in market gardening it is said that 50% of the population may be interested in market gardening.

However, in a community of 20 households, a 10% sample would be only 2 households and the findings could be easily skewed by the opinions or experience of 1 household. For example, if only 1 household said they were interested in market gardening this would be 50% of the sample. A sample of 50% or even 100% of this community would provide much more representative information. In order to be confident that the results of the sample can be legitimately applied to the whole population, it is important to have an adequate sample size. Sampling is a complex topic, and anyone who is planning on conducting a large survey should consult with an experienced statistician (Box 4.2).

A *random sample* is one in which all members of the population have an equal opportunity to be selected. This ensures that the resulting sample will be repre-

---

Box 4.2
Statistics and Probability

Statistics has two broad functions.[7] The first is to describe something by summarizing information about it. This helps us to see important characteristics and makes the information more usable. Giving the percentages of households in a village survey that eat fresh fruit once a day, once a week, and less than once a week is an example of how statistics can be used to describe something.

The second function of statistics is *inductive*. That is, it allows us to make generalizations based on a sample or to compare two groups to see if they are really different in regard to the characteristic we are interested in, or if they can be considered as the same. Surveys of a whole community provide data that can be used to describe the community. If only a sample of the community is surveyed, statistical tests can be used to decide what inferences can be made about the whole population based on the sample. If comparisons are being made between different samples from groups, for example, households with and households without gardens, then statistics can be used to decide if any differences between the groups are *significant*, that is whether they can be accepted as real differences, or whether they are due to chance and do not reflect differences between the groups from which the samples are drawn.

Inductive statistics depends on the theory of *probability*, which allows us to tell whether the patterns being observed in the data occur by chance, or whether they are "really" there, that is, whether they are significant. (Section 11.4.1 gives an example of the use of probability in predicting rainfall.)

---

sentative of the range of persons or households in the area being sampled. For example, a sample chosen from one location in a village will not be representative of the whole village if residence in a community is itself nonrandom. That is, the section of the community a person lives in may depend on her social status, economic level, caste, or ethnic group. These different groups within the community may be distributed according to features like roads, schools, markets, mosques, shrines, pumps, or the best garden land.

An easy way of selecting a random sample is to give each household or person in the population a consecutive number, beginning with "1," writing each number on a slip of paper, folding the slip in half, placing all the slips in a container, mixing them up, and drawing out the slips one at a time without looking at them until the desired sample size is reached (Figure 4.3).

Random samples may not be appropriate in communities with more than one distinct economic, cultural, or ethnic group. In this case a *stratified sample*, one which intentionally selects a specific number of representatives of different groups, is better. A stratified sample can be selected in two ways. Let us say a survey is being done on a sample of 100 households in a community where 25% of the households are Moslem and the rest follow local religious practices. In this case a *proportional* stratified sample would randomly select 25 households from the Moslem part of the community and randomly select 75 households from the non-Moslem part of the community. This provides a more representative sample than random sampling for a population composed of distinct subgroups.

If comparing the differences and special needs of each subgroup is a goal of the survey, then a *disproportional* stratified sample can be used, selecting 50 households from each subgroup. For example, in parts of Burkina Faso, villages may have both Moslem and non-Moslem residents. This religious difference may be reflected in social differences that have a significant effect on household income and food supply.[8] Since Moslem women cannot make or sell sorghum beer due to religious prohibitions on alcohol, other income-earning activities including gardening may be of greater importance to those women than to their non-Moslem neighbors.

It is a good idea when selecting a sample to make it somewhat larger than will actually be needed, so that if some people or households do not participate for any reason, they can be easily replaced by others. In addition, some extra people or households should be selected to use when pretesting the survey.

4.5.4 Pretesting

Before the interview is given, it should be pretested with people from the same population who are similar to those in the sample. Pretesting identifies problems with the interview— if questions are unclear or inappropriate, or if the interview is too long it can be changed before giving it to the sample.

Figure 4.3 Selecting a Random Sample

4.5.5 Conducting the Interview

Whether they are community members or not, interviewers should be able to listen patiently, have a good sense of humor, respect those being interviewed, be interested in the project, and have neat handwriting.

Even when an assessment is being carried out in collaboration with the community, many may not have participated directly in the planning. Therefore, the first step when conducting interviews is to discuss the purpose directly with those being interviewed. This should be done before each interview and any questions should be answered at that time. The confidentiality of interview responses should be explained and maintained at all times. Each person or household interviewed can be given a code number and only that number needs to appear on the form. The key to the code should be kept in a safe place, separate from the interview forms.

The interview should be timed to least interfere with the schedules of those being interviewed. Every effort should be made to minimize the disruption caused by the interview and to show how it will contribute to improving the interviewees' situation. If this is not done people will be reluctant to be interviewed and some may refuse. If someone cannot be convinced to participate in a survey they should never be forced to do so. Instead, another person or household should be selected from the population.

When conducting interviews, the values of the community and its members should be respected. For example, in many Moslem countries permission of the male household head is required before interviewing women. This permission may only be granted if the interviewer is a woman, if the interview is conducted in the presence of a male household member, or both. It may be culturally appropriate to offer a gift when visiting a household. In northern Ghana we gave *kulikuli* (fried peanut balls) to children and kola nuts to adults.

4.5.6 Coding, Checking, and Analyzing

Finished questionnaires should be reviewed as soon as possible after completing the interview. This means coding all answers that can be coded, and checking for any obvious errors, misunderstandings, or missed questions. Returning to check an answer with an interviewee should be done as soon after the interview as possible.

Tabulating the information is one of the simplest methods of analysis. It involves counting the number in the sample with one response to a given question and comparing it to the number with different responses, often expressed as percentages. For example, a sample of 30 mothers of young children were interviewed to discover if they used DGLVs in weaning foods. Twelve said they did and 18 said they did not. These results showed that 40% (12/30) of the mothers in the sample used DGLVs in weaning foods and 60% (18/30) did not. More sophisticated analyses of the data can be made, and someone experienced in statistics should always be involved (Box 4.2, section 4.5.3).

The survey results should answer the questions with which the community and the field worker started. These results are then the basis for further discussion and, combined with any other information gathered, will help the community and field worker decide what actions they will take, and whether gardens will be included in a project.

4.6 Seasonality

The marked seasons that characterize drylands mean that availability of resources changes through the yearly cycle. In an assessment it is very important to find out how the situation differs from season to season. For example, changes in the availability of water have a big effect on the need for, and supply of, food and income. In many drylands, food is in shortest supply during the rainy season before the harvest. This is also a time of increased demand for agricultural labor, and water-borne diseases such as malaria are common.

Gardens also change with the seasons. For example, rainy season gardens may be located near the home compound, while in the dry season they are grown in dry streams and depressions where soil moisture is highest. Cool season crops can differ substantially from those grown in the same garden during the warm season.

A good way to understand seasonality is to make an annual calendar showing variations through the year. Figure 4.4 is an example of an annual calendar for northeast Ghana. It is best to begin by quickly gathering preliminary information, entering it on the calendar, then deciding what additional information is needed (Box 4.3). This is a good safeguard against wasting time making a calendar much too elaborate for practical use. A number of smaller calendar forms can be used for taking notes, and the information can be put together later on a larger sheet of paper.

Scales on the vertical axis of the calendar can be absolute values, for example, 0-300 mm (0-12 in) of rain, or total household income, for example, 0-500 rupees/month. Or the vertical axis may illustrate relative values such as no rain, some rain, much rain, or no income, less income than needed for basic needs, or more income than required for basic needs. Estimates of a probable range of values are better than nothing. As more information is gathered these estimates can be adjusted.

As the project becomes more focused, calendars addressing specific people or crops can be made. For example, if women are interested in expanding gardening into the hungry season a calendar of women's activities could be created. This would help field workers understand and discuss the idea with local women.

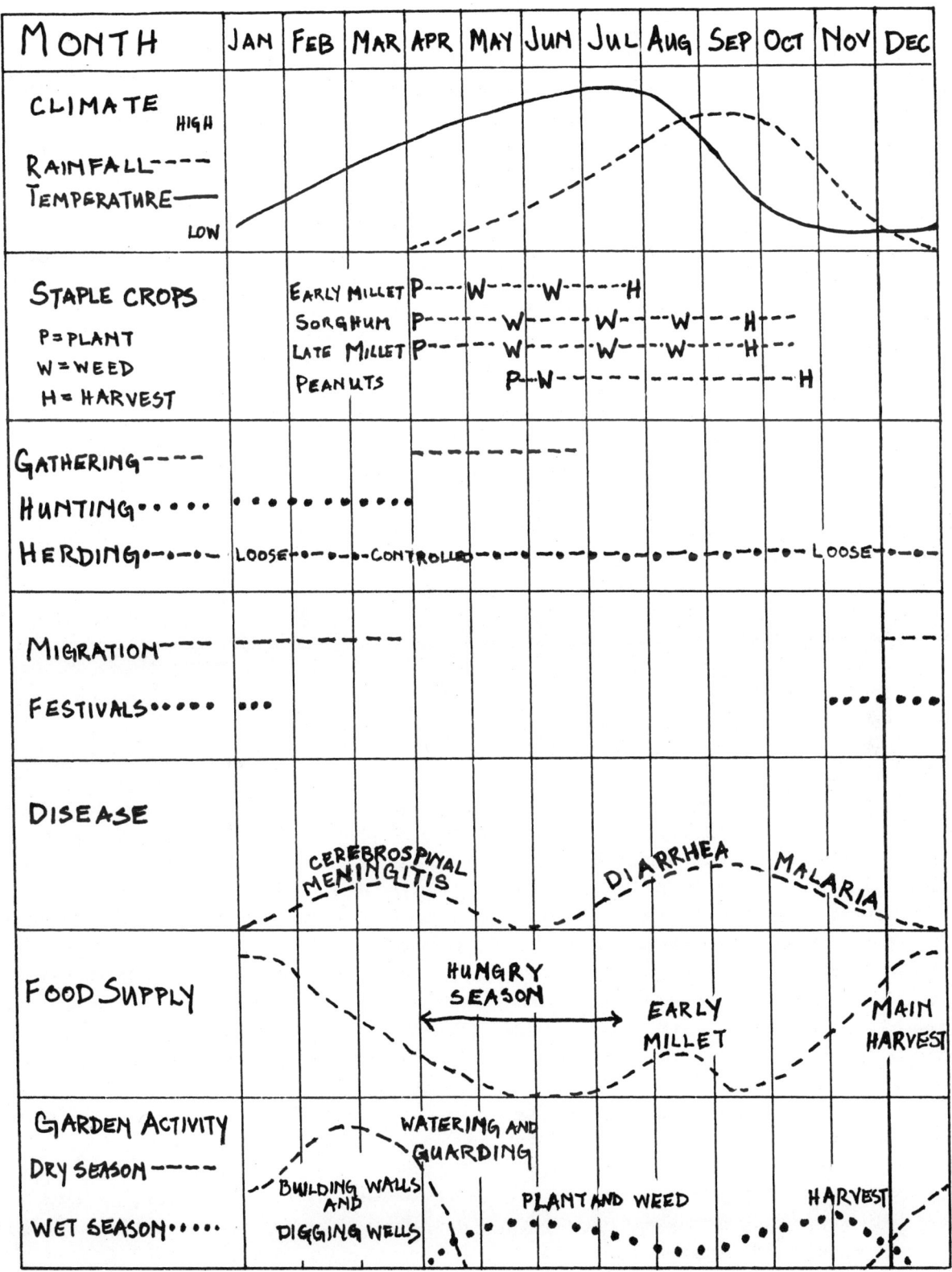

Figure 4.4 *An Annual Calendar for Northeast Ghana*

> **Box 4.3**
> *Possible Topics for an Annual Calendar*
>
> I ENVIRONMENT
> - Rainfall
> - Temperature
> - Availability of water for irrigation
> - Major agricultural pests and diseases
> - Availability of material for fencing, shades, trellises, mulches
> - Availability of soil amendments: manure, sand, compost
> - Availability of land, as influenced by cropping pattern
>
> II PEOPLE'S ACTIVITIES
> (These can then be divided into women's and men's activities)
> - Field crops: sowing, weeding, pest management, harvesting, processing
> - Other food-producing work: gardening, gathering, hunting, fishing
> - Wage labor
> - Animal herding, including seasonal migrations
> - Market activity
> - Labor migration
> - Ceremonial activities
> - School attendance
>
> III NUTRITION AND HEALTH
> - Actual or potential availability of garden foods
> - Availability of staple foods
> - Availability of wild, gathered foods
> - Availability of market foods
> - Hungry season
> - Disease incidence
>
> IV MARKETS
> - Actual or potential availability of garden produce for marketing
> - Shortages of income for the household and individual members
> - Path and road conditions
> - Availability of transport
> - Market demand for garden produce

4.7 Food Distribution and Consumption

Food supply and consumption patterns offer insights into nutritional needs. Market surveys listing the kind, quantity, and cost of foods in the market at different times of year can be done quite easily and are important for understanding local food supplies and for assessing the potential for marketing garden produce. However, they are just part of the food system and should never be used by themselves as indicators of nutritional status. Food in the market is often not available to those who need it most because they cannot afford to buy it. In dryland West Africa we have seen markets full of food while villagers a few kilometers away were hungry but unable to translate their needs into buying power. If they could, the markets would have soon been emptied.

Eating with a household gives insights into what they eat, how food is prepared, and how it is distributed. However, the field worker is often considered an honored guest and, at least the first few times, larger quantities and more special foods may be served (Figure 4.5). These are unusual circumstances and should not be used in an assessment. In addition, the field worker must be careful not to overburden the household which may feel obliged to provide these special meals.

Understanding local diets and nutritional needs means finding out what people eat, how often they eat it, and how this changes at different times of the year. Interviewing a sample of households to find out their source for major types of foods at different times of year can help a field worker understand the seasonality of diets in the community. Adding this information to the annual calendar (section 4.6) will show where dietary change fits in with other seasonal fluctuations.

Collecting information on the frequency with which different foods are eaten, how they are prepared, and who eats them gives a rough idea of what nutrients might be lacking in the diet at different times of the year. This does not provide precise figures on nutrient deficits, or the quantity of nutrients needed to supplement the present diet, but it can give ideas for the kinds of crops to encourage in gardens.

Often the person responsible for preparing the food

Figure 4.5 Often Field Workers are Served Special Meals

will be able to provide the most reliable information for a food frequency survey. Useful questions include, "What do the people in your house eat in the morning?" (or during the day and in the evening) "Are there special days when the quantities of foods eaten are different?" and "How often do people in your household eat fruits?" (or vegetables, cereal and root staples, legumes, dairy foods, nuts and seeds, meats, and fish). The foods eaten and the ways they are prepared should be noted. It is very important to find out how this information varies for different people in the household.

Some important reasons why people have different diets and eating habits are because of individual preferences, cultural values, and geographic locations. Beliefs or customs about foods may have originated for reasons of health, to ensure distribution of valuable foods, because of local beliefs about the cause of illness, or for other reasons. Special diets that may affect health are often prescribed during sickness, pregnancy, and lactation. Some are beneficial, but others are harmful. Working with people to support their healthful beliefs about food and discussing why some beliefs can be harmful is a difficult process. Werner and Bower discuss ideas of how to do this.[9]

The following patterns in data gathered on food will help to understand the local nutritional situation:
- Patterns of food supply and consumption as they vary between rainy and dry, warm and cool seasons.
- Nutritional quality of the most common meals (Chapters 2 and 15).
- The variety of foods: staples, legumes, fruits and vegetables, dairy, meat.
- Sources of produced and purchased foods: field, garden, gathered, market, friends or relatives, food aid.
- Social and economic access to food: in the community (by class, caste, ethnicity, religion, other group), in the household (by age, sex, relation).

4.8 Maps

Making a sample map of an area with community members is a good way to become familiar with the location of natural and social resources. Most available maps will probably be on too small a scale to be useful for village-level information. Maps on a scale of about 1:50,000 (1 cm = 0.5 km or 1 in = 0.8 mi) are helpful guides for making larger scale maps of areas between 10 km^2 and 50 km^2. However, a simple map can be drawn easily by quickly making a preliminary sketch, and then deciding what, if any, additional information is needed. Later the location of resources can be added. Box 4.4 lists some useful information that can be shown on maps. Figure 4.6 is an example of a village sketch map helpful when doing an assessment.

Box 4.4
Useful Information that can be Shown on Maps

- Compass directions or orientation to major landmarks
- Water sources: rivers, streams, wells, pumps, springs, qanats
- Forests, quarries, clay pits, and other sources of garden building materials
- Prevailing winds (may differ with season)
- Residences, including those of local leaders
- Agricultural lands, including gardens and fields
- Places where wild foods are gathered
- Grazing areas, animal enclosures, sources of manure
- Medical resources: local healers, midwives, clinics
- Roads, markets, stores
- Sacred and ceremonial areas
- Seasonally flooded areas

4.9 Long-Term Trends

Identifying and understanding long-term trends in an assessment helps to ensure that changes made by the project will continue into the future. Awareness of these trends helps the community and its projects foresee and plan for changes.

For example, a common trend in the rural Third World is environmental degradation. In many regions deforestation is a serious problem which leads to soil erosion and desertification. The social implications of deforestation are equally serious and include big increases in time and energy spent collecting fuel wood, destruction of agricultural lands and their productivity, and loss of wild food sources for people and their animals. Frequently women bear most of the increased work burden because they are often responsible for providing their households with both fuel and wild, gathered foods.

Overall this trend may show the need to redistribute resources, lower consumption by some, and find sources of energy and income that will not destroy local resources. In terms of garden projects this may mean people have less time and energy for gardening. However, it could also mean gardens will be increasingly important as a source of fruits and vegetables, and income for purchasing fuel.

Other long-term trends that affect dryland community development efforts, including gardens, are dropping water tables, soil salinization, changes in land tenure such as increasing privatization, out-migration of young people, and changing eating patterns.

4.10 Outside Sources

For many areas information has already been collected that can be useful for an assessment. Frequently this information may be difficult to obtain, for example because it is only available in the capital city, or from people or organizations outside the country. Even so, finding it may be worth the effort, and may save time and resources. As with all other information gathered for an assessment, printed or published information must be assessed for representativeness, accuracy, and objectivity. Box 4.5 lists some types of outside information useful for assessment, and possible sources of that information.

4.11 Resources

Many resources about assessments and how to do them are long lists of "Questions to ask." These can stimulate thinking, but too often discussion of how to go about actually doing the assessment and what to do with the information is lacking. To some extent this is unavoidable because methods and questions must be tailored each time to meet the special circumstances of each community and project.

One of the best books to read when preparing to do

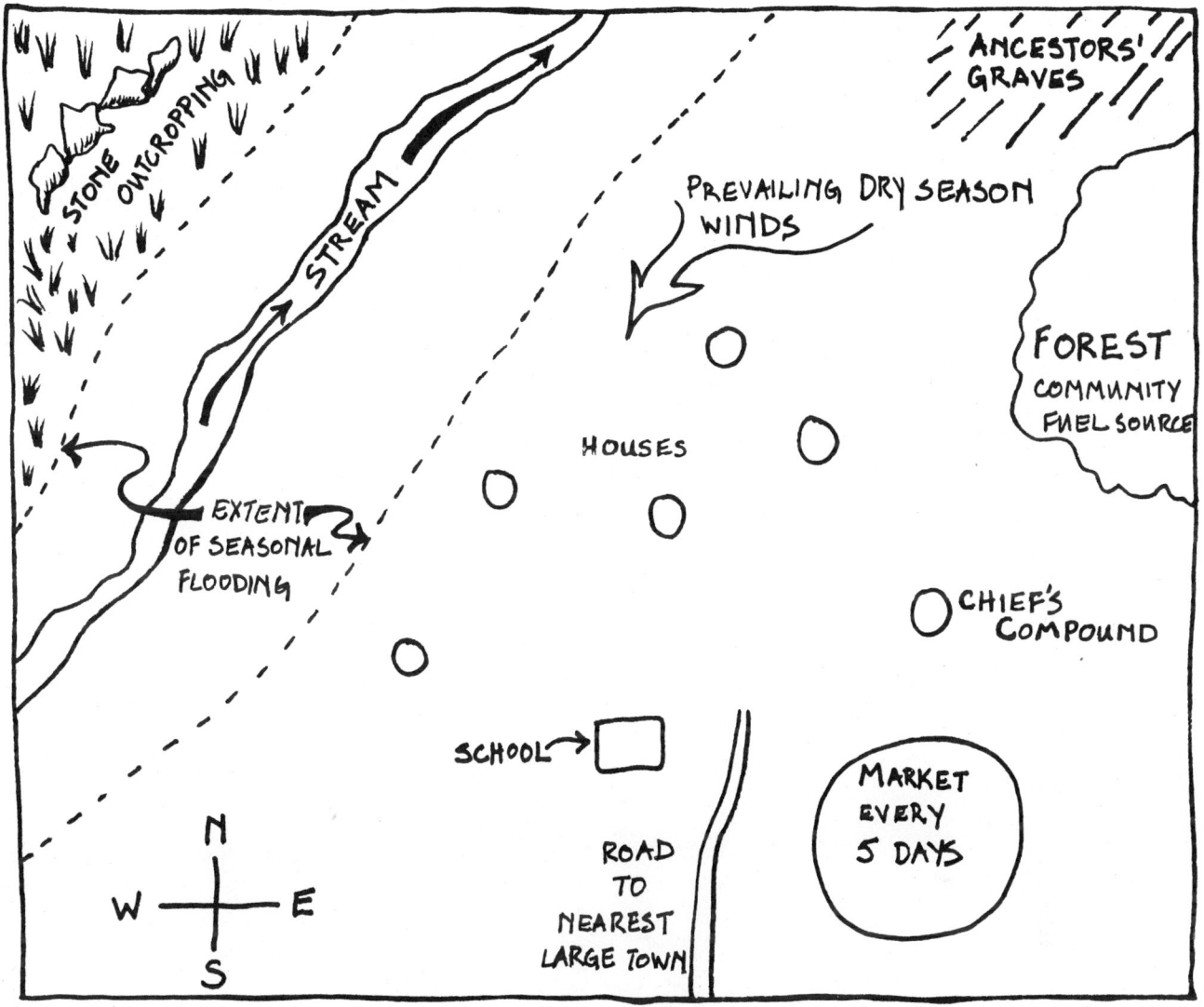

Figure 4.6 A Sample Sketch Map

**Box 4.5
Outside Information**

I *TYPES OF INFORMATION*
- Census reports of both population and agriculture
- Medical and nutritional surveys
- Soil surveys
- Surface and groundwater surveys
- Anthropological, historical, geographical reports
- Botanical surveys
- Agricultural experiment station reports
- Research papers by college and university students
- Project and planning documents and reports

II *SOURCES OF INFORMATION*
- Local and national clinics, health posts, hospitals
- Universities that have carried out local research
- Local, regional and national government departments of agriculture, health, nutrition, community development, census
- Libraries
- People who have previously worked in the area
- Local or international development organizations like ILIEA, Hesperian Foundation, Cultural Survival, and many others (Chapter 19 has brief descriptions of these and other organizations.)

an assessment is *Rural Development: Putting the Last First* (Chambers 1983). This is an easy-to-read discussion of why and how many development activities have tended to overlook those most in need. It also gives some brief suggestions of ways to overcome this problem in projects and project assessments.

We think the best outline for assessment is by David Werner and Bill Bower in their book *Helping Health Workers Learn* (1982). This is an excellent, inspiring book full of information useful for anyone working in community development. Werner and Bower constantly emphasize practical methods that support local control, while keeping in mind the goal of improved well-being for those most in need. The following sections are especially helpful for assessments: Chapter 6, "Learning and Working with the Community;" Chapter 7, "Helping People Look at Their Customs and Beliefs;" and the discussion of evaluations in Chapter 9, "Examinations and Evaluation as a Learning Process."

Anthropologists and other social science field workers have been struggling with the best ways to gather and use data from local communities for a long time. Spradley's books on participant observation (1980) and ethnographic interviewing (1979) are good, practical introductions which emphasize the need to understand the local situation from the people's point of view, and to do research with, and in the best interests of, local people. Bernard (1988) has published a helpful handbook on field methods in cultural anthropology, emphasizing quantitative measurement.

Part three, "Field Methodologies," in OXFAM's *The Field Directors' Handbook* (Pratt and Boyden 1985) provides a general introduction to assessment at the program level.

References

[1] Bunch 1982:61-63.
[2] Werner and Bower 1982:9-18 — 9-22.
[3] Grün, et al. 1989; O'Brien-Place 1987.
[4] Spradley 1980.
[5] Pacey and Payne 1985:210-213.
[6] E.g., Cleveland 1986.
[7] Blalock 1972:3-8.
[8] Saul 1981.
[9] Werner and Bower 1982:Chapter 7.

Part II
Garden Management

The interaction of many factors influences the health and productivity of plants in the garden. Among these are the microclimate, soil, water, microorganisms, neighboring plants, and the genetic inheritance of the plant itself.

There are also many ways to propagate and care for plants, and more than one technique can produce the specific results desired. For example, a gardener wanting to keep her garden free of weeds could spray an herbicide, mulch heavily, or remove the weeds by hoeing. If she wants to start seedlings for transplanting she could purchase specially manufactured, imported seed boxes, use locally made containers, or ones she made herself from free materials. The different techniques in these two examples will all produce the desired short-term results: weeds can be controlled by spraying, mulching, or hoeing; seeds can be started in specialized boxes, local containers, or home-built ones. Because there are many ways to accomplish these and other garden management tasks, the choice of technique must be based on other criteria as well.

If the main goal of garden management is to maximize production and profit, and the required resources are available, then the industrial agriculture model may be effective (Part I). Large-scale, industrial agriculture has been very successful in increasing production and yields by greatly increasing the use of machinery and the fossil fuels to run them, chemical fertilizers and pesticides, and irrigation water. Compared with small-scale, indigenous agriculture it has much higher returns to labor, but much lower returns to energy (section 3.2.1). Industrial agriculture is based on increasing centralization of management and marketing, and on increasing control over nature, rather than working with nature.

Genetic diversity in crops, ecological diversity in fields and regions, and social diversity in management has been drastically decreased in the the drive to increase production (section 14.2). In turn, this lack of diversity results in decreased sustainability, because industrial systems are less and less capable of maintaining their high levels of production when challenged by drought, shortages of irrigation water, a break-down in the fertilizer distribution network, increasing oil prices, or outbreaks of pests and diseases.[1] The typical response of industrial agriculture to such problems is to attempt to increase control over nature and to centralize the system even more.

Experience has shown that the industrial approach to food production often results in increasing inequity because the capital and resource requirements are beyond the means of many Third World households. This approach has also been found to be harmful to the environment and to human health.

The criteria for selecting garden management techniques which we use in this book are self-reliance and local control of the food system; equal distribution of food for improved human health and nutrition; preservation of biological and cultural diversity; and conservation and protection of natural resources. This is why the approach we take to garden management in the Chapters of Part II is quite different than the approach of industrial agriculture, and reflects a growing interest in sustainable agriculture.

The term sustainable agriculture is widely used today to describe agriculture that has the goals of conserving the environment for the future and providing nutritious food for all people equitably (section 1.2). There is increasing awareness in both industrial nations and the Third World of the need to conserve resources for the future. Decreasing profits resulting from environmental degradation such as groundwater depletion and soil erosion, and consumer pressure for a more healthy food supply, are pushing industrial agriculture toward sustainability (section 3.2). For example, the United States' National Academy of

Sciences has published a major book titled *Alternative Agriculture* which advocates moving that country's agriculture away from high inputs of chemical pesticides and fertilizers.[2] Agriculture is increasingly being studied from an ecological perspective.[3] However, because the concept of sustainability has become so popular, it is sometimes used in ways that distort its meaning and make it subservient to production economics.[4]

In the Third World there is also interest in reorienting agricultural development away from the industrial model of the green revolution and toward sustainability.[5] Detailed descriptions of indigenous agricultural systems contribute to a growing appreciation of their ecological (environmental) and social sustainability.[6] However, in many of the World's poorest communities population pressure, social disruption and incorporation into the world economic system have made indigenous agriculture environmentally destructive, socially inequitable, or both. A redistribution of resources from the rich, industrial sector is essential for those poor communities to create sustainable agricultural systems. Ultimately, a sustainable agriculture must be one that supports an end to growth of the human population, to our increasing levels of consumption, and to cultural and environmental destruction.

The Chapters in Part II discuss methods of dryland garden management based on a striving for ecological and social sustainability in its fullest sense. This means making the most of the indigenous knowledge, ecological and social diversity, and locally adapted biological resources which characterize many dryland food systems, while using the knowledge and techniques of Western science to enhance sustainability.

References

[1] Cleveland and Soleri n.d.c.

[2] NAS 1989a.

[3] E.g. Carroll, et al. 1990; Cox and Atkins 1979; Gliessman 1990.

[4] Cleveland 1991; Orr 1988.

[5] E.g., AGRECOL/ILEIA 1988; Dupriez and De Leener 1983.

[6] E.g., Lagemann 1977; Richards 1986: Westphal, et al. 1981, 1985.

Part II Garden Management

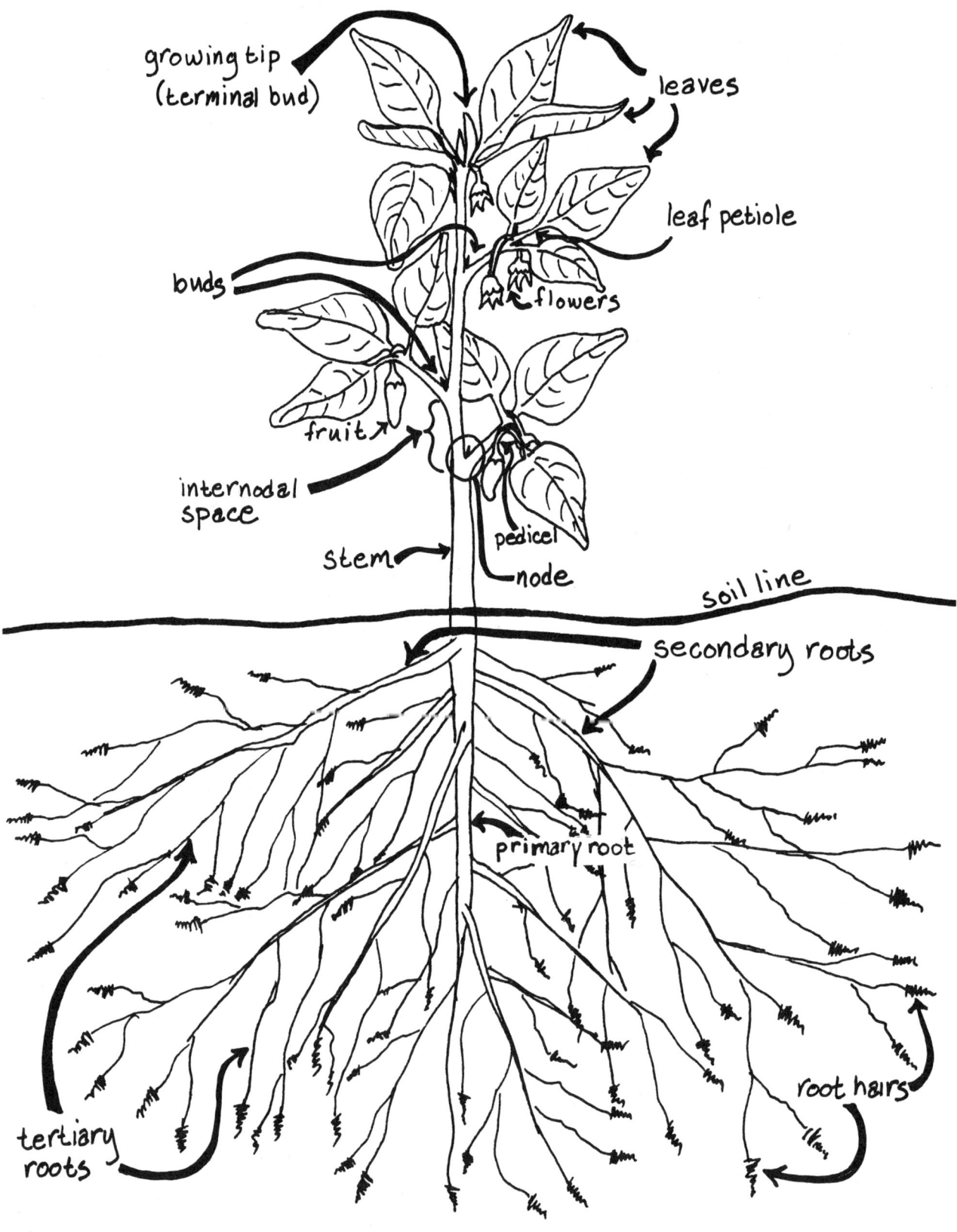

Figure 5.1 Plant Anatomy—the Chili, a Dicot

5
How Plants Live and Grow

Knowing how plants live and grow enables gardeners to adjust management practices according to specific local situations and helps them solve problems in the garden. For our discussion of how plants live and grow we use the terminology of Western science. However, many local systems exist which serve the same purpose, using terms and concepts developed through people's experiences. These local systems are also valid; appreciating and attempting to understand the local system is essential for working with gardeners.

5.1 Summary

This chapter begins with an illustration of basic plant anatomy (Figure 5.1). The vascular system transports food, water, minerals, and other essential substances throughout the plant. Photosynthesis and transpiration provide the plant with the food energy necessary to live, grow, and produce a harvest. Under hot, dry conditions the rate of water loss from the plant increases and can lead to water stress that reduces yields. Plants have evolved a variety of responses to help them survive under these dryland conditions. Some plants also have a tolerance of salty soil, a common problem in drylands. Many crops or crop varieties also have daylength and temperature requirements that can limit their growing seasons.

5.2 The Vascular System in Plants

The *vascular system* is the network of plant cells responsible for the movement of water, minerals, food (sugars), hormones, and other vital substances inside plants.

Water in the soil is taken up by the roots through a combination of osmosis and cohesion. *Osmosis* is the pattern of water movement across a water-permeable membrane such as the cell membrane. If two liquids are separated by such a membrane, water will move out of the more dilute solution, the one with a lower concentration of solutes like salt, and into the more concentrated solution (Figure 5.2). This movement will continue until both solutions have the same concentration of solutes per volume of water. If the concentration of solutes is greater in the root cells than in the soil, water will move into the roots. Water loss from transpiration increases solute concentration in the leaves and so water continues to be pulled up through the plant by osmosis.

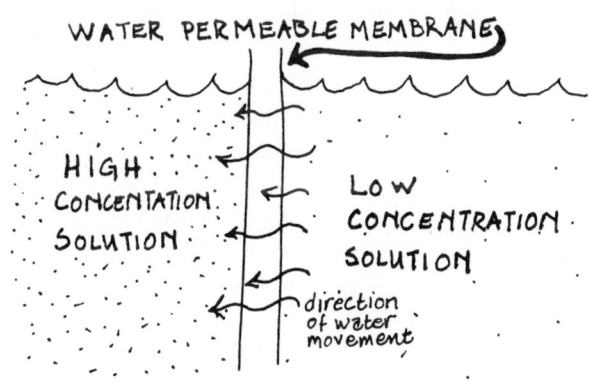

Figure 5.2 Osmosis

Cohesion is the tendency of like substances to stick together. The cohesion of water molecules, together with transpiration and osmosis, causes a continuous flow of water to move up the plant. Once the soil moisture is depleted to the wilting point (section 10.3.1) osmosis and cohesion will no longer be strong enough to move water out of the soil and into the plant.

Dicots and monocots are the two major groups of garden plants. Their vascular systems are arranged differently. *Dicots* are those plants such as beans, cucurbits, amaranths, and many fruit trees which have two *cotyledons*, or seed leaves, in their seeds, and branching leaf veins. *Monocots* have only one cotyledon and usually the veins in their leaves are parallel to each other, running the length of the leaf as in maize, onions, date palms, and most cereals. In larger seeds the difference between a monocot and a dicot is obvious. For example, a bean seed can be easily split into two halves, the cotyledons. A maize seed, however, does not split because it has only one small cotyledon.

The *xylem* is the part of the vascular system that carries water and nutrients from the roots to the leaves. In monocots the xylem tissues are scattered in bundles that run the length of the plant, throughout the leaves, stems, and roots. In dicots the xylem tissues occur in a discrete layer, which in the stem surrounds the pithy center. In dicot roots the xylem is the tissue at the core (Figure 5.3).

The sugars made by photosynthesis (section 5.3) and many growth-regulating hormones produced by plants' growing tips flow through the *phloem*. Osmosis is also thought to be the source of movement for substances in the phloem. As the concentration of sugars produced by photosynthesis increases in the phloem, water from the xylem enters these cells, building up pressure within them. This forces movement of the solution to cells with lower concentrations and pressure until it reaches a place where the sugars are needed or can be stored for later use. Because most photosynthesis occurs on the outer and upper layers of the plant, those leaf areas exposed to sunlight, the movement of solutions in the phloem is primarily inward toward the main stem and downward to the roots where there is little or no photosynthesis. Sometimes the fluids in the xylem and phloem are called *sap*.

In monocots the phloem and xylem tissues are grouped together in vascular bundles running vertically through the plant. In dicots the phloem is a distinct layer separated from the xylem by a thin layer of *cambial tissue* (Figure 5.3). These continuous layers of phloem and cambial tissue make grafting and layering of dicots possible (sections 7.6 and 7.7), whereas with monocots these techniques are not possible.

The outer surface of green plant parts is the *epidermis*. Underneath the epidermis in green shoots and stems lies the *cortex*, tissue that surrounds the vascular system. In dicot trees the outer layer of the trunk and branches is called *bark*, a term that refers to all of the tissue from the cambium and phloem to the outer surface. In bark the cortex and epidermis are replaced by a more rigid, woody tissue called the *cork*, which includes a layer of dead cells on the outer surface.

5.2.1 Roots

Even though they are not usually visible, the roots are one of the most important parts of a plant. Roots provide structural support by anchoring plants in the soil, and they absorb water and nutrients in the soil and transport them to the *shoot system*, the aboveground portion of the plant. *Root hairs* are fine "hairs" that grow out of the root's epidermis, just above the actively growing part of the root and root tip. The root hairs provide much of the root's surface area and so they are very important for the absorption of water and nutrients. Some plants have large, fleshy roots that store energy and water for the plant. A number of these large roots are commonly eaten such as sweet potatoes, carrots, beets, and cassava.

There are two easy-to-identify patterns of root growth: fibrous and tap roots (Figure 5.4). *Fibrous roots* spread out and downward in a mass of fine roots, none of which dominate. Fibrous root systems include many secondary and tertiary roots, or *lateral roots*, those that grow out of an older root and therefore do not tend to grow straight down (refer to Figure 5.1 in section 5.1). Monocots like maize and sorghum commonly have fibrous root systems. Garden crops that are dicots, for example, carrots, okra, chilis, sweet peppers, and amaranth, have a *tap root*, a dominant vertical root with other smaller roots growing out from it. These tap roots can make use of water deep below the soil surface. Many dryland fruit trees such as carob and olive also have a tap root. When the tap roots of mature plants are cut off, for example, in transplanting, the plants may die. Some of these plants can recover by developing alternative roots in a pattern similar to a fibrous root system. However, this will only occur if the plant is young, vigorous and its shoot system is relatively small.

Plants' root systems also vary depending on a number of factors including the soil, irrigation patterns, distribution of nutrients, plant density, and neighboring plants. Root systems have a great capacity for *compensatory growth*. That is, in areas of soil where the conditions are favorable the roots will proliferate, compensating for areas of the root zone that are less favorable. This is important to consider when irrigating young plants, because the root system will develop

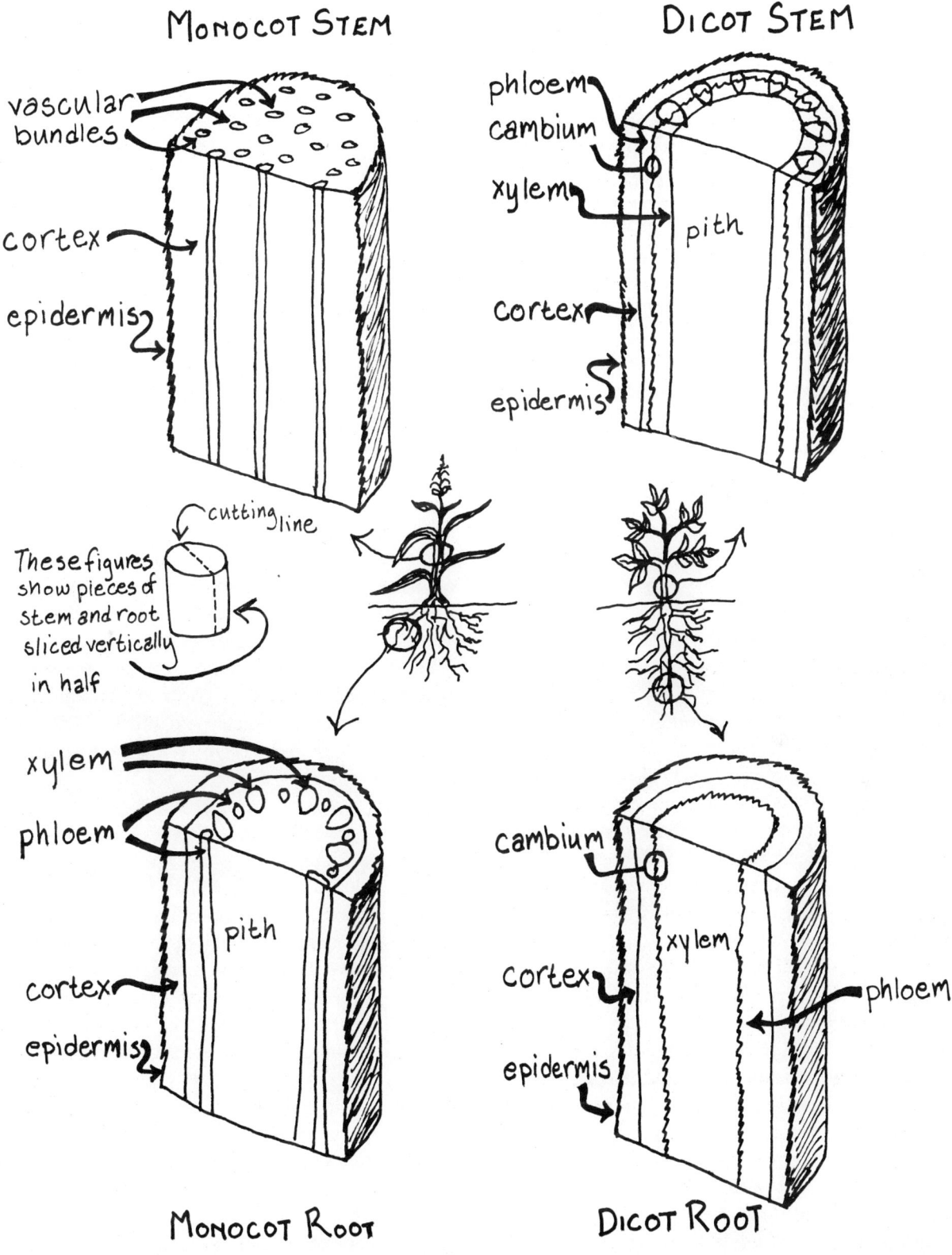

Figure 5.3 Stem and Root Structures of Monocots and Dicots

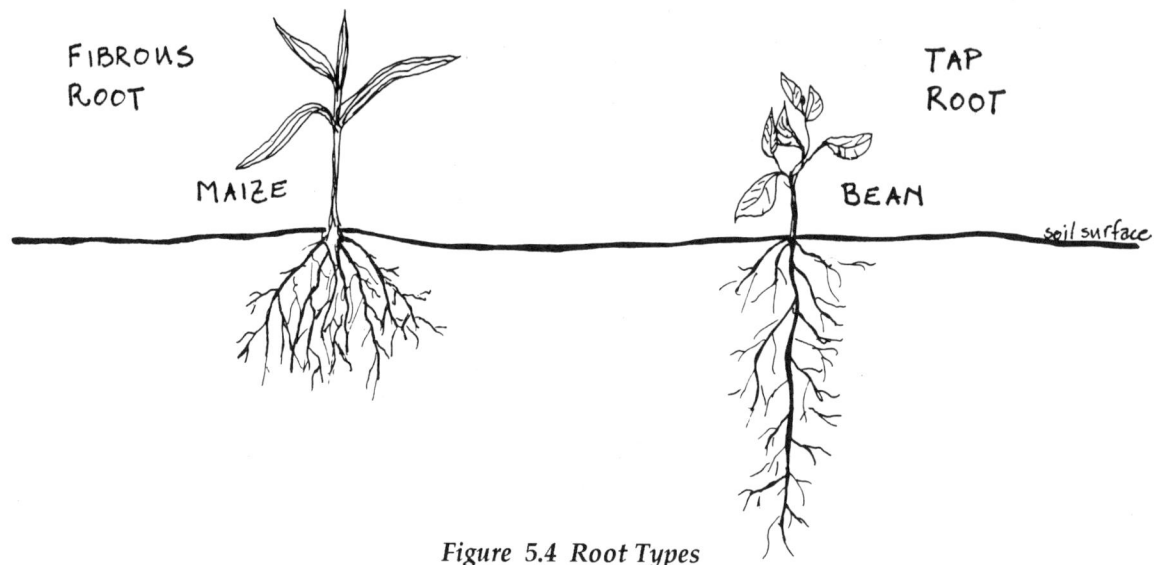

Figure 5.4 Root Types

most strongly where there is consistent moisture. If irrigations are frequent and shallow, for example 10-15 cm (4-6 in), then the plant will develop a shallow root system. Under hot, dry conditions moisture in this surface layer is lost quickly by evaporation. Shallow-rooted plants will require more water applied in more frequent irrigations than plants that have received deeper and less frequent irrigations, encouraging them to develop a deep root system.

Poor drainage and overwatering also cause shallow rootedness as the roots avoid waterlogged soil. Watering patterns that encourage shallow rootedness can lead to other problems such as salinity (section 12.6.2) or roots growing primarily in upper soil layers where temperatures are high, both of which can inhibit growth and kill the plant in severe cases. For these reasons, when watering established seedlings and older plants it is important to wet the soil down to at least 15-40 cm (6-16 in), and below this for trees, in order to encourage deep root growth. However, because compensatory growth is a gradual process, one should not switch abruptly from frequent shallow irrigations to less frequent deep irrigations without a transition phase of deep but less and less frequent waterings.

Root growth is also affected by soil texture and structure (section 9.3.1). Roots will grow where soil conditions are best, for example, where compost and manure have been added and where the soil structure allows easy penetration of roots, air, and water. Extremely heavy, clayey soils with little structure make it difficult for roots to grow and they can become thick and deformed from trying to push through the soil.

From the soil roots obtain nutrients such as nitrogen and phosphorus which are essential for healthy plant growth. In some cases this is made possible through mutually beneficial or *symbiotic* relationships between plant roots and soil microorganisms. Mycorrhizae (Box 9.5 in section 9.5) symbioses enable plants to use more of the phosphorus, zinc, or copper in the soil.[1] Symbiosis between *Rhizobium* bacteria and roots of legumes makes nitrogen in the air available to the plant while also enriching the soil (section 9.5.2).

5.3 Photosynthesis

Photosynthesis is the process by which green plants change the energy in sunlight into energy stored in carbohydrates (CBHs), the food used for growth and reproduction. *Chloroplasts* are the structures in plant cells where photosynthesis occurs. They contain a green pigment called *chlorophyll* which uses sunlight to fuel a reaction with carbon dioxide (CO_2) gas in the air, and water (H_2O) in the plant. The products of this reaction are oxygen (O_2), water, and carbohydrates, such as starches and sugars (Figure 5.5). Any plant part containing chlorophyll can conduct photosynthesis, but the leaves are the main areas of photosynthesis in most green plants.

The carbohydrates produced by photosynthesis are broken down into the simple sugar glucose, which then combines with oxygen to produce CO_2, water, and energy. This process, called *respiration*, provides the energy necessary for the plant to live and grow.

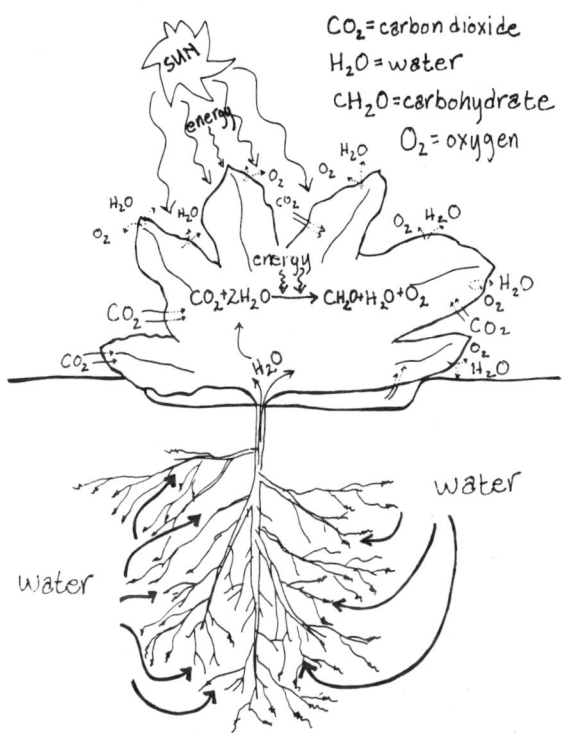

Figure 5.5 Photosynthesis

5.4 Transpiration

For photosynthesis to occur carbon dioxide (CO_2) must enter the chloroplasts, most of which are found in the cells under the plant's epidermis. Most CO_2 enters the plant through the *stomata* (singular is *stoma*), tiny holes in the epidermis which can close (Figure 5.6). When the stomata are open not only can CO_2 reach the chloroplasts, but moisture from the inside of the leaf is able to evaporate into the environment. This movement of water vapor through the plant's stomata is called *transpiration*. As water evaporates from the leaves during transpiration, the concentration of nutrients in the surface cells increases compared with that in adjacent cells, from which water then moves by osmosis. The same process is repeated all the way down to the roots. Because of the great cohesiveness of water and its *adhesion* (the attraction between dissimilar substances) to the cells of the passages along which it moves to the leaves, the water is pulled upward from the roots to the leaves. The energy that keeps this water moving upward is supplied by the sun which causes evaporation of water from the plant during transpiration.

Transpiration is important for two reasons: as just described, it provides the "pull" that keeps water and nutrients moving up through the plant from the roots (Figure 5.7), and, under hot, dry conditions transpiration cools the plant the same way evaporation cools our skin when we sweat. About 90% of all water absorbed by plant roots is released in transpiration. Under stressful (hot, dry) conditions, the amount of water needed by the plant, and thus the amount released in transpiration, increases.

Transpiration rates vary depending on plant types and environmental conditions. Photosynthesis increases with available sunlight, so under sunny conditions the stomata are open longer to supply the necessary CO_2, thus increasing transpiration. Conditions that increase evaporation, such as low air humidity, heat, and wind, also increase transpiration.

The stomata in some plants such as grapes will shut under extreme water stress. However, this may not save the plant. When stomata close to prevent water loss, the cooling effect of evaporation also stops, which can cause problems with high leaf temperatures.

Under sunny, dry, hot conditions transpiration rates are extremely high. If the soil is unable to provide enough water to keep up with the rate of transpiration the plant will wilt. If this water loss is not replaced soon, the plant will die. By shading, mulching, and providing garden plants with protection from drying winds the gardener reduces the need for water and so the amount of water that will be lost to transpiration (section 10.8). Directing water down to the roots, for example, with vertical mulch (section 10.8.2), minimizes the amount lost by evaporation from the soil surface, and makes more water available to meet the plant's needs. Gardeners also use plants with lower rates of transpiration and other characteristics that make them better able to survive and produce under dryland conditions.

5.5 Coping with Heat and Drought

Dryland garden plants must often produce food and other products under hot, dry conditions. For plants, *drought* is a condition in which there is insufficient water available in the soil to meet the plant's needs. A consequence of drought can be *water stress*, or water deficit—that is, insufficient water inside the plant for it to maintain itself and grow. Distinguishing between drought, a condition in the environment (especially the soil), and water stress, a condition in the plant, makes it easier to understand how plants respond to dryland conditions.

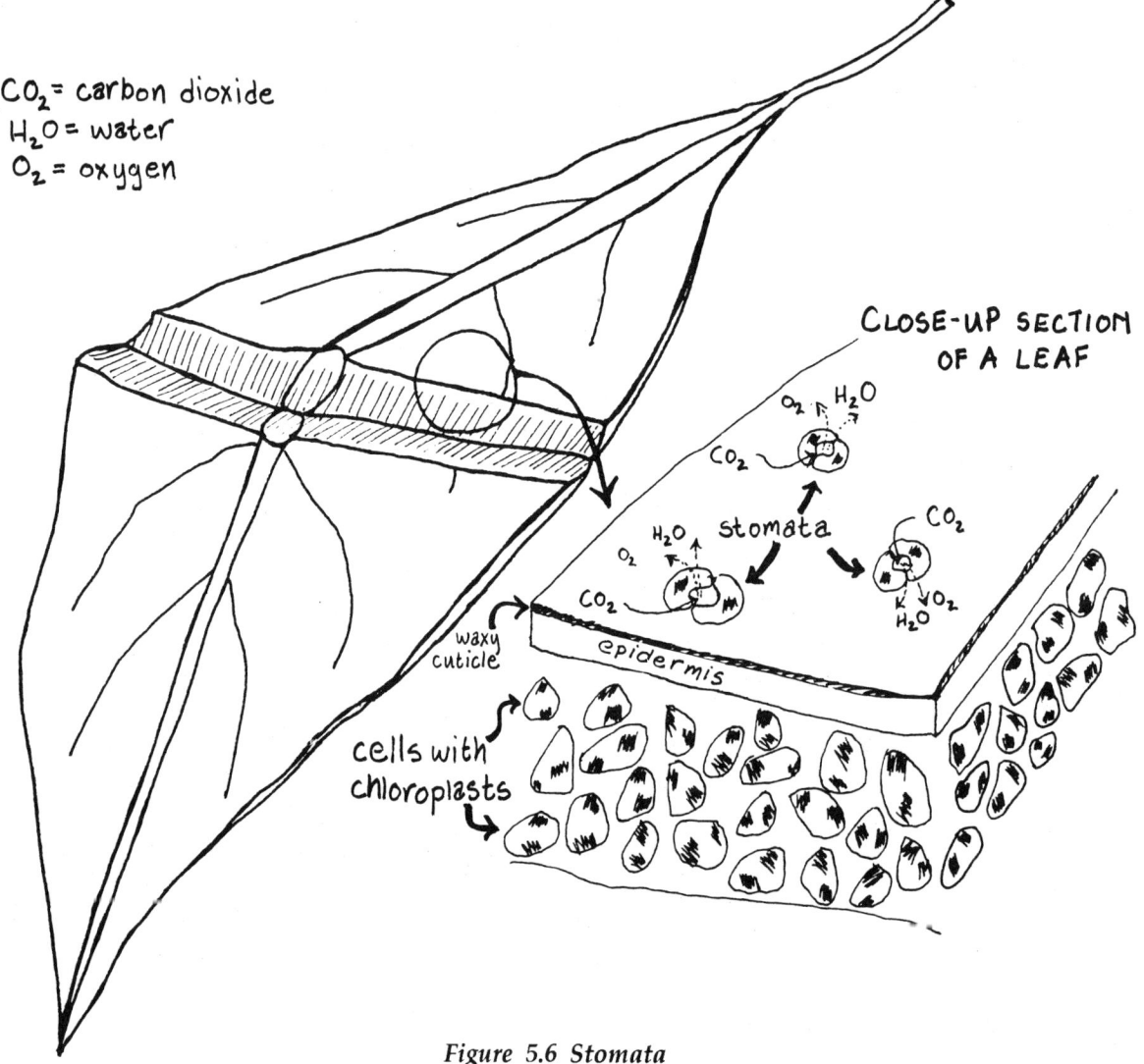

Figure 5.6 Stomata

Drought-adapted plants either escape drought or resist it in some way (Figure 5.8). ***Drought-escaping*** plants have short, rapid life cycles, allowing them to take advantage of brief periods of adequate moisture and decreasing their chance of experiencing drought. Some "famine crops" like short-season millet varieties and tepary beans follow this strategy, maturing before late season drought sets in.

Drought-resistant plants use one of two strategies, either they avoid drought or they tolerate it. ***Drought avoidance*** means more efficient use of water so that the plant will not experience water stress. For example, during periods of drought cowpeas avoid water stress by changing the orientation and movement of their leaves in relation to the sun, minimizing the amount of sunlight and heat they receive. This, in turn, reduces the amount of water lost from the leaves due to excess transpiration.[2]

Physiological differences enable some plants to lower transpiration rates in other ways. C_4 plants, named for the four-carbon molecule they produce and use, are able to use CO_2 more efficiently in a special form of photosynthesis. Because of this these plants do not need as much CO_2, and therefore their stomata need not be open as long as in other plants. Shorter periods with stomata open mean decreased transpiration. Some C_4 dryland garden plants are maize, sorghum, sugarcane, and amaranths.

Another modification of photosynthesis is found in ***Crassulacean acid metabolism (CAM)*** plants. In these plants photosynthesis happens in two stages, one during the day and one at night. The stomata are open only at night when they receive CO_2 and transpiration occurs. Due to the cooler, moister, dark nighttime conditions the rate of transpiration is much lower than it would be during the day. The CO_2 is then stored for use during

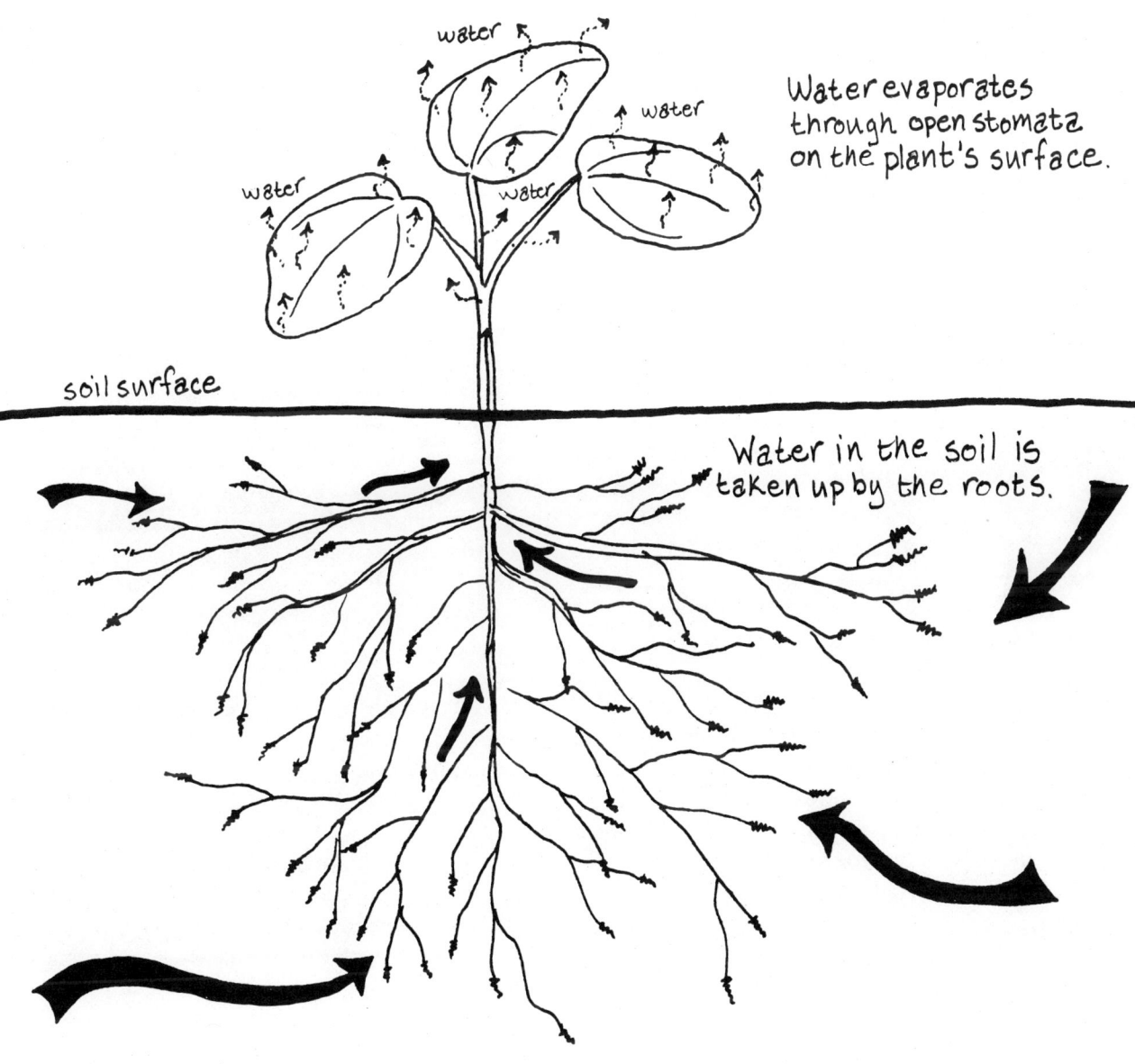

Figure 5.7 Transpiration

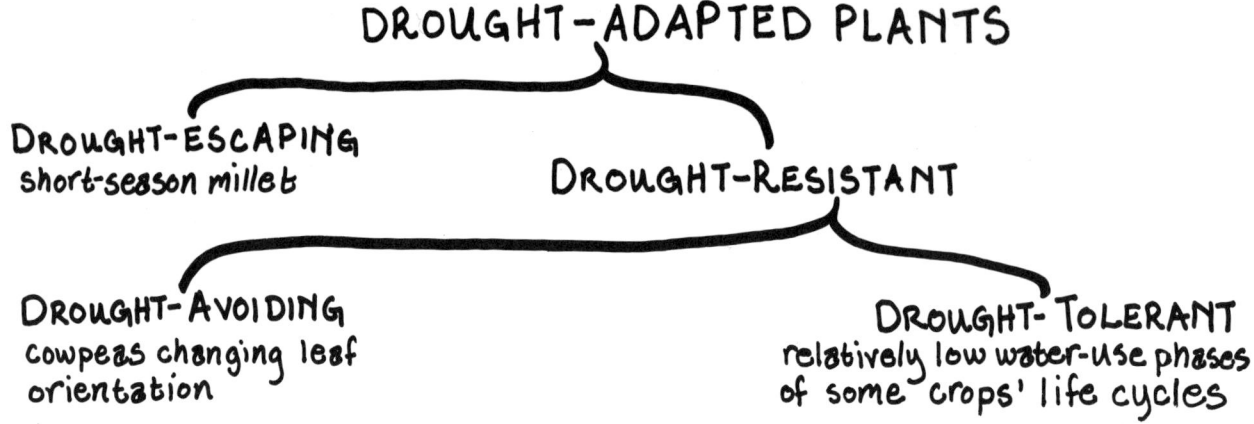

Figure 5.8 Plant Adaptations to Drought

the day when light energy from the sun is available. CAM plants found in some dryland gardens are pineapple, prickly pear cactus, and agave. CAM and C_4 plants are not necessarily the best ones for dryland conditions. For example, maize and sugarcane, both C_4 crops, are high water users.

Other plants may significantly reduce transpiration rates in different ways. Some other physical characteristics that cut down rates of transpiration, making plants better able to cope with drought conditions include:

- Small leaf surface area.
- Small number of stomata per unit of surface area.
- Majority of stomata on the more protected, underside of leaves.
- Thick, waxy or resinous layer or *cuticle* on the leaf surface.
- Light-colored leaves that reflect light, resulting in lower leaf temperatures and therefore less need for cooling by transpiration.
- Hairs on leaves also reflect light and provide additional surface area for cooling the plant and reducing air movement, leading to reduced evaporation.
- Self-shading canopy.
- Deep rootedness.
- Drought deciduousness (section 6.2.1).

Plants that can survive water stress are called *drought tolerant*. The ability to tolerate drought depends on the stage in the life cycle during which drought occurs. For example, if cowpeas experience a water deficit while they are flowering and forming seeds the yield will be significantly reduced. However, if the same water deficit occurs while the cowpeas are forming leaves, before the flowering stage, then the reduction in yield will be much less[3] (section 10.4).

Heat tolerance refers to a plant's ability to survive and produce under hot conditions. Plants commonly respond to hot air temperatures with increased transpiration to cool the leaf surfaces (section 5.4). Cooling through increased transpiration is an example of why heat tolerance and drought adaptation do not always occur together. A plant may be capable of withstanding high temperatures but if it does so only by greatly increasing transpiration it is not very drought adapted. However, a few of the physical characteristics listed above such as leaf orientation to the sun, hairs on leaves, and light leaf color reduce leaf temperature in ways that do not increase transpiration.

Distinguishing between heat tolerance and drought adaptation is useful. In most drylands hot daytime temperatures are very common and so heat tolerance is a desirable characteristic. However, in gardens that receive a regular supply of water, drought adaptation may not be necessary. This is especially true if other varieties or different crops will give a bigger and better harvest with the same amount of water and other inputs.

Gardeners in drylands recognize that heat-adapted crops may differ widely in drought adaptation. We saw an example of this at a new rural settlement in arid Sonora State in northern Mexico where villagers were planting fruit trees. The only source of water for the 30 households in the village was a well 2 km (1 mi) away, and each household had only a few small containers for carrying the water on foot. Orange, pomegranate, papaya, mango, guava, and lime trees were planted. All of these trees were growing vigorously in a nearby town which has a reliable piped water supply. But the harsh, dry conditions in the new settlement were killing all except the lime trees, which were growing slowly. According to the gardeners, lime trees are the best for coping with heat and drought. These villagers also have avocado seedlings in containers that they keep in the shade near their houses. They said they will not plant the seedlings out into their gardens until a more secure water supply is found, because they know that avocados would not survive the heat and sun exposure with the little water they could provide.

5.6 Salt Tolerance

The accumulation of salts can be a serious problem in dryland gardens and agriculture. Discussion of salty soils and water, and related management techniques can be found in section 9.3.1, Box 11.1 in section 11.2, and section 12.6.2. Whatever the source of salts, when they become concentrated in the soil they have an osmotic effect on plants. This results in a slower uptake of water and changes in hormone production leading to lower rates of transpiration and photosynthesis and increased respiration. The browning of leaf edges described in section 13.4.2 is a sign of this (Figure 5.9). Eventually under saline conditions, insufficient energy is available for the plant to grow or even maintain itself, and it will die.

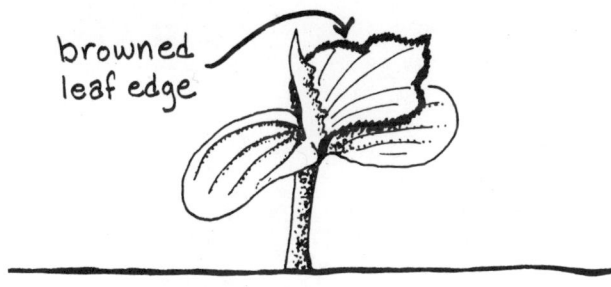

Figure 5.9 Salt Burn

Some plants are less sensitive to salt accumulations than others and are referred to as being **salt tolerant**. *Halophytic* (salt-loving) plants actually like salty growing conditions, producing more as salinity increases to low levels. Some salt-tolerant dryland garden crops are beets, asparagus, cowpeas, spinach, date palms, and some tomatoes.[4] There are also many salt-tolerant indigenous crops and wild plants, and new salt-tolerant varieties of widely grown crops are also being developed.[5] Plants that are particularly sensitive to salinity such as the stone fruit and citrus trees are called *halophobic*.

5.7 Seasonal Constraints to Plant Growth

No matter how carefully the garden environment is improved and managed, there are times when certain plants will not grow. This may be due to their needs for particular daylengths or temperatures that do not occur during some seasons. Local gardeners know the appropriate growing seasons for their crops, but they may be unfamiliar with the needs of newly introduced crops. Understanding seasonal constraints to plant growth improves the chances for healthy, vigorous garden plants.

5.7.1 Daylength Requirements

Some plants have a *photoperiod* requirement for a certain number of hours of darkness before they will grow, flower, and produce fruit. Without this they will not complete their life cycle and will not produce fruit and seeds for gardeners to eat and to plant in the future.

Closer to the equator there is less difference between hours of darkness and hours of daylight, both daily and seasonally. It is not unusual to find crop varieties from the tropics and subtropics with longer darkness requirements than varieties from higher latitude areas. For example, when grown in the northern Sonoran Desert where we live, some beans from central and southern Mexico will not flower until September, even if they are planted in March. This is because we live farther north where the longer nights the beans need to flower do not occur until September, the beginning of the cool season (Figure 5.10). Because beans are warm-season crops they cannot be sown early enough in the year to take advantage of the long nights of late spring. Therefore the only time for us to plant these varieties is in the late warm season, late July or early August, and this does not give some beans enough time to mature before the freezing weather in November.

Onions and sesame are other garden crops whose production is controlled by photoperiod sensitivity, although precise requirements differ by variety. For example, long-night (more than 12 hours, sunset to sunrise) varieties of onions and sesame are required for semitropical savanna West Africa. If short-night (less than 12 hours) onion varieties adapted to higher latitude temperate regions are planted in this area of West Africa they will not form bulbs.[6] Tomatoes are an example of a photoperiod-neutral garden crop.

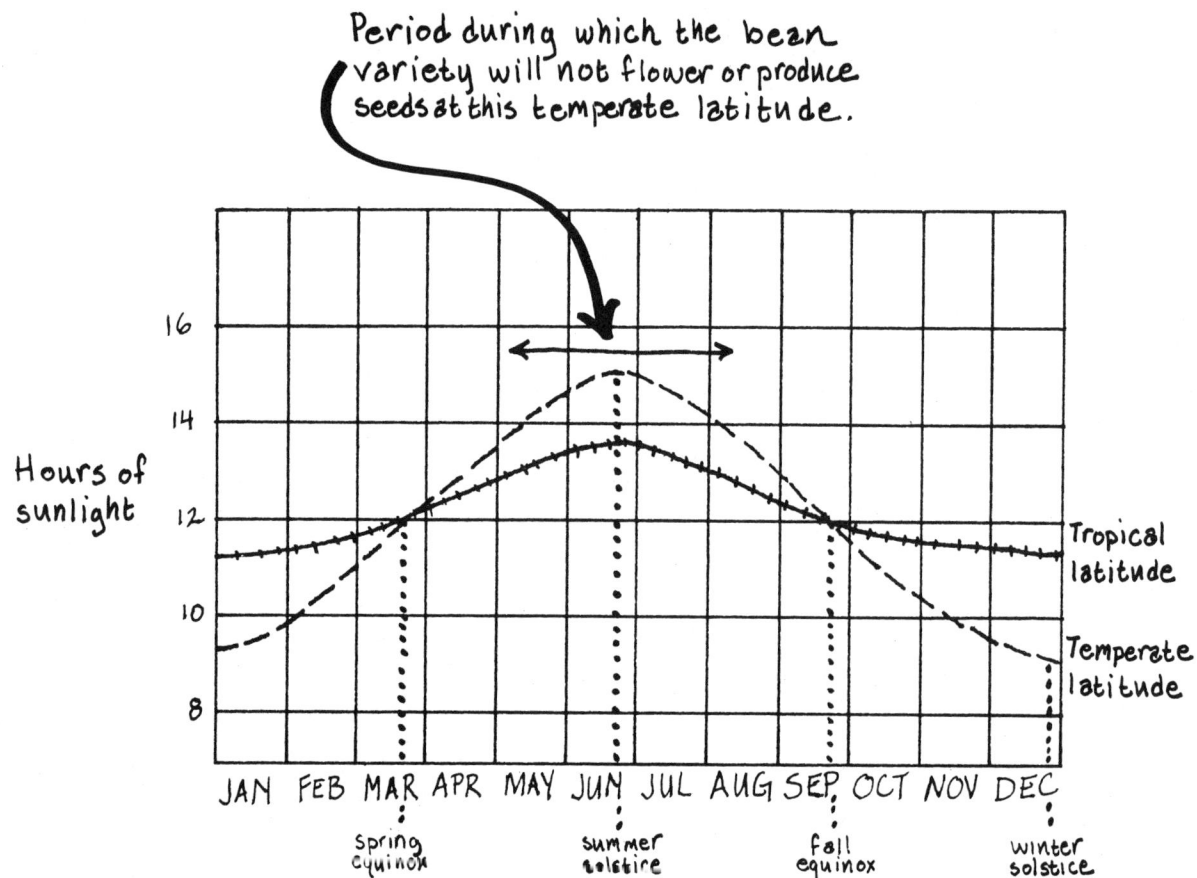

Figure 5.10 Daylength Sensitivity

5.7.2 Temperature Requirements

Like daylength, plants' temperature requirements vary greatly both between and within species. Most eggplants, cucurbits, some pulses and peppers require soil temperatures above 15°C (60°F) for normal germination and seedling development.[7] Some varieties of deciduous fruit trees require a minimum number of days at temperatures below 0°C (32°F) for dormancy in order to produce fruit (section 14.4.1).

A uniform problem among all plants of one variety that cannot be traced to any other cause—such as failure to produce flowers, fruit, or bulbs, or *bolting* (premature flowering)—may be a sign that the plant's daylength or temperature requirements are not being met (Figure 5.11).

5 How Plants Live and Grow

Figure 5.11 A Uniform Problem Such as Bolting may mean Growth Requirements are not Being Met

5.8 Resources

Through careful observation and long experience many farmers and gardeners understand a great deal about their crops. They are the best resource for learning about how local crops live and grow. For a Western science approach, basic principles of botany can be found in many school textbooks. The perspective of a botanist or ecologist is often different than that of an agronomist. While the first two approach the subject with a broad environmental outlook, the agronomist often tends to emphasize production economics. This results in different priorities and concerns, and most importantly, in asking different kinds of questions. Frequently the information in botany or ecology books is more relevant to small-scale, low-input food production, such as household gardens. This is especially true for drylands because agronomy often assumes a modified, optimal environment for crop production, instead of considering how best to cope in a marginal environment with limited resources.

On the other hand, there are many valuable agronomy texts and a growing number of agronomists whose approach to food production is appropriate for small-scale, marginal systems. Lessons 23-27 and 31-32 in *Agriculture Tropicale en Milieu Paysan Africain* (Dupriez and De Leener 1983) describe the needs of plants for resources such as water, air, and light. Part I of *Crops of the Drier Regions of the Tropics* (Gibbon and Pain 1985) includes a section on "Crop Factors" with a good discussion of drought and water use in crops. The *Better Farming* series of pamphlets from the FAO (1976-1977) contain simple discussions of botany and other topics relating to agriculture.

References

[1] Feldman 1988.
[2] Hall, Foster, and Waines 1979.
[3] Hall, Foster, and Waines 1979:156-157.
[4] Ayers and Wescot 1985:31-35; Cox and Atkins 1979:300-304.
[5] NAS 1990:17-39.
[6] Kassam 1976:79, 82, 104.
[7] Hartmann and Kester 1983:147.

6
Growing Plants from Seeds

Most annual garden crops are grown from seed, and so are some perennials. It is easy to grow crops from seeds, and seeds can be traded, transported, or stored. An important reason for using seeds from open-pollinated garden crops is to maintain genetic diversity. The variability that exists for many traits between individual, open-pollinated plants allows gardeners to continually select plants best adapted to changing needs and conditions. In Chapter 14 we discuss genetic diversity and what it means for the gardener, her garden, and for all of us. In this chapter we discuss how seeds are produced and present ideas for planting them in drylands.

6.1 Summary

Many garden crops reproduce sexually when male reproductive cells (contained in pollen) and female reproductive cells (contained in ovules) are joined together during fertilization, producing an embryo that will be contained within the seed. After the seed matures, it will germinate and grow if environmental conditions are right. Appropriate techniques for preparing and planting seeds and for watering, mulching, and shading seedlings conserve water and protect the seedling from the harsh environment. Diagnosis and remedy of seed planting problems may include a germination test to check the health of seeds. Once the seedlings have emerged, thinning them can improve vigor and production.

6.2 Sexual Reproduction in Plants

Seeds and the plants that grow from them are the products of sexual reproduction. Some garden crops can be propagated vegetatively through asexual reproduction, which is discussed in Chapter 7.

Sexual reproduction is the combination of genetic material from the reproductive cells or *gametes*: sperm contained in *pollen* from the male combines with the *ovule* in the female (Box 14.1). The result is a seed that carries characteristics of both parents. *Flowers* are the specialized plant parts where the gametes are produced, and those flowers with female parts are the site of pollination, fertilization, and seed production.

6.2.1 Life Cycles

Plants that produce seeds have two distinct phases of growth. During *vegetative growth* roots, stems, and leaves grow, and during *reproductive growth* the plant's resources are focused on developing flowers, seeds, and fruit. A plant's *life cycle* is defined as the time it takes to produce seeds. How long an individual plant lives is its *life span*. In some plants, life cycle and life span are the same length of time; in others they are not.

Annual plants are those that take 1 year or less to go through their entire life cycle: germination of the seed, vegetative growth, reproductive growth, and seed production, after which they die. That is, their life cycle and life span are equal. This is also true of plants that spend their first year in the vegetative growth stage, and enter and complete their reproductive growth stage and die in their second year. These plants whose life cycle and life span are both about 2 years long are called *biennials*. *Perennials* are those plants that live longer than 2 years, usually going through vegetative and reproductive stages each year after an initial period (1 or more years) of only vegetative growth. That is, their life cycle may be 1 year long, but their life span is much longer as in the case of olive trees, which can live for hundreds of years. On the other hand many agaves, whose swollen leaf bases, roots, and flower stalks are eaten, and leaf fibers used for weaving, have a life span of about 20 years. During this time they go through only one life cyle, producing a flower stalk once and then dying.

Most annuals and many biennials are *herbaceous*, that is their aboveground growth is green, pliable, and tender. Many perennials such as bananas and yams are herbaceous as well. However, some are *woody* in that their stems, trunks, or branches become hard, rigid, and covered with bark, as with olive and peach trees.

Some dryland perennials such as pomegranates, figs, the stone fruits, and jujubes are *deciduous*. That is they have a repeating seasonal cycle of losing their leaves, and becoming dormant, followed by a period of growth, leafing out, and flower and fruit production (Figure 6.1). Nondeciduous perennials are sometimes referred to as being *evergreen*. Both deciduous (e.g., fig) and evergreen (e.g., carob) trees may lose their leaves to reduce transpiration during extreme drought.[1] Because they avoid drought in this way such plants are said to be *drought deciduous*. Cassava is a drought-deciduous, short-lived perennial that loses all but a few leaves on the ends of its stems during drought.[2]

6.2.2 Flowering

In plants male gametes, contained in *pollen grains*, are produced in the *anthers*, and female gametes, contained in *ovules*, are produced in the *ovary*. Some plants like okra have *perfect* flowers which contain both male and female structures. *Monoecious* plants have separate male and female flowers on the same plant as in most squashes and maize. *Dioecious* plants such as the pistachio and date palm bear female flowers on one plant and male flowers on another (Figure 6.2). The papaya is an interesting example of a tree that can be perfect, monoecious, or dioecious. Dioecious papaya plants may even change sex, and in savanna West Africa dioecious male papaya plants are cut back to the ground to encourage female shoot production.[3]

The flowers of many herbaceous garden plants last only a very short time. Squash blossoms, for example, wither and drop off after only 1 day. Stressful conditions such as high temperatures and drought may shorten the flower's life as well, making hand pollination useful (Box 6.1 in section 6.2.3).

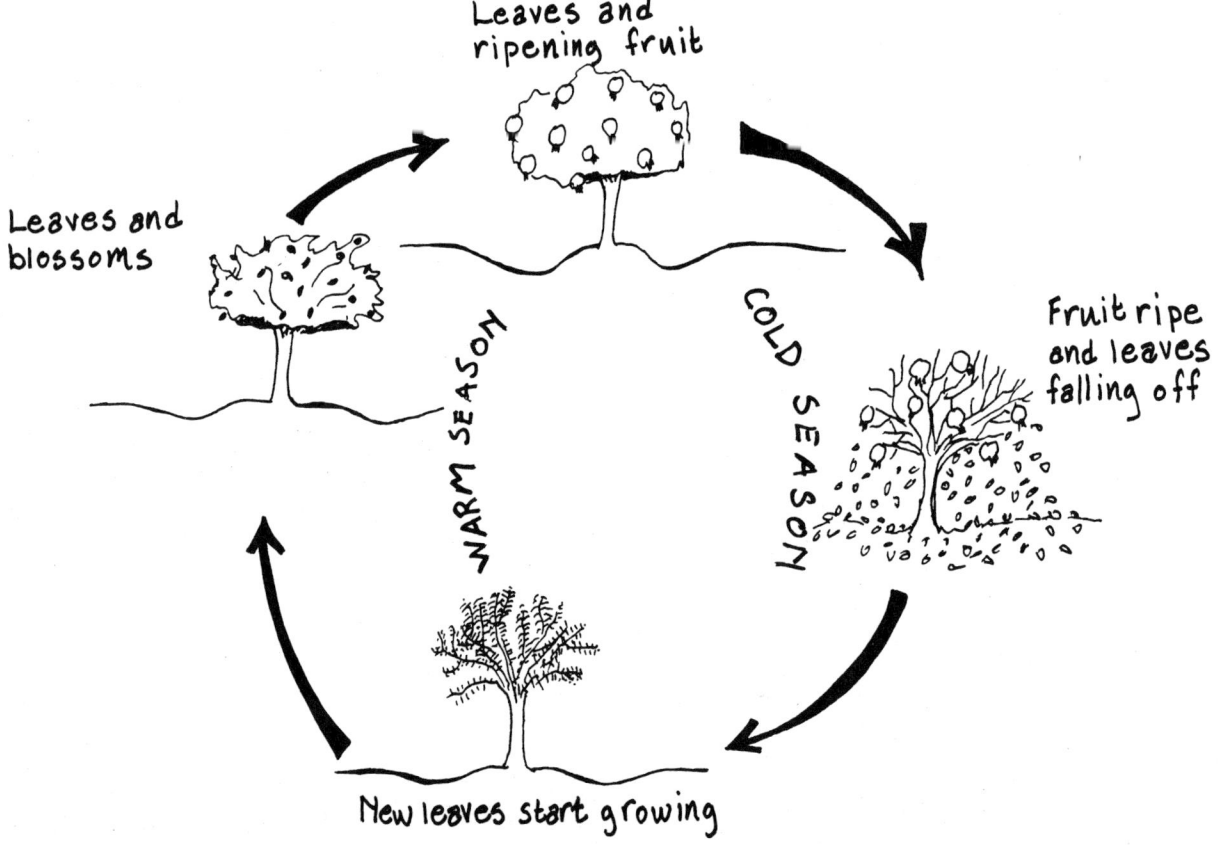

Figure 6.1 Yearly Cycle of a Deciduous Tree—the Pomegranate

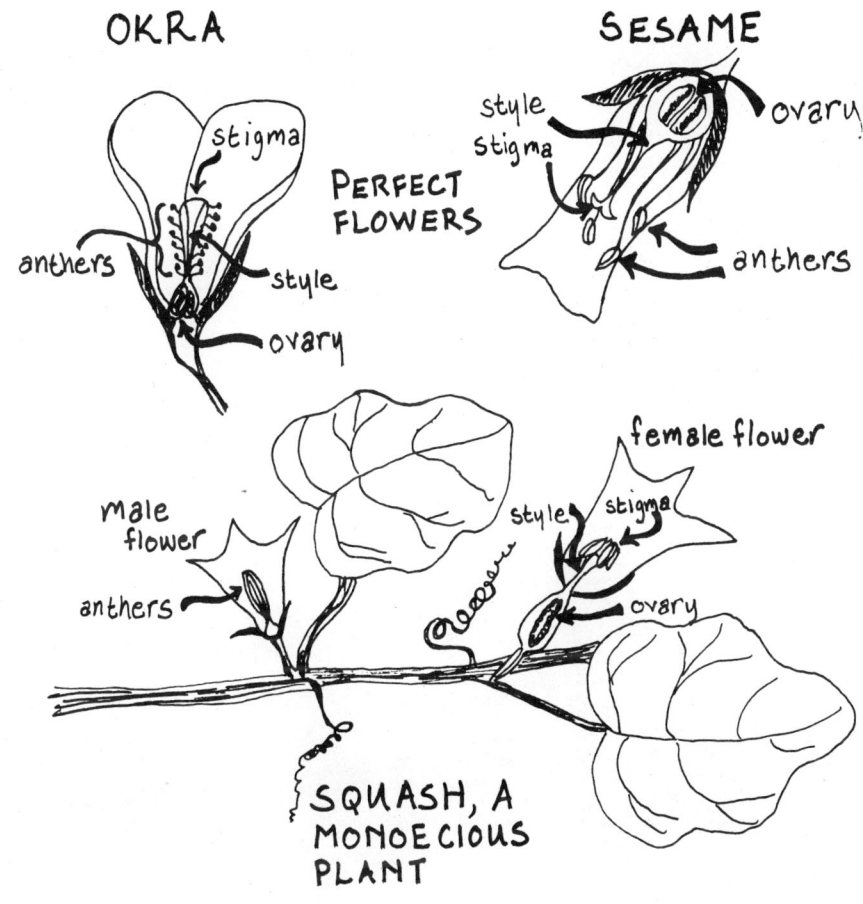

Figure 6.2 Perfect, Monoecious, and Dioecious Flowers

6.2.3 Pollination

Pollination happens when a pollen grain lands on the **stigma**, the receptive surface of the female flower part where the pollen grain germinates, and grows down the style to reach the ovary. Flowers can be cross-pollinated or self-pollinated. **Cross-pollination** occurs when pollen from one plant pollinates the flower of another plant in the same species that is genetically different. When the pollen from a male date palm is blown onto the flowers of a female tree, cross-pollination has occurred (Figure 6.3). An example of cross-pollination of a monoecious plant is the pollination of maize when pollen from one plant is blown to the silks of other plants (Figure 6.9 in section 6.2.4).

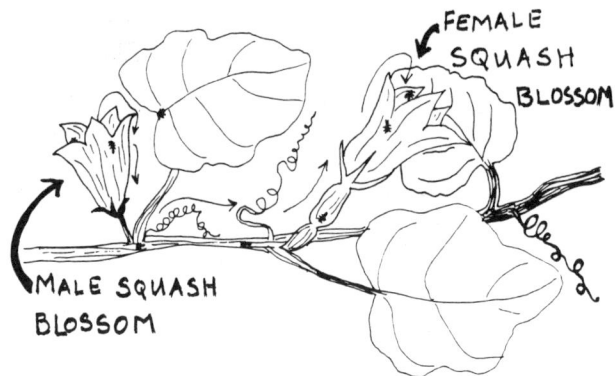

Figure 6.4 Ants Assist the Self-Pollination of a Monoecious Plant

Figure 6.3 Wind Cross Pollinates a Dioecious Plant

Self-pollination refers to the pollination of a flower on a plant that is genetically identical to the pollen donor. Two thyme plants started by cuttings from the same original plant may pollinate each other, because they are genetically identical this is self-pollination, not cross-pollination. Other examples of self-pollination are when a monoecious plant such as a squash or a plant with perfect flowers, like sesame, okra, or tomatoes, pollinates its own flowers, often with help from insects (Figures 6.4 and 6.5).

Knowing how plants are naturally pollinated improves the gardener's understanding of how different genetic combinations occur. It also helps her control pollination for seed production, selecting parent plants with the most desirable traits (Box 6.1).

When pollen is carried by the wind to female flowers, as in Figure 6.3, they are said to be *wind-pollinated*. Examples of wind-pollinated crops are maize, dates, pistachios, olives, and the amaranths. *Insect-pollination* occurs when insects carry the pollen to the female flower parts, as in Figure 6.4. Some insect-pollinated crops grown in dryland gardens are the cucurbits, pulses, tomatoes, garlic and onions, the stone fruits, and mangoes. Box 6.1 discusses how wind- and insect-pollination can be controlled.

There is no absolute rule for distinguishing wind- and insect-pollinated plants, however, the flower is often a good clue. Plants with many inconspicuous, small flowers lacking color or fragrance are often wind-pollinated. Their pollen is relatively dry, light, and easily blown by the wind.

Showy, fragrant, white, or brightly colored flowers usually rely on insect-pollination. Their appearance or fragrance attracts insects such as wasps, bees, ants, flies, and butterflies. Bats, rodents, and some birds also act as pollinators. Because the pollen in these flowers is frequently heavy, moist, and sticky it adheres to the insects or other animals which carry the pollen to another flower, pollinating it when they rub against the stigma.

Figure 6.5 A Perfect Flower Self Pollinated with Help from a Bee

Box 6.1
Controlling Pollination

If successful pollination of a crop by wind, insects, or other natural means is uncertain, then hand-pollination can be done. Examples are when the number of female flowers is limited (as in squash), when pollen supply is limited (as in date gardens), or to take advantage of environmental conditions most favorable for fertilization (as with cool mornings for maize).

When controlling pollination, the first step is to identify the plant's life cycle and flowering characteristics. Flowers that are just about to open are best for hand-pollination. With flowers such as squash, which are usually pollinated by insects, the pollen can be rubbed on the sticky surface of the stigma (Figure 6.6). For wind-pollinated ones, the male blossoms can be shaken over the female flowers, dusting them with pollen (Figure 6.7).

Maize, for example, is a wind-pollinated crop that should be hand-pollinated in the cool of early morning because hot, dry conditions will kill maize pollen. The male flowers or *tassels* are shaken so that the pollen falls on the *silks*, which are the stigmas and styles of the female flowers. Maize should always be planted in clusters or blocks, not in single rows or as isolated plants, since it often needs cross-pollination between plants for good seed production (section 6.2.4).

Pollination may also be controlled to maintain the purity of a specific variety. In these cases steps have to be taken to prevent unwanted pollen from fertilizing the ovaries. In wind-pollinated plants, female flowers can be closed or covered with cloth or paper before and after being hand-pollinated. Wind-pollinated varieties can also be separated from each other in space (e.g., planting different maize varieties at least 0.5-1.6 km or 0.3-1.0 mi apart), and in time (staggering planting times so that different varieties will not be flowering at the same time). The Hopi Native Americans of southwestern North America have maintained a large number of very distinct varieties of maize for hundreds of years by planting the varieties in fields separated from each other.

Surrounding insect-pollinated plants with a frame of sticks, bamboo, or wire covered with a finely woven screen or netting may be enough to control pollination by large flying insects. If ants are pollinators, the female blossoms can be covered and tied shut. If the flowers are perfect, their anthers must be removed so that they will not self-pollinate. In monoecious plants the male flowers on the plant should be removed for the same reason.

Pollination can also be controlled when gardeners want to improve the drought resistance, taste, yield, or other qualities of their crops. This is done by selecting the male and female plants with the desired characteristics and crossing them to produce seeds.

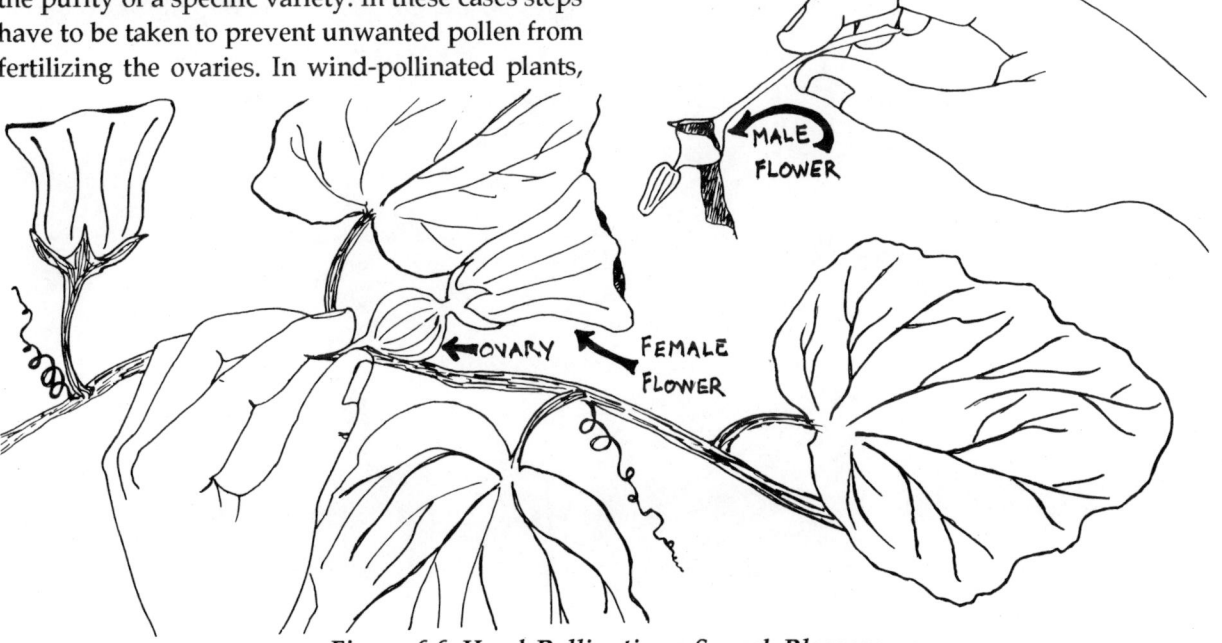

Figure 6.6 Hand-Pollinating a Squash Blossom

Figure 6.7 Hand-Pollinating a Date Palm in Iraq

6.2.4 Fertilization

After pollination the pollen grain germinates and a pollen tube grows from it, down the style, into the ovary and finally the ovule (Figure 6.8). When the male gamete from the pollen grain joins with the female gamete in the ovule, *fertilization* has occurred. The fertilized ovule will develop into a seed, and in some plants the ovary will thicken around the seed or seeds. This thickened membrane is the fleshy part of a *fruit*.

Fertilization is important for two reasons: a) fruit and seed foods such as olives, jujubes, okra, tomatoes, and sesame will only be produced if fertilization occurs, and b) seeds are needed for growing many garden plants, especially annuals.

Fertilization will fail if either the pollen or the ovules are no longer viable. A cell, flower, seed, graft, or cutting is *viable* if it is capable of living. Maize pollen is only released for several hours around sunrise. The pollen is usually viable for about 24 hours but under hot, dry conditions this period is significantly shortened. This is why hand-pollinating maize in the cool of early morning improves chances of fertilization.[4] Similarly, the pollen from tomato flowers may pollinate the stigma but hot, dry weather can kill the pollen during the approximately 50 hours it takes for fertilization to occur.[5] This is why some tomato varieties stop bearing fruit under very hot conditions, and why shading can help increase production.

A few crops, such as some maize varieties are *self-sterile* or *self-incompatible*. This means that even though they are monoecious, flowers on the same plant cannot fertilize each other. While self-sterile plants are incapable of fertilization by self-pollination, genetically different individual plants of the same variety can be fertilized by cross-pollination (Figure 6.9). This is a good reason to have a cluster of plants of one variety growing in the garden.

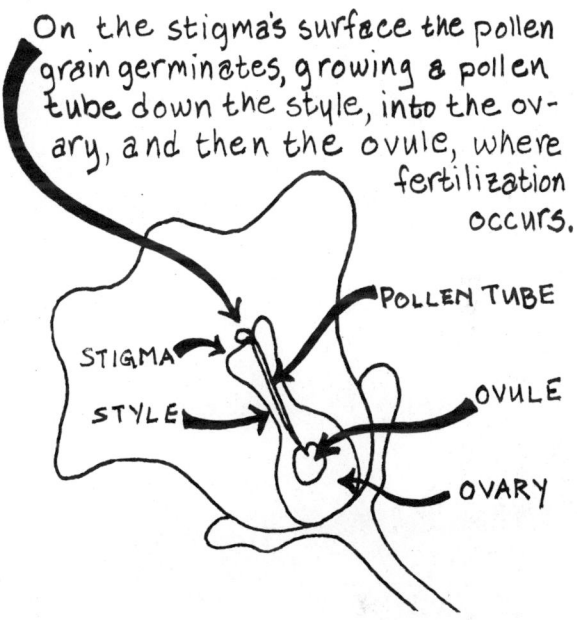

Figure 6.8 Fertilization

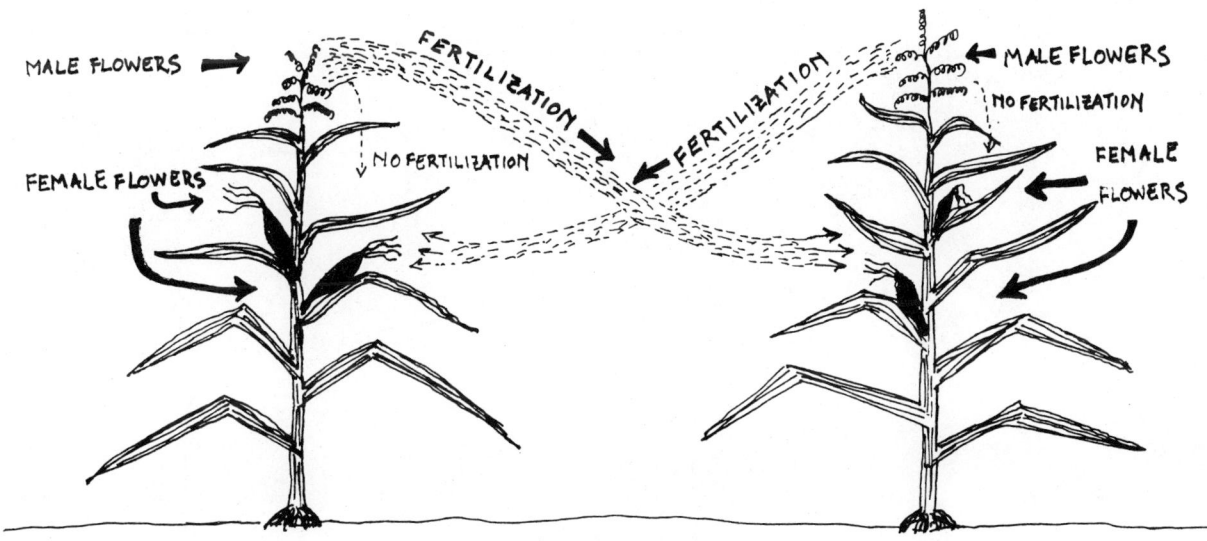

Figure 6.9 Self-Sterile Plants can be Fertilized by Cross-Pollination

6.3 Seed Germination and Dormancy

Once the seed has matured it is ready to be planted and grow into a new plant. A mature seed contains a living plant embryo and food reserves to fuel the seedling's growth until the root and shoot systems take over.

When water penetrates the protective outer seed coat, the seed swells, breaking the coat open. The coat may also be broken by extreme temperatures due to fire or freezing, the growth of soil microorganisms, or because of being eaten and digested by animals.[6] Breaking of the seed coat is a physical process, not a biological one; even dead seed can do this, so it is not a sign that the seed is alive.

When a living seed has swollen with water and produces a root and shoot, it has *germinated*. *Emergence* is when the shoot first breaks the surface of the ground. Seeds with *epigeal* germination, for example, many legumes and cucurbits, push their cotyledons above the soil surface. This contrasts with *hypogeal* germination, for example, in peaches and maize, in which the cotyledon remains underground and the new shoot grows up above (Figure 6.10).

In some seeds a condition of **dormancy** prevents germination for a period after the seed is mature. A dormant seed will not germinate when exposed to water until other requirements are met. Some seeds contain chemicals that inhibit germination for a time after maturation. The fruits of citrus, tomatoes, and some cucurbits contain chemicals that usually keep the seeds inside them dormant. When the seeds are removed from the fruit, and either planted or washed several times, they will no longer be dormant.

Other seeds require environmental stimuli such as low or high temperatures to germinate. The seeds of some high-latitude or -altitude dryland fruit trees, such as peaches, require a period of chilling to break embryo dormancy (section 14.4.1). Some seeds require thorough drying before they will germinate, preventing germination while still on the parent plant.

Overall, compared with wild species, domesticated crops tend to have brief dormancy periods (0-6 months), and high germination percentages which drop significantly as seeds get older. This is because domesticated crops are selected for immediate germination and growth under relatively controlled, improved conditions. In contrast, seeds of wild plants have more requirements for germination, longer periods of dormancy, and staggered germination, reducing the risk that they all germinate and then die if conditions worsen.[7]

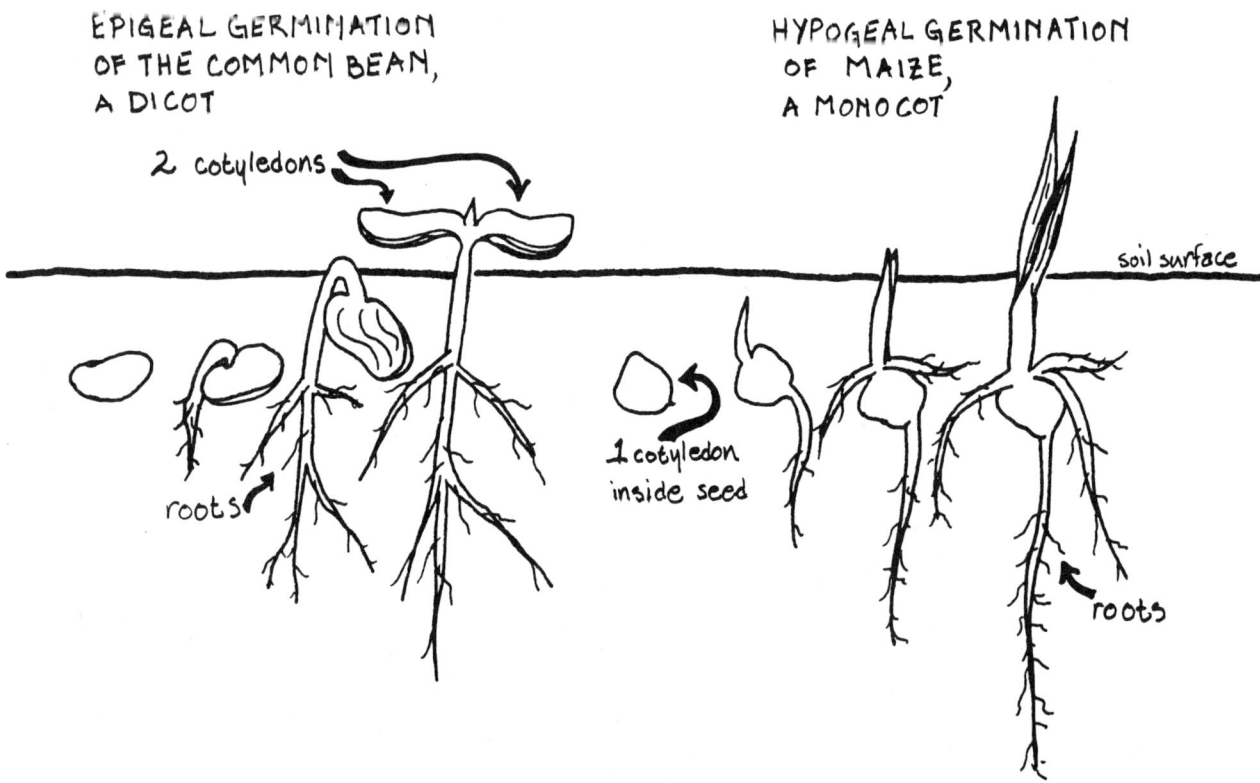

Figure 6.10 Seed Germination

6.4 Suggestions for Planting Seeds under Dryland Conditions

The suggestions given here for direct planting seeds can also be used with the nursery beds and containers discussed in section 8.2.

6.4.1 Preparing the Seeds

Soaking large or hard-coated seeds like maize, beans, and squash in water before planting helps break or at least soften their seed coats. This minimizes the time the seeds are in the ground before germination, when they are most vulnerable to being eaten by birds or insects, or attacked by disease. Presoaking also means that the garden does not need to be watered as much to keep the seeds moist until they germinate, and so saves water. During presoaking bad seeds, which are hollow due to disease or pests, or because the embryo never developed, will float to the surface and can be composted.

Presoaking for too long can kill the embryo. The larger and/or older the seed, the longer it will need to be soaked. Generally, seeds of herbaceous plants need no more than 8 hours of soaking. Large, hard-coated seeds of perennials can be soaked up to, but no longer than 24, hours.[8] Once the seed is softened it must be planted. Seeds for planting should never be soaked and then left to dry out completely because this will kill the embryo.

Seed coats of some plants, like carob, are so tough that scratching them is recommended to speed germination.[9] Scratching can be done easily by lightly rubbing the seed on a rough surface like a rock. The area where the seed was attached to the ovary, known as the *hilum* (Figure 6.11), should be avoided when scratching seeds this way, or the embryo may be injured.

The seed coats of many small seeds can be lightly scratched by putting them in a gourd, can, or jar with some small-sized gravel. Cover the container and shake the contents vigorously (Figure 6.12). In this case damaging the hilum is not a problem because the scratching is so gentle. Hard shells protecting seeds of some crops such as the stone fruits, olive, and many nuts can be carefully cracked just before planting to speed absorption of water and germination.

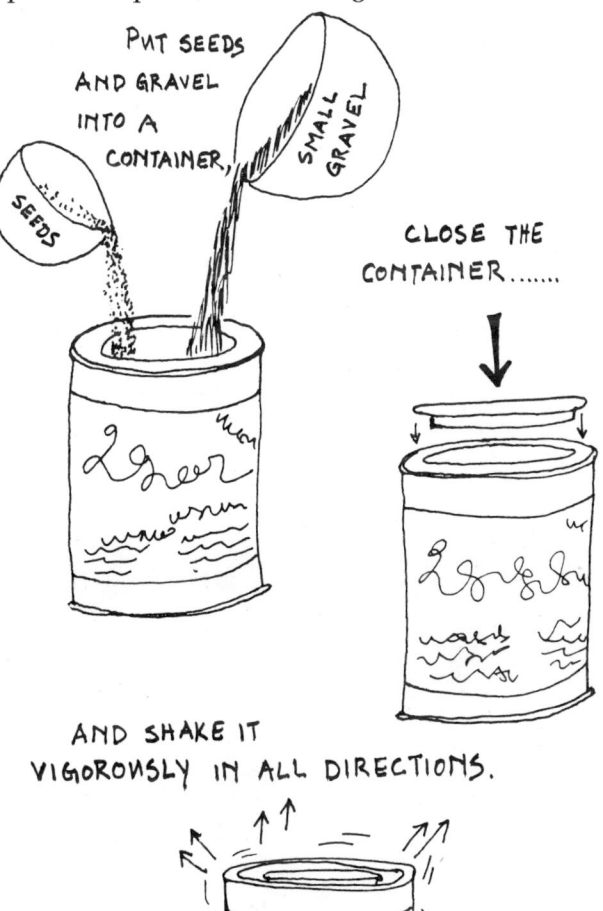

Figure 6.12 How to Lightly Scratch Seed Coats

6.4.2 Preparing the Planting Site

The soil in the garden should be prepared before planting seeds (Chapter 9). Planting seeds in depressions such as furrows or basins concentrates and saves water (Figure 6.13). Seedling emergence and growth

Figure 6.11 The Hilum of a Legume Seed

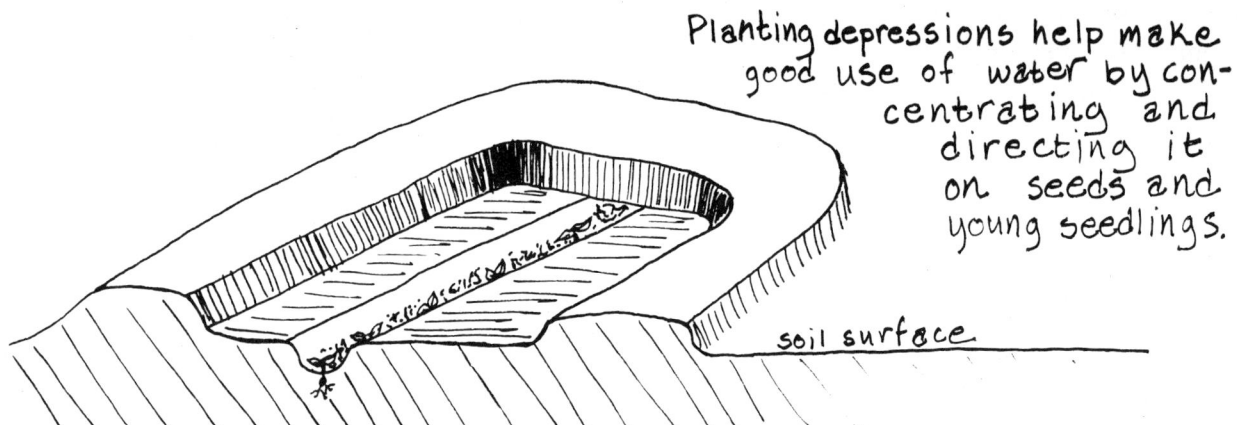

Planting depressions help make good use of water by concentrating and directing it on seeds and young seedlings.

While the young seedlings' root systems are still very small only the planting depression needs to be watered, not the whole garden bed.

Later, as the plant grows, making a ring-shaped trough around it with its stem at the highest point in the center has two benefits:
1) it makes a holding area for water that will soak in and go down to the roots,
2) it keeps the stem out of the water, avoiding disease problems.

Figure 6.13 Planting Depressions

can be encouraged by special attention to the soil immediately around the seed (section 6.4.5). For example, adding some extra compost to soil in the planting depressions makes it rich in nutrients and light textured, encouraging root growth. The compost also breaks up the soil surface, making it easier for water to infiltrate. Good infiltration also helps avoid damping-off fungus (section 13.4). In heavier soils, mixing some extra sand into the soil around the seeds of root or bulb crops such as carrots, beets, or onions helps young roots and bulbs get a vigorous start.

6.4.3 Planting the Seeds

Gardeners can control their seed planting density by sprinkling seeds into the planting depressions with their fingers (Figure 6.14). Pouring them can result in too many seeds being planted (Figure 6.15). This is a waste of seeds and time because many seedlings will have to be thinned later on. Mixing very small seeds such as amaranth or basil with sand or dry soil prevents them from sticking together, making it easier to sow them evenly.

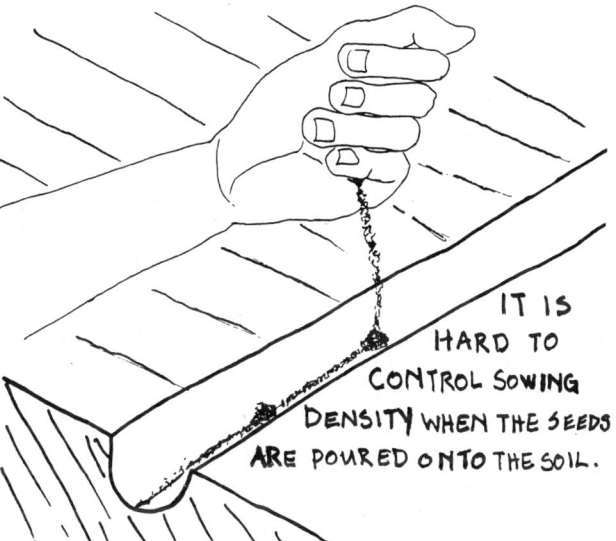

Figure 6.15 Pouring Seeds

Figure 6.14 Sprinkling Seeds

6.4.4 Planting Density

In indigenous, mixed gardens plants often grow much closer together than is recommended in many garden books. This close planting in a well-managed garden can increase production, but planting too closely may reduce production if soil nutrients, water, and sunlight are inadequate to meet all the plants' needs.

There is not a lot known about how planting density affects different crops and crop mixes. It has been found, for example, that closely planted apple trees with overlapping root areas produce many more downward growing roots in contrast to widely spaced trees which produce more horizontal roots.[10] How much some crops can adapt their physical structure to different planting densities and which crops can do this is not really known. Each gardener must experiment with planting densities and combinations that work for them. (See section 8.5 for discussion of intercropping.)

The way plants grow and use space affects how densely they can be planted. Mixing plants of different ages and life cycles decreases competition for resources because they will not have the same requirements and will make use of different levels of space both above and below ground (Figure 6.16). For example, plants with shallow, widespreading roots can be more densely planted with deep-rooted plants than with other shallow-rooted ones. Vines can be interplanted with trees, bushes, and cereals whose leaves occupy different levels above the ground. Annual leaf or fruit crops can often be successfully interplanted with root crops (Figure 6.17). Seeds can be sown among mature annuals which provide shade and wind protection for the seedlings. As the seedlings grow and require more space the older plants are nearing the end of their lives and can be removed or cut back, allowing the younger ones to replace them.

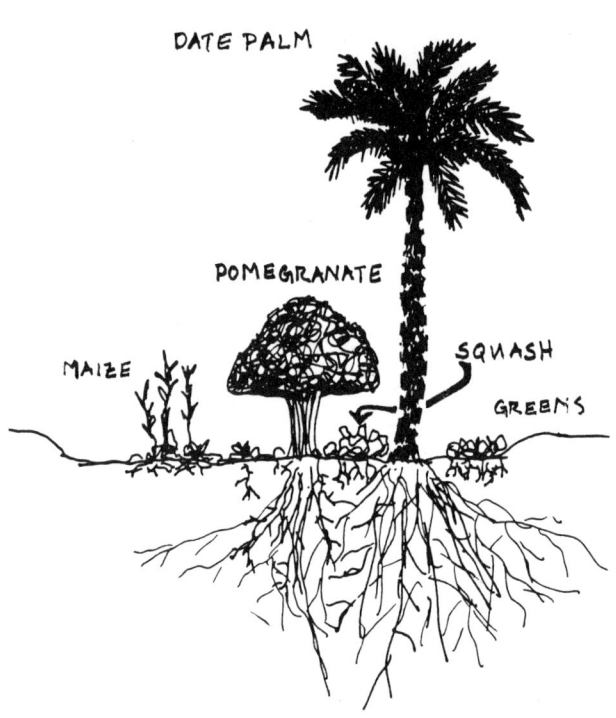

Starting with dense planting allows the gardener to thin, transplant, or harvest as the plants mature and if more space is needed. However, experimentation and careful observation are important because planting too densely can be a waste of seeds and can lead to disease and to competition for scarce dryland resources like water.

We do not feel that charts of precise planting densities are useful and have not provided them here. Such charts are usually based on industrial-style row gardens, not mixed gardens in dryland environments. They do not take into account variables such as the soil quality, water availability, temperatures, and the other plants including the many unique local varieties growing in the garden. The soundest approach is to observe and talk with local gardeners, keep basic principles in mind, and experiment.

Figure 6.16 A Mixed Garden with Annuals and Perennials

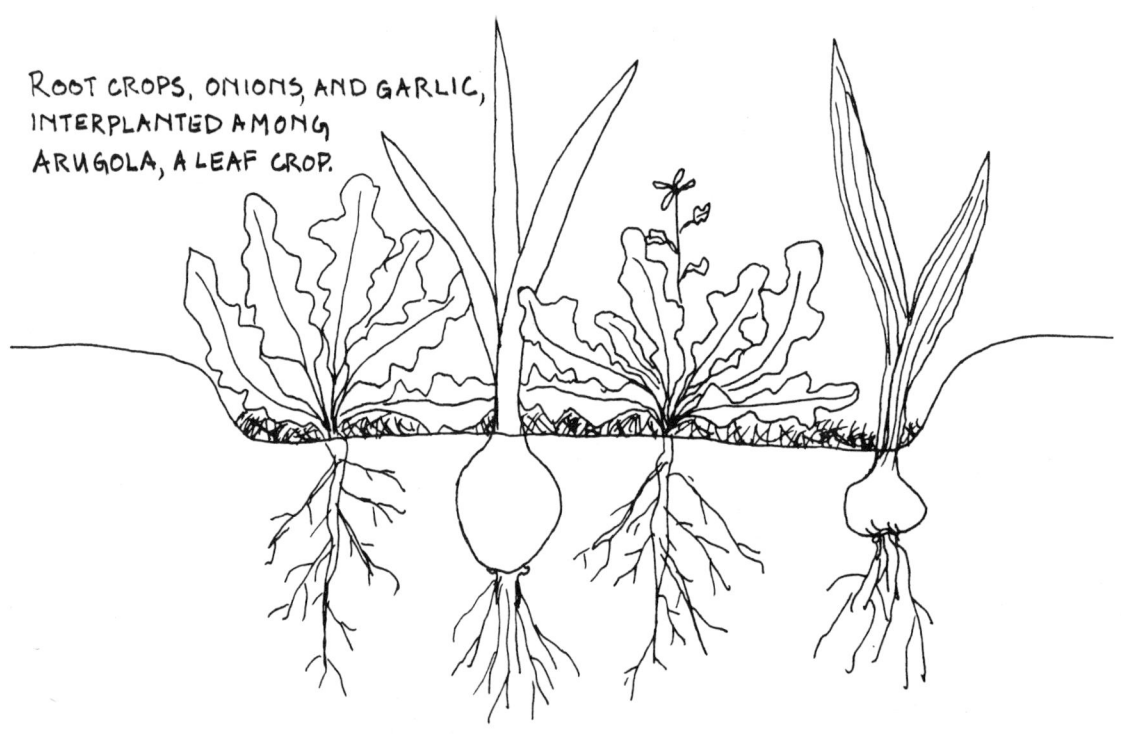

Figure 6.17 Interplanting or Intercropping

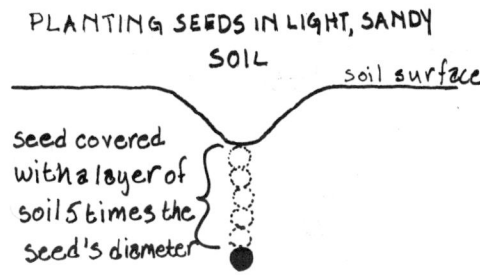
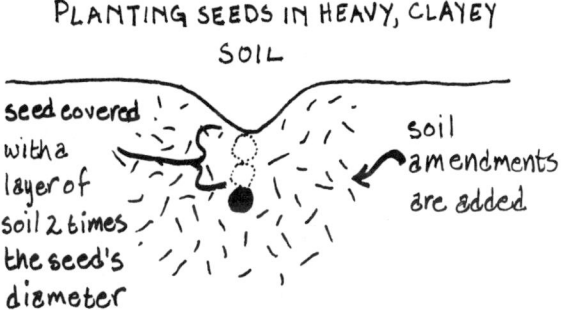

Figure 6.18 Seed Planting Depth

6.4.5 Covering the Seeds

Seeds need to be covered with enough soil to prevent them from drying out, but not so much that the shoot has difficulty emerging. The thickness of this cover depends upon the type of soil and the size of the seed (Figure 6.18).
- In heavy clay soils the covering should be thin (approximately two to three times the diameter of the seed) because these soils dry out more slowly and are harder to penetrate.
- In light, sandy soils the cover should be thicker (approximately four to six times the diameter of the seed) because these soils dry out more quickly and are easy to penetrate.

Seeds with epigeal germination may benefit from being planted slightly shallower than suggested above. If these seeds are planted too deeply and are unable to push their cotyledons above the soil surface they will die.

Diverse local conditions mean there are, however, exceptions to the guidelines given above. For example, in their dry-farmed maize fields the Hopi Native Americans living in the high desert of southwestern North America plant seeds about 25-33 cm (10-13 in) deep (Figure 6.19). This unusually deep planting is an adaptation to local soil and water conditions. In these fields approximately the top 30 cm (12 in) or more of soil is fine sand with a clay-sand loam below. In the spring, water from melted winter snow percolates quickly through the sandy soil, but is held in the clay-sand loam. When the soil has warmed, the seeds are planted in this moist layer. By planting deeply the Hopi take advantage of moisture stored in the soil. The Hopi have selected maize varieties over many years for physical adaptations to this planting method such as a dominant, deep root and a shoot with an underground portion about twice as long as in other varieties.[11] These characteristics and the sandy texture of the upper layer of soil make it possible for Hopi maize to emerge from such a deep planting hole and grow successfully.

Adding sand and organic matter to clayey soils, especially to soil covering the seeds, is a good idea. When ready to plant seeds in their terraced garden beds, Hopi carry up buckets of fine yellow sand from the valley floor below. They explain that they cover their garden seeds with this sand because the heavier, clayey soil in the terraces forms a crust when it dries out after irrigation. This crust can be so hard that it prevents seedlings from emerging.[12]

The soil covering the seeds should be pressed down firmly as it not only helps retain moisture but protects seeds from being removed by wind, insects, and birds. The surface can also be lightly mulched (section 6.5.2).

6.5 Caring for Newly Planted Seeds and Young Seedlings

When caring for newly planted seeds and young seedlings under dry conditions the most important consideration is maintaining soil moisture. Protecting young plants from hot, dry winds and strong direct sunlight helps reduce water consumption and stress on the plant. (See section 13.4 for diagnosing problems with young seedlings.)

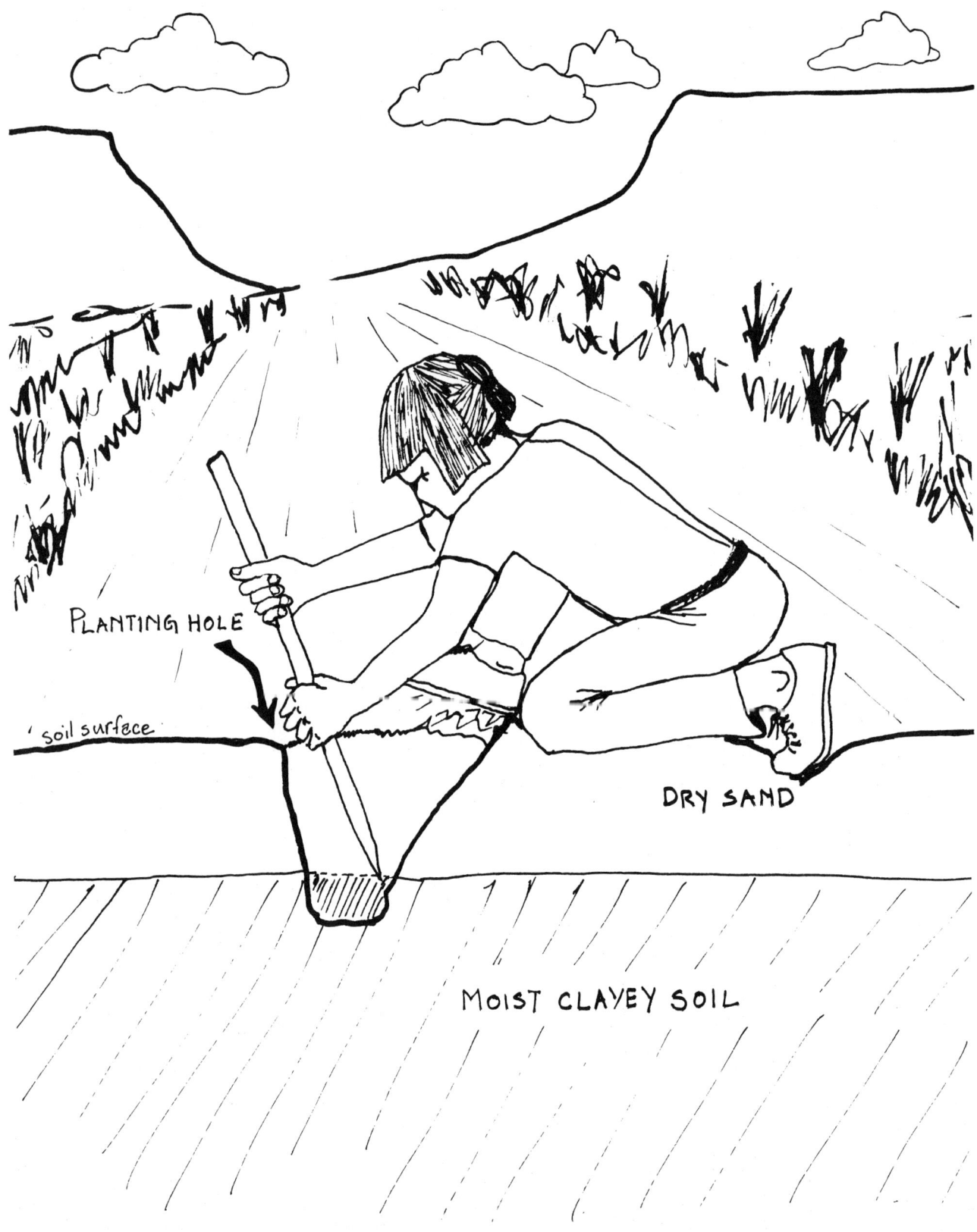

Figure 6.19 A Planting Hole in a Hopi Maize Field

6.5.1 Watering

The soil must be kept moist around seeds and young seedlings because they do not have an extensive root system for gathering and storing water. If the soil dries out they will quickly die. Seedlings experiencing water stress often have retarded shoot development and a greater proportion of root growth than unstressed seedlings. This is because under water stress, development of roots, which provide water, takes priority over shoot development for photosynthesis, which increases transpiration. Overwatering, on the other hand, can lead to soil saturation, forcing oxygen needed by the roots out of the soil (section 10.3.1). Lack of oxygen can kill the seed or seedling and encourages damping-off fungi which can also be fatal (section 13.4). Thus the soil should be kept moist but not saturated. Watering frequency will depend upon weather and soil conditions, planting methods, and plant types, and must be arranged to fit the gardener's schedule.

Breaking the force of flowing water when watering prevents seeds and seedlings from being washed away. This can be done by sprinkling or splashing the water by hand or pouring it through a bunch of leaves or stalks. A sprinkler can be made by putting holes in the bottom of a container such as a calabash or tin can, or making ceramic vessels with holes in their bottoms (Figure 6.20). Whether using a sprinkler or not the water should be poured as close to the soil surface as possible to avoid eroding the soil around seeds and young seedlings (Figure 6.21).

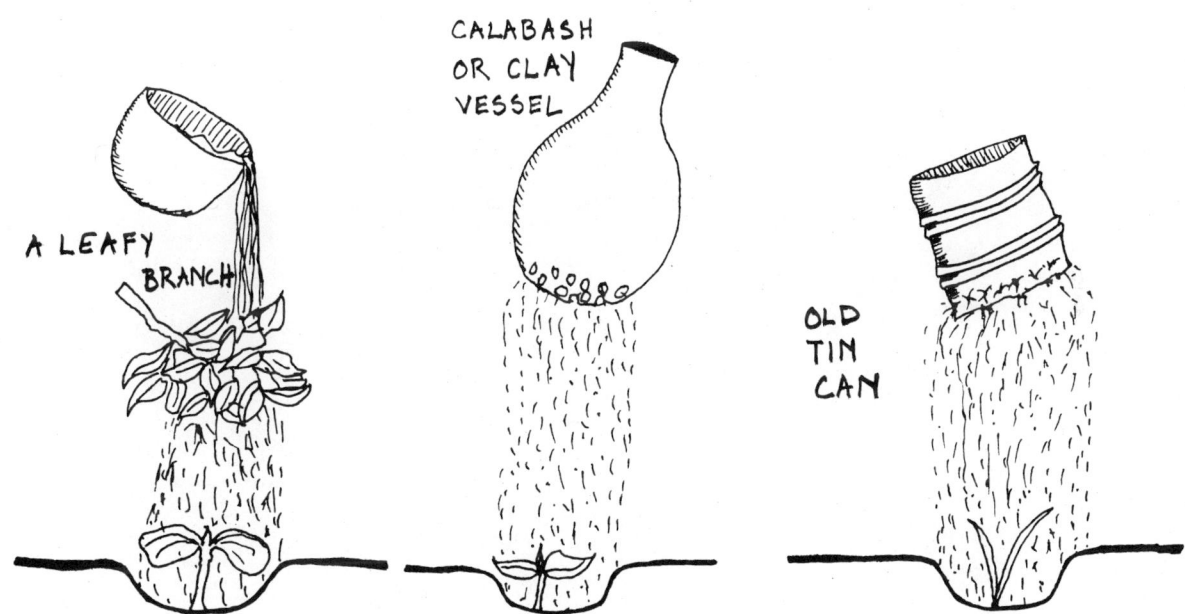

Figure 6.20 Ways to Disperse the Flow of Water on Seeds and Seedlings

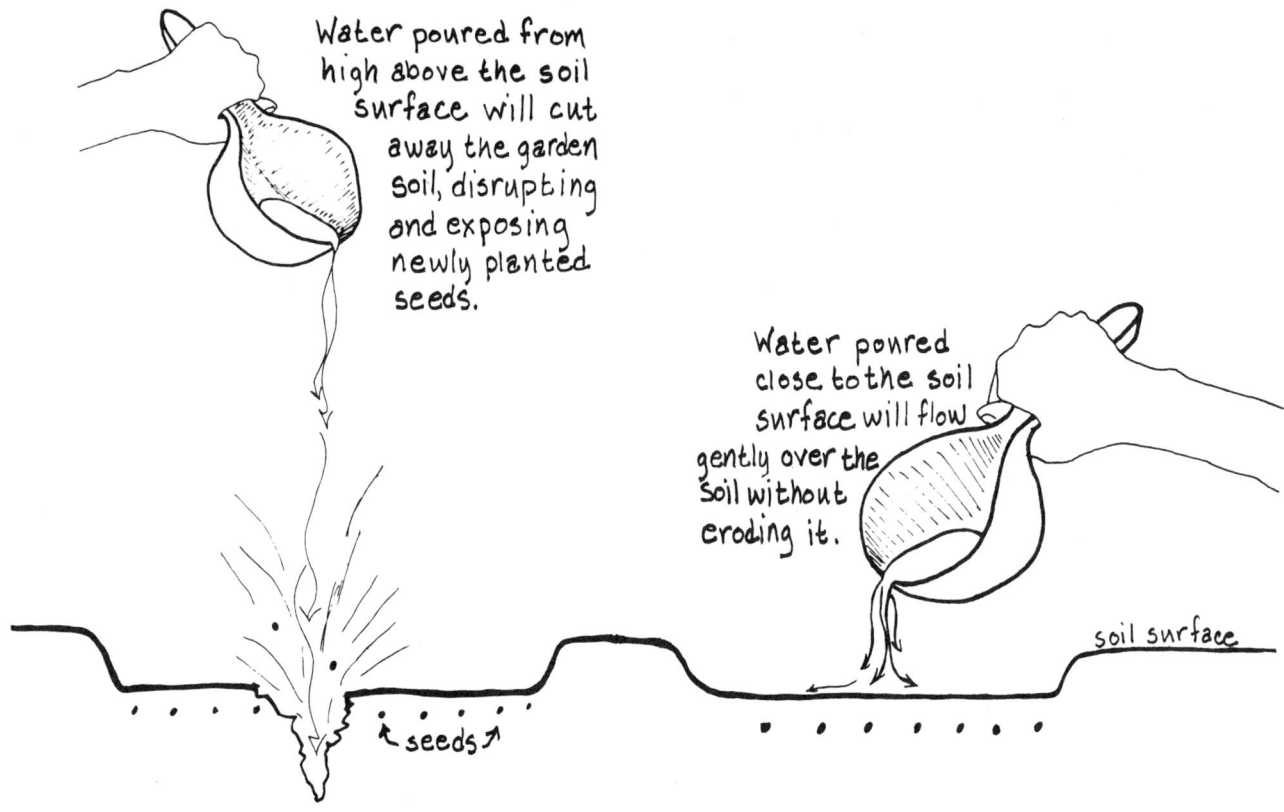

Figure 6.21 Watering Seeds and Seedlings

6.5.2 Mulching and Shading

Mulching and shading conserve moisture around seeds and seedlings when the weather is hot and dry. (For general information on mulching, shading, and windbreaks see section 10.8.) Special points about mulching and shading seeds and seedlings include the following:

- A thick mulch can harbor insects and encourage disease, both of which prey on tender seedlings and seeds. It can also smother and kill emerging seedlings. A light mulch such as a sprinkling of fine organic matter on the soil surface after watering will reduce evaporation. It is a good idea to periodically clear the mulch away and check for problems such as insect or fungal damage.
- Mulching and shading should allow enough sunlight to reach the plants. Pale, spindly seedlings with long stems are signs of insufficient sunlight (Figure 6.22).

Figure 6.22 Too Much Mulch can Harm Seedlings

6.6 Diagnosing Seed Planting Problems

If seedlings fail to emerge the following questions should be asked (Figure 6.23):
- Was it too cold, hot, dry, or wet?
- Were seeds past their after-ripening (section 14.3) or dormancy period?
- Were the seeds old or moldy and therefore no longer viable?
- Was enough time allowed for germination to occur?
- Were the seeds carried away or eaten by insects, birds, rodents?
- Were the seeds buried too deeply for the seedling to emerge? Or were they planted too near the surface, causing them to dry out?

Table 6.1 provides more information on diagnosing and treating seed planting and seedling problems. If the evidence is unclear or indicates a problem with the seeds, the following germination test can be used.

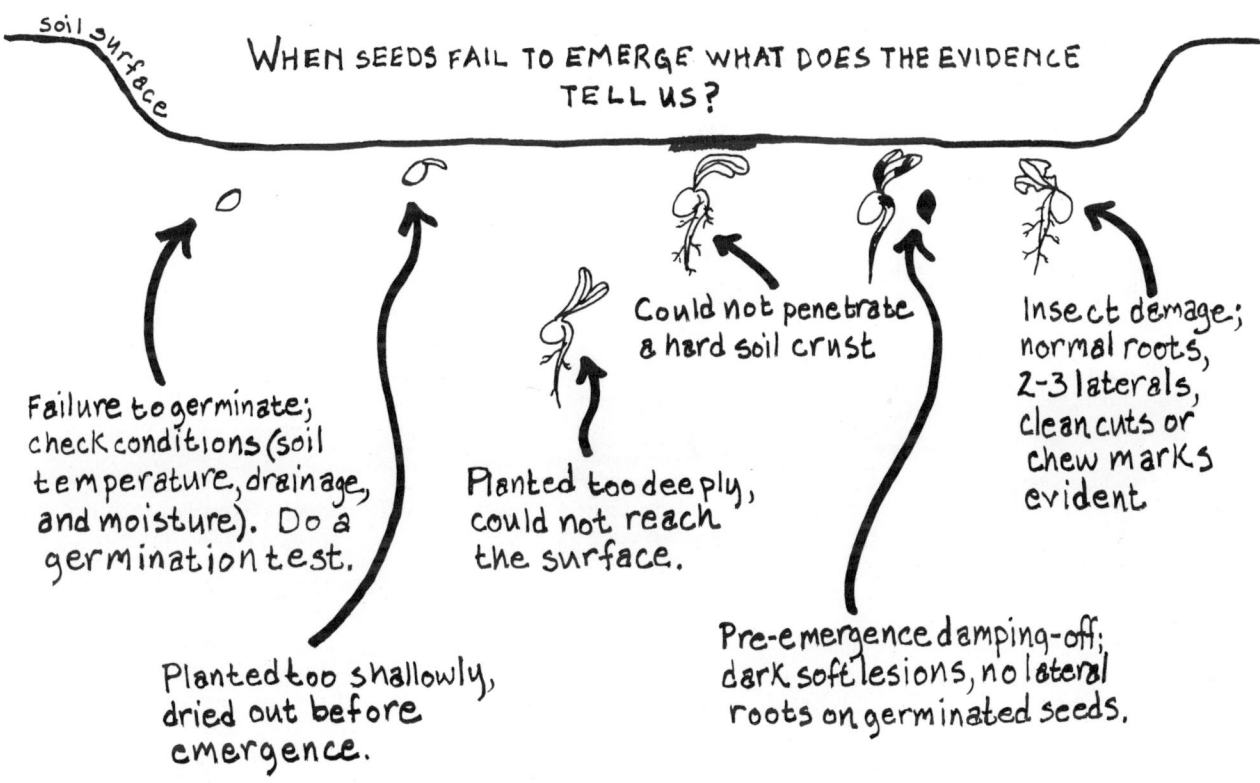

Figure 6.23 Why Seeds Fail to Emerge

Table 6.1 Diagnosing and Treating Seed Planting and Seedling Problems

Observations	Possible causes	Suggested actions
Failure to emerge:		
Seed did not germinate	Environment; old or nonviable seed	Investigate environment; try germination test, look for other seeds
Seed missing	Insects, small mammals, or birds	Plant again, try giving more protection, bird scaring, container planting
Seed germinated but dried up before it emerged	Planted too shallow, did not get enough moisture, got too hot	Plant again, deeper, water more often, mulch
Seed germinated and grew but did not reach soil surface to emerge	Planted too deeply	Plant again, shallower
Seed germinated and grew curled up under soil surface	Soil crust too hard	Add organic matter and sand to improve soil quality around seeds
Seed or seedling dead, normal roots with chew marks or clean cuts	Ants, beetle larvae, and other insects	Plant again, try using containers
Preemergent seedling dead; dark, soft, lesions and dark root	Preemergence damping-off	Add organic matter and sand to improve drainage, use containers with fresh soil
Parts of seedling missing	Probably eaten by insects or birds (wild or domestic), or possibly damaged seed stock	Look on/around plants, under mulch, especially at night, remove mulch, fence, bird scaring, plant in containers, find new seed stock
Spindly, pale seedling	Insufficient sunlight	Increase exposure to sunlight, remove surface mulch
Seedling deformed, abnormal	Damaged or infected seeds	Do germination test to check viability, find new source of seeds

6.6.1 Testing Seed Germination

The *germination percentage* is the percentage of a variety of seeds from a particular source which can be expected to germinate. The standard method for measuring germination percentage on a commercial scale involves four tests, each having a sample size of 100 seeds.[13] For each test the number of seeds germinating within what is considered a reasonable amount of time for that particular crop are counted. Most domesticated annual crops germinate in a few days if in the proper temperature range, although some seeds, for example cilantro and chilis, take longer. The counts for each of the four tests are summed and divided by four to give an average germination percentage. For example, if the numbers of germinated seeds from the tests were 79 + 83 + 81 + 74 = 317, then the germination percentage is 317/4 = 79.3%.

This test is also useful for finding out how strong and healthy the seeds are. Nutritional deficiencies in the seed-producing crop, or poor postharvest handling may result in deformed seedlings. Seeds that are very slow to germinate in this test, assuming the correct conditions are provided, are said to lack vigor. Old seeds lose vigor over time as the embryo becomes weaker; eventually they can no longer germinate and they die. When planted, seeds that lack vigor are more likely to die under stressful conditions or produce weak seedlings.

The seeds of most annual garden plants have a germination percentage of over 70%, so approximately three-quarters of a seed sample should germinate in a test. However, germination in the garden is usually slightly lower than germination tests and this difference increases as the germination percentage drops.[14] For example, if the germination percentage in the test is 95%, in the garden it may be about 90%, but if the test shows 70%, in the garden it may be only 50%.

Household gardeners usually do not have the resources to devote 400 seeds to a germination test, and can use a smaller sample. If a group of gardeners obtain seeds from the same source they could each contribute some seeds for a cooperative test. Using a smaller sample size increases the possibility of test results being affected by sampling error (section 4.5.3). Because of this the results of a smaller test are not meant to compare with those of the standard germination percentage test described above. However, a small-scale test can still help the gardener check the viability and vigor of the seeds she is using. The goal is not to determine an actual germination percentage, but to see if there is a significant problem with the seeds. In other words, is the germination rate closer to 80% or 10%? If it is closer to 10% there is a problem with the seeds and it probably is not worth using them.

When doing the test, small samples of seeds should be selected from various parts of the storage container, ensuring a representative sample of the seed stock. A piece of cloth with each side measuring about two hands in length (35-45 cm, 14-18 in), is wet and the seed samples scattered on it, keeping them about 1 cm (0.5 in) apart. The cloth is carefully rolled around a stick and tied, making a snug but not tight roll[15] (Figure 6.24).

Keeping the roll moist is critical; sprinkling it with water and protecting it from sun and wind will prevent drying. However, if the roll is kept dripping wet the seeds will rot. If they are seeds normally planted at

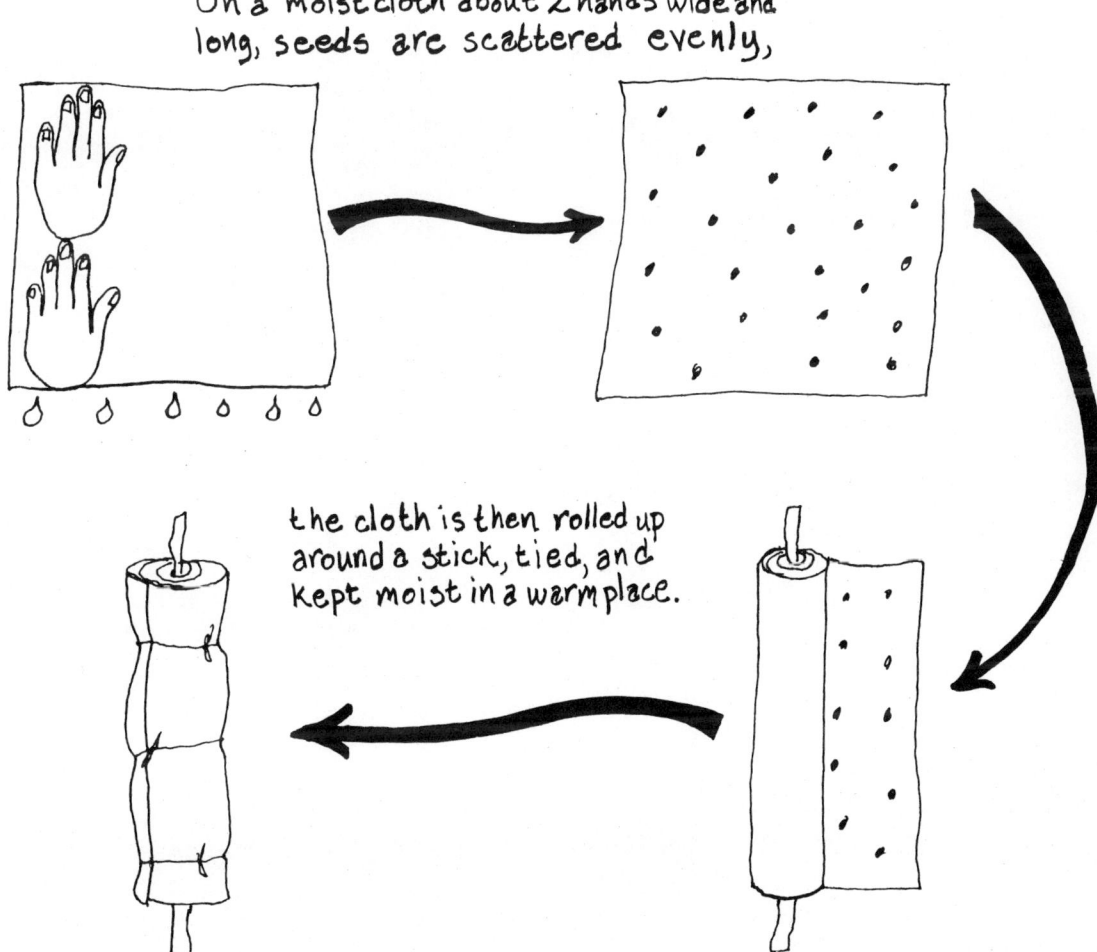

Figure 6.24 Testing Seed Germination

the time of year when the test is done the temperature in the shade should be fine.

The cloth is left rolled for the amount of time germination is thought to take for that particular crop. For most annual garden crops seven to ten days is plenty of time. The cloth is then unrolled to check for signs of germination. If no germination occurs within the expected length of time it is possible that certain conditions were not appropriate. Did the seeds dry out? Did they drown? Were they too hot? If there are no obvious problems the cloth and seeds can be rolled back up and the test continued. If that does not succeed the test can be tried again. If there is still no germination, or it is poor, or seedlings are abnormal or diseased it is best to look for another source of seeds or try a different crop or variety. But if normal germination does occur then the problem must be sought in the seeds' environment (refer to Table 6.1).

6.7 Thinning

Thinning is done so that plants have enough space to obtain sunlight, water, and nutrients for vigorous and productive growth. Thinning is common with plants grown from seed. Vegetative propagation involves larger plant parts which are usually properly spaced on first planting.

Ideally, plants should be left dense enough so that when healthy and mature their leaves will shade the soil, reducing soil temperature and evaporation. But, if plants are too crowded they compete for water, light, and soil nutrients; they lack vigor and cannot produce nutritious, good tasting fruit, leaves, or roots. Because these plants are weaker they are also more vulnerable to stress and disease.

We describe three thinning techniques; pulling, cutting, and transplanting, because each is more appropriate for certain situations. Many seedlings that are pulled or cut can be eaten, including leaf crops and legumes like beans and peas. But some, like tomatoes, are poisonous. Usually local gardeners and others can identify which seedlings can or cannot be eaten. Because transplanting is an important technique used not only for thinning it is described in the Chapter on plant management, section 8.4.

Pulling and cutting are done when transplanting is not desired. The less vigorous plants—those that are smaller, paler, or weaker—are chosen. Seedlings should be cut when they have extensive root systems and are growing so closely together that pulling would disturb adjacent plants. For cutting, a knife or other sharp tool can be used to cut the plant off at or near ground level. If plants are small they can be pinched off with the fingers. Cutting is not appropriate for some perennials, bulbs, or other plants like onions and Jerusalem artichokes which can store energy in their roots and send up new shoots from this reserve. These plants must be pulled or transplanted to thin them out. To thin by pulling, plants are grabbed low on the stem and pulled straight up and out. After pulling the soil should be pressed down so no holes or gaps remain which expose the roots of living plants.

6.8 Resources

Local gardeners and farmers have often developed special techniques for controlling the pollination of their crops and planting the resulting seeds. They are the best sources for locally appropriate insights and methods. Botany and biology textbooks are good resources for explaining the process of sexual reproduction in plants. Items #6-2, #7-3, #7-4, #9-3 and #10-4 from Developing Countries Farm Radio Network (DCFRN) cover topics discussed in this chapter such as seed and fruit formation and planting seeds.

References

[1] Evenari, et al. 1982:208.
[2] Purseglove 1974:174.
[3] Samson 1986:256.
[4] Purseglove 1983:315.
[5] Purseglove 1974:534.
[6] Hartmann and Kester 1983:129.
[7] Mayer and Poljakoff-Mayber 1975:25.
[8] Hartmann and Kester 1983:145.
[9] Hartmann and Kester 1983:171-172.
[10] Atkinson, et al. 1976.
[11] Bradfield 1971:5; Collins 1914.
[12] Soleri 1989.
[13] Hartmann and Kester 1983:164-166.
[14] FAO 1961:102-103.
[15] DCFRN #4-1.

7
Vegetative Propagation

Propagating plants from plant parts other than seeds is called *vegetative propagation*. Plants grown from seed are the product of sexual reproduction and contain a mixture of genes from two parents. Plants grown by vegetative propagation are reproduced asexually, and have only one parent. The genetic characteristics of the new plant will be identical to those of the parent plant, although responses to different environments can make them different in some ways.

For example, date palms can be propagated both from seed and vegetatively. Using seed, the sex of the new date tree and the qualities it will have are unknown until it flowers, 8 to 10 years after planting. The seedling will be a unique combination of the characteristics of its two parents. However, if propagated using vegetative methods, in this case an offset, the new seedling will be a *clone*, or a genetic copy of its parent: the same sex, growing characteristics, and fruit qualities.

Knowing the characteristics of a new plant is one of the main reasons for using vegetative propagation. This is particularly important for trees that do not produce fruit or nuts for several years. (Refer to Figure 7.1 to identify plant parts used in vegetative propagation of trees). Some crops such as bananas, sweet potatoes, and cassava are usually propagated vegetatively because it is a fast and easy method, and because they produce few or no viable seeds.

7.1 Summary

There are several methods of vegetative propagation. The choice of which method to use depends primarily on the crop being propagated. Although we list a few examples of dryland garden crops that can be propagated using each method, local gardeners often know methods that work best with crops in their area. The explantions given in this chapter will help the field worker understand how indigenous methods work, how they might be improved, and what new methods might be introduced.

7.2 Cuttings

Cuttings are plant pieces, usually stems or branches, capable of growing new roots, called ***adventitious roots***. To grow these new roots cuttings must rely on stored energy or energy that they can produce. However, the cutting can only provide this energy if it is carefully protected from stress like heat and drought. Some of the dryland garden plants that may be propagated by cuttings are deciduous trees such as the stone fruits, fig, mulberry, and pomegranate. Olive and carob are two nondeciduous trees that can be started from cuttings. Cassava, sweet potatoes, and some perennial herbs can also be propagated from cuttings.

In sections 7.2.1 through 7.2.4 we give some examples of how cuttings are used to propagate different dryland garden crops.

7.2.1 Trees

Depending on the tree, cuttings can be of either green, new growth produced during the current season, called ***softwood cuttings***, or older, woodier shoots from the previous season or earlier, called ***hardwood cuttings***. Shoots that are producing flowers or fruit should not be used for cuttings. In general, cuttings from younger plants are easier to root. However, if the parent plant is old, heavy pruning will encourage lots of new growth that can then be used for cuttings.

Cuttings should not be taken during times of great environmental stress such as drought or very cold or hot temperatures. Otherwise, the best time to take a

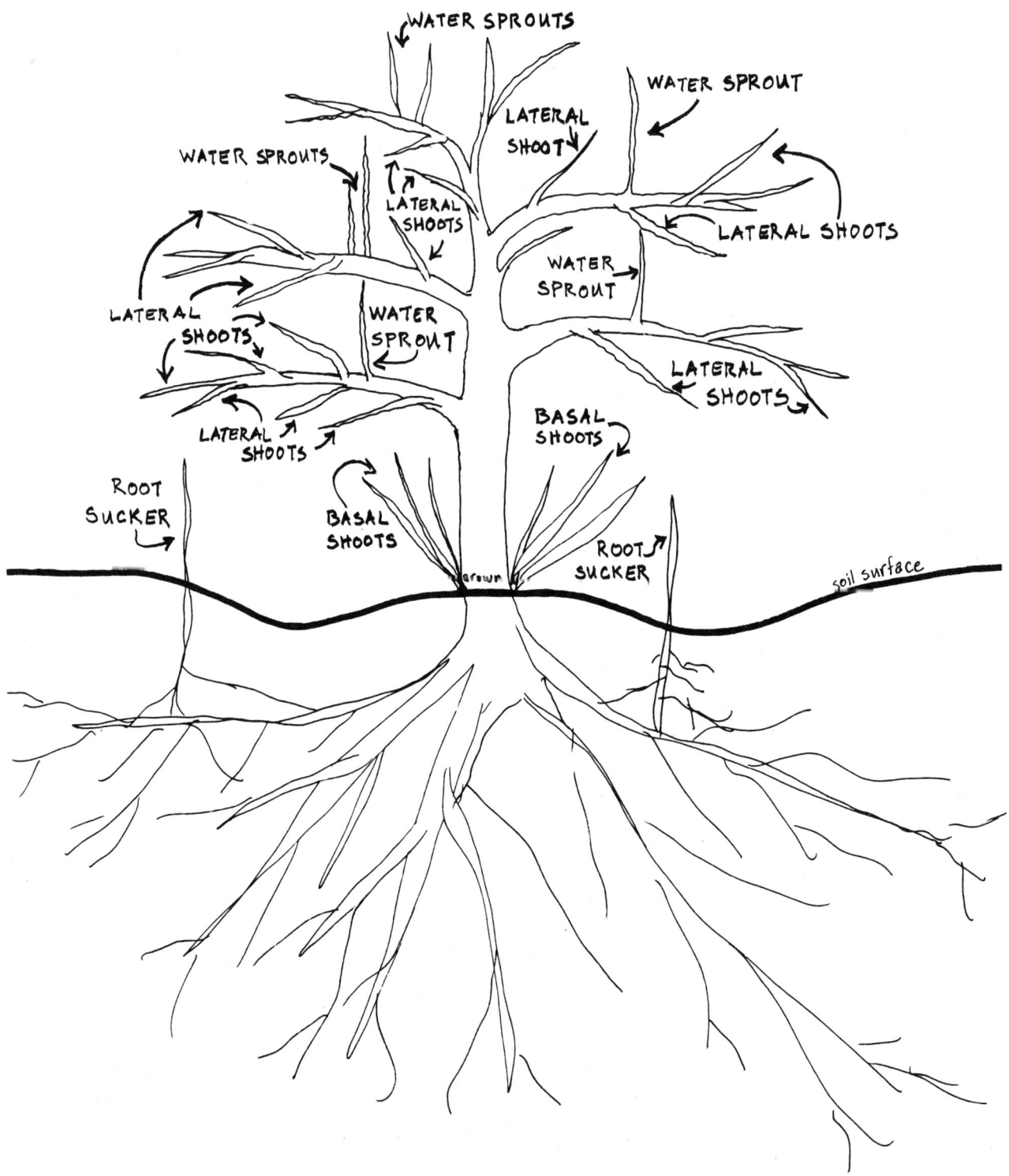

Figure 7.1 Terms Used in Describing the Vegetative Propagation of Trees

cutting depends on the kind of tree and the kind of cutting. In deciduous trees hardwood cuttings are taken during the dormant period before any buds have begun to swell or develop.[1] For nondeciduous trees the time for taking hardwood cuttings is not as easy to predict and experimentation with cuttings taken just before a period of rapid growth may be the best place to start. Softwood cuttings are taken during the early part of the growing season, and are pliable but not woody. They should snap if bent in half.[2] All but the top few leaves are removed from softwood and evergreen hardwood cuttings.

Hardwood deciduous tree cuttings should be dormant and leafless, and the tip is usually removed. Cuttings range in diameter from about 0.6 cm to as much as 5 cm (0.25-2 in). All cuttings should have at least three nodes (Figure 5.1 in section 5.1). A relatively young, vigorous shoot with many leaf nodes that is growing in full sunlight is best. It should be a lateral shoot, one growing horizontally or at an angle in relation to the ground, or a basal shoot from around the base of the trunk (or *crown*). Rapidly growing vertical shoots with large spaces between the nodes are called *water sprouts* and are not good for cuttings (Figure 7.2). The growth habit of a vegetatively propagated plant will often resemble that of the cutting from which it is grown.[3] For example, a vertical water sprout may produce a tall tree with fewer side branches than a tree grown from a lateral shoot which will tend to have fuller, more widespreading growth.

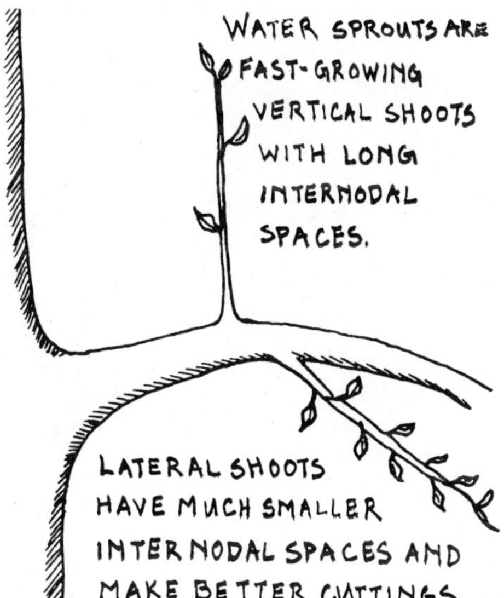

Sometimes shoots are prepared for hardwood cuttings by girdling them several weeks in advance. *Girdling* is the technique of constricting or cutting the stem bark, which blocks the downward flow of carbohydrates, hormones, and other substances through the phloem. These substances accumulate at the base of the cutting where they stimulate increased rooting.[4] Girdling can be done by lightly scoring the bark or tying string or wire around the shoot base.

All cuttings should be cut off from the parent tree just below a leaf node or bud. If girdling is used, the cut is made either on the girdled area or slightly below it, that is, closer to the parent plant. When making hardwood cuttings it is often best to include the shoot base and the area just below it because in some trees this is the area most likely to sprout, since most carbohydrates are stored here. In many cases like that of the olive, the cuttings are made with a small piece of bark and cambial tissue from the main trunk still attached to ensure that the base is included (Figure 7.3). This piece is sometimes called a *heel*.

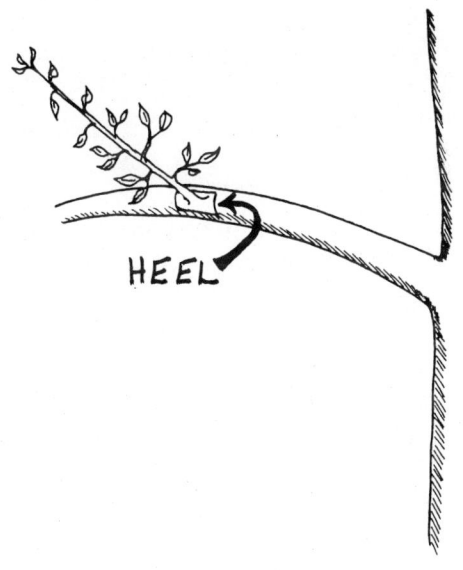

Figure 7.3 A Cutting with a Heel

A sharp tool such as machete or knife makes it easier to get good clean cuts, minimizing the damage to both tree and cutting. If possible the tool should be washed with soap, bleach, alcohol, or vinegar to remove any disease-causing microorganisms. If the wound on the parent plant is large or exposed to harsh conditions, shading the wound will help protect the plant from shock.

Figure 7.2 Choosing Shoots for Cuttings

Softwood cuttings should be planted as soon as possible after being removed from the parent tree. If there is a delay between removal and planting these cuttings must be kept cool and moist, by wrapping them in a damp cloth for example. Putting the cut end in water is not recommended as this can leach out needed nutrients and hormones.[5]

Hardwood cuttings taken during the dormant season should be kept cool and moist until planting time. One way to do this in temperate regions is to bury them outdoors in the sand.[6] The cuttings should be buried horizontally, or vertically with their tips pointing downward. Burying them prevents the cuttings from sprouting in storage and then dying due to lack of water and cold temperatures. If the cool-season temperatures are very cold (at or below 0°C, 32°F) then the cuttings need to be buried in a protected area, or in a box of sand indoors.

Cuttings are planted so that at least two leaf nodes are below the soil surface. The soil around cuttings must be kept moist but not saturated. However, because cuttings—especially softwood ones—are susceptible to rotting good drainage must also be provided. Therefore, the area immediately around the cutting should be sand, or a mixture of sand and garden soil which drains well. If any containers are used they must also have good drainage. (See section 8.4.2 for more suggestions on preparing a permanent planting site for the tree's future growth.)

Once planted, cuttings need to be protected from drying sun and wind with mulch, shades, and windbreaks (section 10.8). Sometimes cuttings are covered with mud or a thin clay and water wash, or wrapped with straw to prevent sunburn and to reduce transpiration.[7]

7.2.2 Perennial Herbs

Young stem cuttings can be used to propagate a number of perennial herbs such as the mints, oregano, marjoram, and rosemary. The best time to take cuttings from these herbs is at or just before the period of most rapid growth. The cutting should be young, healthy, non-woody growth. The length of the cutting will vary with the plants' size and type but should include at least five leaf nodes.

The bottom one-half to one-third of the cutting, including several leaf nodes, is submerged in a container with water. Since root production is sometimes inhibited by light, keeping the submerged portion in the dark may help. This can be done by putting the rooting end in an opaque container and covering the area around the stem with a piece of cloth. The top, leafy part of the cutting should be exposed to normal day and night light cycles because it must photosynthesize to produce energy for root growth.[8]

Roots will develop from the cutting in several days or weeks. Cuttings root most easily when it is warm and humid. When roots have developed, the cutting can be planted in moist soil, shaded, mulched, and watered probably once or twice a day during the first few days.

7.2.3 Cassava

Cassava, grown for its edible roots and leaves, is propagated from stem cuttings called *sticks*.[9] The sticks are cut when the plant is mature and the roots are being harvested. If not planted within a few days of cutting, the sticks can be stored dry and in the shade for up to 8 weeks. If stored, their ends must be recut just before planting. The sticks should:

- Be from a plant free of cassava mosaic virus (see section 13.3.4 for a general discussion of plant viral diseases).
- Have at least three leaf nodes.
- Be taken from the central area of the plant, about 15 cm (6 in) above the ground, to get pieces mature enough that they will not rot[10] (Figure 7.4).
- Be 30-45 cm (12-18 in) long.

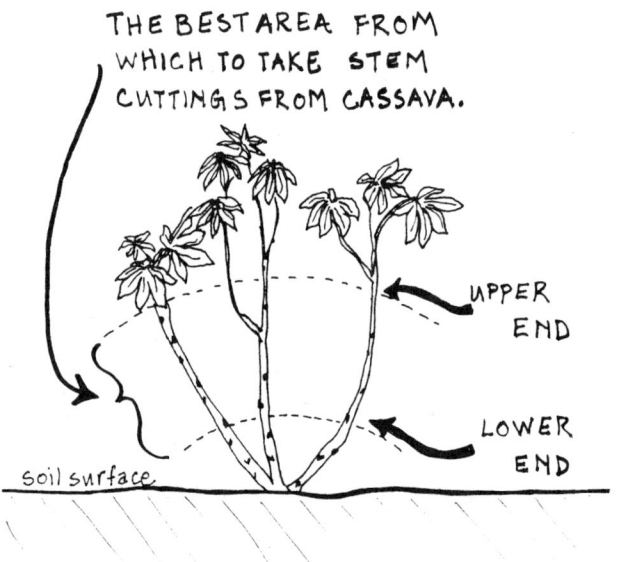

Figure 7.4 Taking a Cassava Cutting

Sticks are planted by burying two-thirds of the lower (older) end in the soil. The buds point up to the upper, younger end of the stick. In parts of both East and West Africa cassava sticks are frequently planted at approximately a 45° angle[11] (Figure 7.5).

7.2.4 Sweet Potatoes

Sweet potatoes are also grown for their swollen tuberous roots. They can be propagated from vine cuttings at least 20 cm (8 in) long, taken from healthy growing tips of mature plants. The cutting should have at least seven nodes and can be planted vertically or at an angle, one-half to two-thirds of its length. If the soil is kept moist, roots can develop from the nodes in as little as five days[12] (Figure 7.6). The tuberous roots can also be planted in a nursery bed and when shoots grow to 23-30 cm (9-12 in) long they can be separated and planted.

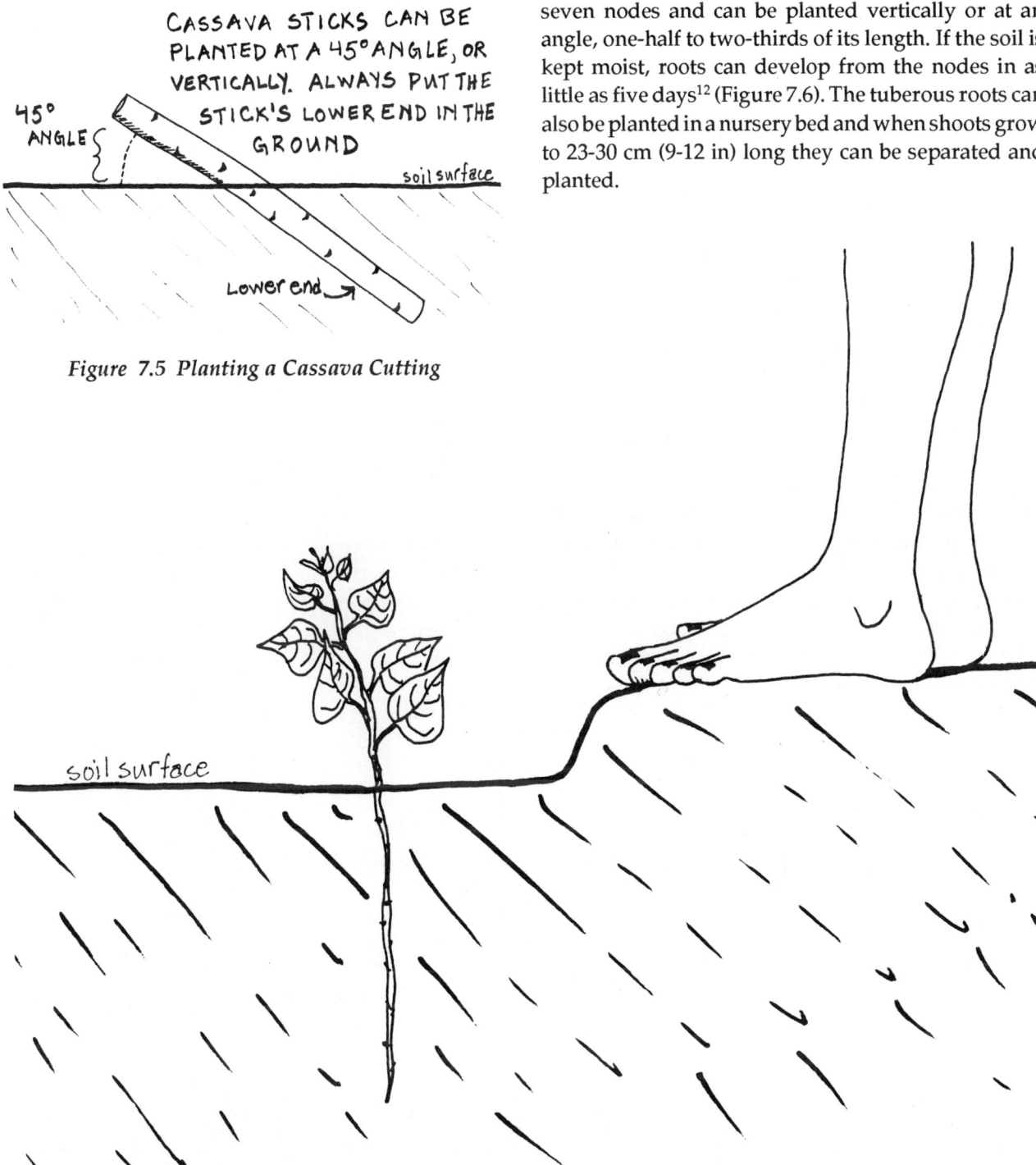

Figure 7.5 Planting a Cassava Cutting

Figure 7.6 Planting a Sweet Potato Cutting

7.3 Tubers, Tuberous Roots, and Bulbs

Tubers are enlarged stems that grow underground. Hausa potatoes, tiger nuts, Jerusalem artichokes, potatoes, and yams are dryland garden plants that can be propagated from their edible tubers. The tubers are selected at harvest time and stored until the next planting season.

Tubers chosen for propagation should be:
- Mature.
- Free of disease (sections 13.3.4, 13.4).
- Not blemished with cuts or bruises.
- Of good eating quality.

Potatoes, Hausa potatoes, and yams are usually grown from *setts*, or pieces of tubers. Setts are made by cutting up the tuber so that each piece has least one or more *eyes* or leaf nodes. Tiger nut tubers are soaked in water for a day and then planted whole.[13] Jerusalem artichokes are also planted whole.

Tubers are usually planted horizontally. In hot drylands setts can be protected from heat, which inhibits shoot or slip production, by planting them with the cut side down. Both whole tubers and setts are covered with a layer of soil three to four times their thickness.

Swollen edible storage roots, or *tuberous roots*, resemble tubers, but are actually part of the root of the plant and not of the stem. Most garden root crops such as beets and carrots are grown from seed, but sweet potato can be grown by planting the root itself, as well as from shoots (section 7.2.4).

Bulbs like garlic and onions are formed by enlarged leaves which grow underground. Each of the cloves of a garlic bulb can be planted to produce new plants. Thus 10 cloves of garlic can produce 10 new bulbs. To grow garlic the cloves of a mature bulb are separated, and each one planted separately, with the narrow tip pointing upward (Figure 7.7). Some onions called multiplier or bunching onions grow in clusters similar to garlic bulbs and are propagated the same way.

Many nonbunching onions are biennials because their full life cycle takes 2 years to complete. In the first year seeds are planted which produce plants with edible bulbs. During the second year a bulb is planted and grows a shoot that flowers and produces seed (Figure 7.8). Because part of this cycle involves the use of seeds, nonbunching onions are not really vegetatively propagated.

In some areas of savanna West Africa nonbunching onions are propagated indigenously in the following way:[14]

a) The selected bulb is cut approximately in half horizontally and the top is eaten.
b) The bottom half is planted with the cut side up in a nursery bed, and side buds or offsets (section 7.4) sprout from the base and grow into shoots.
c) When the shoots are 10-15 cm (4-6 in) tall the bulb is dug up and the shoots separated and transplanted into their permanent growing site.
d) Each of these shoots will flower and produce seeds used to grow bulbs for eating and planting.

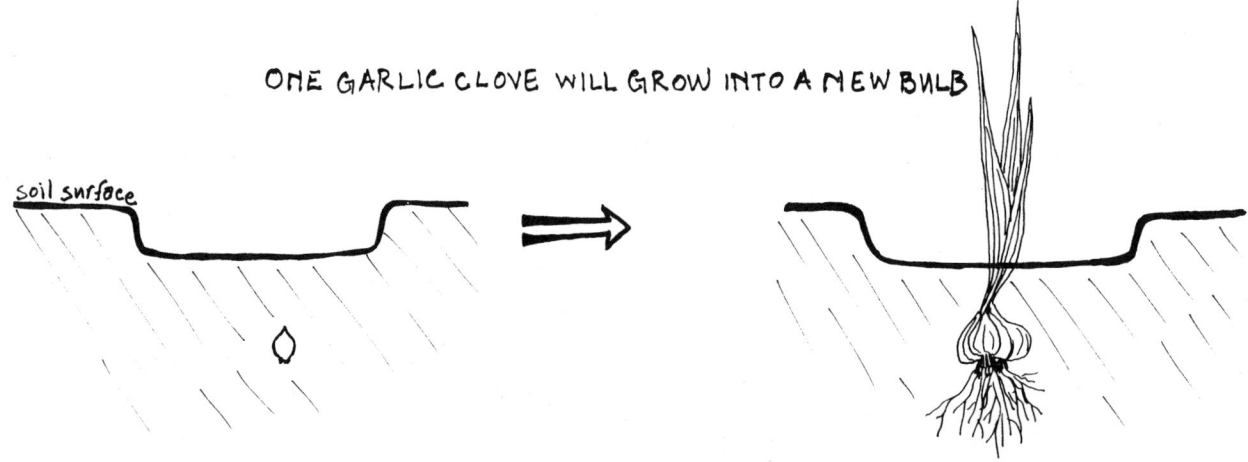

Figure 7.7 Planting a Garlic Clove

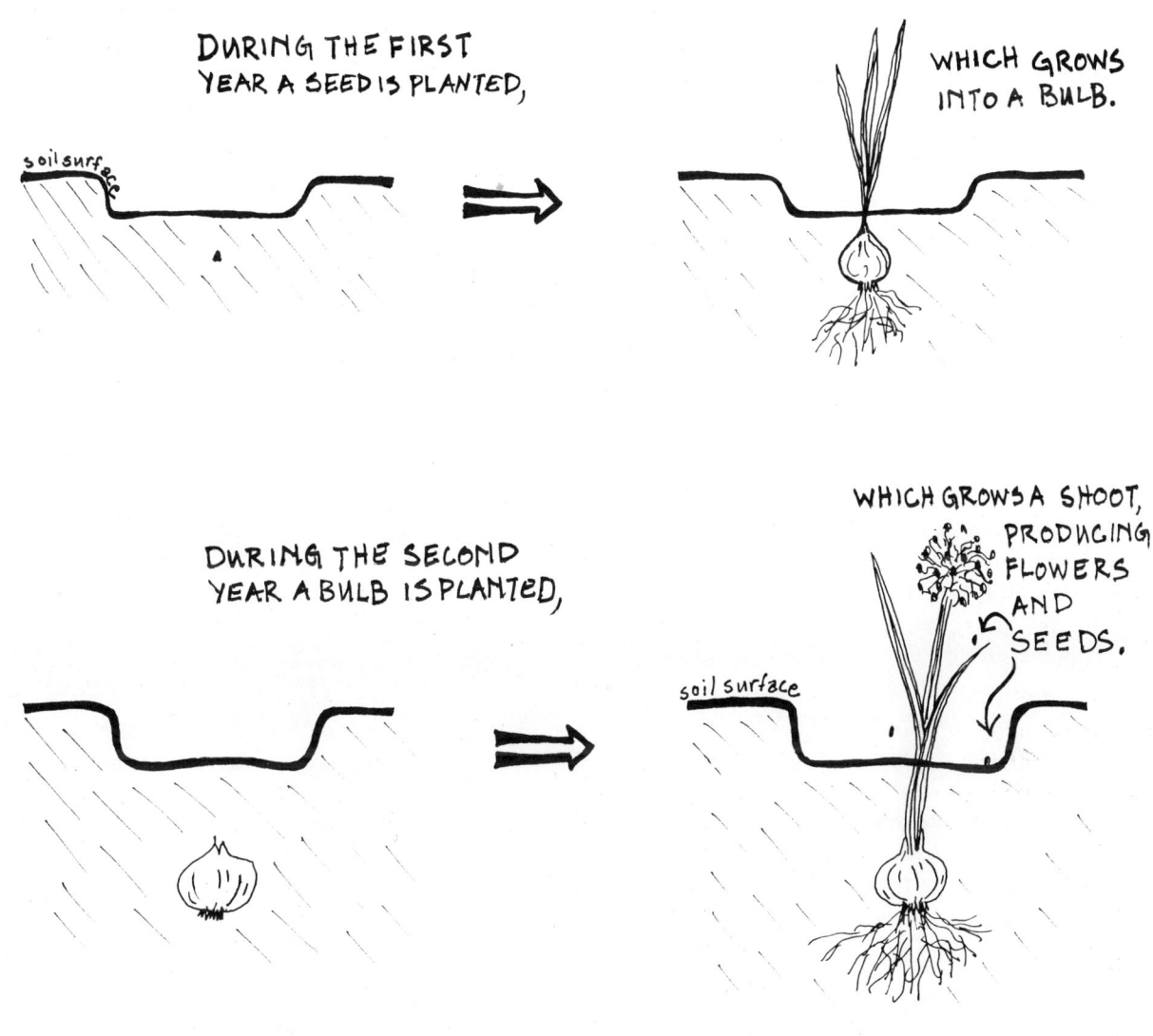

Figure 7.8 Planting an Onion—a Biennial Bulb

7.4 Offsets

Offsets are lateral shoots used for plant propagation. They develop around the base of a monocot's stem or trunk, and are allowed to grow and develop their own root system before being separated from the parent plant and planted elsewhere.[15] When natural production of offsets is slow, cutting back the main stem, as described for nonbunching onions in West Africa (section 7.3), can stimulate their formation. A few dryland garden plants propagated by offsets are certain onions, date palm, banana, and pineapple. In bananas these offsets are called suckers, but because they grow from the *corm*, which is an underground swollen stem, they are in fact offsets. Box 7.1 gives an example of propagation with an offset.

Box 7.1
Propagation by Offset—the Date Palm

The following is an example of indigenous offset propagation of the date palm, an important dryland garden tree.[16]

a) The planting site is prepared for the transplant (section 8.4.2).
b) Date offsets are not removed until they have a root system and some of the dark green foliage characteristic of a mature plant. This usually takes 3 to 10 years after the offset first appears.
c) Thoroughly watering the area around the offset makes it easier to dig up.
d) The offset's leaves are tied and often cut back to as little as one-half of their original length. The remaining leaves are then wrapped in a moist cloth to decrease water loss.
e) The offset's longer roots are dug up and cut to a length that can be transplanted.
f) A chisel or other sharp tool is used to separate the offset from the parent plant (Figure 7.9). In cases of difficult separation it may be best to cut more from the parent plant than the offset because the larger plant is better able to recover from such cutting. However, this should be done very carefully because a large, healthy, productive tree should not be jeopardized to remove an offset. In some areas it is common to coat the exposed surfaces of both plants with wet clay or soil to prevent sunburn and dehydration.
g) The offset is transplanted as soon as possible or temporarily placed in water. Figure 7.10 shows date palm offsets soaking in a small irrigation canal at an oasis in western Egypt.

Figure 7.9 Separating Offsets from a Date Palm

7 *Vegetative Propagation* 107

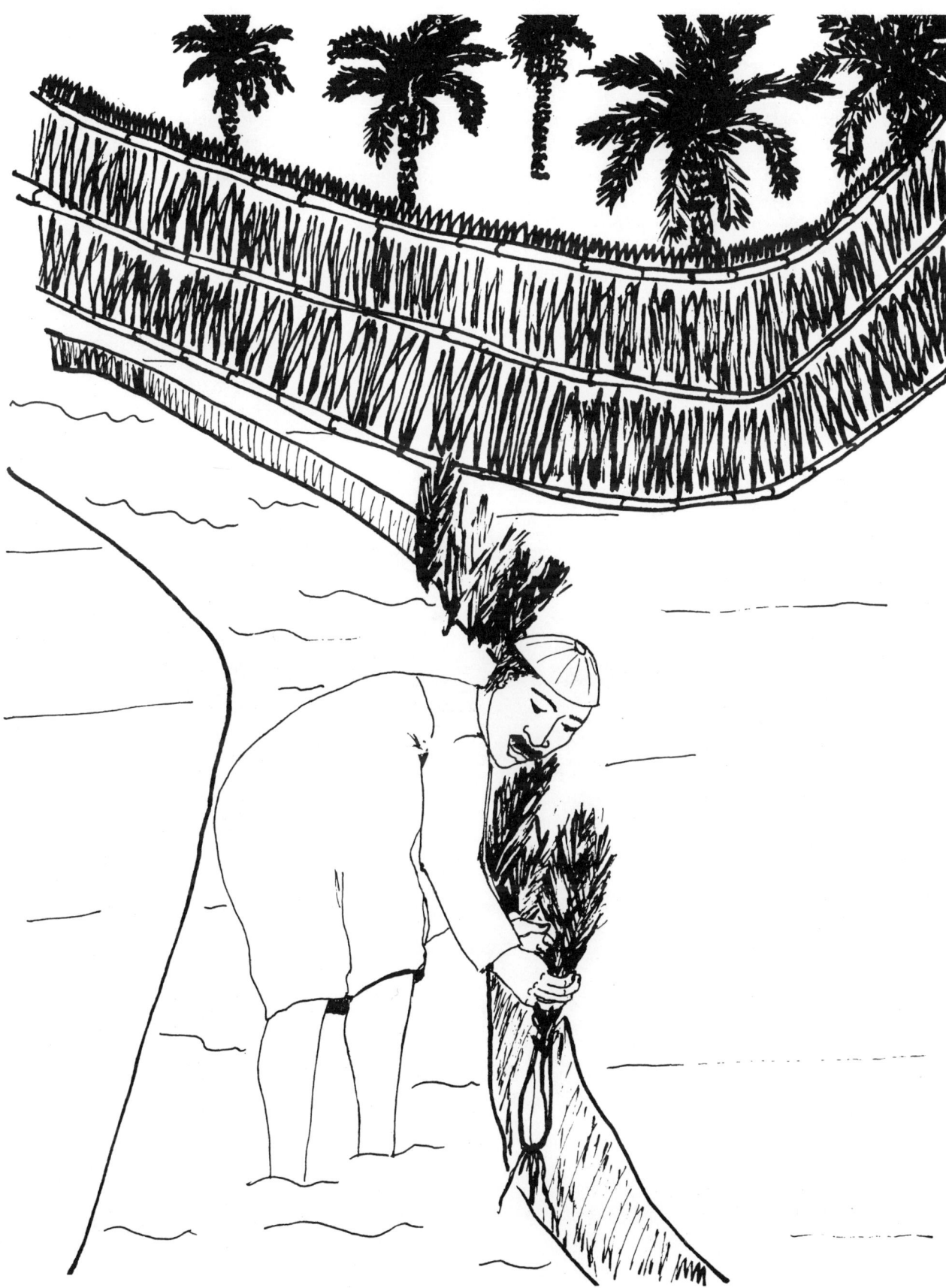

Figure 7.10 Date Palm Offsets Soaking in an Irrigation Canal in Western Egypt

7.5 Suckers

Shoots originating underground from the roots of dryland trees like olives and jujubes are called *root suckers* (Figure 7.11). Drought encourages root suckering in jujube trees. Suckers can be dug up and separated during the dormant or "pregrowth" season and transplanted. Often a small portion of the parent root is transplanted with them. Suckers growing from a tree which was grafted onto root stock will produce plants with the same qualities as the stock, not the scion of the parent plant (Figure 7.12).

7.6 Grafting

Grafting is the technique of connecting pieces from two different plants so that they will develop and grow as one plant. The *stock* is the part used as the base and roots of the new plant. The *scion* is the piece used as the top, fruit-producing part of the new plant. The union between the stock and scion, each from different plants, is formed from the contact of the cambium, a layer between the phloem and xylem in dicots. (Figure 5.3 and the discussion on the vascular system in section 5.2 are helpful for understanding how grafting works.)

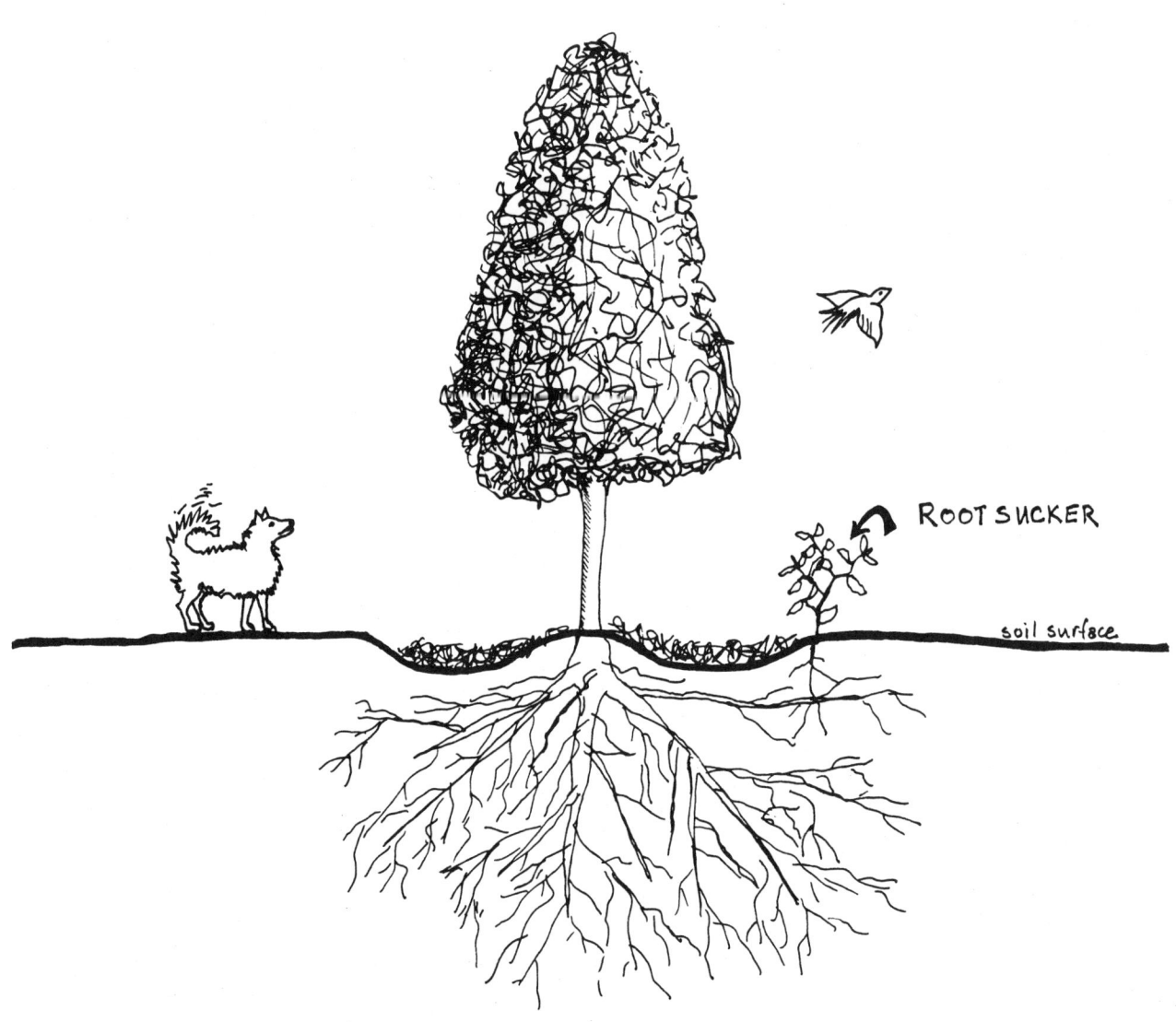

Figure 7.11 A Root Sucker Growing from a Jujube Tree

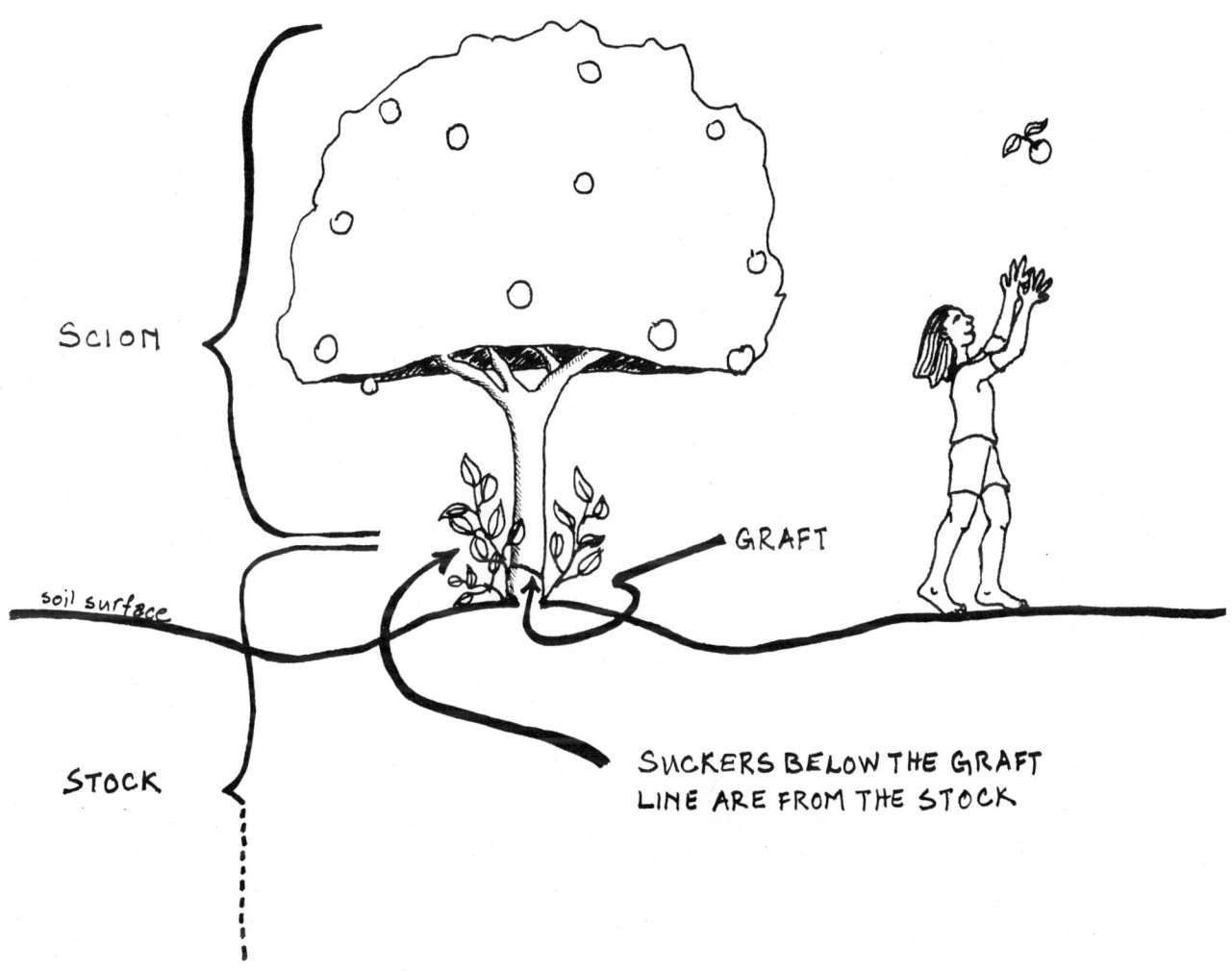

Figure 7.12 Suckers Growing from the Stock of a Grafted Tree

Parenchyma cells are thin-walled plant cells that perform many functions. They are produced in the cambium of both stock and scion to heal the graft wound. The parenchyma cells, which line up with the cambial cells of stock and scion in the healing process, later become new cambial cells. In turn, these new cambial cells form new xylem and phloem cells, establishing connections between the vascular system of the stock and the scion. This connection is only possible in dicots, and is essential if the scion is to receive the water and nutrients necessary for it to survive.

Grafting is used when one variety is hardy and resistant to root disease, but produces low yields of poor-quality fruit, and another variety is sensitive to drought and root diseases but has good fruit production (Figure 7.13). In such cases the scion of the fruit producer may be grafted onto a stock of the hardy variety. Sometimes a hardy, nondomesticated variety is used for stock. For example, the atlantica pistachio (*Pistacio atlantica*), which bears inedible fruits and grows wild in the Negev Desert, is used as root stock for domesticated pistachio scions.[17] Grafting is also useful for changing fruiting varieties of a tree or for repairing a damaged tree.

7.6.1 Compatibility for Grafting

Grafting is limited by the compatibility of stock and scion. There are no simple or absolute guidelines for determining if two plants are compatible, but using plants in the same genus is often, though not always, successful. For example, the stone fruits—almonds, peaches, and apricots—are all in the genus *Prunus*. Almonds and apricots graft well onto peach rootstocks, but an apricot scion is not compatible with almond stock nor is an almond scion with apricot stock.[18]

When the graft fails soon after being made the

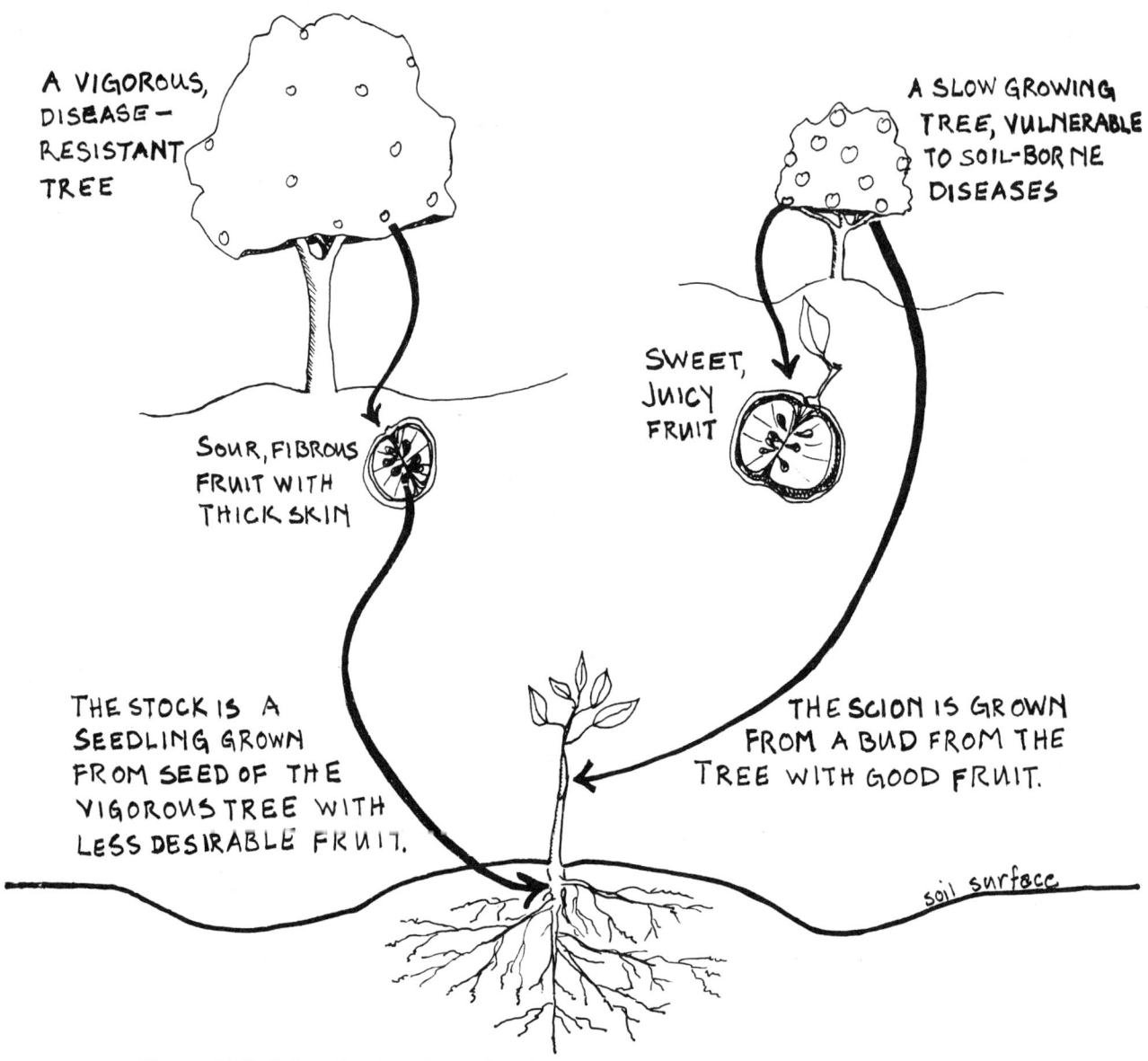

Figure 7.13 Scion that Produces Good Fruit is Grafted onto Vigorous, Hardy Stock

reason could be either incompatibility or a problem with the graft itself, such as infection, pest damage to the graft tissue, drying of the graft wound, or poor alignment of cambial tissues. Incompatibility between scion and stock may be obvious shortly after grafting but it can also take years to show up. It may be due to a failure to produce new xylem or phloem tissues or the production of a chemical or virus by one part of the graft which is poisonous to the other.[19] If, years after grafting the scion breaks off the stock in a clean line where the graft wound was it is almost certainly because of incompatibility. The clean break shows that the graft failed to join into an integrated, strong union. Indicators that a graft combination may be incompatible are:[20]

- Failure to form a graft union.
- Yellowing leaves at the end of the growing season, early seasonal leaf loss in deciduous trees, slow growth, shoot dieback and general lack of vigor.
- Differences in the growth of stock or scion.
- Differences between stock and scion in the time when their new growth for the season begins or ends.
- Premature death of trees.

Experimentation and consultation with local gardeners and other experts in the area are the best ways to determine which grafting combinations are compatible.

7.6.2 Effects of Stock and Scion on the Grafted Tree

The effect of stock and scion on each other depends on their characteristics, as well as on the growing conditions. The stock should be adapted to local soil moisture, salinity, pH, and drainage. Although these conditions affect both stock and scion, the impact on the stock has greater influence on the tree's health. In some cases tree size is determined by the stock and "dwarfing stock" is used to produce small, compact trees.

The influence of the stock on fruiting is complex and variable. Time of blossoming and fruit set, number of years a tree will produce fruit, fruit size, and total yield can all be affected. For example, one experiment found that when scions of orange, grapefruit, and tangerine are grafted to sour orange (*Citrus aurantium*) stock, they produce fruits that have smooth skins, are juicy and sweet, and store well. But the same scion varieties grown on rough lemon (*Citrus limon*) rootstock frequently produce fruits with thick skins, having poor taste, texture, and storage qualities.[21]

Scion varieties are chosen primarily for their fruiting characteristics such as flavor, texture, size, quantity, and storage qualities. In some cases scions may influence the vigor and growth pattern of the stocks. For example, on a grafted citrus tree with a vigorous rootstock and a weak scion the scion will dominate and slow the tree's growth.[22]

There are many ways of grafting using different cuts, some more adapted than others to particular plants and conditions. Gardeners in many areas have a long tradition of grafting their trees and are often very knowledgeable about techniques and the best tree varieties to use. Working with them is the best way to learn about grafting and grafting problems in the local area. In the following sections we give a brief introduction to four basic grafting techniques. The first three are usually used for young trees, the last one for more mature trees.

7.6.3 Approach or Attached Scion

The **approach** or **attached scion** grafting method is useful for many species that are difficult to graft. It is also used for grafting during stressful environmental conditions such as drought. Overall it is the most reliable grafting method although it does require more time and attention than others. With this technique both stock and scion are growing rooted plants during the grafting process which reduces the risk that either will die. Mangoes and guavas are two dryland garden trees that are grafted using this method.

There are many different cuts that can be used for this method; the one shown in Figure 7.14 is the most common, *sliced-approach grafting*. Stock and scion plants whose stems are the same size are selected. Smooth, flat, vertical cuts are made on both stock and scion to expose cambial tissue. The wounds are matched, bound together with string or other fiber, and sealed with mud or wax. It takes approximately two or more months for an approach graft to become successfully joined. When the union has formed, the root of the scion and the top of the stock are gradually cut back and removed. This should be done over several weeks, first nicking the scion about 2.5 cm (1 in) below the graft union. A week later the cut is deepened, and during the third week it is completed. At the same time, between half and all of the stock above the graft union should be removed.[23]

7.6.4 Budding

This is a popular grafting method in which a bud is grafted onto the stock plant. There are many types of cuts used depending on the kind of tree and local conditions and practices. In hot drylands the bud should be placed on the side of the stock plant that is most protected from sun and wind.

Budding is done during periods of active growth for the stock plant, when the cambial cells are rapidly dividing.[24] North of the equator budding is done in spring (March-April), June (late May-early June) and fall (late July-early September). South of the equator budding time is spring (September-October), December (late November-early December) and fall (January-early March). Fall budding is most common and is done before the end of the growing season when the rootstock seedling has grown large enough to support a bud. In the spring, rootstock growth above the graft is removed and the bud produces the new top. Spring budding is done right after the rootstock starts growing again. About 2 weeks after grafting the rootstock growth above the bud is removed and the new top grows from the bud. In places with long growing seasons December (or June in the northern hemisphere) budding may be used. A rootstock seedling started in the spring of the same year is used, as are buds of the current season's growth. Soon after grafting, the rootstock growth above the bud is removed to allow the bud to produce the new top growth.

Some dryland garden trees that can be bud grafted

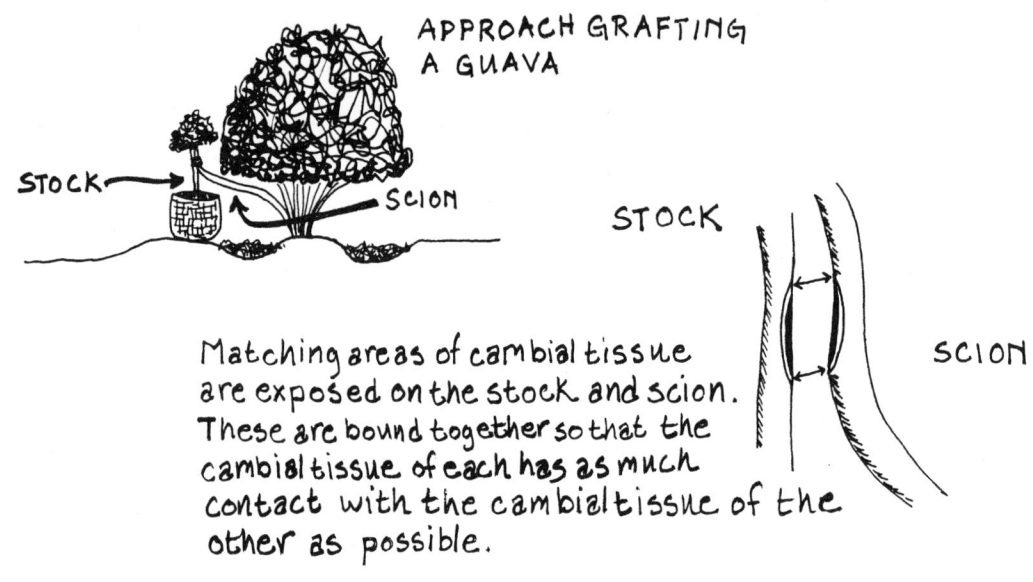
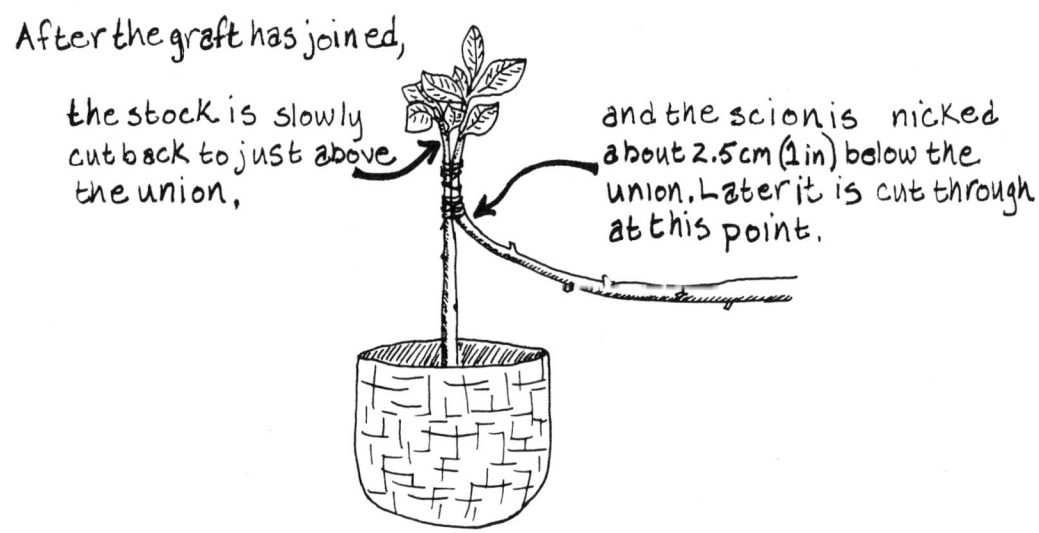

Figure 7.14 Approach Grafting (After Dupriez and De Leener 1987:234)

are mango, guava, citrus, jujube, and the stone fruits.

T-BUDDING The *T-bud* or *shield* is one of the quickest, easiest, and most reliable budding methods. However, if the stock plant is stressed by drought, pests, or other problems it will not work. This is because it will not be possible to separate the bark from the cambium and xylem (often referred to as the *wood*), as is necessary for this method.[25] If this is the case, chip budding may be more appropriate because it does not require separation of the bark and wood.

In T-budding the stock is prepared by making a vertical cut about 3 cm (1.25 in) long through the bark and a horizontal cut, extending about one-third of the distance around the tree, intersecting the vertical cut to form a T (Figure 7.15).[26] The scion bud is cut out by making a horizontal cut through the bark and into the wood about 2 cm (1 in) above the bud. Then an upward slice is made to meet the horizontal cut, starting about 1.5 cm (0.5 in) below the bud and slightly into the wood to ensure that the bud is included. This bud piece is then pushed down, under the flaps of stock bark until the horizontal upper edges of stock and scion are flush. The bud piece should fit snuggly and be covered by the two flaps of bark. Then the wound is bound and tied shut, avoiding any direct pressure on the bud.

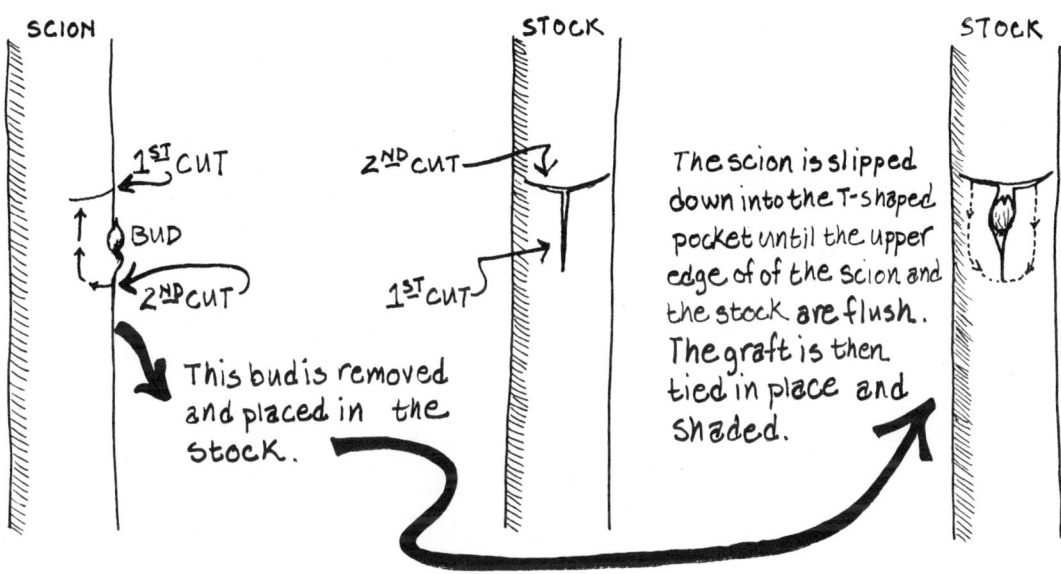

Figure 7.15 T-Budding

CHIP BUDDING   As already mentioned, *chip budding* does not require separation of the bark from the stock, and so can be used when the bark will not separate, for example, when the weather is dry. Instead, this method relies on the contact of two flat surfaces. The stock seedlings used are frequently still quite young; mangoes, for instance, are only three weeks old.

The stock is prepared by making a horizontal slice below the bud that is to be removed. A second slice starting above that bud, running around and behind it, and meeting the first cut will free the piece from the stock. This piece can be composted. This exposes an area of woody tissue ringed by cambial tissue and bark. Continuing the vertical cuts slightly below the horizontal one on the stock will create a lip on the lower end of the stock cut. A scion bud can be cut out following these same steps and should be of matching size with the exposed area on the stock. The scion is placed on the stock so that their cambial tissues are in contact, with the lower lip of the stock helping to keep the scion in place (Figure 7.16). The graft is then tied and carefully wrapped to cover all exposed cuts.

CARE FOR THE BUDDED SEEDLING   The budded seedling should be shaded and watered during the healing process. Tying some leaves above the grafted bud will give it protection from the hot sun (Figure 7.16). About 10 days after grafting, whatever material has been used to wrap the graft should be cut to avoid constricting the bud. Once the graft has been successfully joined, which can take about 2 to 4 weeks, the focus of the plant's growth must be shifted from stock to scion.

In many plants a condition of *apical dominance* exists; that is, the top or *terminal bud* is the center of growth, supressing growth in any lower buds. To overcome apical dominance the top of the stock may be bent over and tied or pegged at a level lower than the graft. This focuses growth on the scion while still using the stock as a source of food until the graft is more established, at which time the stock can be cut off just above the graft. Other methods are to bend and partially break the stock, or simply cut it off above the graft immediately after grafting.[27]

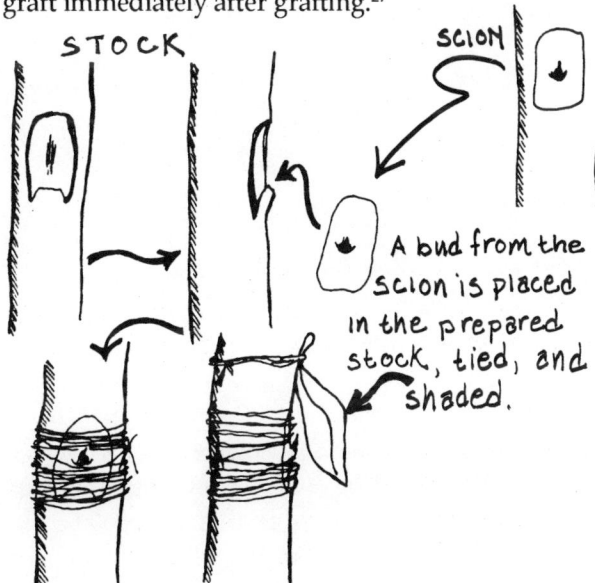

Figure 7.16 Chip Budding

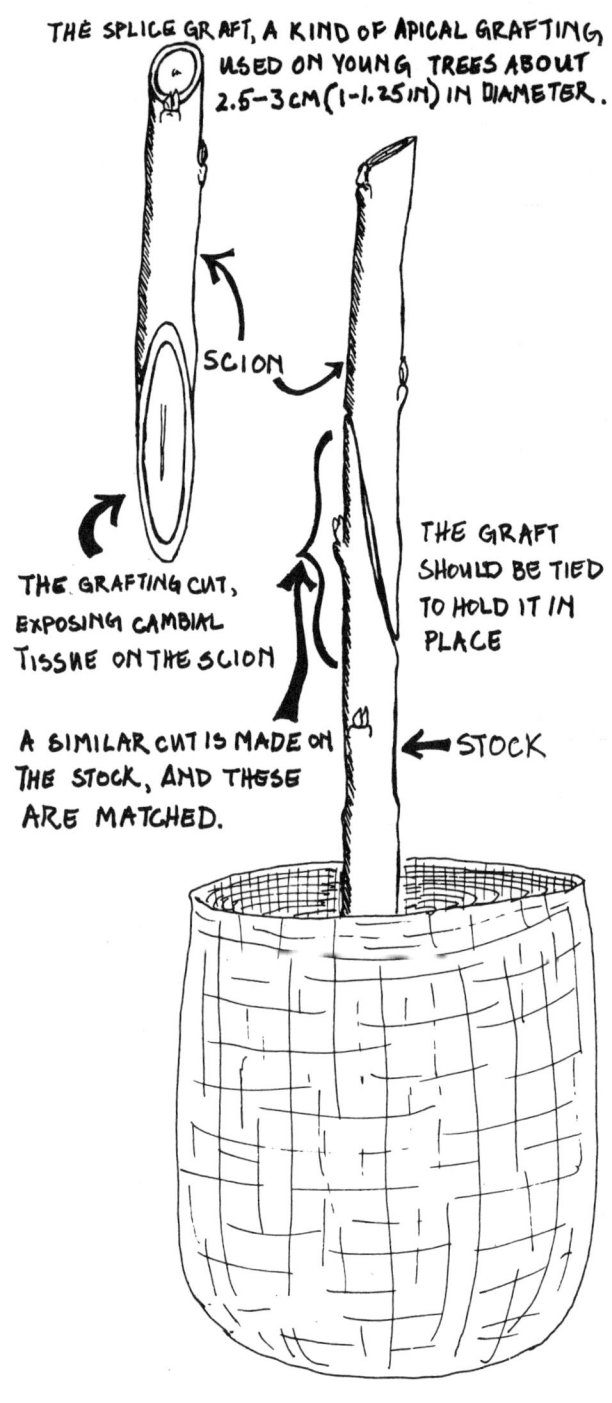

Figure 7.17 Apical Grafting

7.6.5 Apical Grafting

Apical grafting is a technique in which there is no stock above the grafting point. It is used on dryland garden trees such as olive, peach, and citrus. For apical grafting the scion should include several buds. Matching areas of cambial tissue on stock and scion are joined (Figure 7.17). If the stock is larger than the scion, only a portion of the cambial tissue can be joined (Figure 7.20 with Box 7.2). The graft is then tied into place. Because of the relatively large amount of surface area exposed in this kind of graft it is important to cover it with a protective seal, such as the clay and dung mixture described in Box 7.2.

7.6.6 Topworking

Topworking is done on trees that are more mature than those used in the three grafting methods already described. Citrus, olive, avocado, and mango are examples of dryland garden trees for which topworking may be appropriate. There are many different ways to do topworking. One method used when younger trees are the stock is to remove all branches except the structural limbs and lateral branches 7.5-10 cm (3-4 in) in diameter. Scion cuttings with at least three buds and with lower ends cut into a wedge shape with one side longer than the other are inserted into cuts made near the base of the stock tree's lateral branches (Figure 7.18). The ends of the stock branches on which a graft has been placed are then cut back at an angle just above the graft. The tips of other stock tree limbs are cut back so they are not higher than the highest scion.

The advantage of topworking is that the established stock tree provides the root and branch structure; this allows the scion to start fruit production more quickly than in other forms of grafting. Shading is very important for topworking as with other types of grafting, though it may be more difficult because it involves protecting a whole tree. Sometimes a thin coating of a clay and water mixture is applied to protect the scions from sunburn (Box 7.2). Leaves or palm fronds tied around the main limbs also provide shade. As soon as the scions start to leaf out and grow they will provide some shade. After the graft is well joined, the branch angle spreading technique described in section 8.7.2 can be used. To ensure the establishment of the grafts, all new stock growth such as suckers and other shoots should be removed as soon as they appear.

7.7 Layering

Layering is a method of vegetative propagation that encourages the growth of adventitious roots from branches or shoots, eventually producing a viable plant that may be separated from the parent and transplanted elsewhere.

7 Vegetative Propagation

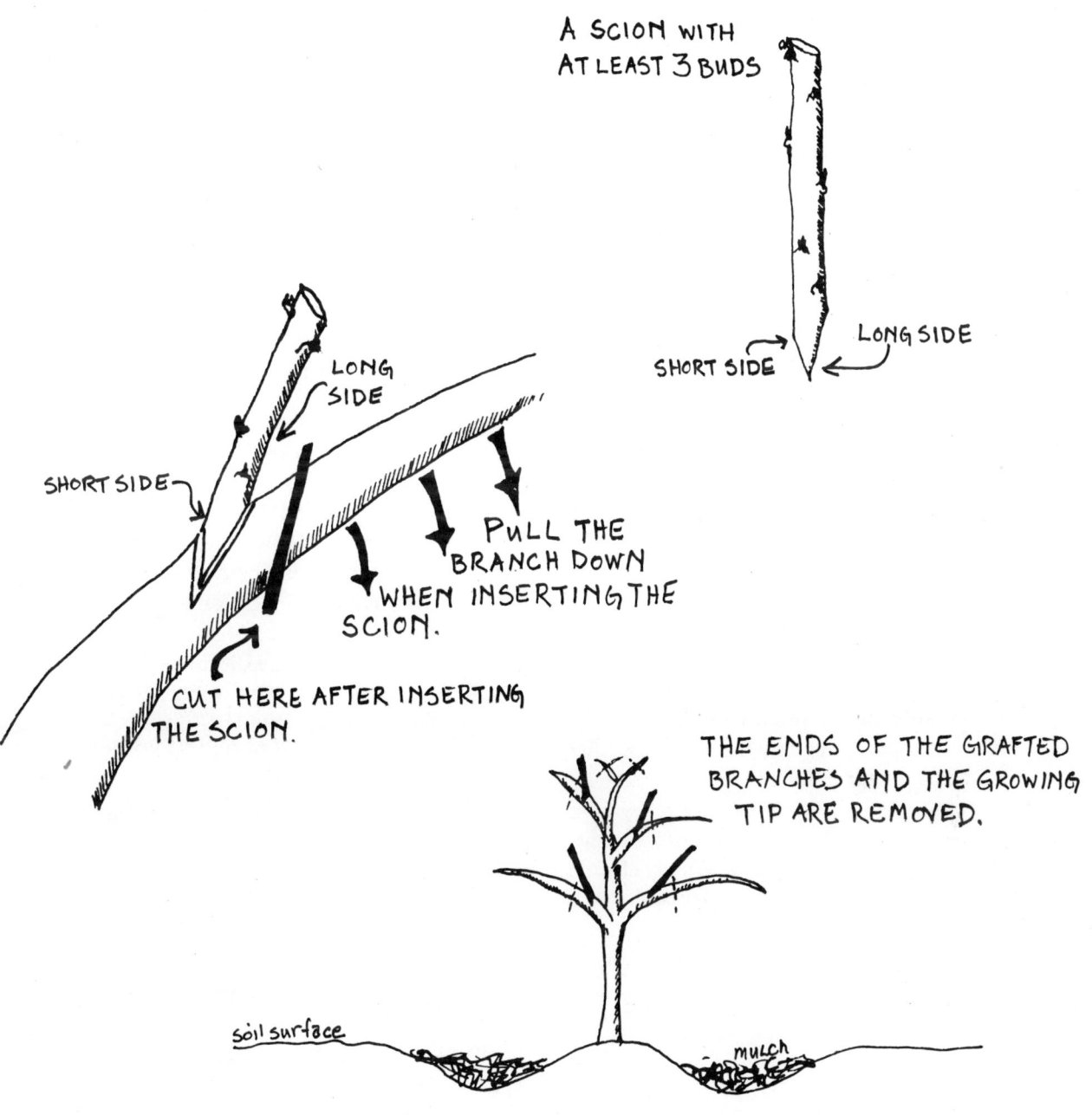

Figure 7.18 Topworking a Tree with Stub Grafts

By bending, girdling or cutting, the movement of food and hormones through the phloem down to the roots is interrupted. Root development is encouraged where these substances accumulate, just above the cut or bend. The flow of water and minerals from the roots to the layered part continues through the intact xylem. Rooting is also encouraged when the stem is hidden from light by being wrapped or covered with soil.

If simple layering is to be done, layering material such as shoots and low-hanging branches should be allowed to grow. Severe pruning one year before

Box 7.2
Guidelines for Grafting

- When choosing stock and scion keep in mind that:
  - Plants in the same species are usually but not always compatible.
  - Differences in maturity between stock and scion are not usually a problem, although young seedlings for stock and growth less than 2 years old for scion, graft more readily.[28]
  - Both stock and scion should be alive and healthy.
- The best scion material:[29]
  - Is young, less than 2 years old, with short internodes.
  - Has healthy vegetative buds, not flower buds. Vegetative buds are narrower and more pointed than flower buds. The buds should not be in their most active growth phase; if they are the scion will leaf out immediately and the leaves will transpire and use up the water available to the scion rapidly, causing it to dry up and die.
  - Is from the upper part of the tree but includes only the bottom two-thirds of the shoot.
- The best stock are young, vigorously growing plants. For topworking however, they are often older than for other grafting techniques. Although rooted cuttings or layered plants can be used, seedlings produce the strongest rootstock plants.
- Both stock and scion should be washed clean of dirt and debris before the cuts are made.
- For surface cuts in bud grafting, making the starting and ending slices before doing the entire cut avoids a sloppy cut which can strip off bark (Figure 7.19).
- The cutting tools should be sharp to give a controlled, clean cut without crushing the plant's tissues and leading to infection.
- Cleaning cutting tools with soap, bleach, alcohol, or vinegar between cuts avoids spreading diseases.
- The scion pieces should be oriented properly. The upper (farther from the roots) and lower (closer to the roots) ends of the scion should be identified and kept in the same position in the graft. This is easy to do because the buds always grow up, with their pointed ends aimed away from the ground. If the scion is turned upside-down the graft will fail.
- Matching the cambial tissue of stock and scion of similar size is relatively easy. If stock and scion are of different sizes or stages of development the inner edges of their bark must be matched to ensure contact of cambial tissues (Figure 7.20).[30]
- Once the graft is in place, it can be secured with plant fibers, leather strips, rubber bands, pieces of plastic, or string. These ties must be removed when the union has healed so they will not constrict the growing tree.
- Under hot, dry conditions it is particularly important to see that the graft and any exposed cut surfaces do not dry out, killing the cells and preventing a union. A mixture of two parts clay and one part fresh cow dung can be used,[31] and adding some plant fibers or animal hair to the mixture will make it stronger and less likely to crumble away as it dries. Heated beeswax, painted over the ties that hold a graft in place, can be an effective seal. Melted paraffin wax can also be used although it has a lower melting temperature than beeswax. However, melted wax should be applied carefully to the area of the wound to avoid burning the graft tissues. In some extremely hot drylands wax seals may not work because they will melt off. Other useful materials include plastic, which is wrapped around bud grafts on mango trees in Mali, moist cloth, and possibly local tree resins.

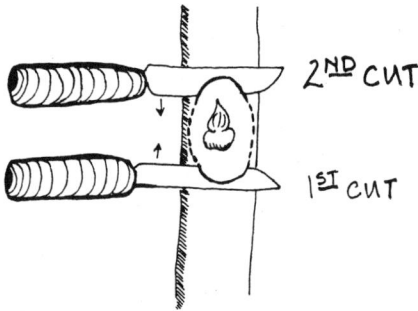

Figure 7.19 Cutting Grafts to Avoid Tearing the Bark

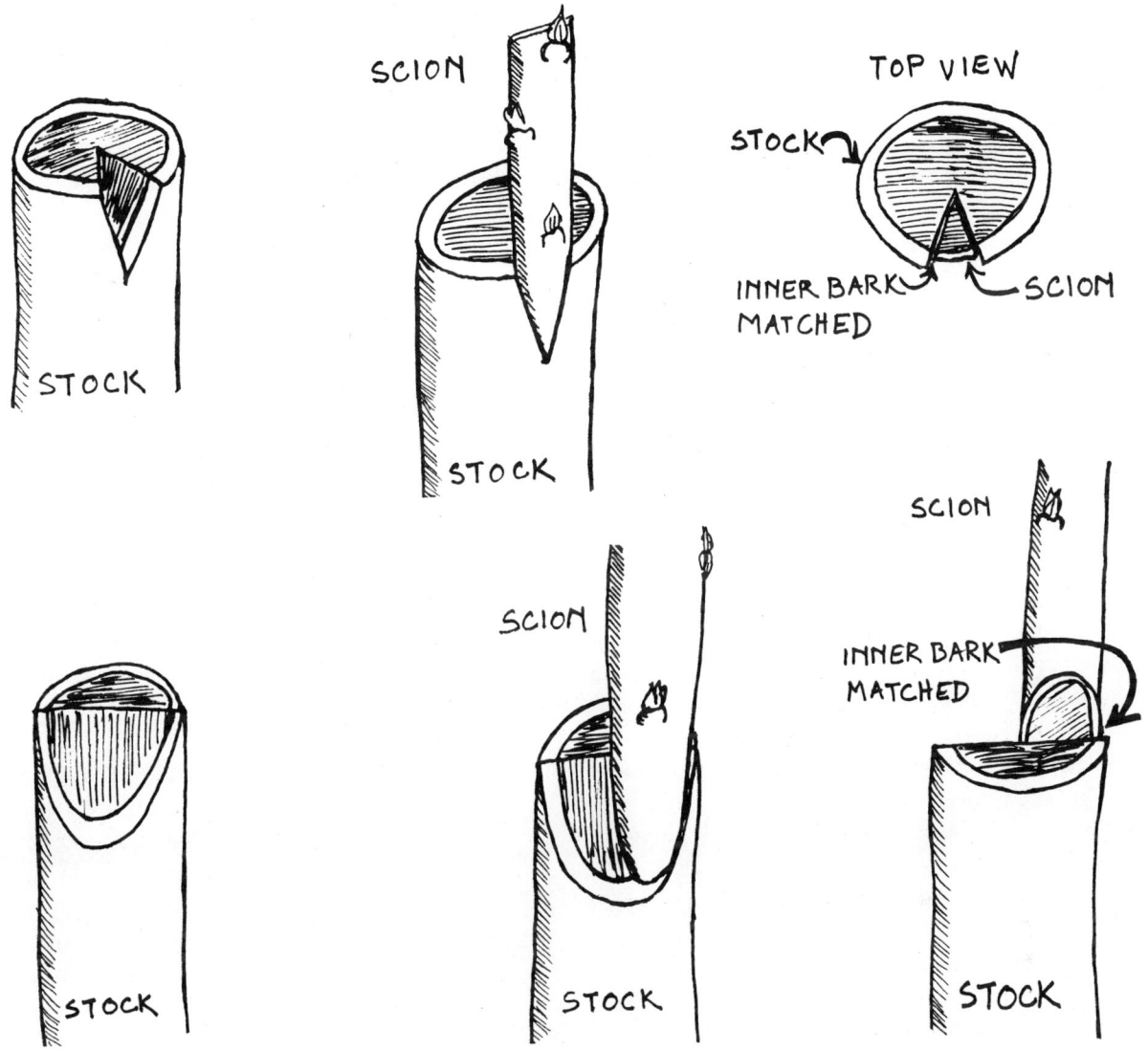

Figure 7.20 Matching the Cambial Tissue at the Inner Edge of the Bark
(After Garner, et al. 1976:86)

layering encourages the growth of young shoots around the plant's base which can be used for layering.[32]

Deciduous plants such as figs and grapes are layered during the dormant season and so are prepared to take advantage of the increase in vegetative growth that follows. Evergreen trees like the cashew, carob, and mango can be layered just before or during a period of rapid vegetative growth. Perennial herbs such as mint, thyme, and oregano can also be propagated in this way.

7.7.1 Simple Layering

For simple layering (Box 7.3), vigorous shoots from around the parent plant's base work well because they bend easily and tend to root quickly. The shoots used should be about 1 year old and sufficiently long to produce new roots far enough away from the parent that digging them out will not harm the parent's own roots. Branches are also used but because they are older tissue than shoots, they are stiffer to bend and slower to produce roots.

7.7.2 Air Layering

Another kind of layering called *air layering*, or *marcottage*, is based on the same principles as simple layering but is done on shoots that are not buried in the soil (Figure 7.22). Young shoots, about 1 year old with leaves and leaf buds, growing in an exposed part of the

> **Box 7.3**
> *Guidelines for Simple Layering*
>
> The following are guidelines for simple layering in drylands.
>
> a) Vigorous shoots with numerous leaf buds that can be easily bent to the ground are selected.
> b) The area around the parent plant should be cleared of weeds and other debris.
> c) Organic matter is worked into the soil where the layering will be done. Adding sand to clayey soils improves drainage.
> d) A small depression is dug in which to place the shoot and to concentrate moisture.
> e) All leaves are removed from the layering material except those on and around the tip, which is gently bent to the ground, and all buds on the upper surface are removed. In the case of trees, some sort of wounding is usually done to encourage the production of roots in the buried portion of the shoot. This can be done by girdling or making several cuts through bark and cambial tissue on the shoot's lower surface. Filling these cuts with soil prevents the wound from growing shut (Figure 7.21).
> f) The layering material is staked or weighted down so that it forms a U-shape or at least a right angle, with the growing tip extending up vertically. If layering trees, wooden or metal stakes, rocks, or other heavy objects, can be used to hold down the branches or shoots. Freshly cut wooden stakes can sometimes encourage the growth of fungi and other materials are better. The shoot is buried with compost-enriched soil, leaving the tip exposed above ground.
> g) The layered shoot should be watered generously, and the area kept moist. However, it must not get waterlogged as this will cause the shoot to rot. Under hot, dry conditions the shoot should be mulched and shaded to protect it from drying and high temperatures, both of which inhibit rooting.
> h) After the growing season or when the shoot tip appears to be growing, well-established and vigorous, the shoot can be cut off from the parent plant. This can be done all at once if the shoot is very vigorous, has well-developed roots, and the environmental conditions are not stressful. However, making the separation gradually, first nicking the parent branch, cutting more and more and then completing the cut over a few weeks minimizes shock to the shoot. Cutting as close as possible to the original rooting point without damaging the new shoot's roots is best (Figure 7.21). Once cut, the new plant is left in place two to three weeks before transplanting, allowing it to recover from the separation.

plant are best. The shoot is girdled, removing a ring of bark approximately 2.5 cm (1.3 in) wide near the base. Care should be taken not to injure the xylem as this will weaken the shoot and interfere with the flow of water and nutrients to the shoot tip. The girdling wound is surrounded with a light rooting mixture which can be made of compost, sand, sawdust, and soil. This mixture is moistened and then wrapped in plastic, cloth, or bark and leaves, and bound in place. During the rooting period the soil mixture must be kept moist; shading can help. Rooting can take from 30 to more than 100 days after which the shoot should be slowly cut from the parent plant and transplanted, preferably during a period of slow growth. Air layering is frequently used with avocado, cashew, and mango trees.

7.8 Resources

For information on grafting, talk with local gardeners, farmers, and extension workers who are experienced grafters. For a more formal, printed, source of information on grafting, layering, and other means of propagating fruit trees, *The Propagation of Tropical Fruit Trees* (Garner, et al. 1976) is a useful book. While it focuses on the commercial production of a limited number of common tree crops, it still provides some useful basic information about propagation. However, the technical information is often not relevant for the gardener or small farmer. *Plant Propagation* by Hartmann and Kester (1983) is a widely used source on plant propagation in the English language. The background infor-

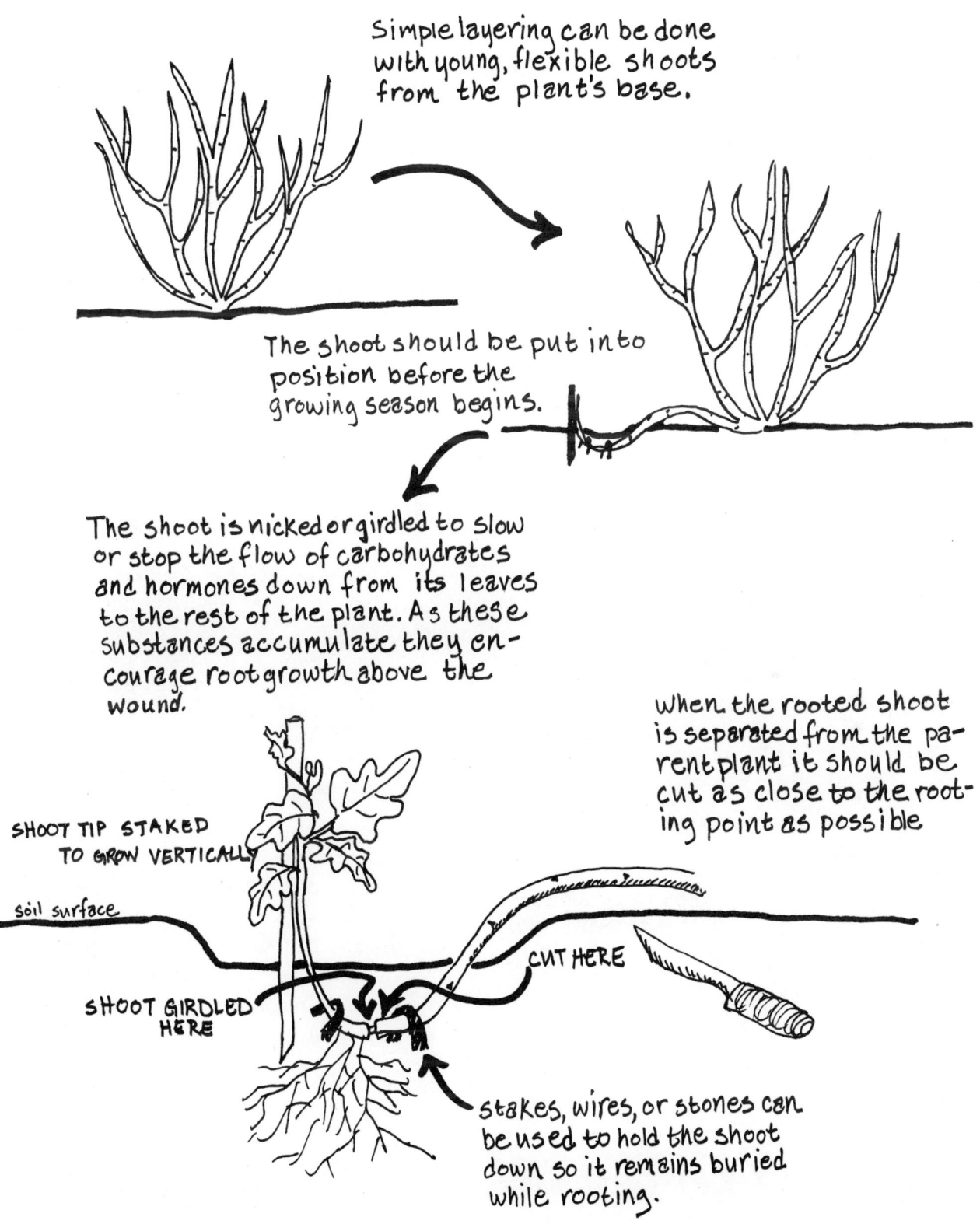

Figure 7.21 Simple Layering of a Fig

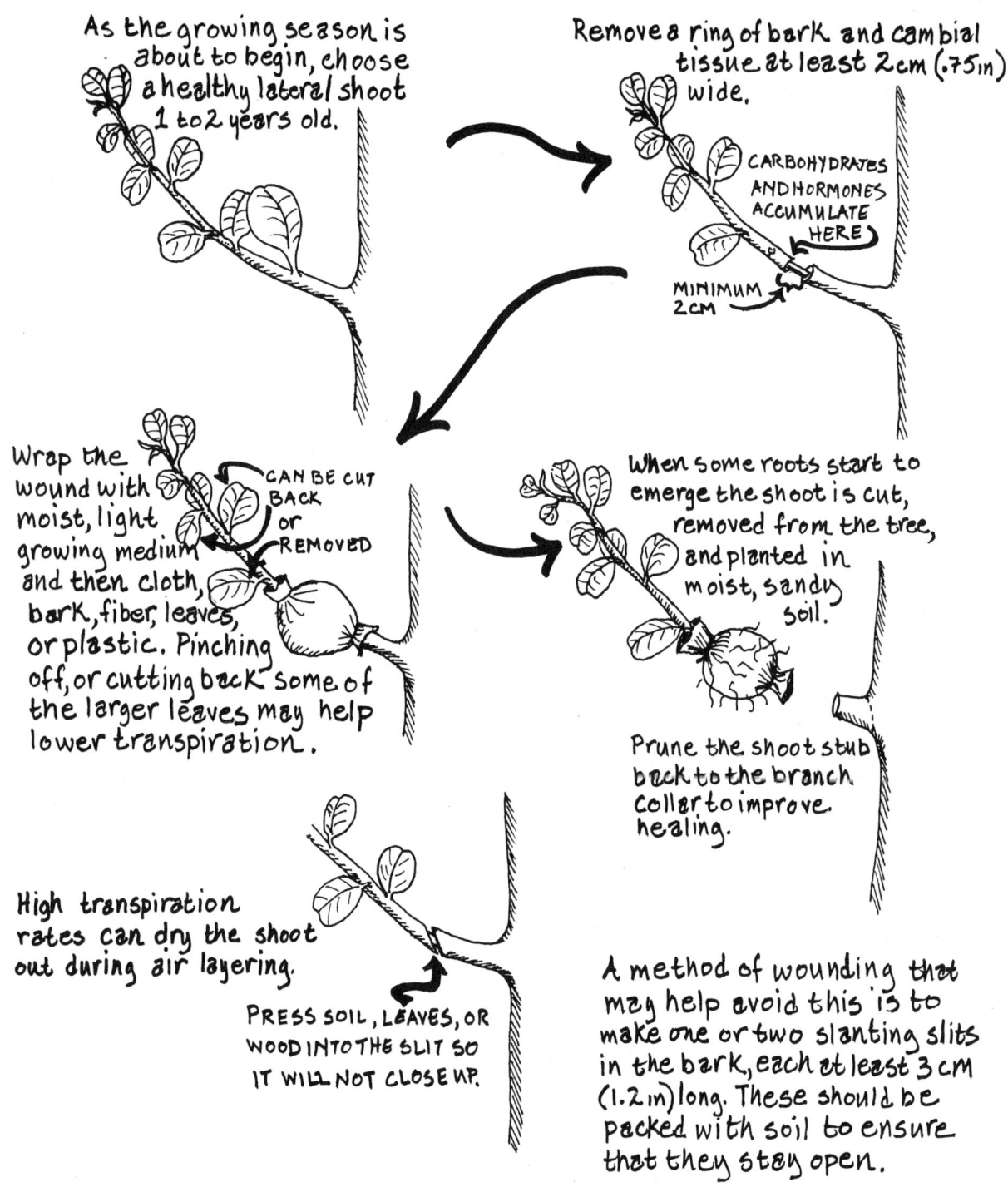

Figure 7.22 Air Layering, or Marcottage, of a Cashew

mation on the anatomy and physiology of different propagation methods is its most valuable contribution to gardeners. The methods described are frequently only relevant for large-scale commercial propagation and many of the specific examples are of temperate-region ornamentals. Aumeeruddy and Pinglo (1989:36-48) give a number of very brief examples of indigenous techniques of vegetative propagation.

References

[1] Nokes 1986:38.
[2] Nokes 1986:42.
[3] Hartmann and Kester 1983:205-206.
[4] Hartmann and Kester 1983:260.
[5] Garner, et al. 1976:61.
[6] Hartmann and Kester 1983:304.
[7] Garner, et al. 1976:65.
[8] Hartmann and Kester 1983:278.
[9] Kassam 1976:54.
[10] Acland 1971:36.
[11] Kassam 1976:56.
[12] Kassam 1976:66.
[13] Irvine 1969:185-187.
[14] Kassam 1976:83.
[15] Hartmann and Kester 1983:485.
[16] FAO 1982b; Popenoe 1973.
[17] Evenari, et al. 1982:364.
[18] Hartmann and Kester 1983:368.
[19] Hartmann and Kester 1983:374-376.
[20] Hartmann and Kester 1983:371.
[21] Hartmann and Kester 1983:380.
[22] Hartmann and Kester 1983:382.
[23] Garner, et al. 1976:108.
[24] Hartmann and Kester 1983:449ff.
[25] Hartmann and Kester 1983:448.
[26] Hartmann and Kester 1983:456-459.
[27] Garner, et al. 1976:124.
[28] Garner, et al. 1976:84.
[29] Hartmann and Kester 1983:427.
[30] Garner, et al. 1976:85-80.
[31] Garner, et al. 1976:90.
[32] Garner, et al. 1976:39.
[33] Garner, et al. 1976:42-48.

8
Plant Management

Young plants are tender and especially vulnerable to stress and a plant of any age needs special protection for a period after being transplanted, especially under hot, dry conditions. However, once a garden plant becomes established in its final growing site it needs less attention. As the plant matures and the root system grows, it is better able to withstand heat, drought, disease, and pests, and will not need to be watered as frequently. At this point the main work in the garden is maintaining the plant's health to ensure a good harvest.

8.1 Summary

Seeds or plant parts for vegetative propagation can be planted in nursery beds, containers, or directly in the garden. Starting plants in places other than their permanent growing site can offer such advantages as convenience for the gardener and special conditions that encourage and protect young plants. Eventually all plants started in temporary growing sites must be transplanted. The choice of a permanent growing site for garden crops depends on the gardener, the crop, and the growing environment. Transplanting requires care especially where hot, dry conditions add to the stress that the plant experiences. Weeding, pruning, and trellising are management practices that help maintain and improve the health and productivity of the garden.

8.2 Nursery Beds and Container Planting

Seedlings and recently planted cuttings require extra care and attention. To make this work easier they are sometimes started in nursery beds and containers, temporary growing places from which healthy young plants can be transplanted into the garden. Whether nursery beds or containers are useful or not depends on the type of plant being propagated, convenience for the gardener, and environmental conditions such as temperature, wind, rainfall, pests, and disease.

8.2.1 Nursery Beds

Nursery beds are garden beds devoted entirely to starting plants, both from seed and by vegetative propagation. They may be in the garden itself or elsewhere, but wherever they are located, nursery beds are planned and planted as the temporary location of the plants, not their final growing site (Figure 8.1).

One indigenous example comes from central Mexico where a system of canal-fed gardens is used and special nursery beds are constructed for starting grain amaranth seedlings during the dry season.[1] A rectangular bed measuring approximately 2 m x 15 m (6.5 ft x 50 ft) and 4-5 cm (1.5-2 in) deep is filled with rich, muddy soil scooped up from the canal bottom. The next day as the soil dries it is cut into 3 cm x 3 cm (1.25 x 1.25 in) squares with a special slicing rake called a *cuchilla*. Using a small stick or their fingers, the gardeners make a small hole 1 cm (less than 0.5 in) deep in the middle of each square and then drop amaranth seeds in. The bed is sprinkled with manure which is then swept off so only that which falls into the planting hole remains, covering the seeds. Twenty to 30 days later the seedlings in their soil cubes are transplanted into the garden beds. Where clayey soil is available gardeners can try making soil cubes to start transplants (section 8.2.2).

Nursery beds allow the gardener to concentrate and save resources such as water and time, and to take

Figure 8.1 Nursery Beds for Starting Seedlings for the Garden

advantage of favorable microclimates for early- or late-season germination. For example, in temperate drylands warm-weather plants can be started in sunny, protected areas before the end of the cool season. Moving nursery beds seasonally or yearly helps avoid problems with nematodes which can build up in continuously cultivated soil (section 13.3.2).

Steps in selecting and planting a nursery bed are:

a) A site is selected that is convenient for daily care, easy to protect from wind, sun and pests, and has enough room for plants to grow to the stage at which they can be transplanted.

b) Soil and bed are prepared as described in Chapter 9 except that the soil need not be as deep because plants will not grow to maturity here. For most annuals 15-20 cm (6-8 in) is deep enough. When establishing a nursery bed for trees, the length of time required for them to become strong enough to transplant needs to be taken into account. This will affect where the nursery bed is located and how deeply the soil should be prepared.

c) Planting depressions are made (see section 6.4.2).

d) Seeds may be sown more thickly than in a permanent bed because plants will be removed and dispersed before maturity. However, enough room must be left so that individual plants can be removed for transplanting without major damage to the roots.

e) The sown seeds are covered with soil and possibly mulch. The soil is kept moist by gentle watering as needed. Shades, windbreaks, and bird-scaring devises are often used.

8.2.2 Container Planting

Containers offer many of the benefits of nursery beds with the added advantage of being easily moved to adjust to changes in sunlight, shade, temperature, and the gardener's schedule. Containers are often used for starting cuttings or seedlings for transplanting. Some plants, whose leaves or small fruit are continually harvested in small quantities for seasoning or medicinal purposes, are kept permanently in containers. Examples include mint, basil, oregano, marjoram, rosemary, epazote, and chili peppers.

Containers may be the only option for urban gardeners with little space and no access to land. For them, permanent container plantings on rooftops, in windows, or on balconies can provide herbs, condiments and some greens for the household meals (Figure 8.2).

CONTAINERS A wide variety of locally available and free or inexpensive containers can be used for planting including calabashes, pots and pans with holes in the bottom, baskets, steel cans, wooden boxes, trays, cardboard cartons, plastic bottles, and other made, found, or discarded materials (Figure 8.3). Galvanized steel containers are not good because this material gives off zinc salts toxic to plants.[2]

Baskets made of leaves or other plant fibers make excellent containers for eventual transplants. Often there is no need to remove tree seedlings from a basket container when transplanting, because the basket will eventually decompose in the soil, thus avoiding disruption of the tree's root system. However, the basket fibers should be cut through and pulled open in several places before planting the tree. This is especially true if the basket is tightly woven or made of a tough, woody material. Planting containers should not have been used for paint, fuel, pesticides, or other substances that can harm the seedling, contaminate the garden where it is transplanted, or poison people and animals.

Soil cubes are good for starting transplants because the seedling's roots are not disturbed by removing the soil from a container. The soil cubes described in section 8.2.1 are slices of a special type of nursery bed, but soil cubes or blocks can also be made and used like containers. One way to make the blocks is to press the wet soil into a form such as a container or even a hole in the ground. Then as it dries, it is sliced into squares which will be cubes when removed from the form. However, as already mentioned, the soil used must have the right clay and humus content to hold together

Figure 8.2 Container Gardening in Mexico City

Figure 8.3 Containers

enough so that the gardener can move them without difficulty. However, very small containers, such as a cup, are not appropriate for most plantings as they offer little room for the plants to grow and insufficient soil from which to receive nourishment. Also, the larger the container the more soil there is for holding water, and the less frequent the waterings need be. Containers should be deep enough to allow the seedling's roots to develop to the stage when they can withstand being transplanted. To allow transplanting, the top opening must be as big, but preferably bigger, than the bottom, making it easy to remove seedlings without damaging their roots (Figure 8.4). Vertical ridges on the inner surfaces of containers prevent roots from growing in a spiral pattern which results in root-bound seedlings (section 8.4.4). A community garden or tree nursery project that includes local manufacture of containers can design them with a cone shape and vertical ridges on the inner surface (Figure 8.5).

in blocks until it is time to transplant. Timing of transplanting is also important, because if the seedlings' roots grow out of the soil block they have no protection and will dry up or be injured when transplanted.

CONTAINER SIZE Containers can be of any size and may hold one or more plants, but should be small

TEMPERATURE REGULATION The relatively small amount of soil in containers can dry out rapidly and the soil temperature can rise and fall to extremes of hot and cold, all of which are harmful to plant growth. Containers with thick walls provide better insulation, reducing temperature fluctuations. Under sunny, hot conditions light-colored containers are good because they reflect sunlight, and thus heat, away from the soil. In the cold season or in cold areas, dark-colored containers that absorb sunlight and heat may be better. A container can be shaded and insulated by wrapping it with cloth, or by mounding soil, compost, or leaves around it.

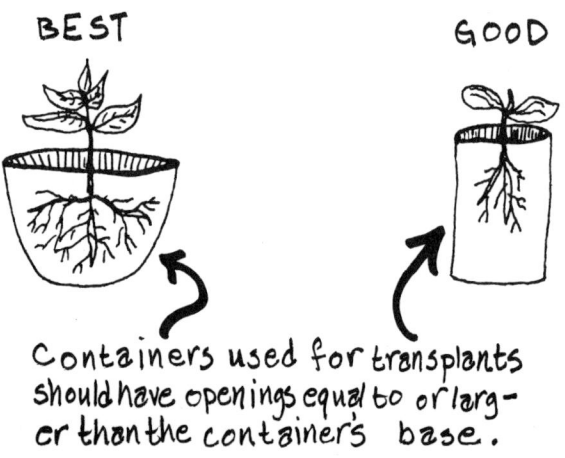
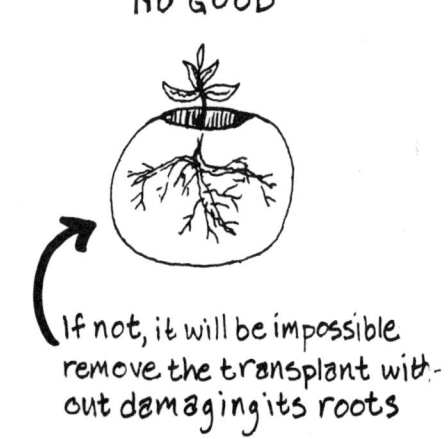

Figure 8.4 Container Openings

8 Plant Management 127

WATERING AND DRAINAGE Containers must allow drainage of excess water or else waterlogging (section 12.6.1) and damping-off (section 13.4) will occur. Holes can be made in containers of most materials except glass and fired clay, which are so brittle that holes cannot be made in them without breaking. In addition to holes, a few small stones or pieces of broken clay pots in the bottom of containers can improve drainage.

Salts in water or soil can accumulate on unglazed ceramic containers, appearing as a white patina or coloring on the outside surface. If this occurs, the empty container can be submerged in water for several days and scrubbed to remove the salts.[3] Using rainwater will reduce salt buildup in locations where this is a problem with other water sources.

SOIL MIXTURE Filling the container about three-quarters full leaves room for adding enough water to soak down to the plant's root zone (Figure 8.6). In Chapter 9 soil qualities good for the garden are discussed. The same qualities are good for container soil mixtures but with an even greater emphasis on good drainage. The planting mixture must be able to hold some moisture but both soil and container must allow good drainage because water-saturated soil encourages disease problems. How the mixture is made depends on the local soil. Two helpful guidelines are:

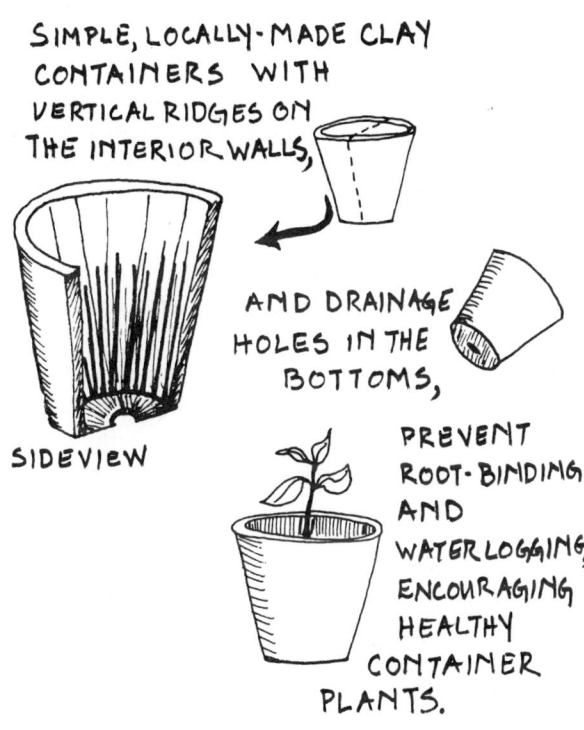

Figure 8.5 Ridges on the Inside of Containers

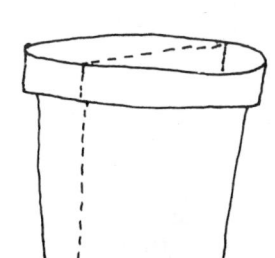

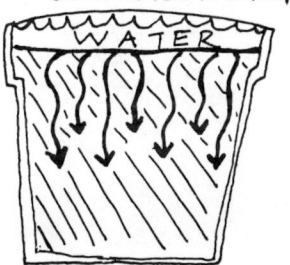

Figure 8.6 Soil Depth in Containers

- Well-composted organic matter is the best soil amendment both to open up the structure of a heavy clay soil and to improve water-holding capacity of sandy soil.
- Keeping the top 2-3 cm (1-1.5 in) of the soil a sandy texture allows quick infiltration and keeps water from gathering around the stem base where it encourages disease. In heavier soils, adding some sand as well as organic matter improves drainage.

If soil-borne diseases are a problem, heating small amounts of soil, called *soil sterilization*, can help. Heating moistened soil to 71°C (160°F) for about 30 minutes kills most bacteria and viruses.[4] In areas with high daytime temperatures, moist soil could be put in a covered metal pot or closed plastic bag and left to heat up in the sun for several days. Because heating the soil also kills beneficial microorganisms it should only be used when absolutely necessary.

Before planting, the soil in the container should be leveled and wet throughout. After this, drainage can be tested by adding water up to the container's rim. If there is still water standing on the soil surface after about 15 minutes, the drainage needs to be improved. The drainage openings in the container may need to be enlarged or more organic matter or sand added to the soil.

PLANTING In large containers the gardener can make planting depressions for seeds using her fingers or a stick. Small containers concentrate water on seeds so there is no need for planting depressions. In small containers seeds can simply be sprinkled across the soil surface (Figure 8.7), covered with a layer of dry garden soil (section 6.4.5), pressed down firmly, and the planting area gently watered.

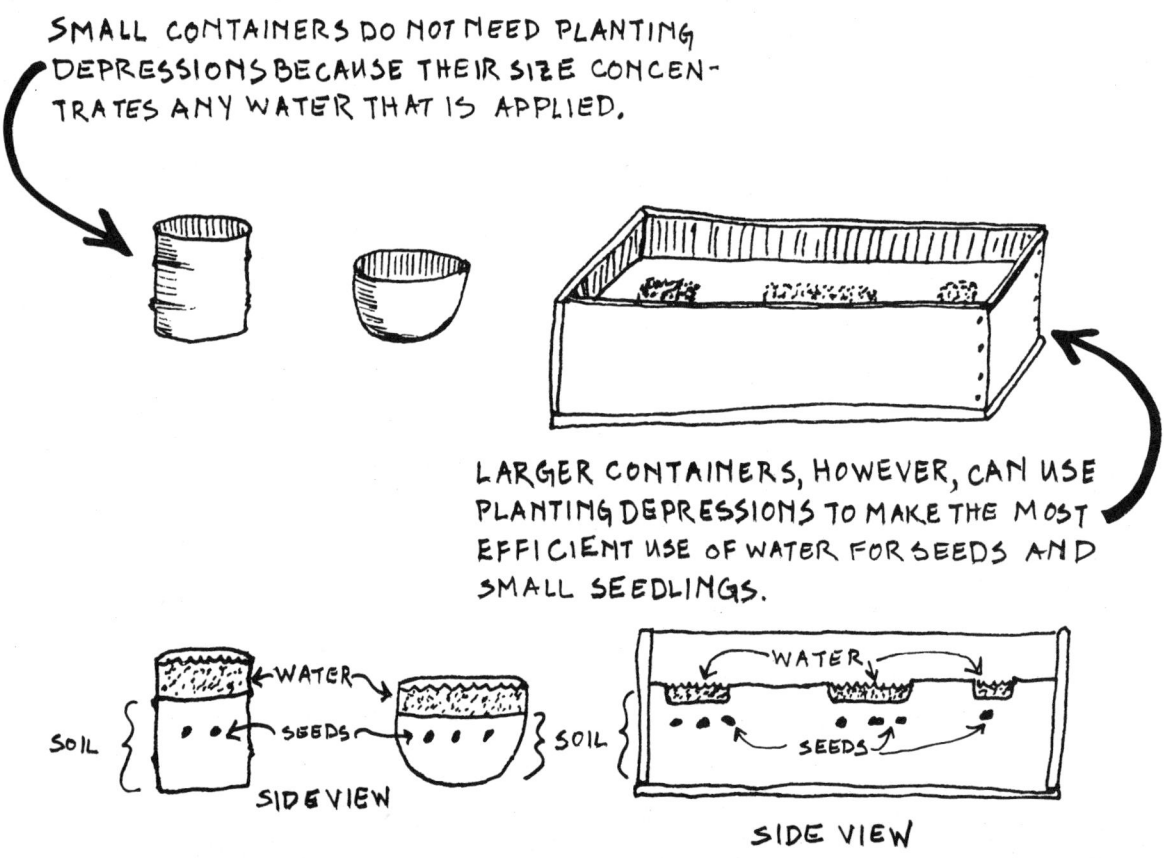

Figure 8.7 *Planting in Containers*

PLACEMENT AND CARE Containers should be kept in a convenient location with adequate shade and sunlight. Sometimes the drainage holes become clogged by the surface on which the container is resting. If this happens putting gravel under the container can help by making air spaces that water can pass through to drain out of the wet soil in the container. The soil surface should be allowed to dry slightly between waterings (section 6.5.1).

8.2.3 When Direct Planting is Better

Sometimes planting seeds directly in the garden is more appropriate than container or nursery bed planting. The primary advantage of planting directly in the garden is that the plants will not be moved, and therefore their growth will not be disturbed by hardening-off (section 8.4.4) or transplanting shock. For example, seedlings of cucurbits and some root crops may die or their growth may be severely slowed by transplanting. Direct planting also eliminates work for the gardener. The gardener's schedule or the garden location may make it convenient to care for the seedlings right in the garden with less work, and the garden environment may be good for starting seedlings.

These advantages must be weighed against the possible disadvantage of direct planting: increased vulnerability to pests and the elements because management and resources are not as concentrated as with nursery beds or containers.

8.3 Planting Sites and the Sun

The location of plants in relation to the sun should be considered when direct planting or transplanting. Garden plots and crop mixes can be planned to protect tender plants from the harsh midday and afternoon sun (section 10.8.3). For example, sunflowers, sorghum, okra, or other heat-tolerant, tall crops can shade melons, tomatoes, chilis, and other more sensitive crops. Some garden crops like prickly pear cactus do better when transplanted with the same orientation to the sun that they had before being moved, so that the north-facing side is still facing north.

Trees and other perennials will shade larger and larger areas as they grow bigger, and this should be planned for. For example, in subtropical drylands the winter sun warms the garden, house, and work area. If a nondeciduous tree such as a carob, olive, or loquat is planted in the southern part of a northern-hemisphere garden or the northern part of a southern-hemisphere garden, it may block the low winter sun, keeping surrounding areas in shadow. A deciduous tree would lose its leaves in the cool season, allowing the sun to warm the area, but then leafing out and providing shade in the hot season (Figure 8.8).

8.4 Transplanting

Transplanting can be done while thinning plants, or to move seedlings from containers or nursery beds to permanent locations in the garden. It may also be used for moving mature plants. The chances for successful transplanting can be improved if timing, the transplant site, water availability, and preparation of the transplant itself are planned.

8.4.1 Timing

Deciduous trees like figs, the stone fruits, pomegranates, and jujubes are best transplanted when they have lost their leaves and are dormant. This minimizes stress and takes advantage of the surge of growth after dormancy to help the transplant become established. If moved during this time, young deciduous perennials can be **bare-root** transplanted, that is, they can be transplanted without their root ball covered with soil.

Nondeciduous trees are best transplanted before a period of growth, for example, at the start of the rainy season. Avoid transplanting any plant just before or during a time of environmental stress such as the middle of the hot, dry season. Time of day is also important for transplanting. When it is hot and dry the evening is the best time for transplanting because it allows the plant a cooler, dark adjustment period when transpiration rates are lowest. Cloudy days are also good times for transplanting.

Of course, the most important consideration for timing transplanting is the gardener's schedule. Food preparation, field work, and marketing may take priority over gardening, so transplanting must be planned around those activities.

8.4.2 The Site

Planning ahead for the growth of garden plants helps in choosing the location for a transplant. A sketch of the garden layout is a good tool for thinking about different transplant sites and for anticipating the effect

Figure 8.8 Planting Sites and the Sun South of the Equator

on space, shade, and sunlight. This is especially true for perennials. When the site is chosen, holes are dug according to the needs of each transplant. The bigger the hole the better, because the less energy the plant must use to develop a healthy root system and obtain nutrients, leaving more energy available for producing edible parts.

Seedlings of annuals such as tomatoes or peppers are often transplanted into a spot in an existing garden bed. In this case, the hole need only be made large enough to receive the transplant's root ball because the soil is already prepared deeply enough for the plant's future growth (Figure 8.9). The main considerations in this situation are the other plants growing in the same bed or nearby (see section 6.4.4 on planting density).

For transplants not placed in existing garden beds, or seedlings of trees and other large plants, the soil at the planting site will have to be prepared (Figure 8.9). Annuals need a hole at least 45 cm (18 in) deep and 30 cm (12 in) in diameter. The exact size will depend on soil quality and depth (Chapter 9). If the soil quality is poor the hole should be made bigger than it would be in better soil to protect the plant from negative effects of the poor original soil. This hole is then refilled with compost and good soil just as would be done when preparing a garden bed (section 9.8).

When transplanting it is best to use no more than one-third compost or other organic matter for refilling the hole. The remainder should be soil, or soil and sand. Exact proportions are not as important as making sure that there is not an abrupt and extreme change in soil texture between the soil in the transplant's root ball and the soil in the planting hole. For example, mixes containing much more organic matter may be so much looser than the transplant's root ball that this major change in soil texture slows the flow of water (section 10.3.2) and discourages the transplant's roots from growing out of the root ball, causing it to become root-bound after transplanting. Planting holes should not have smooth sides, as this also encourages root binding. The sides of the holes should be loose and rough textured. Any manure used in the planting hole should be well rotted or composted or it may burn the transplant's roots.

For trees, the size of the hole depends upon the soil and the type of tree being planted. For example, in dense, hard soil with thick layers of caliche or ironstone, a hole 1.5 m (5 ft) deep and 1 m (3.3 ft) in diameter is a minimum size. Once again, the deeper the hole the better because more soil is loosened and prepared making root growth easier. If its roots cannot penetrate hard, dense soil a tree will be stunted and may lack the hardiness needed to survive harsh dryland conditions. In addition stunted shallow roots do not anchor a large plant well and so cannot prevent it from being blown over. Holes dug in sites where soil is of better quality and less dense need not be as big because the roots will be able to penetrate beyond the planting hole more easily.

Planting holes should also provide good drainage, because plant roots will not survive long periods in water-saturated soil. An impermeable soil layer like dense clay or caliche can prevent water from draining below the root zone. Roots become suffocated and

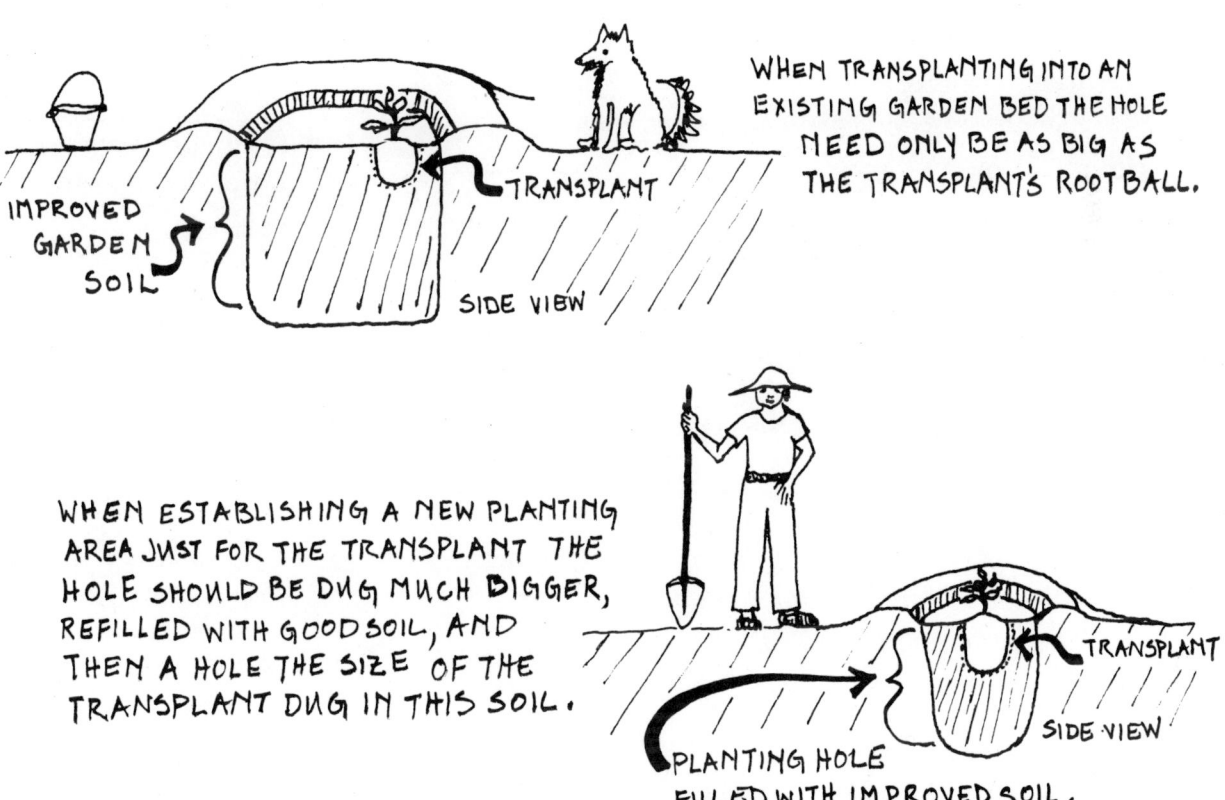

Figure 8.9 Transplanting Holes

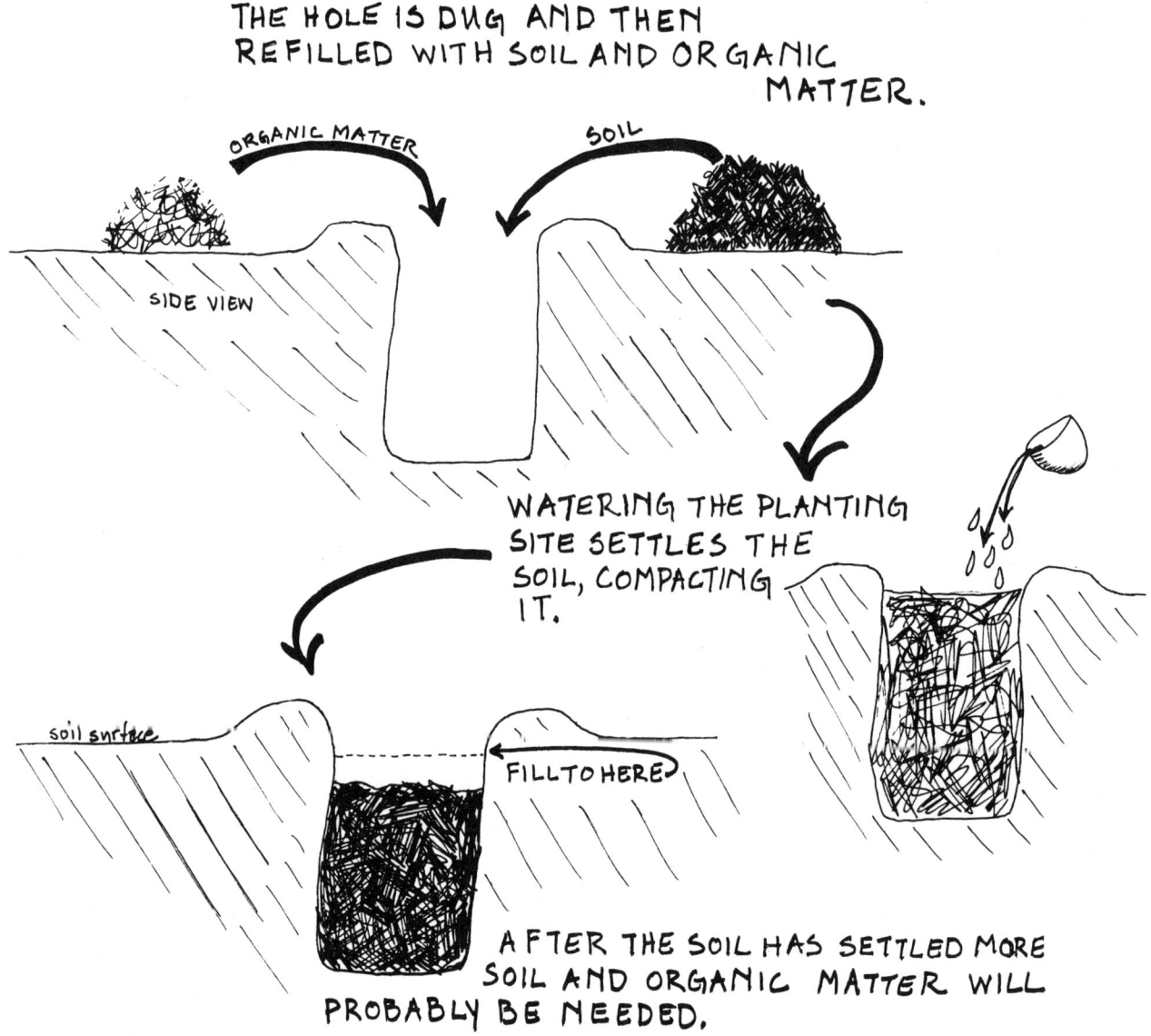

Figure 8.10 Watering a Transplant Hole to Settle the Soil

salts that would otherwise be washed below the root zone accumulate (Figure 12.5 in section 12.6.1). (A test for drainage in a tree hole is described in section 9.4.)

A large hole is refilled with good quality soil and organic matter using coarse organic matter near the bottom and finer material near the surface. The hole is then watered thoroughly to settle the soil. When it is watered and the organic matter is compressed, the soil level in the hole may sink significantly. Allowing a day or two for this to happen before planting will avoid later problems with sinking or shifting soil. Before planting more soil can be added (Figure 8.10).

8.4.3 Water

There must be enough water at the planting site to wet the refilled hole before transplanting and to thoroughly water the transplant afterward. In addition, enough water should be available for daily watering during the period of adjustment after transplanting. A basin is made with a diameter about two to three times larger than the root ball to hold water with the plant in the center on a slightly raised area which protects the trunk or stem from standing in water (Figure 8.11).

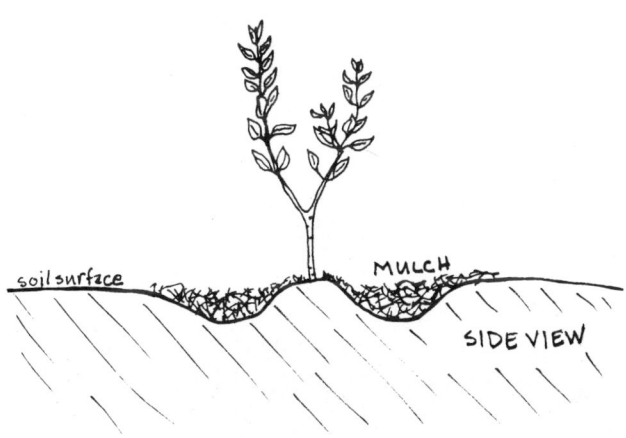

Figure 8.11 Protecting the Transplant's Trunk from Water

An example of indigenous transplanting basins comes from the dry Oaxacan valley of southern Mexico. Large vegetable plots are carefully laid out with planting basins or *cepas* prepared for each chili or tomato transplant.[5] Hand-dug wells among the *cepas* in each plot provide water for hand irrigation.

For the people of Dar Masalit in central Sudan, the major work in establishing fruit tree transplants (mango, guava, and lemon) in seasonally dry stream beds and terraces is watering while the trees become established. This must be done daily at first, tapering off to about once a week after 3 years.[6] The trees also have to be protected by thorn fences against livestock; and thorn branches are no longer easy to find. One-half of the fruit trees are lost to termites, flooding, and other causes. However, it seems that in good locations (those with a high water table yet protected from flooding), the trees require little work once they begin bearing fruit.

8.4.4 The Transplant

Preparing plants for transplanting begins about 3 to 7 days in advance for annuals, and even longer for perennials. It is done by gradually exposing the plants to more sun and heat and reducing their water intake. This process of exposing them to controlled stress, called *hardening off*, reduces the rate of transpiration and photosynthesis and causes the plant tissue to become more dense because it contains less water. Hardening off also encourages food storage in plant tissues because growth is slowed.[7] Some plants may wilt slightly when the hardening process begins but usually recuperate at night. As long as their central stalk and growing tip remain green and firm they are not being harmed. After several days the plant will stop wilting, unless the hardening is too severe. Just before and then immediately after transplanting the plant should be well watered. Drought-hardened plants seem to be better able to cope with subsequent drought and are more productive under dry conditions than plants that have not been hardened.[8]

Younger plants have more vigor and resilience and are generally better for transplanting. For example, tomato transplants 5 weeks old were found to be far more productive than 7-, 9- or 11-week-old transplants.[9] Transplants should not have flowers or fruits because these use the energy they need for surviving and becoming established after transplanting. If the plant must be transplanted when it is flowering or fruiting, flowers and fruit should be removed. Annuals are transplanted after the first two to four true leaves (not including the cotyledons) have developed. Perennial transplants can have more aboveground growth than this, although younger plants have a better chance of surviving, and they are smaller and easier to move.

Before planting, the transplant should be checked to make sure it will not introduce pests or diseases into the garden. It should have a vigorous root system with no *galls* (unusual swellings or growths due to pathogens, sections 13.4.1 and 13.4.3) or soft brown lesions and no harmful insects or their eggs on the stems or leaves. Plants weakened by disease, predators, or other sorts of stress will also have more difficulty surviving transplanting than those in good health.

Except for bare-root transplanting of deciduous perennials and untangling a root-bound transplant, the roots should be disturbed as little as possible. This is because damaged roots reduce a plant's ability to obtain water and nutrients, and make it more susceptible to disease and environmental stress.

A seedling should be handled as little as possible, and then by gently holding its soil-covered root ball. The plant should not be held by its stem because young, tender stems bruise easily and diseases often develop on these bruises.

No matter how carefully the transplanting is done, some root hairs will be damaged. Though small, these root hairs constitute the majority of the total root surface area and are vital for the intake of water and nutrients. Loss of root hairs, and thus a reduction in root surface area and water absorption, causes the wilting common in transplants. This is why extra watering and protection are needed after transplanting, especially under hot, dry conditions.

The roots should be pointing downward when the plant is placed into the transplanting hole. If pushed upward during transplanting roots will be closer to the soil surface where higher temperatures and less moisture will slow growth (Figure 8.12).

A seedling that has been left too long in its container will not have sufficient room for its roots, or adequate soil from which to obtain water and nutrients. The roots spiral around each other in the restricted space, eventually choking the plant (Figure 8.13). These plants are root-bound, often appear unhealthy, and frequntly wilt, even with regular watering. This problem occurs frequently with perennial transplants left in the container for 1 or more years. Vertical ridges on the inside of containers as described in section 8.2.2 help avoid this pattern of spiraling root growth.

The root ball of root-bound plants should be gently loosened and the roots untangled. Briefly soaking the root ball in water helps loosen the soil. If untangling

A TIGHT MASS OF ROOTS GROWING IN CIRCLES ALONG THE CONTAINER WALLS SHOW THAT THIS JUJUBE SEEDLING IS ROOT-BOUND.

ROOTS WERE EVEN GROWING OUT THE POT'S DRAINAGE HOLES.

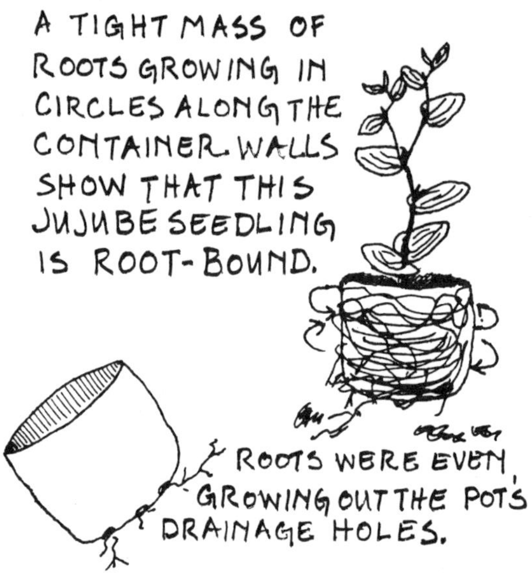

Figure 8.13 A Root-Bound Transplant

them is impossible, the gardener may try cutting them in a few places to pull the spiraled roots apart and direct them out and away from the root ball. Obviously all this handling and cutting make a root-bound transplant extremely vulnerable to water and heat stress and it will need shade and a lot of water until it becomes established. However, if the roots are not untangled and slightly spread, the plant will remain root-bound even after transplanting because its roots will continue to follow the same spiraling pattern. A root-bound plant is poorly anchored in the ground and easily blown over.

Tomatoes can be transplanted slightly deeper than

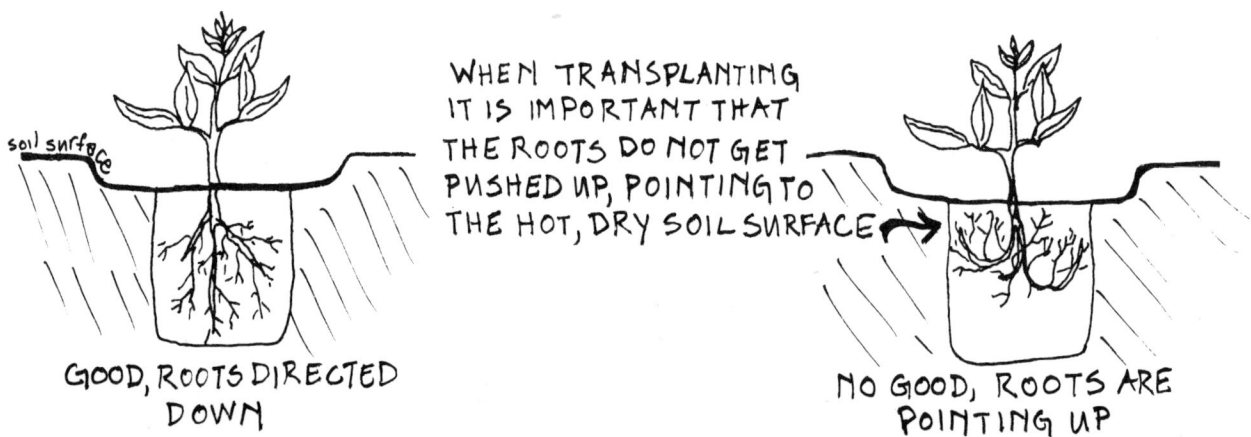

Figure 8.12 Transplant Roots

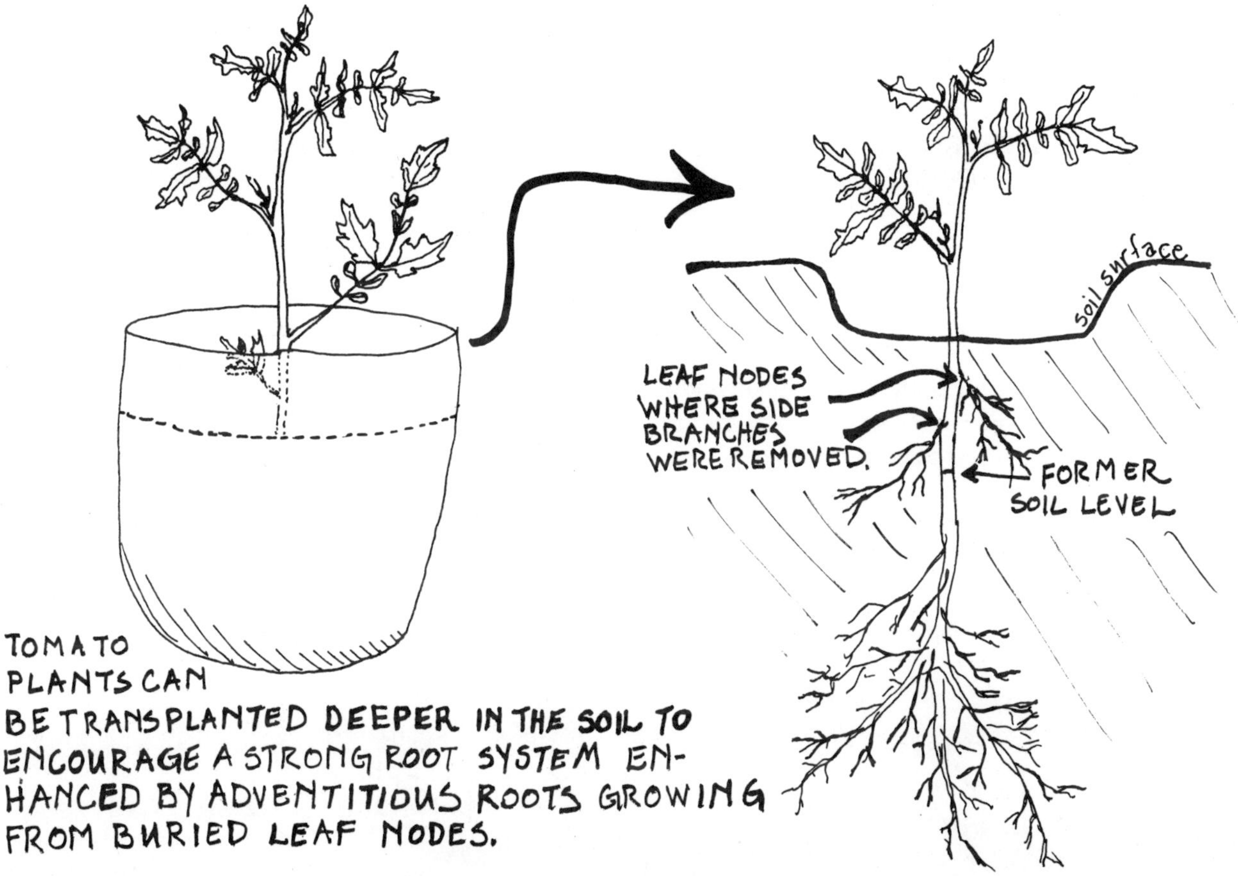

TOMATO PLANTS CAN BE TRANSPLANTED DEEPER IN THE SOIL TO ENCOURAGE A STRONG ROOT SYSTEM ENHANCED BY ADVENTITIOUS ROOTS GROWING FROM BURIED LEAF NODES.

Figure 8.14 Transplanting Tomatoes

they had been growing because adventitious roots will grow from nodes on the stem, strengthening the plants and improving their ability to obtain water and nutrients (Figure 8.14). All other plants including perennials should be transplanted to the same depth at which they were already growing.

When transplanting a grafted seedling (section 7.6), it is best to keep the graft union at least 20 cm (8 in) above the soil surface. If close to or under the surface the union may remain wet when the seedling is watered, encouraging the growth of microorganisms which could destroy the graft union and rot and kill the plant.

Under hot, dry conditions some perennials and mature annuals are pruned to decrease the total leaf surface area and thus the amount of water lost through transpiration (Figure 8.15). When transplanting woody fruit trees their aboveground growth should be pruned back about one-third to compensate for root damage. By diminishing total plant size with pruning, the plant can focus valuable resources such as water on survival. Drought-deciduous plants self-prune under extreme drought for the same reason.

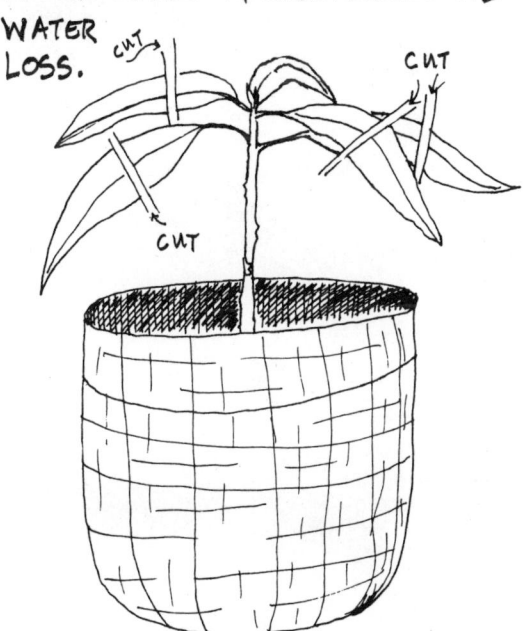

Figure 8.15 Pruning the Transplant's Leaves

Box 8.1
Steps in Transplanting

The basic steps for transplanting in drylands are reviewed below. Figure 8.16 illustrates transplanting an annual.

a) The transplant is hardened off and pruned if needed.
b) A hole is dug corresponding to the soil conditions and the transplant's needs.
c) Before transplanting both the transplant and the transplanting hole are watered thoroughly.
d) Transplanting is best done in the evening or on a cloudy day.
e) Transplanting is done quickly, leaving the plant out of the ground for as little time as possible. If the transplant is not in a container but must be moved some distance to the planting hole the root ball can be wrapped in moist cloth. This prevents it from falling apart, protecting the roots. Once at the site a root-bound root ball can be carefully loosened; the roots untangled and cut.
f) The transplant is placed in the hole at the same depth at which it was growing before, with the roots directed downward and slightly spread so that they will not bind each other. The hole around the transplant is filled with soil which is packed firmly as it is added.
g) A ring-shaped trough around the plant makes a basin in the soil to hold water. The trunk or stem is in a central raised area which helps keep it dry.
h) The transplant is watered deeply, mulched, and shaded if necessary.

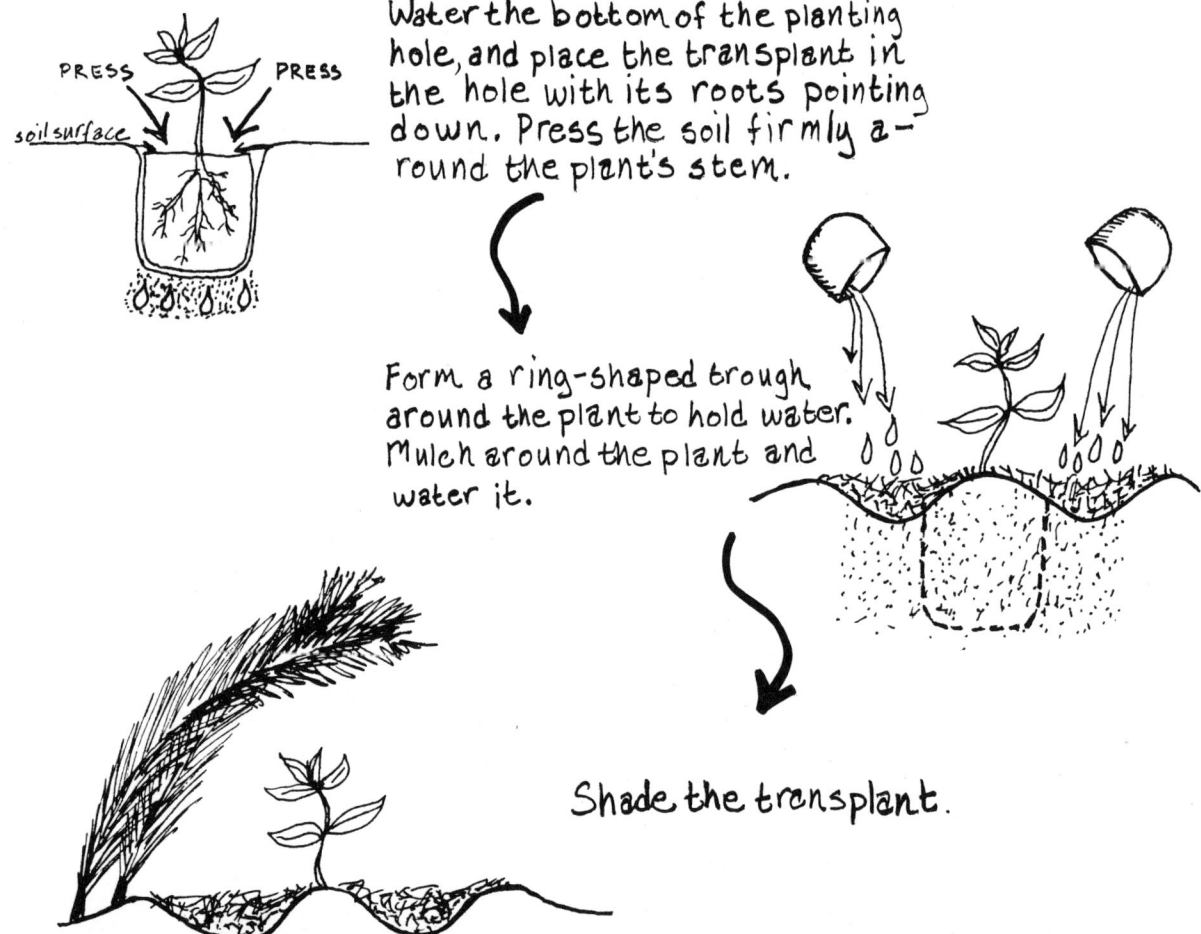

Figure 8.16 Transplanting an Annual into a Garden Bed

Figure 8.17 A Mixed Garden in Northern Mexico

8.5 Plant Interactions

Plant interactions in the garden can affect how well the plants produce and how much care they will need. Many combinations of crops seem to improve garden production, and in indigenous dryland gardens a wide variety of crops and crop varieties are often grown in mixtures (section 14.2.2).

8.5.1 Mixed Planting

Mixed planting in a household garden may take many forms. It can be a combination of various trees and plants such as the garden in Durango, Mexico, in Figure 8.17. It can be fruit trees surrounded by squash vines or garden beds containing alternating rows of different crops. However it is organized, the goal in a mixed garden is a greater average harvest of diverse garden produce for the least amount of labor and resources.

Many gardeners recognize the interactions between different crops in their mixed gardens. These interactions change over time, even from year to year when large perennials are mixed with annuals. For example, a gardener we visited in Durango, Mexico, pointed out that soon she would not be able to plant annual vegetables under some of her peach trees because it was getting too shady. For her the peaches were more valuable and so the trees were not thinned. Also she was finding other areas such as the edges of her garden where annuals could be grown, and vines like chayote were encouraged to climb up fences and over rooftops where they got plenty of light.

In addition to the variety of goods produced, mixed planting takes advantage of limited resources such as space, water, soil fertility, and the gardener's time (section 6.4.4). Some other benefits of mixed planting are discussed below.

DIVERSITY Species and varietal diversity can help reduce pest and disease problems and their impact on the garden's productivity (section 13.2.2). Growing only one or a few crops or varieties encourages the growth of pest populations and disease. Having different plants with diverse forms and life cycles discourages this.

There are a variety of ways in which crop mixes can reduce pests: 1) by providing a habitat for birds or insects like parasitic wasps which prey on insect pests in the garden, 2) by providing alternative host plants for pests, and 3) by protecting vulnerable crops by visually or chemically hiding them.[10] Several dryland garden crops are thought to reduce populations of damaging nematodes in the garden (section 13.3.2).

REDUCING WEEDS Dense crop mixtures create a living mulch that reduces weed growth by shading emerging weed seedlings from the sun. Mixtures that combine crops with upright growth forms with those that sprawl or vine control weeds especially well. This is because these two plant forms can be sown together densely enough that the soil is covered by the mature plants, but not so densely that they are harmed by competition for soil, water, and light. For example, it has been shown that planting cowpeas and mung beans among sorghum or pigeon peas suppresses weed growth just as much as two hand weedings would; and watermelons and sweet potatoes planted among yams or among a yam, maize, and cassava mixture suppress weed growth as much as three hand weedings would.[11] The squash plants commonly grown mixed with maize and beans in southern Mexico send out rapidly growing vines with large leaves that quickly cover any exposed soil.[12]

CREATING MICROENVIRONMENTS Some plants can be used to create an environment favorable for other plants. Trees or other large plants in indigenous mixed gardens provide shade and wind protection to smaller, more tender ones. This mimicks a situation frequently found in nature. For example, in the Sonoran Desert of North America large "nurse" plants appear to protect smaller, more sensitive plants such as wild chiltepines (*Capiscum annuum* var. aviculare or glabriusculum) from frost or extreme heat. By the way they lay out their gardens and mix different species of garden plants, some gardeners have created environments that mimick this pattern found in nature.

In the Middle East, intercropping patterns in the smaller date gardens illustrate the effect of the date palm tree on its environment. When the trees are young alfalfa is often grown around them. As the trees grow they create a partially shaded area protected from desert winds. The alfalfa is then replaced by vegetables and trees such as citrus, fig, mango, pomegranate, and jujube.[13]

Research on mixed cropping and its effects on plants, pests, and resource use is limited and focused mostly on large-scale agriculture in temperate and humid tropical areas. A better understanding of mixed cropping systems, especially under the conditions of small-scale, low-resource agriculture would be extremely valuable. However, the very complexity characteristic of these systems makes them difficult to examine using conventional methods. Perhaps this dilemma will force researchers to adopt a new approach to understanding agriculture.

8.5.2 Allelopathic Plants

Many crops produce chemicals that in some way inhibit the growth of other plants. This is called *allelopathy*.[14] For example, sunflowers, asparagus, eggplant, and sorghum produce allelopathic chemicals that affect other plants of the same species and sometimes other species.

The squash (*Cucurbita pepo*) grown in southern Mexico that covers the soil surface discouraging weeds (section 8.5.1), may also reduce weed growth through allelopathy. The leaves of these squash plants appear to suppress growth of many common local weeds, but not that of the maize and bean crops among which they are planted.[15]

Sometimes allelopathic plants are obvious because no other plants will grow around them, and they are often well-known among local gardeners and farmers. Plants that inhibit the growth of others need to be identified and taken into account when planning a garden. For example, placing a garden in the shade of a eucalyptus tree (*Eucalyptus* spp.) would be a waste of the gardener's time and energy because volatile oils in the leaves of many eucalyptus varieties inhibit germination and growth of some plants. Hot, dry conditions stimulate the release of these oils.[16] In Nigeria *Eucalyptus* spp. are popular trees to grow as boundary markers for fields and as dooryard ornamentals.[17] Their leaf litter was found to suppress germination and early growth of tomatoes and chilis, but affected maize only slightly and cowpeas not at all.

Leaf litter from tamarisk trees (*Tamarix aphylla*) creates such salty soil conditions that most garden plants cannot survive.[18] Walnut (*Juglans* spp.) and pine (*Pinus* spp.) trees also have allelopathic qualities. The extent of the root system and the area that would be affected by leaf drop of these trees should be noted when establishing a garden. Root pruning, described in section 8.7.1, may be helpful.

Walnut wilt is a disease affecting garden plants, like tomatoes, that are growing in the root area of a walnut tree. The woody pith tissues in the garden plant's stems turn brown, the plant wilts, and eventually dies. The cause of walnut wilt is an allelopathic chemical produced by the walnut roots. This poison remains in the soil even after the tree is dead, so putting a garden near where a walnut tree is or was growing is not a good idea.

In addition, parts of some plants may not be useful as mulch or compost because of their chemical properties. However, thorough composting (section 9.6.2) is thought to eliminate allelopathic qualities in amaranth weeds (*Amaranthus retroflexus*) and goosefoot (*Chenopodium album*) which otherwise may have negative effects on garden crops.[19] These "weeds" are also widely favored as green vegetables. Local farmers and gardeners are often aware that particular plants are poisonous and should not be used in the garden or compost. If there is some uncertainty experiments can be conducted, being careful not to contaminate the whole garden or compost pile.

8.5.3 Crop Rotation

In addition to interacting with each other directly, crops can be affected by what was previously grown in the garden. Gardeners can take advantage of the positive effects of this interaction by rotating the location in the garden or field where different crops are planted.

Some soil-borne pathogens and some disease-causing nematodes feed on or parasitize certain crops or crop families and do not harm others (sections 13.3.2 and 13.3.4). Crop rotation can help control or eliminate these diseases and pests by removing the crops they feed on for 1 or more years. For example, if clubroot fungi (which only affect members of the crucifer family) infest cabbage growing in one garden bed, kale planted in that bed the following year will probably also be affected. Rotating crops from other families, such as chilis from the solanaceous family, will control the pathogen.

Soil improvement can be another benefit of crop rotation. By rotating crop mixes dominated by nitrogen-fixing legumes with those emphasizing nonleguminous crops it is less likely that nitrogen will be depleted (section 9.5.2).

8.6 Weed Management

Plants growing in the garden other than those that were purposefully planted are called *weeds*. Weeds are hardy, establishing themselves and surviving without any attention. They quickly take advantage of disturbed habitats such as a garden, which in drylands is an oasis of favorable growing conditions.

Even though weeds are not intentionally planted they are not necessarily unwanted. It is now recognized that weeds are often useful to the gardener and her household.[20]

Some uses for dryland weeds:
- food and condiments for people and animals
- medicine
- mulch
- green manure and compost
- insect traps
- bedding for animals
- erosion control
- crafts.

Some weeds are plants that people are slowly domesticating. In the high-altitude drylands of Chihuahua, Mexico, the Tarahumara Native Americans allow the jaltomata to grow among their maize, on the edges of their maize plots or in patches of fertile soil such as where animal manures or refuse are dumped.[21] The jaltomata produces dark, sweet edible berries. The plants allowed to grow among crops and other disturbed areas had a higher number of berries per plant and a lower number of seeds per berry than those growing in the wild. This may be because the Tarahumara have selected for plants that have the best fruit and that respond to improved growing conditions.

Weedy relatives of garden crops can contribute new genetic material to the crops through cross-pollination.

In the short-run this may reduce garden yields somewhat, but the long-term effect is to increase the genetic diversity and thus the adaptability of the garden crop. An example of this is the crossing of the "weed" teosinte (*Zea* spp.) with maize which is carefully managed by farmers in an area of Jalisco, Mexico.[22]

However, weeds may also have a negative impact on the garden if they attract or spread disease and pests and compete with garden crops for essential resources. In some cases, weeds like leaf amaranth are hardier than most garden crops, easily outgrowing them. There are solanaceous weeds that carry diseases harmful to solanaceous garden plants. For example, in southwestern North America *Datura* spp. weeds can introduce curly top virus (section 13.4.1) to gardens, harming the tomatoes and chilis growing there.

There are few documented examples of traditional weed management. Even though the following example is about farmers in the humid tropics it is worth mentioning because it is rare documentation showing the complexity of a traditional weed management system. Small-scale farmers in Tabasco, in humid southern Mexico, classify weeds as either "good" or "bad" plants, "*buen*" or "*mal monte*".[23] More than 40 weeds are classified this way based on their direct use to people and animals, their effect on the soil and crops growing there, and how difficult they are to control. Many different management methods such as living mulches, burning, cutting, and cultivation are used depending on the particular field or plant.

Which plants are weeds and how they are managed will depend on many things including the crops in the garden and the usefulness of the weed. Yet, often it is not the plant itself which is either good or bad but rather the timing of its emergence and growth in relation to garden resources and crops.

8.6.1 Resource Use

The gardener must decide whether the benefits provided by a weed are worth the resources it will use. For example, under dry conditions all available soil moisture and nutrients may be needed by garden crops and yields may be reduced by weeds that outcompete them. If there is little competition, however, weeds may produce food while helping shade garden crops and the soil.

If products provided by the weeds are desirable enough, the resources they use can be an acceptable loss. For the Tarahumara, *quelites*, or wild greens that emerge in their maize fields, are an essential food source.[24] The *quelites*, amaranth (*Amaranthus retroflexus*), *Brassica* spp., and goosefoot sprout and grow with the rains in May. This is during the hungry season (April to July) when food reserves are low or depleted. Leaves from *quelite* seedlings between 2 to 6 weeks old are harvested and cooked, and make a significant contribution to the Tarahumaran diet. The *quelites* are removed from the field when they are about 6 to 8 weeks old, by which time their roots start to compete with the deeper-rooted maize crop and their leaves become tough and unpleasant tasting.

8.6.2 Effects on Pest Populations

As with other crops in a mixed garden, weeds can have an effect on insect populations in and around the garden (section 8.5.1). Weeds can mask the presence or smell of crops from insects, offer these insects more desirable food sources, and provide an environment in which beneficial, predatory insects will become established.[25] In southern Zimbabwe the common weed, wild marigold (*Tagetes minuta*), has been found to significantly reduce populations of root knot nematodes in soil where it grows, and is left to grow among crops[26] (section 13.3.2).

Some weeds are trap plants for insects keeping them off crops, while others may encourage beneficial insects. Managing these weeds, for example, cutting them at a particular time, can move those beneficial insects onto the crops. However, weeds may also attract or encourage a concentration of harmful insects and provide places for them to reproduce, as happens in the fields of the Tohono O'Odham Native Americans of the Sonoran Desert (section 8.6.3).

The effect of weeds on pest populations depends on many changing factors such as the garden crops, the garden microclimate, pest life cycles, migrations, and predator-prey relationships between pests and other animals. Local experience and knowledge combined with experimentation are the best ways to learn about the relationship between weeds and pest populations in a particular area.

8.6.3 Timing

Timing is the key to easier weed management. Weeding the garden can consume a lot of time, and may compete with other tasks, for example, weeding fields. The time demands of weeding may influence when gardens are planted, garden size, when weeds are harvested for food, and when they are removed.

The Tohono O'Odham Native Americans of the Sonoran Desert manage weeds in a way similar to that of the Tarahumara described in section 8.6.1.[27] Amaranth seedlings are allowed to flourish for several weeks, providing tender, leafy greens. When the plants get taller than about 30 cm (12 in) they are removed to avoid competition with crops and because they start attracting pests such as cucumber beetle larvae and grasshoppers.

Once garden plants are well established they will shade the soil surface, discouraging new weeds from growing due to lack of sunlight. In some date gardens of the Middle East, Bermuda grass (*Cynodon dactylon*, Pers.) is an aggressive and harmful weed around young trees.[28] However, when the trees are larger and more established some gardeners encourage the grass as a fodder for animals. The shady conditions created by the tall trees keep the sun-loving grass from getting too vigorous and competitive.

8.6.4 Methods of Weed Control

Knowing the life cycle of weeds and how they reproduce is essential for controlling them. Removing undesirable weeds before they produce seeds or spread vegetatively will save work in the future. Putting seeds in the compost or mulch could spread the weeds throughout the garden. Pre-irrigating empty garden beds to germinate weed seeds, and then removing the weeds, is effective if there is plenty of water. Weeds easily propagated by vegetative means like Bermuda grass should not be used for compost or mulch.

There are four methods of weed control useful for dryland gardens: heavy mulching, cultivation, hand weeding, and burning, each of which has advantages. The weeding method chosen will depend on a number of factors including the gardener's experience and schedule, the garden layout, the time of year, and the soil type (Table 8.1).

- Heavy surface mulches control weeds by blocking sunlight from the soil. Use of these mulches is described in section 10.8.1. As described in section 8.5.1, plants like squash vines, which cover the soil surface, act as living mulches and are also effective for discouraging weed growth. The main drawback to this method of weed control is that mulches may hide pests that are harmful to young plants.
- Hoeing or some other method of cultivation is relatively quick and kills most weed seedlings by disturbing their roots. However, cultivation may be awkward in a mixed garden and is not appropriate where wind or rain erosion can be a problem.
- Weeding by hand is effective but requires more labor than the other three methods.
- Controlled burning before planting or of weeds removed by other methods is quick, effective, and does not require a lot of labor. Burning is not appropriate in areas with alkaline soil where ashes would raise the pH, or in very small, mixed gardens where fire would harm plants already growing.

Table 8.1 Controlling Weeds in Dryland Gardens

Method	Timing	Positive features	Things to think about
Heavy mulching	Best after crops established to avoid pest problems.	Low labor, added benefit of controlling water losses, long lasting; pulled weeds can be used as mulch.	Mulches can harbor pests harmful to young plants.
Cultivation with hoe, stick, or other tool	Before planting and/or while plants growing, best before seed production.	Kills seeds and seedlings by burying, cutting, or exposure to sun; breaks up soil surface reducing evaporation; weeds can be used in compost.	Be careful not to damage garden crops; if windy can expose topsoil to erosion; difficult in a dense garden or under trees with surface roots.
Hand weeding; pulling and cutting	Before or while garden plants growing; before weeds produce seeds.	Thorough and selective, works well in mixed garden; saves weeds for eating, composting, or other uses.	Most labor-intensive; timing essential to eliminate weeds before seed production.
Controlled burning	Before planting, or burn weeds removed by other methods.	Relatively quick, can be low labor, kills both plants and seeds.	Ashes increase soil alkalinity; lose organic matter useful as compost or mulch.

8.7 Pruning

Pruning is the selective removal of plant parts to promote different patterns of growth and development. Pruning may also be done as part of transplanting or grafting to reduce water loss through transpiration. The products expected from the plant and the way the plant grows determine which parts are pruned. For example, flowering is undesirable in crops such as basil or leaf amaranth because it diverts energy and resources away from leaf production. On these crops growing tips are harvested before they flower, or flower heads are pruned off to encourage more leaf production (Figure 8.18). In fruit-bearing plants such as tomatoes and eggplants, however, flower production is essential for producing the fruit, so some vegetative growth may be pruned to encourage flowering.

8.7.1 Reasons to Prune

Following are a variety of reasons to prune plants in dryland gardens:

- To delay flower and seed production and encourage leaf production. Examples: leaf amaranth and mint.
- To reduce vegetative growth and produce an earlier, more concentrated harvest. Example: tomato.
- To improve the quality and size of fruits by removing some of the immature fruits. For example, on papayas pruning out some of the fruit makes room for those remaining to grow. On trees like olives, lemons, and peaches an extremely heavy fruit crop one year will be followed by a very small one the next, and in some cases no crop at all. Thinning the fruit in the year of heavy production reduces the demands on the tree and stimulates flowering for next year's crop, evening out the amount of fruit borne year to year. Examples: stone fruits, loquat, olive, and citrus trees.
- To selectively thin out branches of perennials creating a strong structure and reducing the risk of wind damage. Examples: stone fruits, olive, fig, and pistachio trees.
- To thin out dense foliage that inhibits air circulation, thus reducing conditions that encourage fungal disease. Examples: tomato and basil.
- To shape the plants for specific purposes, for example, to provide space for other crops (Figure 8.19). If a gardener plans to use layering as a method

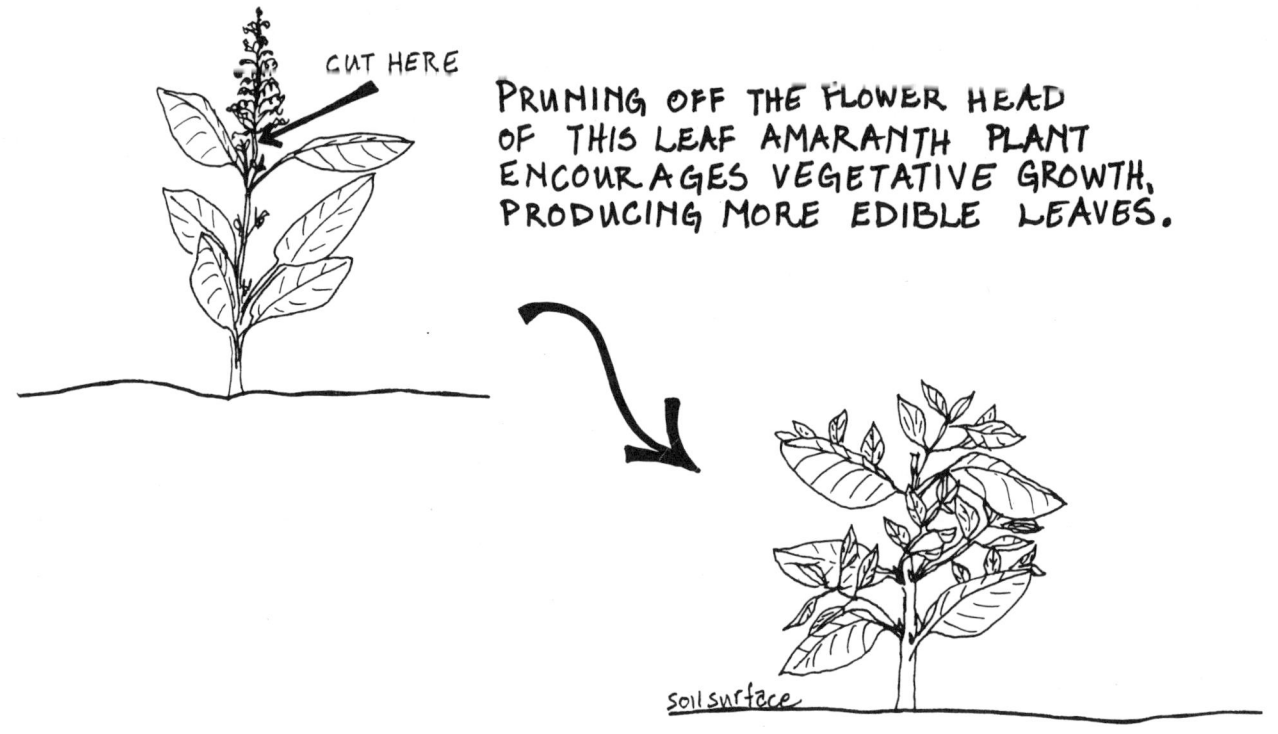

Figure 8.18 Pruning to Encourage Leaf Production

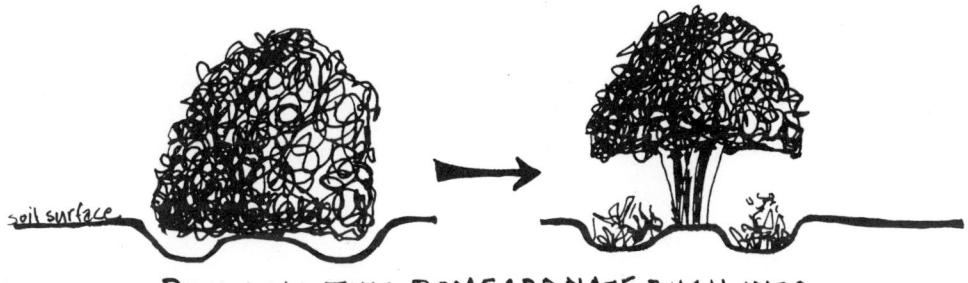

Figure 8.19 Pruning to Shape a Plant

of propagation, pruning should be done to leave some shoots or low-lying branches appropriate for this. In West Africa, trees whose leaves are collected are sometimes pruned into a ladder-like form with short branches making it easy to climb up the tree and reach the leaves growing on its higher branches.[29] Example: most perennials.
- Remove flowers or fruit on very young plants to reduce stress and direct the plant's energy into becoming established and vigorous. Example: most flowering plants.
- To diminish leaf surface area therefore reducing water loss and concentrating resources during times of stress such as grafting or transplanting. In some cases with perennials like mangoes, when transplanting a young seedling the leaves are cut halfway back (Figure 8.15 in section 8.4.4). Examples: many perennials and annuals.
- Harvesting of nonedible plant parts. Examples: date fronds for shading or building, banana leaves for shading and wrapping foods, pollarding of neem trees in West Africa to collect wood for building materials.
- To control tree height by removing apical dominance and encouraging lateral growth, and so making it easier to harvest fruits from the tree. Examples: stone fruits and jujube.
- Root pruning of trees to stimulate deeper, downward growth and discourage root growth into the area of other trees, into garden beds, or under buildings. Root pruning is done by digging a trench where the roots are to be cut. The trench should be at least 1 m (3 ft) away from the tree's trunk, or 50 cm (20 in) beyond the drip line (section 13.4) if this is farther. If root invasion is a recurring problem it may help to leave the trench open without refilling it, taking precautions so that people or animals do not fall into it. Examples: trees with invasive root growth, including nongarden trees like *Eucalyptus* spp. whose roots may be intruding into the garden.
- Perennial herbs that have become woody can be pruned back severely to encourage fresh, full growth and greater leaf production. Examples: oregano, marjoram, mint, and lemon verbena.

8.7.2 Guidelines for Pruning Trees

This section discusses some simple suggestions for pruning applicable to all fruit trees. Table 8.2 summarizes pruning advice for some widely known dryland garden fruit trees.

SHAPING THE TREE Many trees do not need pruning since their natural form is strong and productive. However, pruning these trees is still useful for removing dead or diseased wood, and branches that are crossing or rubbing each other. Branches in a direct vertical line, one above the other, block each other's light and create an unbalanced, weak tree. Pruning one off makes a healthier, more stable tree (Figure 8.20).

Pruning some garden trees to a particular shape can make them stronger and more productive. A shape for the tree should be chosen before pruning begins to guide decisions about which growth to remove. Two basic forms for pruned trees are open center and central leader (Figure 8.21).

Table 8.2 Pruning Suggestions for Some Dryland Garden Trees

Fruit tree/vine	Deciduous	Pruning suggestions
Almond	yes	Pruning to remove secondary or smaller side branches stimulates formation of new fruiting spurs.
Apricot	yes	Prune as recommended for almond.
Avocado	no	Minimal pruning to control height and open up dense growth for improved air circulation and sunlight.
Carob	no	Prune to shape.
Cashew	no	Prune when young to shape and remove shoots around base of trunk unless these are desirable for layering.
Citrus	no	Prune to shape, to open up dense canopy, remove basal suckers; occasionally thin fruits.
Fig	yes	Prune to shape, to open up dense canopy, reduce chance of wind damage; occasionally thin fruits.
Grapes	yes	During dormancy prune heavily leaving trunk and two to four major branches only to stimulate new growth for fruit bearing during the coming growing season.
Guava	no	Prune to shape and remove basal suckers.
Jujube	yes	Prune to shape and control height.
Loquat	no	Minimal pruning to shape, and reduce wind damage; occasionally thin fruits.
Mango	no	Minimal pruning when young to shape or remove heavy flower set.
Olive	no	Prune to shape and open up structure; occasionally thin fruits.
Peach	yes	Prune heavily to stimulate lots of new growth to bear next year's crop.
Pistachio	yes	Prune to shape.
Pomegranate	yes	Prune to shape; occasionally thin fruits.

Open center trees have any central trunk above 0.5-1.0 m (1.5-3.3 ft) removed when the trees are very young. This eliminates apical dominance and produces an open, bowl-like form. As a result fruit can be more easily reached. Spreading the tree's canopy allows light to penetrate into the inner branches, encouraging fruit production and ripening.

The central leader form allows the tip of the main trunk to continue growing and dominating the tree. All branches grow out from this trunk giving the tree a cone or cylinder shape. This shape can be very strong and vigorous and its canopy will take up less room than an open center tree.

BRANCH ANGLES The angle at which branches grow out from the trunk or larger branches affects their strength. Branches growing at small angles (less than 40°) are weaker and more likely to split or break than those with angles of 45° to 65°. This is because there is usually a fissure or crack in the wood's growth pattern when the branch and trunk meet in a narrow angle (Figure 8.22). If it is not possible for an adult to press her index finger into the place where the branch joins the trunk then the angle is too small.

If a branch on a young tree is growing at a narrow angle it can be widened by wedging something between the branch and trunk to spread the angle (Figure

8 Plant Management

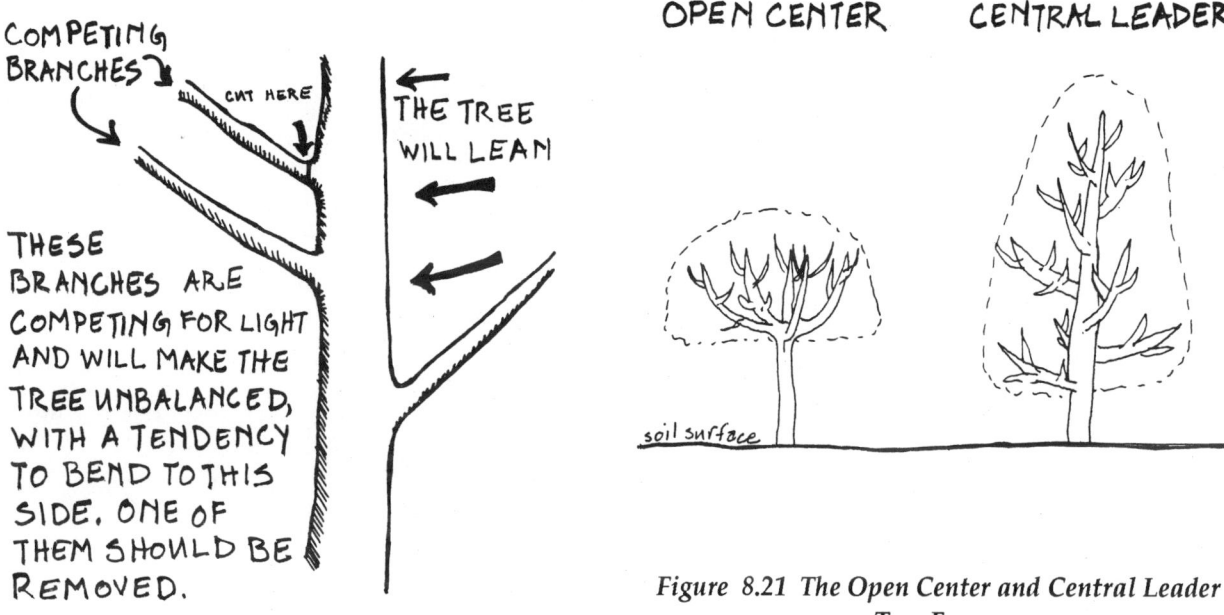

Figure 8.20 *Pruning Out a Competing Branch*

Figure 8.21 *The Open Center and Central Leader Tree Forms*

8.23). Ropes and weights can also be used to pull the branch down, into a more open angle. The spreading should be started at the beginning of the growing season and continued until the desired angle can be maintained by the tree. Putting some cloth padding between the tree and the rope or spreader will prevent

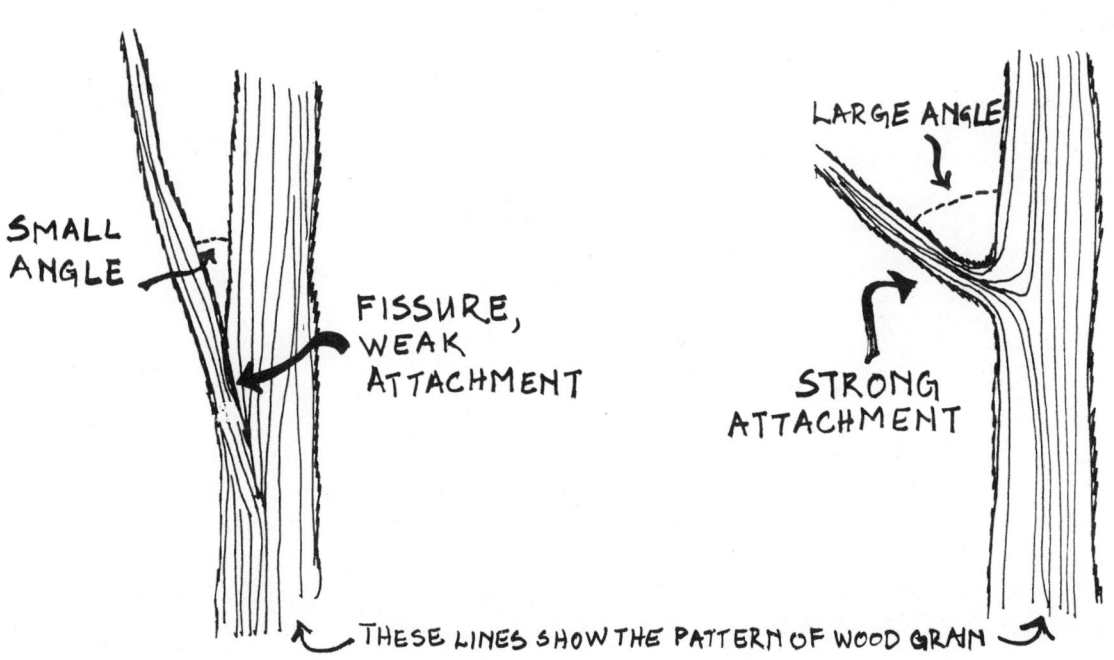

Figure 8.22 *Branch Angle and Strength*

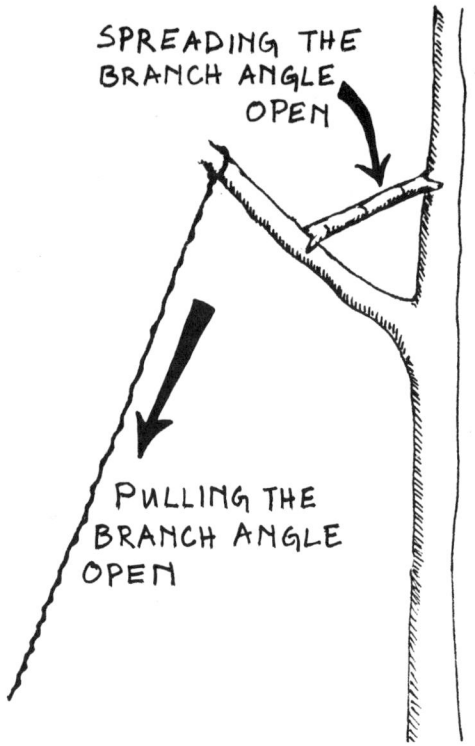

Figure 8.23 Spreading a Branch

damage to the tree. Farmers in Italy use lengths of cane with a V-shaped notch cut in each end to spread branches of fruit trees.

In addition to forming a weak connection, branches growing at a narrow angle from the trunk are more vertical, leading to apical dominance, and vegetative growth is stimulated more than reproductive growth. When the branch's angle of attachment is wider, the tip will soon hang down, overcoming apical dominance (Figure 8.24). This encourages the new growth at the high part of the branch to be vegetative, while reproductive growth is favored at the lower tip.

HOW FRUIT IS BORNE BY THE TREE An important consideration when pruning is how it will affect the harvest of fruit, leaves, or nuts. Some trees, like the mango, bear fruit only on the tips of the current year's growth at the ends of branches; this is referred to as *terminal* bearing (Figure 8.25). Trees like peaches produce fruit only on shoots grown the previous year (Figure 8.26). Trees like apricots and almonds bear fruit on short flowering shoots called *spurs* which grow laterally from branches and last 3-5 years (Figure 8.27).

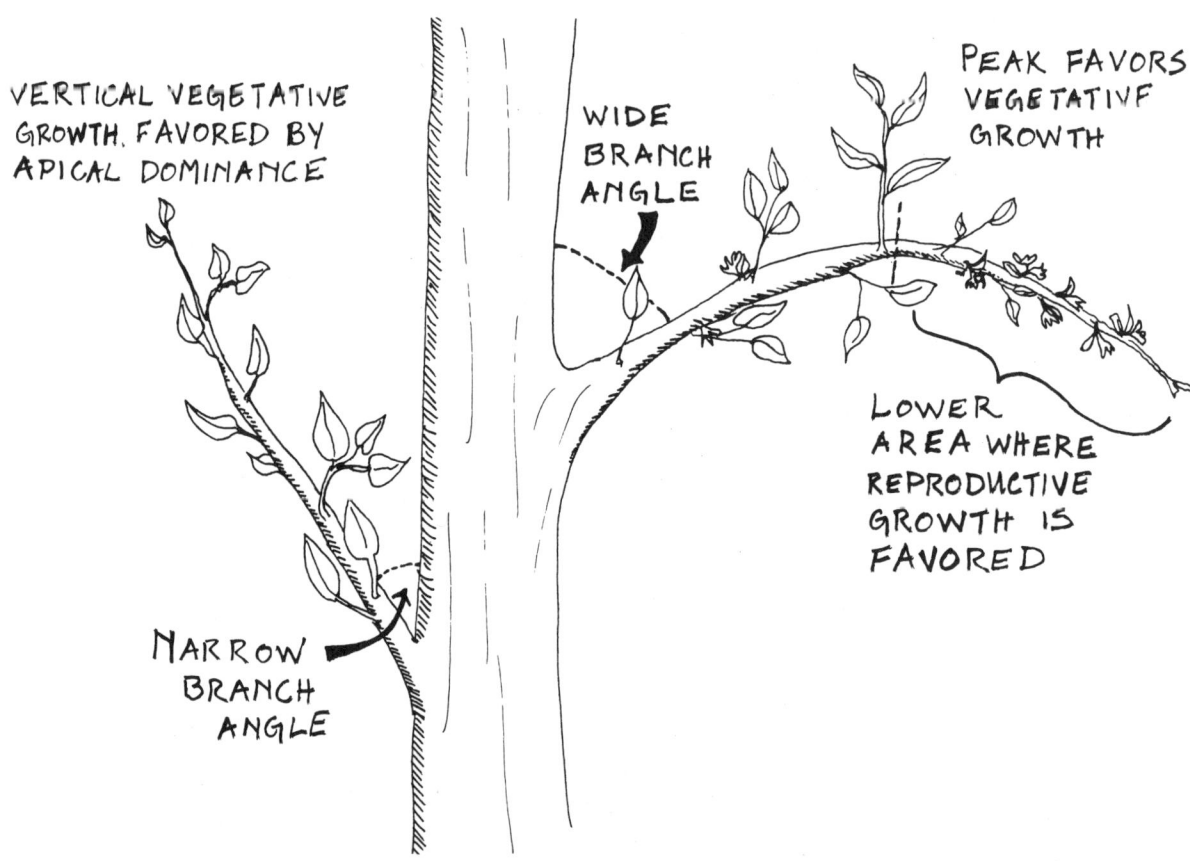

Figure 8.24 Branch Angle and the Type of Growth it Encourages

Figure 8.25 Terminal Fruit Bearing in the Mango

Pruning is most important for trees bearing fruit on this or last year's growth. Heavy dormant-season pruning will stimulate new growth, providing more sites for fruit production. For example, heavy dormant-season pruning is important for grape vines because they bear fruit only on the current year's growth. Fruiting spurs are productive for a limited time, depending on the kind of fruit tree. For example on apricot trees, fruiting spurs are productive for about 3 years and on almonds for about 5 years. Pruning secondary or side branches during the dormant season helps to stimulate new fruiting spurs.

Figure 8.26 Peaches Bear Fruit on Last Year's Shoots

WHEN TO PRUNE Pruning tree branches is best done when the shoots or branches are small, unless, of course, particular products are wanted. In temperate regions with cold winters, deciduous trees should be pruned at the end of the cold season. Pruning tends to increase cold sensitivity and so pruning at the beginning or during the cold season could make the tree more vulnerable to damage and disease. Root pruning is best done in the dormant season or at the start of the rainy season so as not to stress the tree.

Planning the desired form and anticipating the plant's growth lets the gardener prune before consuming resources that could be used in other, more desirable forms of growth. Compared with branches, small shoots and buds are easy to remove and these wounds heal quickly with little danger of disease.

Figure 8.27 Apricots are Borne on Fruiting Spurs

MAKING THE CUT The plant part being pruned should be as completely eliminated as possible. At the same time, it is important not to damage the plant as this can disturb growth and leave a wound vulnerable to infection. In some older branches the **branch collar** or area of tissue surrounding the branch's base is evident. The collar should not be cut off because this tissue forms a rapidly growing, disease-resistant callus over the pruning cut. If a branch collar is not evident, the pruning cut should be made away from the main branch (Figure 8.28).

When pruning many plants the gardener can easily pinch or pull the part off with her fingers. If the branches or shoots are too big or there is a danger of

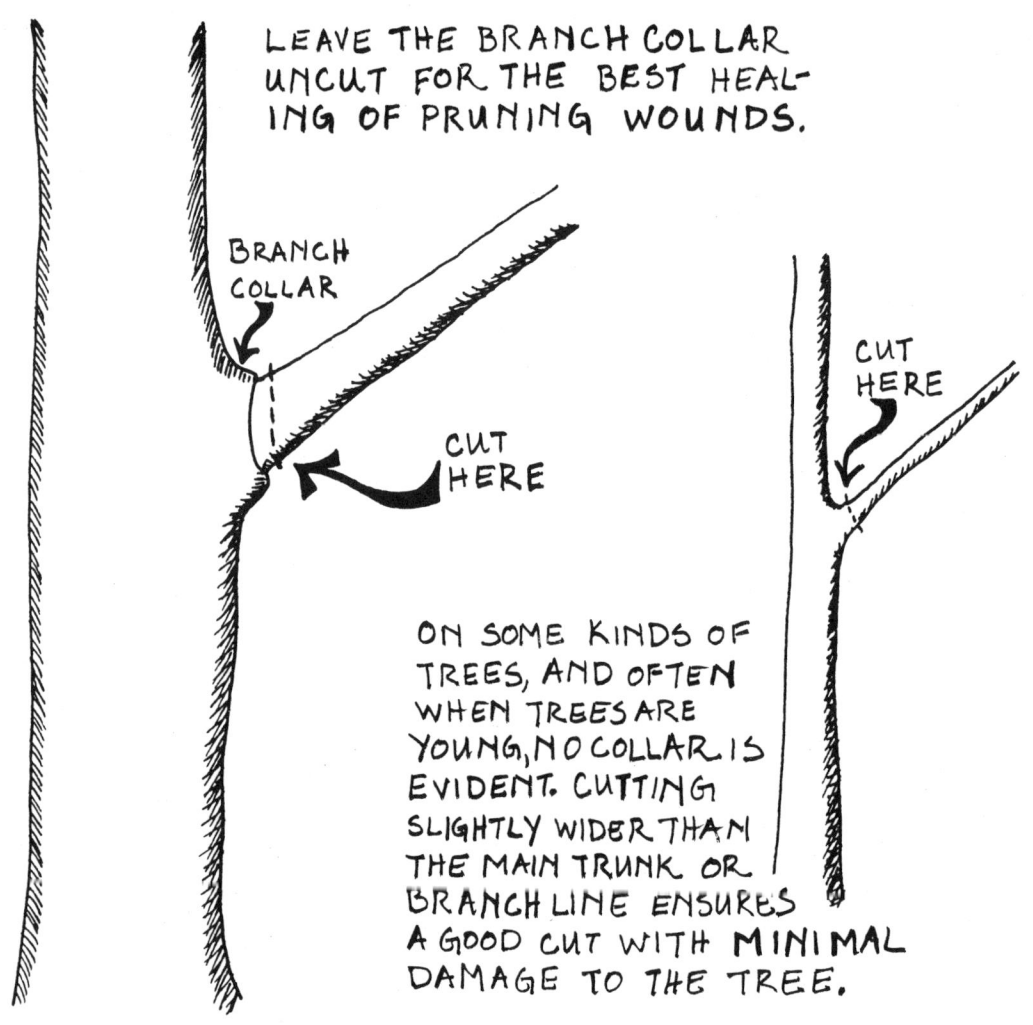

Figure 8.28 Where to Prune a Branch

ripping the remaining tissue then a machete, knife, saw, pruning shears, or other sharp tool can be used.

When cutting large branches it is easiest to start by removing most of the branch so only a small stub remains. Cutting the underside of the branch first will prevent tearing the bark and avoid pinching the tool blade. After most of the branch has been removed the final pruning cut can be made without having to handle the awkward, long, heavy branch (Figure 8.29).

When pruning the ends of branches, making the cut at a slight angle with the high point just above a desirable shoot or bud will promote the growth of that shoot or bud (Figure 8.30). All cuts should be made cleanly, leaving a smooth surface and without tearing the bark. If a diseased area is pruned either by hand or with a tool, washing the hands with soap or thoroughly cleaning the tool with soap, bleach, alcohol, or vinegar before touching another plant avoids spreading the disease.

For best healing, pruning cuts should not be covered or treated with anything. Exposure to the air promotes healing, but under hot, dry conditions it is a good idea to shade large wounds.

NOTCHING *Notching* refers to the technique of wounding the phloem tissue above or below a dormant bud to encourage a particular kind of growth. Notching is based on the same principle as girdling discussed in section 7.2.1. Wounding the tissue interrupts the flow of carbohydrates down the phloem

8 Plant Management

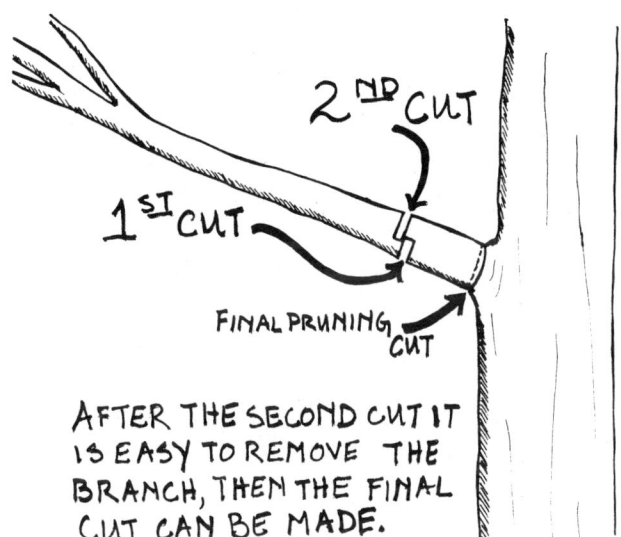

Figure 8.29 Pruning a Large Branch

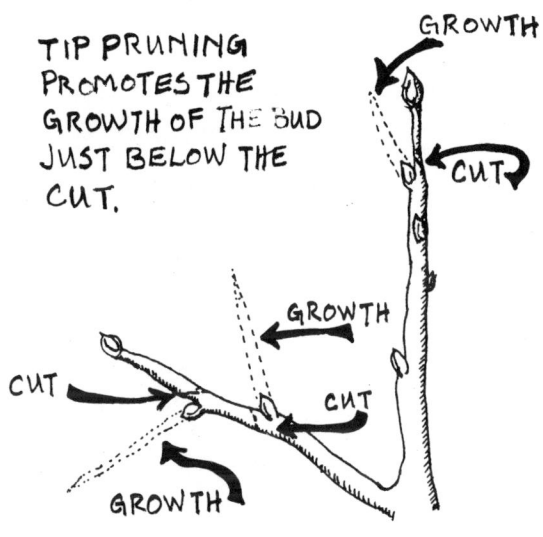

Figure 8.30 Tip Pruning

from the leaves back to the roots, thus concentrating it above the cut. If the tissue is broken above the bud not only is this flow cut off but so is the downward movement of hormones from the terminal bud which supresses growth below it. Therefore the bud will develop into a shoot. Notching below a bud encourages development into a flowering bud because of the concentration of food.

Notching is usually done on the young shoots and branches of deciduous trees at the end of dormancy. It is best to make the cut at least 2 cm (0.8 in) wide to prevent it from closing up rapidly. The notch should only penetrate the bark into the cambial tissue, and should extend about halfway around the branch or shoot (Figure 8.31).[31]

8.8 Trellising

Trellises support the growth of plants up and away from the ground. Plants like passion fruit, grapes, and many cucurbits, peas, and beans, are climbing vines whose growth and production can be improved by trellising. Many tomato varieties also benefit from trellising. Another advantage of trellises is that by encouraging vertical instead of horizontal growth,

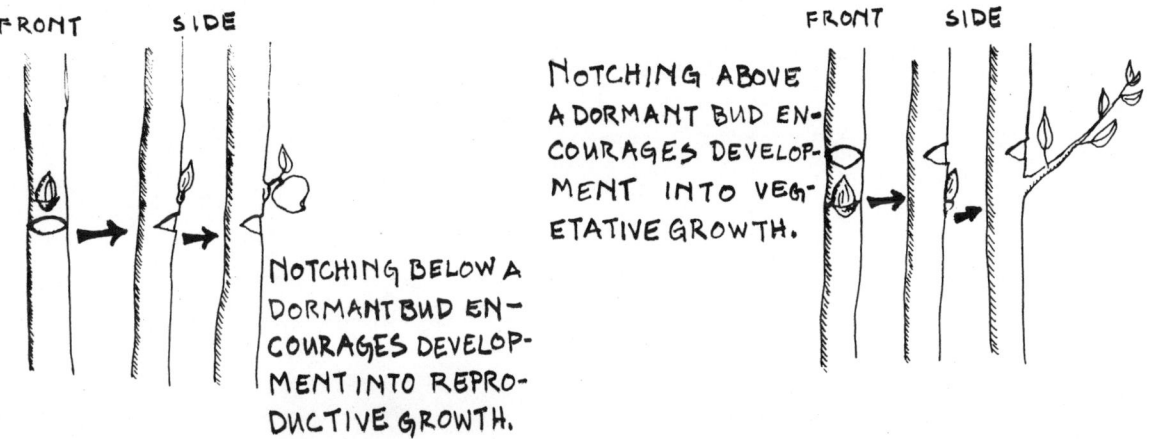

Figure 8.31 Notching to Influence Growth   (*After Kourik 1986:206*)

more room is made available in the garden. Trellises keep fruits such as tomatoes, cucurbits, and peas off the ground, making them less vulnerable to many pests and to rotting from contact with moist soil. Trellised plants can also shade people and other plants. However, where there are strong, drying winds, trellising may not be a good idea because it exposes more leaf surface area, leading to increased transpiration.

Trellis designs take into account special climatic conditions. For example, in high-altitude or high-latitude drylands with cold nighttime temperatures during part of the year, planting beans, tomatoes, or other cold-sensitive plants against a south-facing rock pile is a good idea. This will serve as a trellis for these plants and the heat stored in the rocks during the day create a warm microclimate for the plants as the air temperature drops at night.

Some types of trellises are:
- Other trees and plants in the garden which are living trellises (Figure 8.32),

Figure 8.32 Maize Provides a Living Trellis for Bean Plants

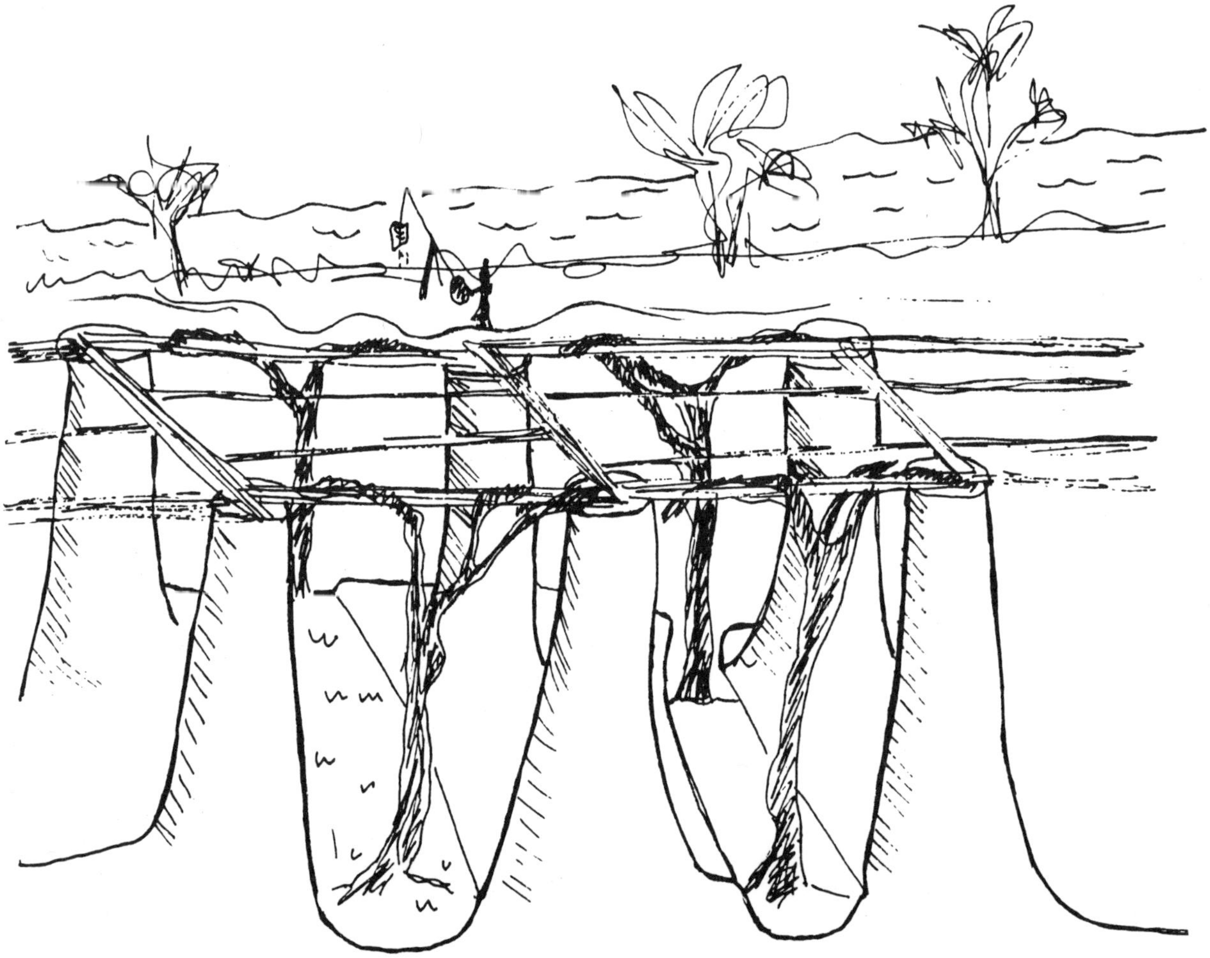

Figure 8.33 A Constructed Trellis for Grape Vines in Egypt

- Constructed trellises made of adobe, wire, string, sticks, cane, branches, and stalks (Figure 8.33). Plant fibers, vines, leather, string, wire, or cloth are useful for tying the parts together and tying the vines to the trellis.
- Existing structures such as houses, walls, or rock piles. In savanna West Africa pumpkin and luffa vines often cover thatch-roofed houses, providing extra shade and cooling from their transpiration,

8.9 Resources

Local gardeners and farmers are the most knowledgeable sources of information on plant management in their areas. Observing, working, and talking with them is the best way for outsiders to understand local techniques, resources, and problems.

Developing Countries Farm Radio Network (DCFRN) has brief scripts for radio programs about plant management topics such as nursery beds (#7-4), transplanting (#7-5, #10-5, #10-6), intercropping (#3-6, #7-1a), and weeds (#5-1a).

Regional books on wild or native plants such as *Traditional Food Plants* (FAO 1988), *The Weed Flora of Egypt* (Boulos and el-Hadidi 1984) and *La Culture des Légumes Indigénes au Nigéria* (van Epenhuijsen 1978) can help an outsider learn to identify important local plants and can be very useful when working with gardeners on weed management. Many gardening books have some information on pruning and there are also some specialized publications on that subject.

References

[1] Early 1977:42-44.
[2] Hatmann and Kester 1983:166.
[3] Garner, et al. 1976:10.
[4] Hartmann and Kester 1983:37-38.
[5] Wilken 1987:164-167.
[6] Tully 1988:130-131.
[7] Hartmann and Kester 1983:192.
[8] Arnon 1975:36; Larson 1975:159.
[9] Horticultural Abstracts 1984 55(5):283.
[10] Altieri and Liebman 1986.
[11] Altieri and Liebman 1986.
[12] Gliessman 1986.
[13] FAO 1982b:97-98.
[14] Rice 1984:1-2.
[15] Gliessman 1986:89-90.
[16] Rice 1984:173-178.
[17] Igboanugo 1986.
[18] Duffield and Jones 1981:158.
[19] Bhowmik and Doll 1984.
[20] Ruthenburg 1980:84.
[21] Davis and Bye 1982.
[22] Benz, et al. 1990.
[23] Chacón and Gleissman 1982.
[24] Bye 1981.
[25] Altieri and Liebman 1986.
[26] Rice 1983:52.
[27] Nabhan 1983.
[28] FAO 1982b:140.
[29] Dupriez and De Leener 1987:90.
[30] Kourik 1986:206.

9
Soils in the Garden

Loss of productive land to the growth of cities and roads, flooding from dams, waterlogging, salinity, soil erosion, and overcultivation is a serious problem in industrialized countries as well as in the Third World. The social organization of production, not merely technology, is a major factor in soil degradation.[1] An emphasis on short-term production, mechanization, reliance on chemical fertilizers, and extensive monocropping contribute to serious losses of soil resources in industrialized agriculture. Many indigenous production systems which at one time conserved soil resources no longer do so because of disruption of their social organization, population pressure, and environmental changes.

Sustainable production from dryland gardens should be based on management techniques that improve and protect the soil using local resources, and that are appropriate for the local social and economic situation. In this chapter we emphasize the importance of reducing wind and water erosion, and of maintaining adequate soil organic matter to ensure fertility and water-holding capacity.

9.1 Summary

Dryland soils can differ a great deal from one region to another and even from one side of a village or garden to the other. The valuable contributions that gardens can make to the household mean that locating the garden where there is good soil and/or improving the soil is worth the effort.

There are many different ways of classifying soils; some based on the way the soil can be used, others on specific physical properties of the soil. Texture, structure, porosity, permeability, color, temperature, profile, depth, pH, salt, and nutrient content are characteristics that help in choosing and managing soil. Often local systems of classification based on gardeners' experience will be the most useful.

Soil is made up of minerals (in sand, silt, and clay), water, air, and organic matter from dead plants and animals. In addition, there are usually many living things in dryland soils, including plant roots, algae, fungi, bacteria, and animals such as moles, beetle grubs, earthworms, and nematodes.

Plants use carbon dioxide and sunlight to make energy, but must obtain water, nitrogen, potassium, phosphorus, and other nutrients from the soil. The best way to make sure these nutrients are available is by regularly adding organic matter such as compost, animal manure, and green manure. Organic matter also improves the ability of the soil to hold water. Increasing and maintaining organic matter and water in the soil are the most important goals of soil management in drylands. Even most poor soils can produce abundantly when improved and managed.

Wind and periodic heavy rains can cause severe erosion problems in drylands where vegetation is sparse. Simple ways to prevent loss of fertile garden soils are decreasing rainwater runoff, modifying the slope, and improving the soil's resistance to erosion.

Garden beds concentrate improved soils into growing areas that can be easily managed and maintained. Beds may be raised or sunken, but for most drylands sunken beds are best because they help moderate soil temperatures and make the most of scarce water.

9.2 Soil and Land-Use Classification

Past and present climate, the rocks from which the soil developed, and topography all affect soil characteristics. Soils in much of the drylands are geologically young and relatively fertile, but are alkaline, and can have a high calcium content with layers of caliche (Box 9.3 in section 9.4). Soils in other parts of the drylands are geologically old and have been leached of basic mineral nutrients such as potassium (K$^+$), calcium (Ca^{++}), magnesium (Mg^{++}) and other cations. They tend to be acidic and may contain impermeable horizons of plinthite (Box 9.4 in section 9.4). These soils cover large areas of sub-Saharan Africa, South Asia, Southeast Asia, and South America.

There is a great deal of local variation in dryland soils. For example, in valleys the soil deposited by seasonal river flooding has a much higher clay and nutrient content, and higher water-holding capacity than adjacent soils (Figure 9.14 in section 9.7.1). Because they are often near a water supply in the dry season, these soils are ideal for gardens (section 11.6). Local variation in dryland soils can also be seen in the patches of younger, more fertile volcanic material pushing up into older, leached soils.

People use *soil classification systems* to help themselves organize different soil types for food production, and most gardeners and farmers have a system that they use in selecting and managing soils. Soils can be classified in many different ways based on their physical properties, profile and depth, plant nutrients, organic matter content, and how easily they are eroded. These topics are covered in following sections of this chapter. The diverse systems for using these characteristics to classify soils reflect the culture, environment, and needs of those who created them.[2] Soil classification systems used by soil scientists in the Third World are often influenced by those of former European colonial rulers. For example, there are differences in the systems used by neighboring Anglophone and Francophone countries in West Africa.[3]

Land-use classification includes not only soil type, but other factors that affect how the land can be used by people, such as tenure systems, social preferences, proximity to water supply, exposure to sun and wind, and yields under different cultivation systems.

For selecting or improving soil in garden sites, understanding local classification systems is the most appropriate place to begin. These local classification systems probably exist in all communities that grow crops, except perhaps the newest or most transient.

9.2.1 Indigenous Classification Systems

Indigenous systems often combine soil and land use classification. The system in southwestern Tlaxcala state in Mexico, for instance, is based on both inherent characteristics and how the soil responds to management.[4] Color and texture are the main characteristics used, but structure, saltiness, or depth may also be included. Inferior or superior ratings reflect anticipated yield, although farmers feel that the true nature of the soil can be known only after working it for several seasons. The ratings take into account the difficulty of the different management practices needed, including irrigation, fertilization, and cultivation, all of which depend on the characteristics listed above.

In semiarid areas of northeast Brazil farmers classify land using 15 specific terms for land types referring to location, soil type, and crop history, as well as many finer distinctions that govern planting of crops.[5] These terms are organized according to two major criteria: strong (fertile) versus weak (infertile) and hot (dry) versus cold (wet).

People often use the distribution and growth of plants and animals as indicators of soil quality, since these are affected by the water-holding capacity, water content, and fertility of the soil. For example, in northern Kenya, Turkana gardeners use the presence of small termites and certain species of trees as indicators of good garden sites.[6]

An excellent place for an outside worker to begin is by developing an understanding of indigenous knowledge including local soil and land-use classification systems (section 4.3.4). Compiling a list or catalog of the names and terms of the system and their meanings helps a field worker from outside the community learn the local system and the criteria on which it is based. This may also be valuable for giving field workers who are community members a new perspective on, and appreciation of, the local classification system.

9.2.2 The USDA Classification of Soils in Drylands

The United States Department of Agriculture (USDA) is promoting its system, the "Seventh Approximation,"[7] for classifying soils in the Third World. The USDA system organizes all soils into 10 different categories, or orders, based on how the soil is formed and under which climatic conditions. The orders are broken down into suborders, the suborders into groups, the groups into families, and the families into series, of which thousands have been defined. Because it is so widely used, we briefly describe the main soil orders found in drylands according to the USDA system: Aridisols, Alfisols, Entisols, Vertisols, and Oxisols.

Aridisols occur primarily in arid areas and are the most abundant USDA soil order in the warm drylands including the Sahara and part of the Kalahari Desert in Africa, deserts of southwestern North America, and from the eastern Mediterranean to India. They are the most common of the 10 major soil orders worldwide, accounting for 18.8% of the land surface.[8] During most of the time when the soil is warm enough for plant growth, water is unavailable to many plants (tension <-15 bars, discussed in section 10.3.1), is too salty, or both.[9] Because they developed under arid conditions, nutrients have not been leached out, and many Aridisols are relatively fertile, although they are alkaline with salt and caliche deposits. The Aridisols also have the sparsest natural plant and animal populations.

Next in abundance to Aridisols in the warm drylands are Alfisols and Entisols. Alfisols are slightly to severely leached, acidic in upper layers, and occur throughout much of savanna Africa, India, and northeast Brazil. Entisols, which characteristically lack distinct soil horizons, comprise most of the Kalahari Desert. Significant areas of Vertisols exist in western India, eastern Sudan, and eastern Australia. They have high clay content, especially montmorillinite, and so shrink and swell a great deal, making them difficult to cultivate. In drylands the low pH, intensely weathered, infertile Oxisols occur mainly in southern Tanzania and northern Mozambique.

9.3 Physical Properties of Soils

Physical properties of soils include texture, structure, porosity, permeability, color, and temperature. Understanding the physical properties of soils helps in selecting new garden sites, and in improving soils in existing gardens.

9.3.1 Soil Texture and Structure

The proportion of different-sized mineral particles (sand, silt, and clay) in the soil is known as *texture*. Texture is important in the garden because it influences movement and storage of water in the soil, the ease of bed preparation and cultivation, the amount of air in the soil, and soil fertility. Knowing the soil texture helps in deciding how the soil can be improved.

The smallest soil particles, less than 0.002 mm (0.00008 in) in diameter, are called *clay*. Clay plays a big role in soil fertility and structure (Box 9.1). The other categories of soil particles are *silt* (0.05-0.002 mm; 0.002-0.00008 in), and *sand* (2.0-0.05 mm; 0.08-0.002 in). *Gravel* and *cobbles* are larger than 2.0 mm (0.08 in) and are not considered in determining texture. *Loam* is the name given to a soil containing sand, silt, and clay (0-40% clay, and 10-80% each sand and silt). Most soils are mixtures, and are referred to as sandy clay, silt loam, sandy clay loam, and so on. In general, sandy soils are those with 70% to 80% or more sand, and clayey soils have more than 40% clay.

Box 9.1
Clay

Clay is the most important mineral portion of the soil for plant nutrition (section 9.5). It is formed in the soil from dissolved minerals resulting from the breakdown of bedrock. Clays are primarily crystalline sheet-like structures made up of oxygen (70-85%) with aluminum (Al) and silica (Si). Most clays also contain some iron (Fe), zinc (Zn), magnesium (Mg), and/or potassium (K). Most clays have a negative charge and attract positively charged cations which plant roots can remove and use as nutrients (Box 9.5 in section 9.5).

The type of clay in a clayey soil can be partially identified by the extent to which the soil shrinks and cracks when drying. Swelling, sticky clays such as *montmorillonite* expand when wet, and shrink and crack during drying. They have a high capacity to store nutrients, and are the common clays in the more arid drylands. Bentonite, a type of montmorillinite, is used in the bottoms of ponds because it seals them as it swells.

Kaolinite expands and shrinks very little with wetting and drying, and is the clay used for ceramic pottery. *Sesquioxide* clays are common in the older soils of the subhumid drylands, where intensive weathering has led to the leaching out of all Si, Fe, Mg, and Al. These soils are extremely fragile. They do not shrink and swell and are infertile and subject to irreversible hardening when the soil horizon is exposed to air through cultivation.[10] These layers are sometimes referred to as plinthite (Box 9.4 in section 9.4). Both kaolinite and sesquioxides have little ability to store nutrients in the soil and their presence indicates a need to improve the soil by adding organic matter.

Loams are considered the best soils for gardens because they share the properties of sand, silt, and clay. Sandy soils are low in nutrients and have a poor water-holding capacity. Clayey soils have poor drainage and are hard when dry, and sticky when wet, making them difficult to work. Both sandy and clayey soils can be improved by adding organic matter (section 9.6). It would be too much work to change the texture of the soil in most fields because of their large size, and lower productivity/area of land compared with gardens. Changing the texture of the soil in a garden, however, is practical and worthwhile, especially if suitable materials are nearby, such as sand from a streambed.

For most purposes, an adequate idea of soil texture can be obtained by simply looking at the soil and rubbing a sample between the thumb and finger.[11] The hard particles of sand can be distinctly felt, and if the sample is dry the crunching sound it makes can be heard when held close to the ear. Silt cannot be felt as individual particles but gives a smooth or soapy feeling. Clay is hard when dry, and slippery and sticky when wet. If a more precise idea is needed, the tests in Box 9.2 can be used.

Soil structure is the way in which sand, silt, and clay are grouped together in the soil to form *aggregates* or clusters of soil particles. Humus (section 9.6) is made up of material from the breakdown of organic matter and clay, two of the most important sources of cementing materials that help create aggregates. A good soil structure for gardens is one with many aggregates that create many spaces in the soil for the movement of the water and air needed for healthy roots and plants.

Large quantities of sodium, sometimes found in arid soils, lead to a loss of soil structure by causing negatively charged soil particles to repel each other, breaking apart aggregates. Clays and small organic molecules then plug soil pores (section 12.6.2). Because of this, salt (NaCl) is sometimes used to seal rainwater catchment areas or the bottoms of ponds.

9.3.2 Soil Porosity and Permeability

Soil *porosity* is a measure of the amount of spaces or *pores* in the soil that can be filled with either air or water. *Permeability* is a measure of the ease with which water and air move through the soil pores (section 10.3). A combination of texture, structure, and organic matter content determine the porosity and permeability of soils. For a good balance between small pores that retain a lot of water, and large pores that allow easy movement of air and water, medium-textured loam soils with good structure and plenty of organic matter are best.[14] While clayey soils have higher porosity, and so can hold more water than sandy soils, they also have lower permeability because the pores are so small. Soils with a small pore size (less than 0.03 mm; 0.0012 in), characteristic of clay soils, retain water by attraction forces and if most pores in the soil are of this size waterlogging can occur. Large pore size, and thus poor water-holding capacity, is a problem only in very sandy soils. Both situations can be improved by adding organic matter to the garden soil.

Soil porosity can also be improved by animals such as earthworms or termites living in the soil. In northern Burkina Faso farmers make depressions in the soil which collect plant debris blown by the wind, and to which they may add manure.[15] The organic material in the depressions attracts termites that make tunnels in the soil. The termites are usually gone by the time the farmers plant their crops and the tunnels increase water infiltration.

Cultivating the soil, that is, hoeing or plowing it, increases exposure to air, speeding the breakdown and loss of organic matter. This reduces soil structure, porosity, and fertility, as well as encouraging wind and water erosion, and eventually leading to lower garden production. However, clayey soils need to be cultivated to loosen the soil and to mix much needed organic matter into the root zone. Whether or not cultivation in the garden is needed or advisable depends on local conditions. The more the garden soil is cultivated, the more organic matter needs to be added, and the more attention must be given to controlling erosion (section 9.7).

9.3.3 Soil Color

The color of a soil is an indication of its mineral, humus, and water content.[16] Soil color can help in choosing a garden site and in understanding what soil improvements are needed.

- Reddish soil: location is generally dry with seasonal wetting, darker red means higher iron content.
- Grey, blue, or green soil: location is poorly drained and if waterlogged for long periods of time, spots (or mottles) of blues, greys, and greens often result, indicating lack of oxygen.
- Dark brown soil: within a small geographical area, darker brown often means higher organic matter content, although darker colors do not always mean a fertile soil.
- White soil: lime or salts are present, pH is high.

Box 9.2
Tests for Soil Texture

THE RING TEST   For the ring test a small sample of moistened soil is shaped and its qualities observed.[12] This test should be done the same way each time so that samples from different sites in the garden and from different gardens can be compared. The sample should be from the root zone (15-45 cm; 6-18 in), and gravel and cobbles should first be removed by screening. If available, window screen or mosquito netting (which often have openings with a diagonal measuring about 2 mm [0.08 in]) work well for screening the sample. If these are not available the most obvious stones and gravel can be removed by hand.

Slowly add water to about 1 tablespoon of soil, enough to form a ball approximately 2.5 cm (1 in) in diameter. Try to roll the ball into a cylinder about 15 cm (6 in) long (Figure 9.1). Sand cannot be shaped into a ball; loamy sand will form a ball, but cannot be rolled into a cylinder. Loam can be rolled into a cylinder; with a clay loam the cylinder can be formed into a ring, though it breaks easily. Clay can be formed into a ring without cracks, which holds together with handling. This test is used frequently by soil surveyors in West Africa and if done carefully gives a good idea of how the soil will behave under cultivation.[13]

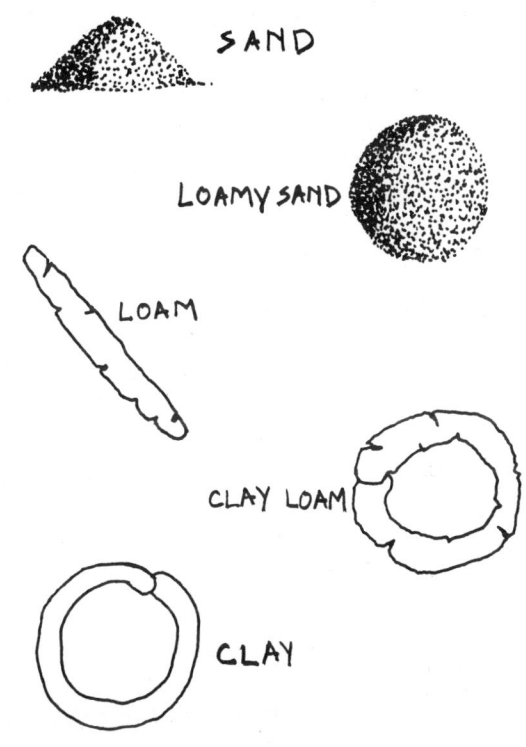

Figure 9.1 The Ring Test

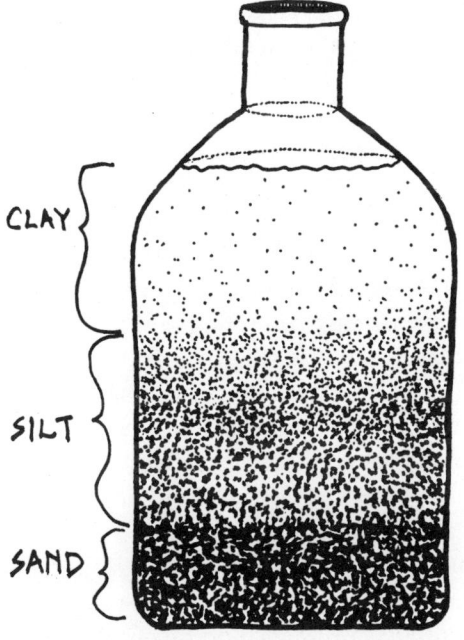

Figure 9.2 The Settling Out Test

THE SETTLING OUT TEST   Another way of finding out about the texture of garden soils is by taking a sample of soil (as with the ring test) and shaking it up well in a straight-sided, clear glass bottle filled with water (Figure 9.2). To help break up soil particles, several pinches (about 20 cc) of table salt (sodium chloride) and an equal amount of baking soda (sodium bicarbonate) can be added. After shaking, the sand settles out in about 1 minute into a layer extending just beyond the point where the particles become invisible to the eye. The silt will settle out in 1-2 hours, and then the clay begins to settle. Because clay particles are so small it may take several days for them to settle out and leave fairly clear water, although it would take months for all the clay to settle out. The relative widths of the different layers gives an idea of the composition of the soil texture.

9.3.4 Soil Temperature

Soil temperature affects seed germination and root growth, and is the result of many factors:
- Direction of slope: determines exposure to the sun. In the northern hemisphere north-facing slopes receive less sun than south-facing ones; in the southern hemisphere the opposite is true (section 8.3).
- Depth: with increasing depth below the soil surface, the less the soil temperature changes daily and seasonally, and the more changes in soil temperature lag behind changes in air temperature.
- Color: dark-colored soils absorb more solar radiation (heat) than light soils.
- Water content: water in the soil moderates temperature; large amounts of heat are used to evaporate water, keeping soil cool on hot days, and, large amounts of heat are given off when water freezes, keeping soil warm on cold nights.
- Surface cover: covering the soil surface with a mulch slows heating and cooling both seasonally and daily, and can be used to regulate soil temperature according to the time of year and crop (sections 5.7.2 and 10.8).

When cold soil slows seed germination or plant growth the following steps can help: planting on slopes exposed to the sun, encouraging deep root growth; planting in dark-colored containers or covering the soil with dark colored stones or other mulch to absorb heat during the day; mulching during cold nights to prevent heat loss; and watering regularly. In many mountainous or hilly areas with temperate climates, gardeners locate their gardens mid-slope, avoiding both valleys where the heavier, cold air settles, and high elevations where winds can blow warm air away.

When hot soil slows seed germination or plant growth the following can help: planting on slopes shaded from the sun; encouraging deep root growth; planting in light-colored containers or covering the soil with light-colored mulch; and watering regularly.

9.4 Soil Profile and Depth

A verticle section through the soil displaying its layers is called a *soil profile*. Soils are divided by soil scientists into three layers or *horizons* within a profile, referred to as A, B, and C (Figure 9.3). The A horizon is sometimes referred to as the *topsoil* and contains most of the organic matter in the profile, which often gives it a darker color. However, in arid areas there may be very little organic matter, and the topsoil can be a light color. The depth of the topsoil is important because this is where most of the roots of annual plants grow. Shallow topsoils have to be deepened by removing or improving of the underlying subsoil, or by making mounds or raised beds of improved soil.

At the point in the profile where the color, texture, or composition of the soil changes, the B horizon begins. The B horizon, commonly referred to as *subsoil*, accumulates material washed out of the A horizon, such as clay, salts, and iron. Any change in soil porosity between the A and B horizons will slow or even stop movement of water downward and tends to increase the water content of soil above it (section 10.3.2). In alkaline soils of arid and semiarid areas a white, nonporous layer composed mainly of calcium carbonate and known as caliche, can form in this horizon, and if it is too close to the surface it may cause problems with root growth and drainage (Box 9.3). In semiarid and subhumid areas with acidic soils a rock-like layer called ironstone can form (Box 9.4).

The C horizon is formed by decomposition of the parent material or *bedrock* over which it lies. It contains no organic matter and very few mineral nutrients available to plants.

In the garden, good quality soil should be deep enough to allow plenty of room for root growth, and good drainage of any excess water. For most annual vegetables 45 cm (18 in) deep is probably adequate (section 9.8). Many trees will need deeper, well-drained rooting areas to prevent stunting or waterlogging. Gravel or rocky layers like caliche and ironstone impede root growth and drainage and should be removed from the root zone and replaced with good soil. After digging a planting hole the drainage can be checked by filling it with about 15 cm (6 in) of water. If the water drains out overnight, drainage is probably adequate. If not, the hole may have to be dug deeper to reach a better draining part of the horizon, or a new planting site found. Salt buildup, waterlogging, or nutrient deficiencies in the B horizon affect plant growth, and are problems that must must be solved if deep-rooted plants, especially large perennials, are to be grown (sections 9.5, 12.6.1, 12.6.2).

9.5 Soils and Plant Nutrients

All living cells on earth, whether in animals, plants, or microbes, contain about the same elements in approxi-

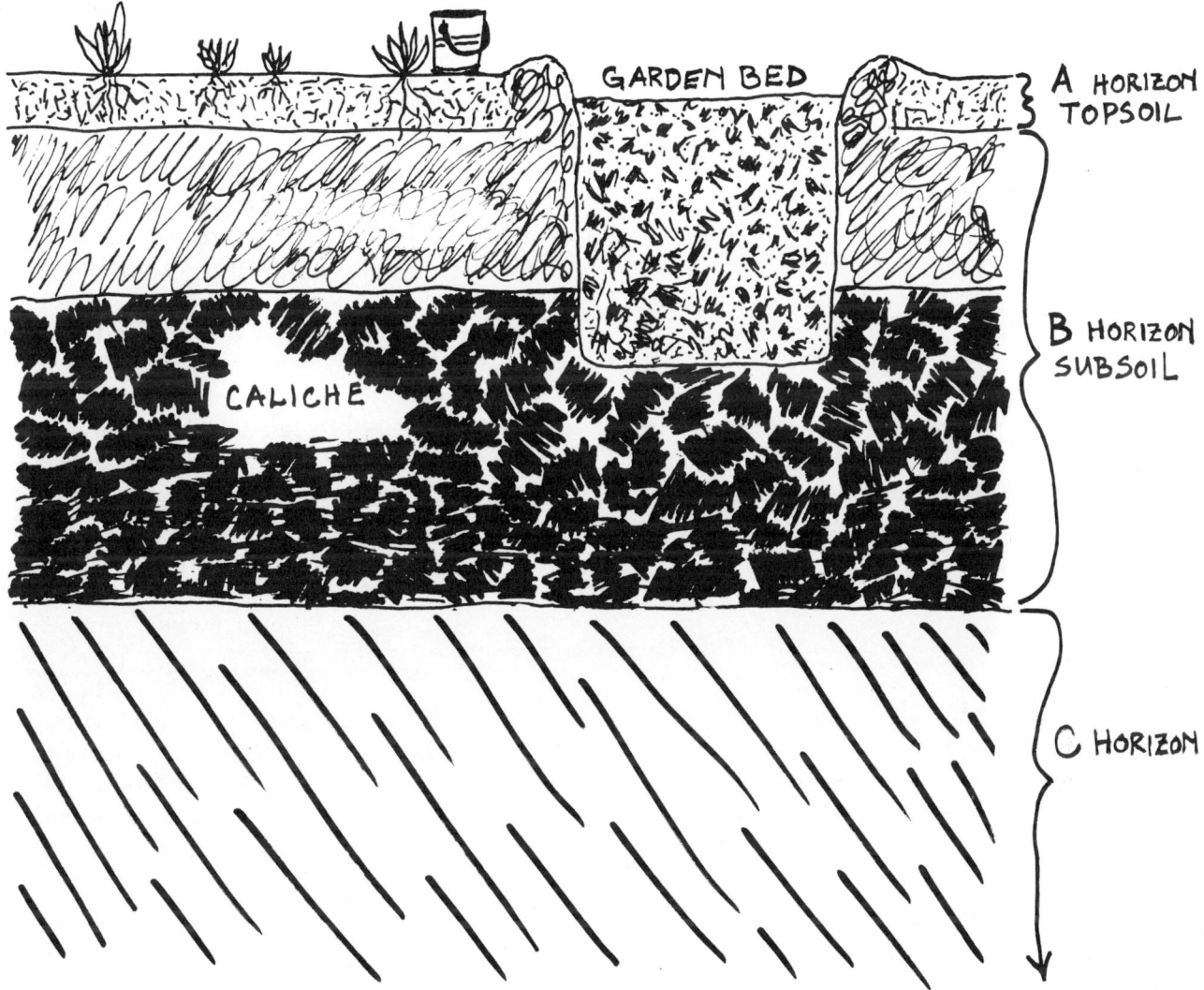

Figure 9.3 A Soil Profile in an Area with Aridisols

mately the same proportions. Sixteen elements are essential nutrients for plants. The six macronutrients, required in large amounts, are carbon (C), hydrogen (H), and oxygen (O) which are all obtained from air and water, and nitrogen (N), phosphorus (P), and potassium (K) which are absorbed from the soil. The secondary nutrients calcium (Ca), magnesium (Mg), and sulfur (S) are required in smaller amounts. The micronutrients chlorine (Cl), copper (Cu), boron (B), iron (Fe), manganese (Mn), molybdenum (Mb), and zinc (Zn) are required in very small amounts. Secondary nutrients and micronutrients are obtained primarly by absorption through the roots from the soil (Box 9.5). With the exception of boron, people also require these same elements (section 2.7). In addition, people need sodium (Na), chromium (Ch), selenium (Se) and iodine (I). Plants need an adequate supply of water, air, sunshine, and soil nutrients to be healthy and to produce a good harvest. When any one of these, including a specific soil nutrient, is in short supply, plant growth slows or even stops.

The soil nutrient supply available to the plant depends on:

- The ability of the crop species, the crop variety, and

> **Box 9.3**
> **Caliche**
>
> **Caliche**, or a calcic horizon, is common in arid and semiarid regions.[17] It is a hard, whitish alkaline layer resembling limestone and can be many meters thick. Caliche is composed mostly of calcium carbonate ($CaCO_3$), but magnesium carbonate ($MgCO_3$) is usually also present. It is the same composition as the "lime" that is often applied to acid soil to raise the pH. Caliche usually occurs in the B horizon, but in very arid areas, or where there has been severe erosion, it may extend to the surface.
>
> The sources of $CaCO_3$ are rainwater, wind-blown dust, and soil layers above the caliche. Carbonate from these sources in the upper soil is dissolved by carbonic acid (H_2CO_3) formed by a combination of water with carbon dioxide (CO_2) from plant roots and microbial activity, and moved into the lower layers by rainwater. When evapotranspiration is much greater than rainfall, $CaCO_3$ becomes more concentrated and forms solids. A decrease in CO_2 concentration below the root zone is also a factor. As the caliche layer builds, it may form a barrier to water penetration and cause further buildup toward the surface.
>
> Caliche layers in the root zone should be removed in order for most garden crops to do well. However, some trees such as olive can tolerate shallow, alkaline soils, and so grow sucessfully where deep caliche layers are near the surface. To remove very hard caliche it may have to be chipped away with a pick or hoe. Caliche can sometimes be softened by soaking with water. Dry-season garden beds can be started during the rainy season and excavated as they are moistened with rain.

> **Box 9.4**
> **Plinthite and Ironstone**
>
> Soil horizons rich in iron and aluminium oxides, usually leached from upper slopes, are known as *plinthite*. These soils tend to have low levels of plant nutrients, harden irreversibly when exposed to repeated cycles of wetting and drying, and form nodules or rock-like layers called *ironstone* or *laterite*. This process is speeded up when at or near the surface, and therefore often seems caused by excessive grazing or cultivation leading to erosion.[18]
>
> The climate under which these soil layers are found may have little or nothing to do with the climate when they were formed. Ironstone and laterite are found in some areas that today are subhumid, semiarid, and even arid, including the Sahel, parts of India, and Western Australia. One soil scientist estimates that there are as much as 250 million ha of soil in semiarid West Africa with laterite 5-25 cm (2-10 in) below the surface.[19]
>
> Like caliche, ironstone gravel or solid layers limit root growth and cause waterlogging by slowing drainage. Horizons with ironstone gravel have very low water-holding capacity (section 10.3.1) as do sandy, gravelly, or rocky soils in general. Soils with ironstone horizons should be avoided as garden sites unless the overlying soil is deep enough to support plant growth and the slope and topographic position allow drainage. Ironstone outcrops tend to occur in the middle of slopes where there is the most erosion (section 9.7.1). If ironstone is present, a better garden site may often be found in a nearby low-lying area that is less eroded, and where there is also likely to be more fertile soil, and better access to water, whether from wells or runoff.

the individual plant to obtain nutrients in relation to other plants with which it is growing.
- The chemical composition of nutrients and their relative amounts.
- Soil conditions such as the temperature, moisture, pH, porosity, and cation exchange capacity (CEC) (Box 9.5).

Nutrient deficiencies often show up as changes in leaf color, including *chlorosis*, a change from darker

> **Box 9.5**
> **Nutrient Uptake by Plant Roots[20]**
>
> Most soil nutrients are in the form of electrically charged particles called *ions*. Positively charged *cations*, for example ammonium (NH_4^+), calcium (Ca^{++}), potassium (K^+), ferrous and ferric iron (Fe^{++} and Fe^{+++}), and zinc (Zn^{++}) are attracted to negatively charged soil particles, primarily clay and humus. Negatively charged *anions*, for example nitrate (NO_3^-), phosphates ($H_2PO_4^-$ and HPO_4^-), and sulfate (SO_4^-), are attracted to positively charged soil particles. The *cation exchange capacity* (CEC) is a measure of the amount of exchangable cations per 100 gm of soil. It is determined by the amount of clay, and of humus from the breakdown of organic matter. Montmorillonite and vermiculite are the clays with the highest CEC, illite has less, and kaolinite and sesquioxides have a very low CEC.
>
> Plants absorb most of the water and minerals they need through the root hairs that grow on younger roots. As water increases to near field capacity, nutrient uptake increases because transport of nutrients through the soil by flow and diffusion increases, as does root growth and transpiration. Most root hairs are only about 0.01 mm (0.0004 in) in diameter and a few mm (0.1 in) long, and are actually extensions of single epidermal cells (section 5.2). The root hair cell walls are coated with a sticky substance that allows them to cling to soil particles, increasing their ability for absorption.
>
> Nutrients are absorbed from the soil in a process that uses energy from plant food created by photosynthesis. Once nutrients and water have been absorbed, they move throughout the whole plant, carried by the flow of water or sap in the vascular system (sections 5.2 and 10.3.4).
>
> Recently, another factor affecting nutrient uptake has been recognized. Symbiotic relationships called *mycorrhizae* can occur between certain soil fungi and the roots of flowering plants including garden fruits and vegetables. Mycorrhizae make nutrients, especially phosphorus, more available to the plant (section 5.2.1), and help the plant resist some fungal diseases. Absence of the mycorrhizae results in stunting and poor plant growth.[21]

green to lighter green and yellow due to a loss of chlorophyll. Care must be taken to distinguish these signs from those caused by pests and diseases (section 13.4.2).

Growing food involves managing plant communities to increase production of fruits, nuts, roots, or leaves that people want. When these products are harvested nutrients are removed from the garden, upsetting the natural cycling of nutrients, and making management of soil nutrients necessary. Figure 9.4 shows the nutrients needed by plants cycling through humans, animals, plants, microorganisms, soil, and the air. Within the soil bacteria, fungi and animals like earthworms and millipedes are constantly recycling organic matter.

Gardeners' understanding of this cycle is shown when they replace nutrients by fallowing, rotating crops, growing cover crops, and returning plant, kitchen, human, and animal wastes to the garden, either directly or after composting. Gardens planted on or around recent animal pens, trash heaps, or latrines take advantage of the enriched soil there. Deep-rooted plants such as many fruit and nut trees can absorb nutrients at deep soil levels and make them available to shallow-rooted plants when their leaves drop to the soil surface and decompose. Nitrogen supplies can be increased by growing nitrogen-fixing leguminous crops (Box 9.7 in section 9.5.2). An adequate supply of nutrients in most gardens can be maintained by adding plenty of organic matter to the soil, and by interplanting and rotating crops.

Persistent problems for which other causes cannot be found may indicate a problem with soil nutrients, and a soil test may be appropriate. Very inexpensive kits that test for pH, nitrogen, phosphorus, and potassium may sometimes be available and are relatively easy to use. In addition, local agriculture departments in some areas of the world provide soil testing to farmers and gardeners.

9.5.1 Soil pH and Plant Nutrition

The *pH* of the soil is a measure of how acid or alkaline it is. pH is important to the gardener because it has a strong effect on the availability of nutrients to both crops and soil microorganisms (Figure 9.5). In most

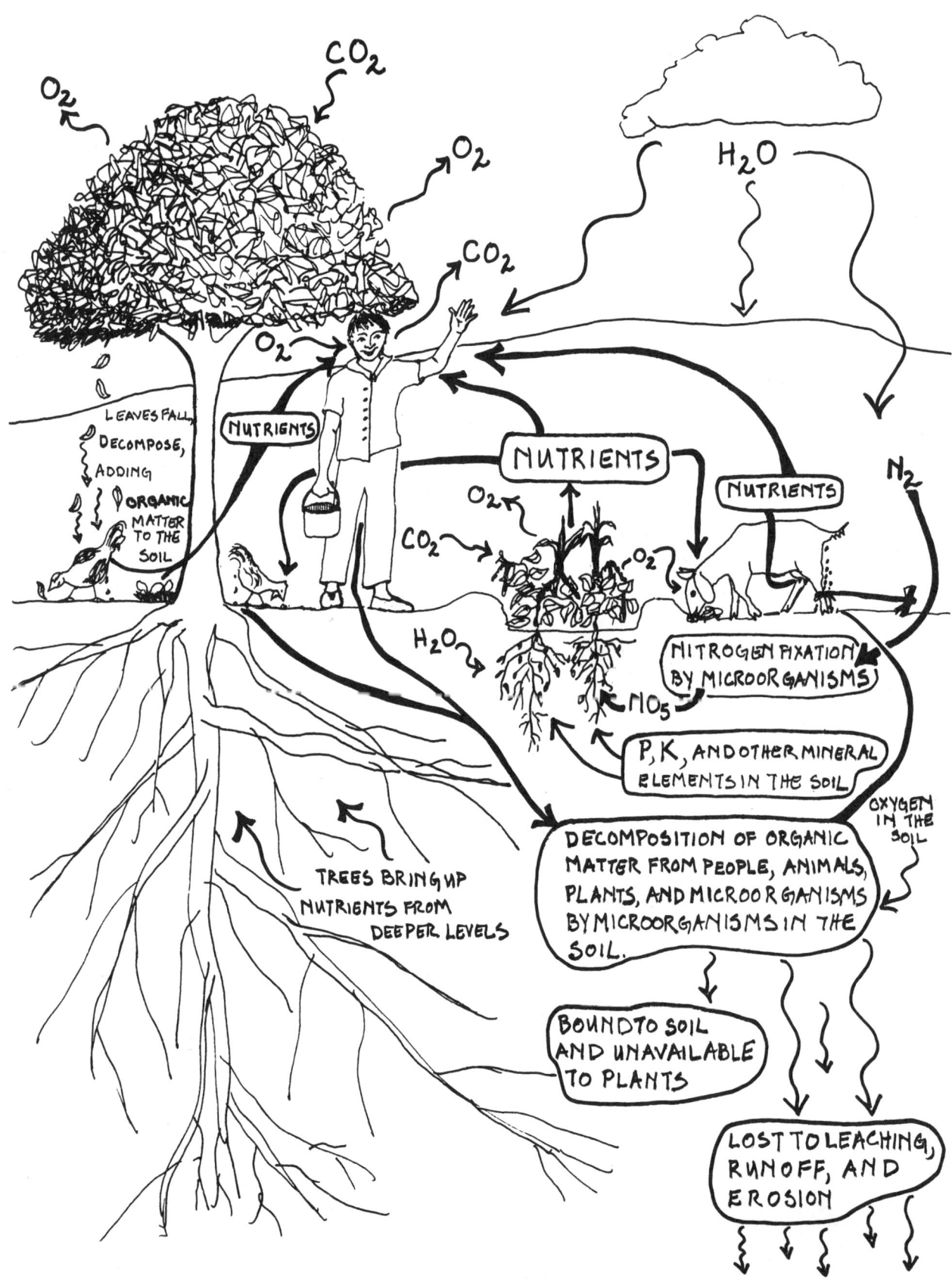

Figure 9.4 Nutrient Cycles in the Garden

soils the pH is between 3.5 and 10. Garden soils should generally have a pH between 6 and 7.5. A neutral soil has a pH of 7, an acid soil below 7, and an alkaline or basic soil above 7. On the pH scale an increase of 1 means that there is a tenfold increase in alkalinity, and a decrease of 1 means a tenfold increase in acidity (Box 9.6). For example, a soil with a pH of 7 is 10 times as alkaline as a soil with a pH of 6, 100 times as alkaline as a soil with a pH of 5 and so on. A pH of 6 to 7 in garden soil makes many nutrients easily available, but a pH up to 8 is still fine for the majority of crops.

Soils in arid areas tend to be neutral to alkaline (pH 7 or higher) especially in the topsoil, because of the lack of rain to wash out basic cations and release acidic anions from the clay.[22] The more arid the region, the higher the pH. In areas with less than 600 mm (24 in) of annual rainfall pH is often as high as 8-8.5. Topography also affects pH, and the middle of slopes tend to be better drained and therefore more leached and more acidic than adjacent bottomlands or valleys.

Basic or *alkaline soils* have a pH greater than 7 and often contain relatively large amounts of exchangeable (available to plants) calcium and magnesium, with lesser amounts of potassium and sodium. As soils become more alkaline the availability of some nutrients decreases (Figure 9.5). Gardeners can reduce the pH of alkaline dryland soils by the addition of organic matter. Sometimes elemental sulfur is also added. Sulfur lowers pH by first being oxidized to sulfate by bacteria, and eventually becoming sulfuric acid. To increase bacterial action, the surface area of the sulfur should be maximized by applying it as a powder.

Acid or *acidic soils* have a pH of less than 7 and also occur in drylands, for example, when parent material is an acidic rock like granite, or when a former wet climate leached the soils. In much of the West African savanna underlying granitic bedrock and many years of weathering have resulted in acidic soils.

When the pH falls below 6, some nutrients become less available (Figure 9.5). The pH of acidic soils can be increased through the addition of alkaline substances like lime ($CaCO_3$) or ashes. Ashes are an indigenous soil amendment in many parts of the world. Their value to the farmer and gardener is shown by the fact that, like animal manure, they are applied to individual plants, for example in West Africa,[23] Mexico, and Central America.[24] Ashes from home fires also reach gardens and fields through trash heaps and compost piles (Box 9.10 in section 9.6.2). While burning crop residue and weeds leads to a loss of most of the nitrogen and sulfur, other nutrients, especially potas-

Box 9.6
pH

The pH scale is a *log scale*, with pH expressed as the negative log of hydrogen ion (H^+) concentration in the soil. When water molecules come apart ($HOH \longrightarrow H^+ + OH^-$) both cations ($H^+$) and anions ($OH^-$) are at a concentration of 10^{-7} moles (molecular weights) per liter, therefore a pH of 7 is neutral. A soil with a pH of 8 has one-tenth as much H^+ and is 10 times as basic (alkaline) as a neutral soil; a soil with a pH of 6 has 10 times as much H^+ and is 10 times more acidic than a neutral soil; a soil with a pH of 5 has 100 times as much H^+; and so on.

sium, become more concentrated, easier to turn into the soil, and wash more quickly into the root zone than bulky plant residues.

9.5.2 Nitrogen

Nitrogen (N) is one of the most important nutrients required by both plants and animals. Even though the air we breath is 78% nitrogen, we cannot obtain

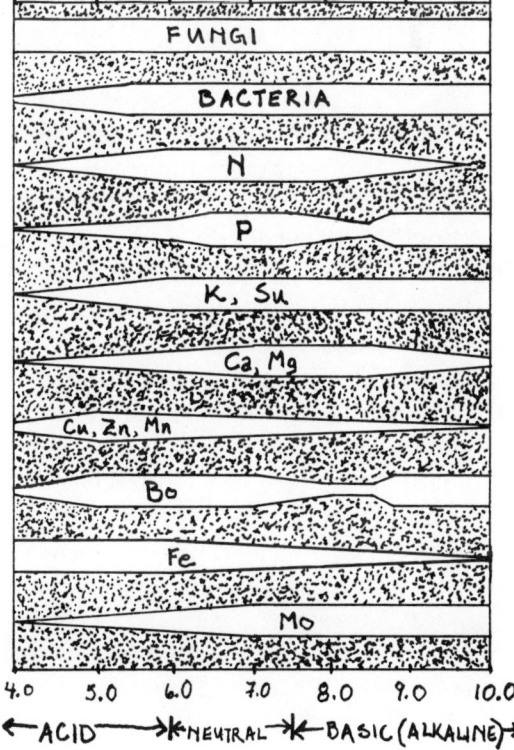

Figure 9.5 Soil pH and Nutrient Availability

nitrogen directly from the air. We depend on microorganisms in the soil to make nitrogen in the air available for plants, and then we eat plants (or animals that eat plants) to obtain the nitrogen we need (Figure 9.6). Plants must have nitrogen to produce amino acids, which are then used to make plant proteins. When eaten by humans, the protein in plants is broken down into amino acids and resynthesized into human protein (section 2.5). Nitrogen is also an important component of DNA.

A key symptom of nitrogen deficiency in plants is whole leaf chlorosis, especially in older leaves (section 13.4.2). Deficient plants may also show slow growth and drying out along the edges of leaves.

Nitrogen is made available to plants by microorganisms in two ways: by fixing nitrogen from the air, and by releasing nitrogen by breaking down organic matter. Some microorganisms fix nitrogen in the air into ammonium (NH_4^+) and nitrate (NO_3^-) which can be taken up by plant roots. Some of these are bacteria that grow in nodules on the roots of legumes (Box 9.7). Bacteria quickly convert most NH_4^+ to NO_3^-, which is very soluble and quickly leached from the soil. The majority of nitrogen in the soil is stored in living and dead organic matter. Therefore, plants depend for nitrogen on the continual decomposition of dead microorganisms, and of all sorts of other organic material such as dead plant roots and soil organisms to make nitrogen available. Gardeners increase this supply of soil nitrogen by adding animal manure, *green manure* (fresh green plant matter), or compost, and by planting legumes. Chemical fertilizers are discussed in Box 9.8.

Box 9.7
Nitrogen Fixation

A critical link in our food web are the microorganisms that convert nitrogen in the air into forms that can be used by plants. Symbiotic bacteria and actinomycetes (another type of bacteria) in root nodules, and free-living bacteria and cyanobacteria (formerly known as blue-green algae) in the soil, all fix nitrogen.

Many garden crops in the legume family, such as cowpea, tepary bean, fava bean, African locust bean, and carob, fix nitrogen in root nodules inhabited by rhizobia, bacteria of the genus *Rhizobium*. Such crops can supply nitrogen to the soil and so to future crops that do not have rhizobia, thus reducing the need for nitrogen fertilizers in the garden. The bacteria enter the root hairs when the plants are young, causing the legume root to form tumor-like growths known as *nodules*. These nodules are different than the knots formed by root knot nematodes (section 13.3.2), because the nodules can be rubbed off without breaking the root apart. The legume protects the rhizobia and supplies it with food for energy, while the rhizobia fixes atmospheric nitrogen into forms that the legume can use. This is a symbiosis, a relationship in which both organisms benefit.

Specific *Rhizobium* species induce nodules in certain species of legumes. Local legume varieties are more likely to become colonized by local rhizobia than imported legumes. Many commercially available legume seeds will need to be inoculated with the appropriate variety of rhizobia that colonize them, since the rhizobia may not be present in local soils.

Leguminous garden crops can supply nitrogen to other crops by rotating the legumes with nonlegumes, or composting leaves, stems, pods, and other parts for later use in the garden (for example carob pods or peanut leaves). In addition, legumes can be interplanted with other crops as in the familiar example from dryland Mexico of corn, with a high nitrogen requirement, and beans. However, most nitrogen is probably contributed by legumes in the season after they die, when their leaves and roots decompose in the soil. It is a good idea to leave the roots of annual legumes along with their rhizobia in the soil where they can serve as a source of bacteria for later plantings as well as provide nitrogen.

Rhizobial growth and metabolism is affected by the carbon to nitrogen ratio (C:N) in the plant in which it is living, and the amount of phosphorus, calcium, magnesium, molybdenum, and boron in the soil. The number of nodules is not a good indicator of nitrogen fixation. The bacteria may even become parasitic and remove nitrogen from the plant. An easy test for gardeners is to cut some of the nodules open. If they are reddish they contain leghemoglobin and are fixing nitrogen; if they are hard and greenish inside they are not fixing nitrogen and may actually be removing it from the plant.[25]

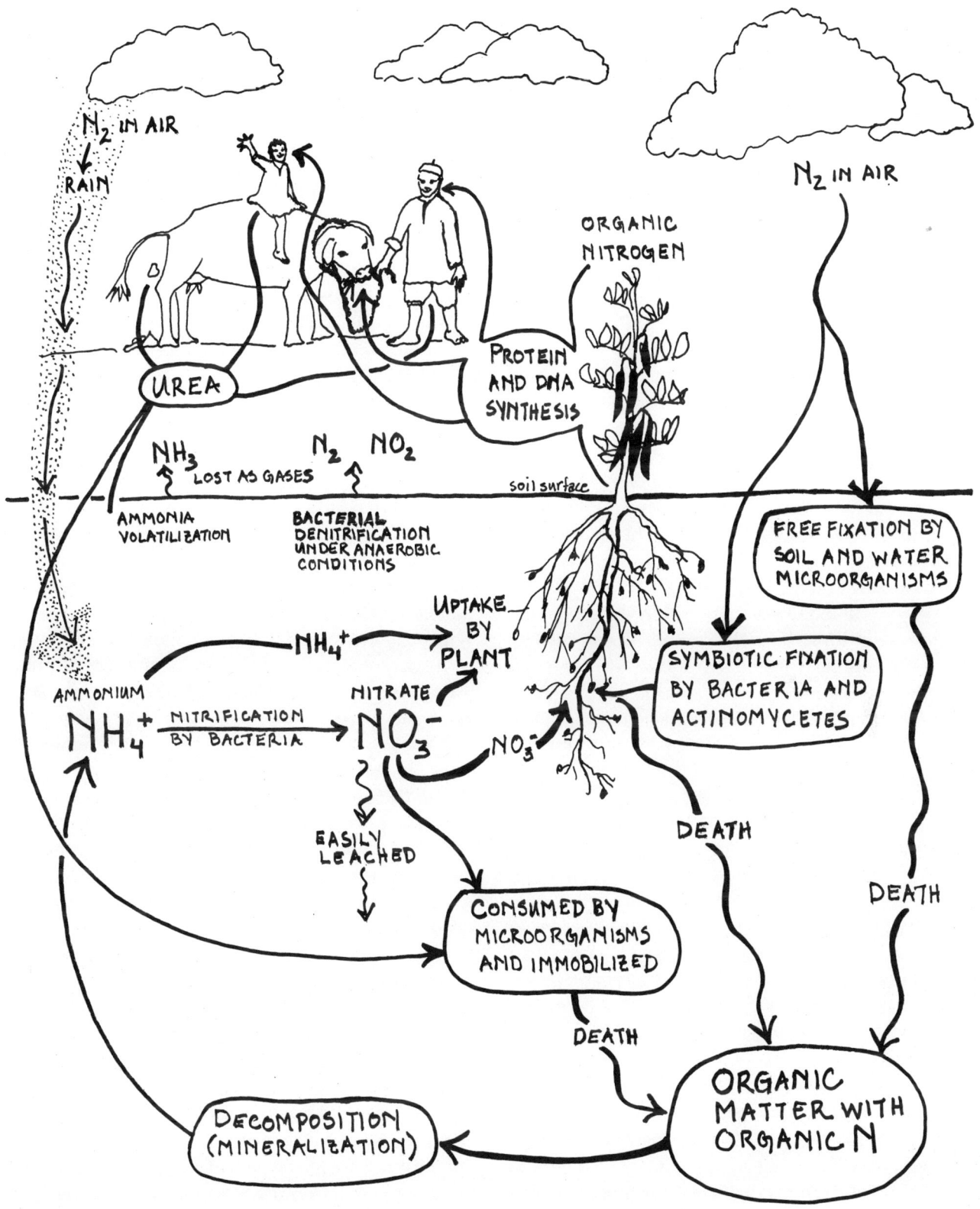

Figure 9.6 The Nitrogen Cycle

9.5.3 Phosphorus and Potassium

Phosphorus (P) is a component of DNA and proteins in plants and animals. Bones and teeth are mostly compounds of phosphorus. It is especially important for young plants which absorb it rapidly, and for fruit and seed production. Phosphorus deficiency causes slow growth and stunting and can lead to a reduction in normal opening of the stomata and thus to an increase in leaf temperatures and overheating (section 5.4). A reddish purple color in the leaves may be another indicator, although this color is also caused by other factors in some plants.

Soluble phosphate (H_2PO^{4-}) reacts rapidly with soil ions to form insoluble compounds inaccessible to plants. Therefore most phosphorus used by plants comes from organic matter broken down in the soil by mycorrhizae (Box 9.5 in section 9.5). At a carbon:organic phosphorus ratio of 200:1 or less, phosphorus is readily released, but when the ratio is 300:1 or greater the

---

Box 9.8
Commercial Chemical Fertilizers

"Chemical" is a word popularly used to describe synthetic fertilizers and pesticides that are commercially manufactured. We will use it in this way because it has become accepted, even though everything on the earth is made of chemicals. Chemical fertilizers contain high concentrations of a limited variety of nutrients in simple chemical compounds, compared with organic matter which has lower concentrations of a wider variety of nutrients in much more complex chemical compounds.

Chemical fertilizers are relatively expensive and their supply is beyond the control of the gardener. In addition, their manufacture is energy intensive and often harmful to the environment. Fertilizers available to project officials, frequently at subsidized prices, may not be available to gardeners when the project is turned over to them, or ended. Therefore, using chemical fertilizers at the beginning of a project because they are easier to obtain and apply, and provide quicker results than organic fertilizers, may jeopardize the project's long-term success.

Chemical fertilizers can also have a bad effect on soil quality. The high concentrations of nutrients in chemical fertilizers encourage rapid plant growth, using up the organic matter in the soil. If the organic matter is not replenished the soil structure will be destroyed.

Gardeners who recycle garden and kitchen refuse, use nitrogen-fixing legumes, and have trees in their gardens do not need to use chemical fertilizers. We give a brief overview of how chemical fertilizers work because they are often promoted as part of garden projects, although we advise strongly against the use of these and other manufactured agrochemicals in the garden.

Grades or nutrient content of fertilizers are usually given as percentage by weight of their elemental nitrogen (N), phosphorus pentoxide (P_2O_5), and potassium oxide (K_2O) content, in that order (Figure 9.7). However, fertilizers are commonly manufactured to include other nutrients such as sulfur, and the effect of a given macronutrient fertilizer may actually be a response to another component of the fertilizer which is the limiting nutrient in the soil. For example, in northern Ghana, peanut yields increased in response to the sulfur in single superphosphate fertilizer and not the phosphorus.[26]

Because the nutrient content is so high compared with organic fertilizers, chemical fertilizers can create an excess of a particular nutrient, upsetting the balance of nutrients in the soil. This makes it impossible for plants to take up the needed amount of another nutrient; this condition is referred to as an *induced deficiency*.[27] To determine deficiencies, before chemical fertilizers are applied, a soil test should be done and then interpreted on the basis of the actual responses of specific crops already growing there.

Nitrogen in chemical fertilizer is supplied either as the ammonium cation (NH_4^+) or the nitrate anion (NH_3^-). Urea (NH_2CONH_2) is rapidly converted to ammonia after application to the soil. Manufactured urea is a solid, water-soluble fertilizer. Nitrate fertilizers do not lower pH, but ammonia-containing fertilizers do because the ammonium cation replaces other cations like Ca and Mg, which are then leached out. For this reason ammonia should not be used as a source of nitrogen in gardens with acid soil. Ammonia will be held in solution as NH_4^+ for a short time, especially in soil with high clay and organic matter. Nitrogen fertil-

phosphorus is immobilized in the bodies of microorganisms. As is also true with nitrogen (section 9.6.2), too much high carbon material in garden soils or compost makes these important plant nutrients unavailable, reducing production.

Potassium (K) is very abundant in plants and important in many plant processes including cell division, carbohydrate formation, movement of sugars, some enzyme actions, and disease resistance in some plants. Uptake of potassium is especially high during early plant growth. The potassium-containing soil minerals are micas and feldspars which dissolve slowly. These sources supply half of the potassium available to plants, while decomposing organic matter supplies the other half. A pH of 6.0 to 6.5 is best for potassium availability.

9.5.4 Other Nutrients

Sulfur, calcium, and magnesium are minor nutrients

Box 9.8, continued

izer, especially nitrate, is easily washed into the soil, but is also easily leached out of the root zone, and so multiple applications are often recommended.

Powdered phosphate-containing rocks can be applied to supply phosphorus, but they are not very soluble. The finer they are and the more they are mixed with soil, the more quickly they will release phosphorus. The most common water-soluble phosphate fertilizers, single superphosphate and triple superphosphate, are manufactured by mixing rock phosphate with an acid. After application the phosphate in these fertilizers is quickly fixed into insoluble forms unavailable to plants. Therefore, it is usually recommended that these fertilizers be applied in concentrated areas near the seed or growing plant. Phosphate is often applied as *side dressing*, that is, close to plant roots such as along the side of a row, so plants can use it quickly.

Most potassium fertilizers are water-soluble, and are produced by mining large deposits of soluble potassium salts. Potassium chloride, or muriate of potash, is a common potassium fertilizer.

The oxides, hydroxides, and phosphates of iron, zinc, copper, and manganese are insoluble but the sulfates are soluble and can be sprayed on leaves where they are absorbed through the stomata. *Chelates*, metal ions bonded to complex organic molecules, are also used. Chelates, which also occur naturally in organic matter, resist fixation in the soil but are still available to plants.

Figure 9.7 Commercial Fertilizer Labels

Figure 9.8 The Living Soil

since they are needed in smaller amounts than nitrogen, potassim, and phosphorus. Chlorine, copper, boron, iron, manganese, molybdenum, and zinc are micronutrients, and are needed by plants in very small quantities.

Symptoms of sulfur deficiency include chlorosis and retarded growth. Soil organic matter is a major source of sulfur. Most soils contain adequate calcium, especially alkaline soils. Calcium is needed in large quantities for cell division, and therefore a deficiency affects the rapidly growing root and leaf tips, causing browning, rotting, and deforming. An example is leaf tip burn in lettuce and cabbage. Blossom end rot of tomatoes and peppers is caused by a calcium deficiency, and is often caused by dry soil, which makes the calcium in the soil unavailable to the plants.

Iron and zinc deficiencies, showing up as chlorosis between the veins of new leaves, are common in alkaline soils or when there are high levels of caliche (calcium carbonate, $CaCO_3$).

9.6 Organic Matter

Soil is full of living organisms. A small handful (1-5 gm; 0.05-0.15 oz) of most garden soils contains over 1 million bacterial cells, hundreds to thousands of fungi, hundreds of nematodes, and large numbers of other microorganisms (Figure 9.8). Larger organisms living in the soil include insects, millipedes, earthworms, and mammals like moles, mice, and rabbits. Roots of growing plants are another living part of soils. When these organisms die, microorganisms decompose them, releasing their nutrients for further plant growth. Harvesting garden produce removes nutrients which must be returned to the garden. The most convenient source of nutrients, and the best for the garden, is additional *organic matter*, the remains of plants and animals of all sizes. The addition of fresh plant remains (green manure), animal manure, and compost to the soil is a widespread indigenous practice of farmers and gardeners in drylands.

In an intensively cropped dryland garden organic matter decomposes rapidly because of high temperatures, high moisture, strong structure with plenty of soil air, pH between 6 and 8, and an abundance of nutrients. In addition to decompostion, organic matter in dryland gardens is lost through harvesting of plant parts, and cultivation of the soil, exposing it to oxygen in the air which increases the loss of organic matter through oxidation. This is why adding organic matter to the garden soil is probably the most important part of dryland soil management. Table 9.1 shows the approximate nutrient content of some different types of dryland organic matter. Box 9.9 lists some of the advantages of adding organic matter to dryland gardens.

Availability of organic matter may be the greatest limitation on its use in dryland gardens. Organic matter content of as little as 3-8% of the soil weight or higher improves plant growth.[28] When first making garden beds or planting holes, 25-50% of the soil volume can be compost. (See section 8.4.2 for discussion of why holes for transplants may be an exception.) Organic matter is much lighter than the mineral components of soil, therefore, a volume of soil would weigh much more than an equal volume of organic matter.

Humus consists of the small particles resulting from the breakdown of plants and animals. Humus continues to be decomposed by microorganisms that use the nutrients to multiply; as they die, their bodies are decomposed by other living microorganisms. As humus and microorganisms break down they make nitrogen and other plant nutrients in forms that are available to plants while releasing H_2O and CO_2. In effect, the soil humus and living microorganisms act as a nutrient storage system. Humus also improves soil structure by acting as a bond to hold together soil particles in aggregates (section 9.3.1), creating space for air and water in the soil, and increasing infiltration rates.

Gardeners often use different kinds of organic matter in their gardens. For example, in Zimbabwe manure from cows and goats, compost, and soil from termite hills is used in irrigated valley gardens, along with some chemical fertilizer.[31] Because of the high use of organic matter the quality of the soil was maintained, whereas it would not have been if only chemical fertilizer was used. Tests found that the organic matter content in gardens (3.5%) was almost the same as in noncultivated areas outside of the gardens (4%), showing that gardeners were replacing the organic matter being lost through cultivation and harvesting.[32]

9.6.1 Animal Manures

Gardeners and farmers are usually well aware of the value of animal manure (dung or feces, and urine) for productive crops. Most often manure is from domestic animals such as cows, horses, goats, sheep, chickens, ducks, and pigs. However, bats, wild birds, or other wild animals may also be a source. In savanna West

Table 9.1 Approximate Nutrient Content of Some Dryland Organic Matter [a]

Source	Nitrogen (N) (% dry weight)	Phosphorus (P) (% dry weight)	Potassium (K) (% dry weight)
Poultry droppings	2.1	1.1	0.37
Brewery waste[b]	3.2	0.6	0.37
Farmyard manure	1.7	0.4	0.38
Cocoa husks	1.1	0.24	4.4
Rice straw	0.5	0.16	1.3
Cowpea husks (pods)	1.4	0.14	2.0
Yam peelings	0.2	0.13	2.7
Sugercane trash	0.08	0.06	0.4
Groundnut husks	0.3	0.08	0.6
Maize waste (cobs)	0.7	0.07	0.5
Plantain peelings	0.8	0.07	2.2
Orange peelings	0.75	0.04	0.30
Maize stem	1.1	0.07	1.20
Melon waste	0.6	0.05	0.2
Cassava peelings	0.65	0.04	1.6

[a] From Titiloye, et al. 1985.
[b] "spent grain"

Box 9.9
Advantages of Organic Matter

Organic matter from many different local sources is a high-quality, low-cost resource for maintaining dryland garden soil fertility (Figure 9.9). It provides the following benefits to garden soils as it decomposes to humus:[29]

- Is the source of 90-95% of soil nitrogen, including that which is cycled through microorganisms.
- When it makes up more than 2% of soil, it can be the major source of available phosphorus and sulfur.
- Is a major source of the cements necessary for aggregate formation to create strong soil structure (section 9.3.1) with a higher proportion of larger pores, which improves water-holding capacity and water and air movement.
- May furnish 30-70% of the negatively charged sites that hold nutrient cations plants can use (Box 9.5 in section 9.5). This electrical property also gives organic matter the ability to act as a *buffering agent*, moderating the tendency to change pH when acid or alkaline substances are added to the soil.
- Acts as a chelate, that is, it forms compounds with metal nutrients (usually iron, zinc, copper, or manganese) increasing their solubility and availability to plants.
- Supplies carbon for energy to many soil microorganisms that perform beneficial functions such as nitrogen fixation.
- In cities where garden soils contain lead from exhaust fumes of vehicles or lead-based paints, soils containing 25% or more organic matter significantly reduce the uptake of the poisonous lead by garden crops.[30]
- Acts as a mulch on the soil surface (section 10.8.1).

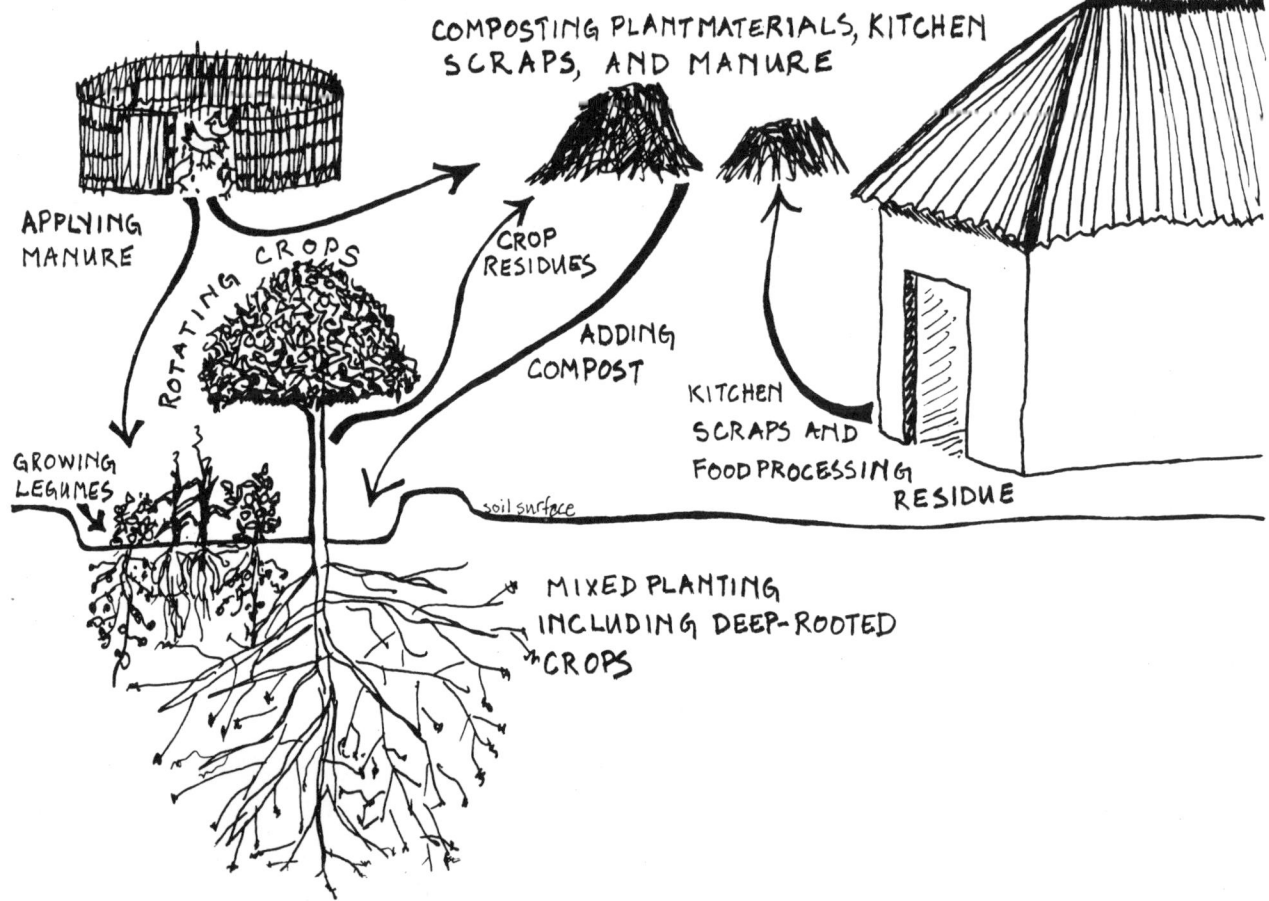

Figure 9.9 High-Quality, Low-Cost Methods of Maintaining and Improving Soil Fertility

Africa the fields that are closest to the house and are most intensively cultivated contain the rainy-season gardens. Here precious animal manure is applied at the end of the dry season to improve the soil. The Yatenga Mossi of northern Burkina Faso use the manure from penned animals in their gardens.[33] It is the richest manure because in addition to dung, it contains urine which is higher in nitrogen and potassium than dung. In southeastern Burkina Faso, Mossi gardeners, especially when growing some of their produce for market, purchase manure from Fulani herders.[34]

Manure can lose many of its nutrients if not properly handled when fresh. Left in piles, nutrients in dung may be washed away and nitrogen lost as ammonia gas. Animal pens should be covered with dry organic material such as straw from cut grasses or other plants. This provides carbon and air for microorganisms which can use the nitrogen in urine and dung for growth, thus storing this nutrient in their bodies.

Weed seeds or cuttings in manure can sometimes grow in the garden. Before applying large amounts of suspect manure, it can be tested on a small area. If lots of weeds grow where the manure was applied but not elsewhere, it should be composted to kill the seeds. This is less of a problem with dung from cows, water buffalos, goats, sheep, and other animals that chew a cud because they have a two-stomach digestive system that kills most seeds. However, horses, donkeys, and birds have less thorough digestive systems that some seeds can pass through unharmed.

Manure from humans can also be used for gardens, however, great care must be taken to avoid spreading disease. The best way to do this is to make use of human manure without actually handling or transporting it, for example by letting tree roots extract nutrients from old latrines. In parts of West Africa, latrine sites are moved each year and the second year after being abandoned the old sites are filled in and bananas are planted on them.[35]

9.6.2 Composting

Compost is a soil amendment made from decomposed organic matter. It can be made solely from plant remains or from a combination of plants and manure. Composting fresh plant remains and animal wastes is often more convenient than adding them directly to the garden, and helps gardeners get the most soil improvement from organic matter. Making compost piles or pits prevents the organic matter from drying up or blowing away, two ways in which nutrients are often lost. If applied directly to the garden, high-carbon organic matter can tie up nitrogen during decomposition, and high nitrogen organic matter can produce considerable heat which can burn the roots of garden crops.

Like other aspects of dryland garden management, understanding the basic principles of composting allows local gardeners or projects to develop or improve techniques best suited to their needs.

Moisture is needed for the decomposition process in composting. However, moisture requirements can be kept low if the compost pile is covered with earth or woven mats to provide shade and retain moisture when the weather is hot and dry.

The size of the compost pile should be small enough and the organic matter coarse enough that the pile does not become packed down or compressed, eliminating oxygen. This leads to smelly, *anaerobic* (without oxygen) decay, as happens when only fresh green weeds are used. On the other hand, organic matter particles should be small enough and the pile large enough that the pile is able to hold heat and moisture, and to provide optimum surface area for microbial growth (Figure 9.10). Corn, amaranth, okra, sunflower, and other stalks are best chopped with a machete into pieces less than 15 cm (6 in) long.

Almost any organic material can be composted, including crop or food-processing residues (e.g., malt left from making beer, seed pods, grain chaff), weeds, and animal bedding with manure and urine. Leaves and stalks of younger plants contain more readily available water-soluble nutrients in their sugars and amino acids, but as the plant ages these become less available woody materials like cellulose and lignin, the most abundant compounds in cell walls. Salted foods or plants that tolerate salt such as tamarisk, which may have high salt concentrations in their leaves (section 5.6), should not be used in the compost pile. This is especially true if the soil or water used for irrigation is already salty. Where soils are alkaline, ashes or other materials that would raise the pH should be kept out of the compost. Where soils are acid ashes and other alkaline materials can be good soil amendments, but should only be mixed into finished compost or directly into the soil. Adding ashes while building a compost pile, and thus raising its alkalinity, interferes with the decomposition of the organic matter (Box 9.10).

When the compost pile is first built the carbon to nitrogen ratio (C:N) should be about 20:1 to 30:1 by weight, which is the ratio at which microorganisms use these nutrients. If it is less than 20, there will not be

Figure 9.10 The Effect of Particle Size on Composting

enough C in relationship to N, and microorganisms will use protein for energy and give off ammonia gas (NH_3) which has a uniquely pungent, sharp smell. This smell is an indication that valuable nitrogen is being lost, and high carbon materials should be added to the pile. If the C:N is greater than 30 the composting process will slow down since there is not enough nitrogen for the microorganisms, and new ones cannot grow until old ones die and release the nitrogen in their protein. If the pile does not heat up and start shrinking, more high-nitrogen material (to lower the C:N) should be added. After some experience observing and working with compost piles it becomes easy to estimate the proportions of different types of local material best for making compost. When the composting process is over the final C:N will be about 10:1 to 12:1, as carbon dioxide (CO_2) is lost through the metabolism of sugar for energy. Table 9.2 gives the approximate C:N ratios of some common dryland sources of organic matter, and can be used as a rough guide to combining organic matter with differing C:N ratios in a new compost pile, trying for an overall ratio between 20:1 to 30:1. Box 9.10 gives some tips for a fast compost pile

The compost pile contains millions of bacteria, fungi, and other microorganisms which feed on the nutrients in the organic matter. The fungi become dominant toward the end of the composting process and may form a grayish-white powder on the outer 10-15 cm (4-6 in) of the pile. Large numbers of these fungi occur in the soil, manure, and other compost ingredients. For this reason there is no need to inoculate the compost pile with commercial inoculants or other material.

A method called **trench composting** or **trench bed gardening** is used in some areas of southern Africa and the Philippines (Figure 9.11).[36] Alternating layers of organic matter and soil are used to fill up a trench and make a slightly raised bed that is then planted. As the organic matter decomposes it releases nutrients for use by the growing plants. This is a method worth experimenting with although care must be taken that the heat of decomposition does not burn the roots of garden plants.

9.7 Preventing Soil Erosion

Erosion is the movement of soil from one place to another by wind or water. Erosion usually leads to a loss of productivity because the finer clay and organic matter which contains most of the fertility are lost. In addition, clay and organic matter are the major source of binding for soil aggregates, and so soil structure is weakened. When erosion is so severe that the subsoil is exposed there is a further reduction in structure.

There is a positive side to limited erosion when it

Table 9.2 Examples of Approximate C:N Ratios of Dryland Organic Matter [a]

Source	Percent organic carbon (C)	Percent total nitrogen (N)	C:N ratio
Organic Matter for Compost			
Alfalfa, young	40	3	13:1
Poultry droppings	27.9	2.0	14:1
Brewery waste[b]	44.6	3.7	15:1
Farmyard manure	33.9	1.5	22:1
Cocoa husks	33.1	0.9	36:1
Rice straw	41.7	0.5	76:1
Yam peelings	51.0	0.3	196:1
Sugercane trash	38.0	0.1	318:1
Soil Microbes			
Soil bacteria	50	10.0	5:1
Soil actinomycetes	50	8.5	6:1
Soil fungi	50	5.0	10:1
Soil humus	50	4.5	11:1

[a] Based on Titiloye, et al. 1985, and Donahue, et al. 1983:146.
[b] "spent grain"

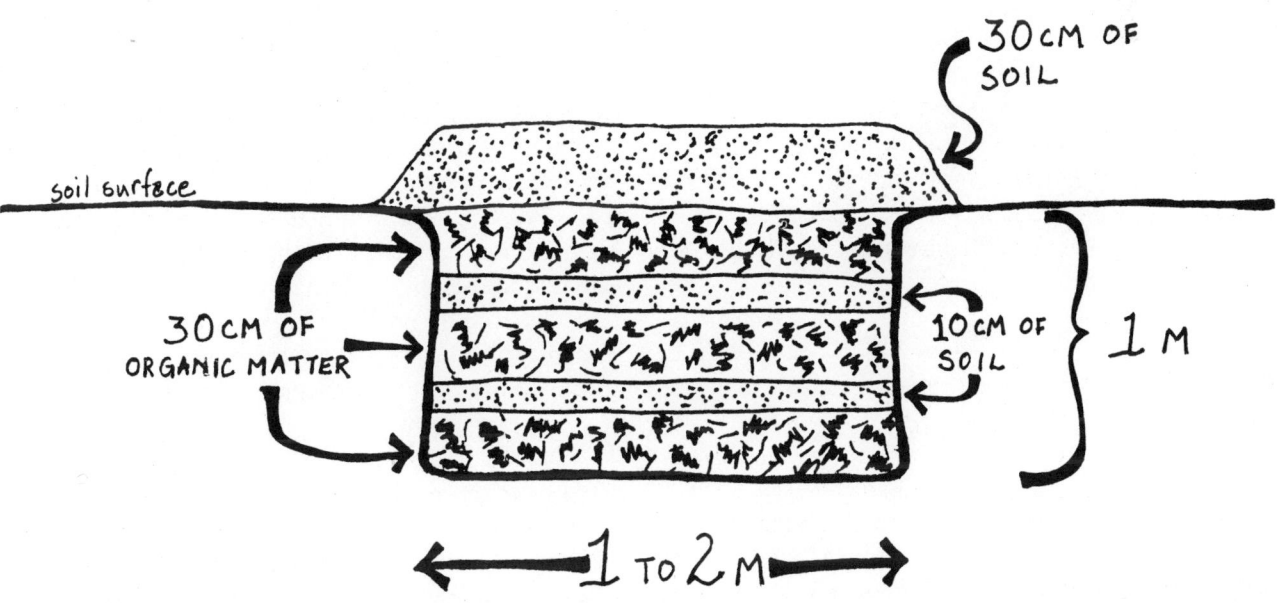

Figure 9.11 Trench Composting   (*After DCFRN #9-7*)

> Box 9.10
> For a Fast Compost Pile
>
> Compost piles can be either fast or slow, that is take a short or long time to finish composting. For fastest piles moisture, oxygen, temperature, and pH should be kept near the optimum levels described below.[37] However, for slow composting controlling these factors is not as critical, because a longer time in the compost pile compensates for lower moisture content and temperature. Slow composting, using two or three piles at various stages of decomposition, is easiest and often most practical for gardeners.
>
> *MOISTURE* Between 50 and 60% moisture is best for optimal microbial growth. This is the point at which water can be squeezed out of the composting material by hand, but will not drip or drain without squeezing. The compost should have a glistening appearance after being wet down. If the pile is too wet the air will be forced out, leading to an increased population of anaerobic, denitrifying bacteria which lose nitrogen by releasing it into the air. In the dry season covering the pile with a layer of clayey earth or woven mats, or putting the compost into a pit dug up to 60 cm (24 in) deep, help to conserve moisture. During the rainy season, however, compost piles in pits can become waterlogged.
>
> *AIR* Oxygen is also required by the microbes, so the pile should not be too big, preventing the center of the pile from "breathing." A good size is roughly a 2.5 m x 2.5 m (8 ft x 8 ft) base and a height of 1.5 m (5 ft). Ammonia and other bad smells are a sign of too little air. Air circulation can be increased by piling branches and other loose, bulky material underneath the pile. Building the pile around poles which are then removed when the pile is finished provides air to the center of the pile (Figure 9.12).
>
> *TEMPERATURE* Heat is released by the microorganisms in the compost pile as they metabolize sugar. If the pile is built correctly it heats up within 2 to 3 days to about 55°C (131°F), then peaks at 60-70°C (140-158°F). The temperature can be tested by digging into the center of the pile, or by sticking a wooden or metal pole into the center and leaving it there for 15 minutes. If it is hot to the touch when it is removed, the temperature is high enough.
>
> *pH* The compost pile usually begins as slightly acidic, then becomes very acidic and ends up somewhat alkaline (pH 7.5-8.5). Ashes should not be used when making a compost pile as they will make the pile too alkaline at the start for the microorganisms required to decompose the organic material.
>
> *Figure 9.12 Creating Air Space in a Compost Pile*

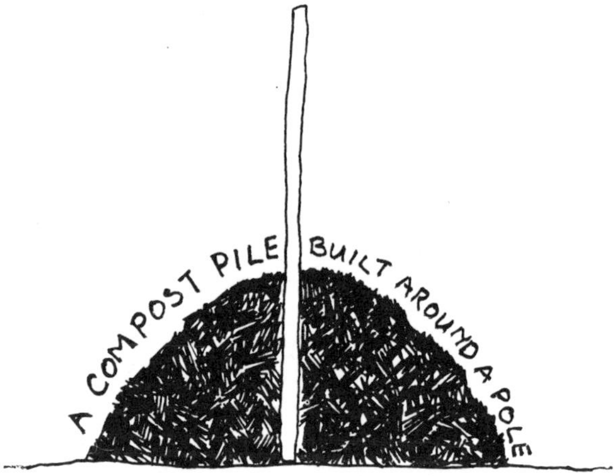

moves and concentrates small soil particles in sites which can then be cultivated. The low-lying areas at the bottoms of slopes and along streams and rivers often have clayey soils rich in mineral nutrients and organic material brought to these sites by rainwater from the slopes above, or by stream flow (section 11.6). The soil deposited by flowing water is called **alluvium**.

However, if erosion continues unchecked, infertile subsoils, rocks, and gravel can be washed down and cover the fertile soil. If streambeds or riverbeds become filled with eroded soil the level of water will be raised and consequently so will the water table, which may cause drainage and waterlogging problems in gardens next to the stream. Gardeners in valleys and other low-lying areas must therefore be concerned about erosion not only in their gardens, but on the slopes above the gardens and in the watershed of the stream that brings them water.

9.7.1 Decreasing Runoff

When rain is falling at a faster rate than the ability of a soil to absorb it, water begins running off the soil surface. In technical terms, when rainfall intensity (mm/hr) exceeds the *infiltration rate* (mm/hr) (the rate at which water on the soil surface enters and continues to move deeper into the soil), then runoff begins (section 11.5). Runoff water flows over and erodes the soil in two ways: a) sheet flow moves as a thin sheet of water removing a thin layer of soil, in a process called *sheet erosion*, b) stream flow cuts into the soil surface, removing soil in larger amounts from the sides, bottom and head of small channels (rills) or large channels (gullies), called respectively *rill erosion* and *gully erosion* (Figure 9.13).

Anything that decreases the amount or speed of water flowing across the soil will help to limit soil erosion. Vegetation and dead plant material on the surface also help to slow water flow and increase infiltration.

Slope is the most important factor causing runoff. Slope is the change in height of the soil surface over a given distance, and is often expressed as a percentage (Figure 11.6 and Box 11.4 in section 11.5.3). The larger or steeper the slope the more quickly water will run off the soil surface, and the more danger there is of erosion. Slope affects soils and growing conditions in

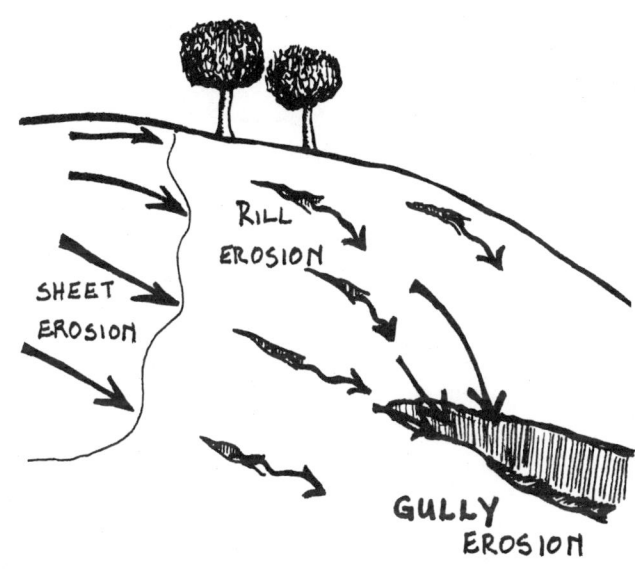

Figure 9.13 Sheet, Rill, and Gully Erosion

many important ways (Figure 9.14). Upland soils are usually well drained, while low-lying ones tend to be poorly drained. The middle sections of slopes are most subject to erosion. Rainy-season gardens do best where there is good drainage, and dry-season gardens

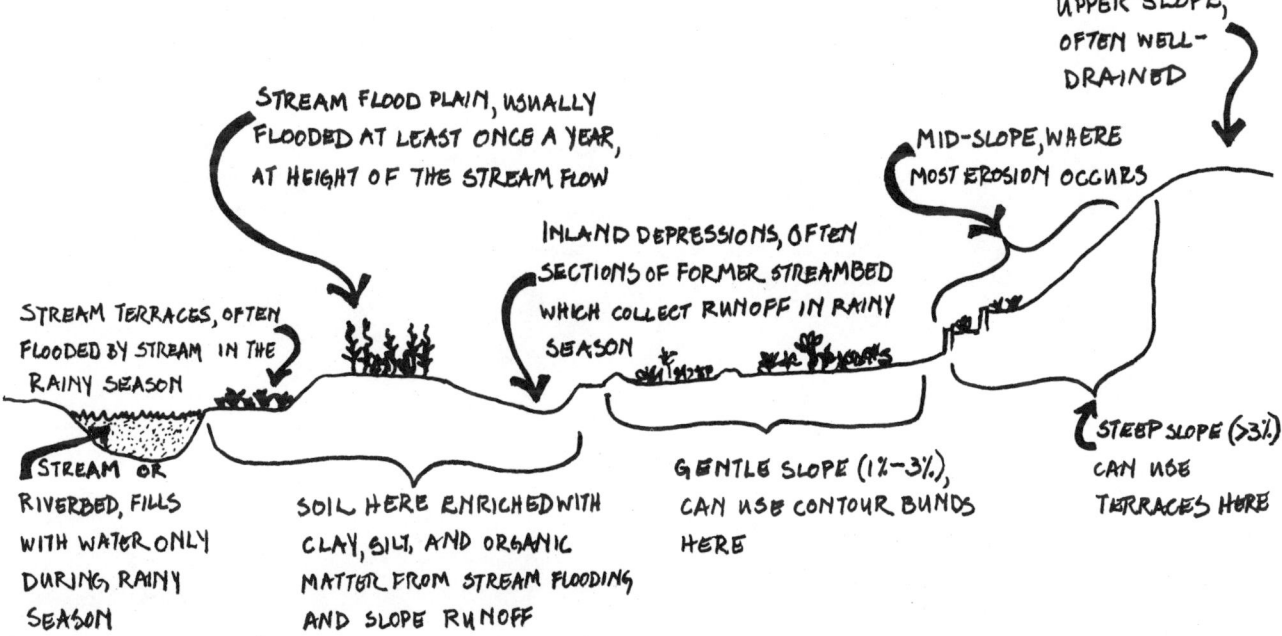

Figure 9.14 Some Typical Effects of Slope in Drylands

benefit from being located in low-lying areas where water is held in the soil, and where they can also collect rainwater. Land with steep slopes should not be cultivated if at all possible, and erosion on intermediate slopes can be decreased by terraces or contour bunds.

Although slopes are not the ideal location for gardens, there are many situations where they are the best place available. Cultivating on slopes greatly increases the risk of soil erosion by rainwater runoff, and the longer or steeper the slope the more danger of erosion. Contour bunds or terraces are often constructed to decrease the slope, slow runoff, increase infiltration, and capture soil eroded from areas further up the slope. Trees, bushes, and grass growing in strips between groups of bunds or terraces further protect the soil. Contour bunds or terraces can also harvest rainwater. Details of calculating runoff and catchment to garden area ratio (CGAR) are found in section 11.5.

CONTOUR BUNDS *Contour bunds* are piles of rocks, soil, or organic matter made in long rows following the contours of the land (Figure 9.15).[38] A *contour* is a line along the ground which is at a constant elevation. Although contours are often constructed to decrease soil erosion, they can also capture and hold rainwater for gardens. In areas where peak rainfall is not heavy or where distance between bunds is small, earthen bunds will retain the most water. However, earthen bunds do not allow excess water to flow through them, so where rainfall and runoff are larger, provision should be made for peak rainfall runoff which can cause much damage. Spillways in the earthen bunds can be built out of rock so that any overflow will not wash away the whole bund (Figure 9.16). Bunds that include rock or dead vegetation such as tree branches are another good solution because they allow excess water to pass through.

If bunds are not constructed accurately on the contours, runoff will tend to flow along the length of the bund, causing soil erosion and damage to the garden (Figure 9.17). One way to correct contour bunds that do not follow the contours is to remove them and start over, but this is a lot of work. An easier solution is to construct short bunds, perpendicular to the existing contour bunds, to stop and hold any water flowing parallel to the contour bunds, thus preventing it from gaining the speed and volume which cause erosion. When the contour bunds are close together the perpendicular bunds can connect them, forming basins.

In theory, information should be collected on runoff potential of the soil and rainfall pattern so that hydrological calculations can be made to design the size, shape, and placement of the contour bunds. In practice such data are often difficult to obtain, persons with appropriate skills are not available, and the time and other resources required may be more useful elsewhere. However, as with many other gardening techniques, using gardeners' knowledge and skills,

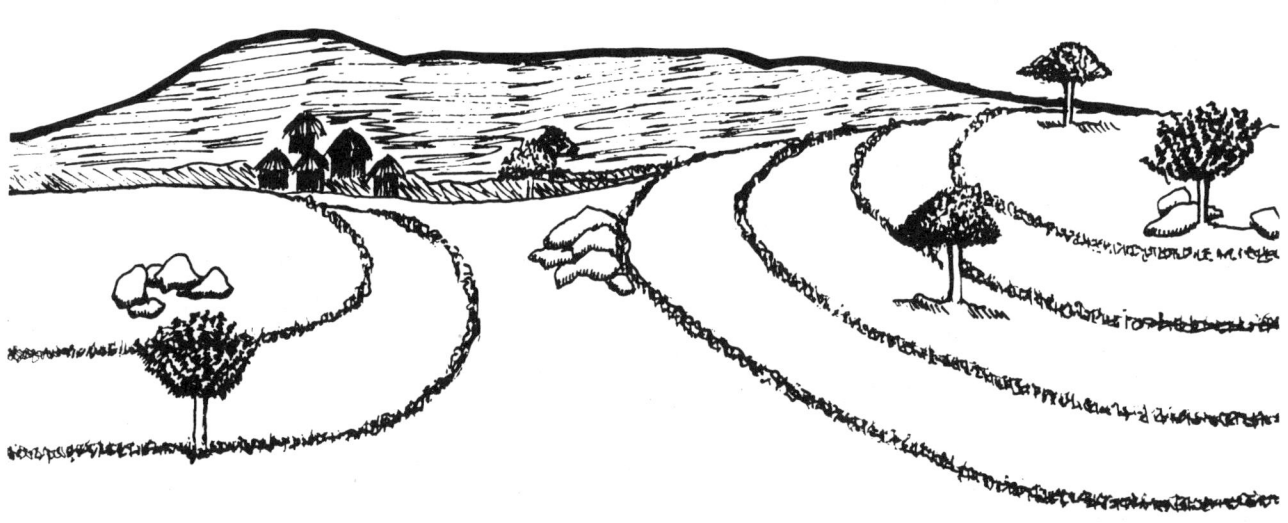

Figure 9.15 Contour Bunds in Northeast Ghana

Figure 9.16 A Stone Spillway in an Earthen Bund

beginning small, and encouraging experimentation and adaptation can be a successful alternative to dependence on outside technical expertise. For example, in Burkina Faso farmers developed successful contour bund systems by starting small, observing the effects of runoff during the rainy season, and then adding more bunds between existing ones when necessary.[39] Box 9.11 describes methods for determining contours.

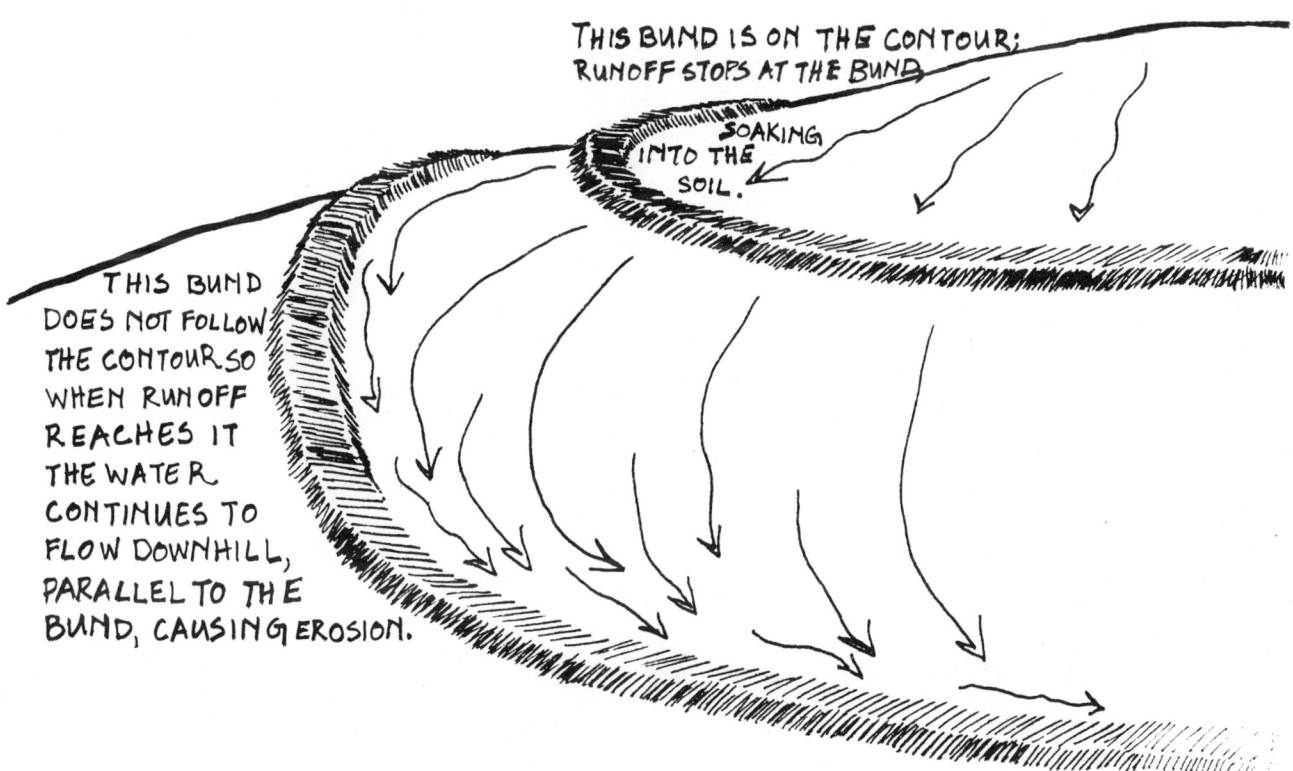

Figure 9.17 Contour Bunds

> ### Box 9.11
> ### Determining Contours
>
> A variety of methods can be used to determine contours, and gardeners can choose the best one for them. One method used successfully by farmers in the Yatenga area of Burkina Faso employs the water tube level.[40] This requires a length of transparent plastic tubing. If long pieces are difficult to obtain, a section about 2.0 m (6.5 ft) long can be fastened to each end of a long piece of nontransparent garden hose. The tube is first stretched out on the ground and then filled with water by pouring or syphoning. The tube ends are fastened to two straight sticks, each at least 2.0 m (6.5 ft) long, standing next to each other. Some water may need to be added to bring the water level to within 0.5 m (20 in) of the top of both ends of the tube, and this point is marked on both sticks (Figure 11.8 with Box 11.4). Then one stick is held still while the second one is moved across the slope and then up and down until the water level on both sticks is at the original mark. This point on the slope is marked and the first stick is then moved to a third point while the second stick remains stationary, and so on (Figure 9.18). If water spills or the tube stretches (as it can when it gets hot) then water will have to be added and the level marked again on the sticks.
>
> Another method for measuring contours is using an A-frame level (Figure 9.19). Two poles about 1 m (3.3 ft) long and a shorter middle bar are lashed together into an A-shape and a weighted string is hung from the center of the top of the A. To calibrate this level the base of each pole is marked, for example, by driving a small stake. The place where the string crosses the middle bar is then marked. Then the A-frame is turned 180°, with each stake taking the place of the other stake. Once again the place where the string crosses the middle bar is marked. The midway point between these two marks on the middle bar is the level point. The A-frame can then be used to mark a contour line along a slope by placing the two legs so that the string hangs on the level mark (Figure 9.20).

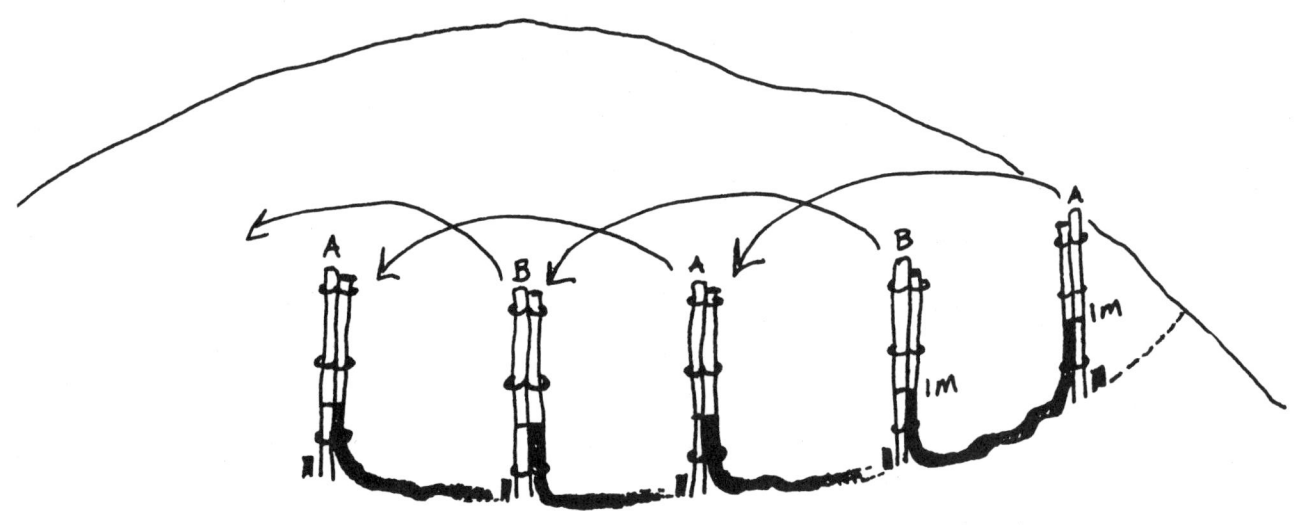

Figure 9.18 Using a Tube Level to Measure Contours

9 Soils in the Garden

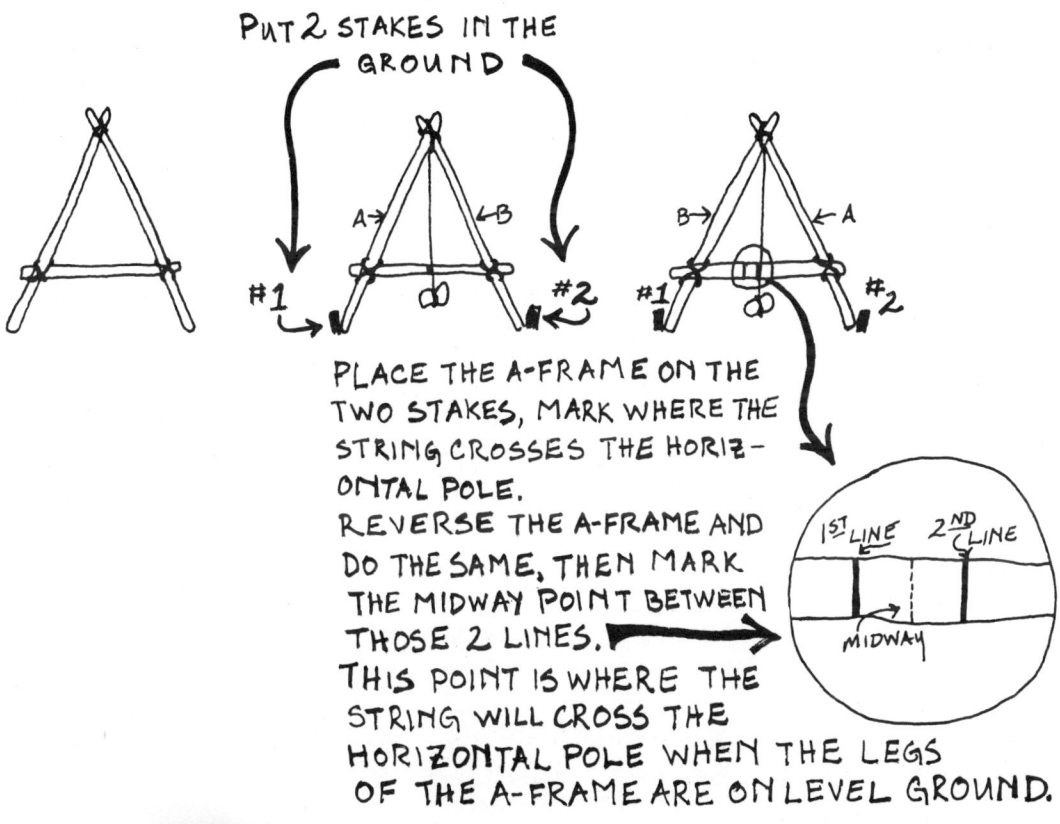

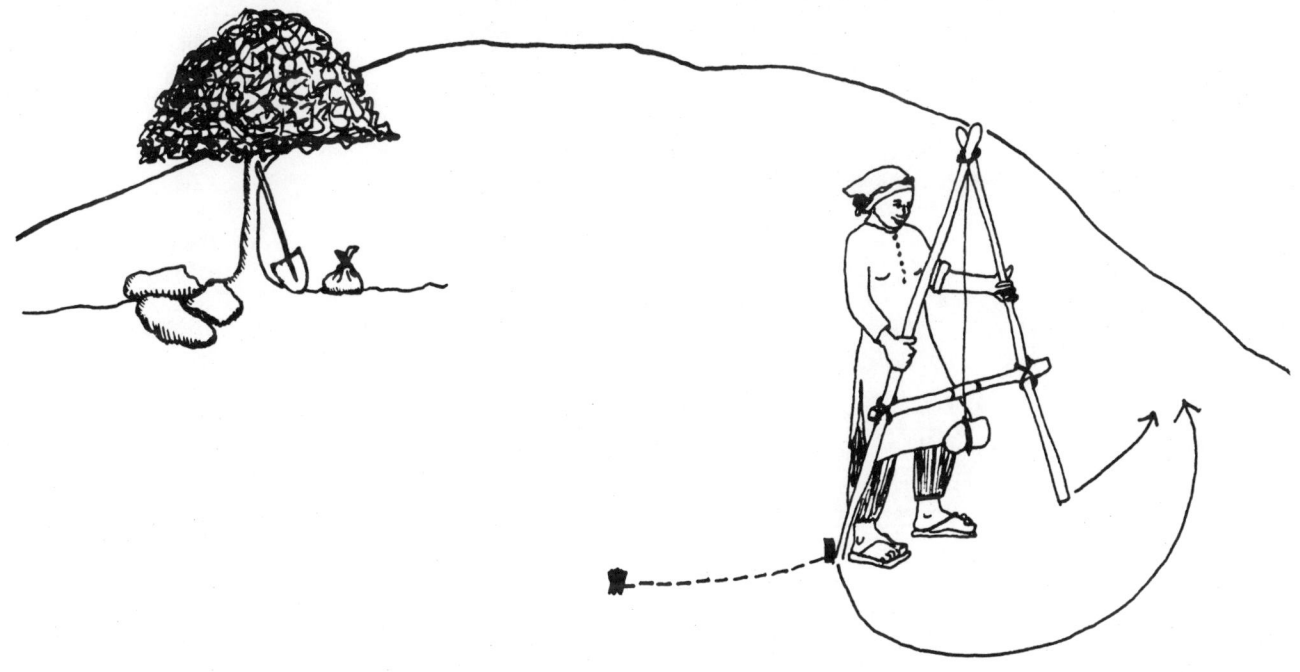

Figure 9.19 Using an A-Frame Level

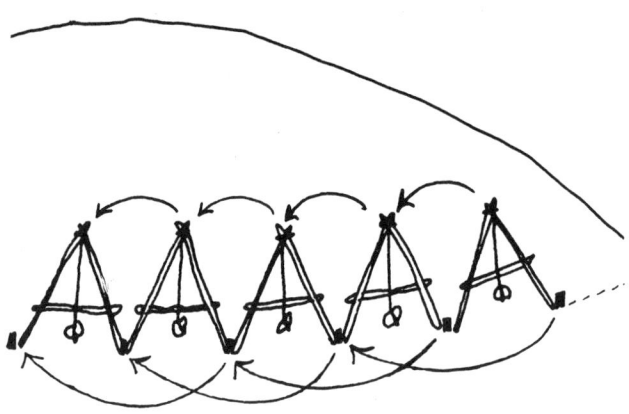

Figure 9.20 Measuring Contours with an A-Frame Level

TERRACES *Terraces* are shelves of land built on slopes to provide level areas for gardens or other uses and are part of indigenous soil management systems in many dryland areas. Like contour bunds, terraces are also built on contours and reduce runoff and erosion, but they are used on steeper slopes (3-30%) than contour bunds. Elaborate terraced systems of agricultural production date back over 2,000 years in dryland South America and Mexico, where most present terrace systems are continuations of ancient ones.[41] During the first century after the European invasion of this region (by 1600), the indigenous population was reduced from 50-100 million people to 5 million people, mainly from epidemic diseases introduced by the Europeans. This reduction in population along with violent social and cultural destruction, led to many terrace systems being abandoned, resulting in increased erosion.

Terraces have also been used for centuries in many other drylands such as those of the Arabian Peninsula and southwestern North America. In the Mediterranean, olives, figs, and grapes are often planted in small stone terraces on slopes that would be too rocky to support many other garden plants. Hopi Native Americans of Arizona, USA, have built up extensive terraces below mesa tops for spring-fed irrigated gardens where they grow chilis, onions, tomatoes, corn, herbs, and other crops[42] (Figure 9.21).

Construction of terraces involves building a retaining wall and leveling soil on the slope above this wall (Figure 9.22). Additional soil will often be needed to fill in the terrace. The steeper the slope and the more shallow the topsoil the closer together the terraces should be (Figure 9.23).

Figure 9.21 Hopi Terraced Gardens

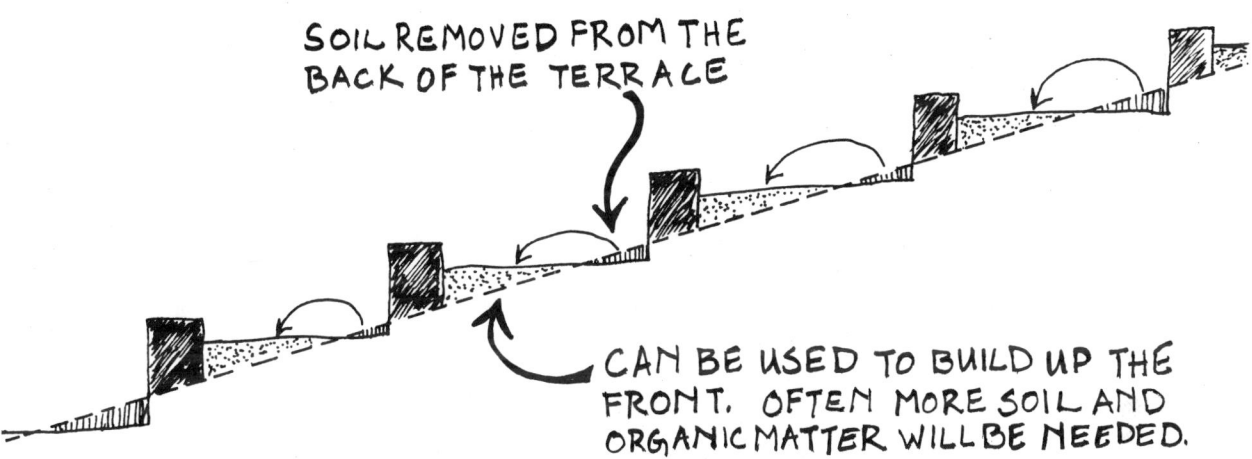

Figure 9.22 Leveling the Soil in Terraces

9.7.2 Decreasing Raindrop Impact

The force of raindrops hitting the surface of the soil can cause as much erosion as runoff. Raindrops break up and detach soil aggregates producing smaller-sized particles which are easier to wash or blow away. Raindrops move soil particles as they splash, and on a slope will move soil downhill, even without runoff. In addition, the impact of raindrops compacts a thin layer of soil on the surface which slows infiltration and increases runoff. Infiltration is also reduced as rainwater carries the detached soil particles into the soil, clogging soil pores.

Protecting the soil surface from the force of raindrops helps decrease erosion. This effect can be seen dramatically where a rock on bare soil exposed to the rain has protected the soil under it from erosion so that eventually it is perched on a pedestal surrounded by an eroded surface (Figure 9.24). The leaves of closely planted garden crops will absorb the energy of the raindrops, slowing their impact. This helps protect the soil from erosion because water that drips from leaves has much less energy than when it first hits the plants.

9.7.3 Increasing Soil Resistance to Erosion

Anything that increases soil permeability or porosity will help decrease erosion (section 9.3.2). A loam tex-

Figure 9.23 The Relationship Between Degree of Slope and Terrace Size

ture is a good compromise between sandy soils, with high infiltration rates but low water-holding capacity, and clayey soils, with high water-holding capacity but low infiltration rates. Loams are also ideal soils in other ways (section 9.3.1).

Vegetation not only protects the soil from raindrop impact, but roots help keep soil in place. An example of the ability of indigenous techniques to slow erosion is the living fences of willows (*Salix* spp.) and cottonwoods (*Populus* spp.) planted by farmers in the Sonoran Desert of northern Mexico.[43] The fences are planted in the winter in riverbeds adjacent to fields to protect them from stream erosion during the summer rainy season. Not only do these trees protect fields in most situations, but they also slow the flow rate of the water upstream. As it slows, the alluvium carried in the water is dropped, forming a deposit of this rich sediment behind the fences. Over the years, farmers have used this soil to enlarge their fields. Unlike outsiders who consider willows and cottonwoods bad because they transpire so much water, local farmers have developed a successful technique for using those trees for soil management and so see them as beneficial.

9.7.4 Reducing Wind Erosion

Wind is a major cause of erosion, especially in drylands where surface vegetation is sparse. It carries fine soil particles away, bounces heavier particles of sand along the surface and into plants, and adds to the erosive power of raindrops by increasing their speed at impact.

A small reduction in wind speed results in a proportionately much larger reduction in erosion. For example, a 13% reduction in wind speed results in approximately a 50% reduction in erosion.[44] Wind erosion of soil and abrasion of plants with windblown soil particles can be reduced with *windbreaks*. Windbreaks can improve yields by protecting a plant, a garden bed, or one or more entire gardens (section 10.8.3). They should be placed perpendicular to the prevailing direction of the strongest wind. If strong winds come from different directions at different seasons of the year, then windbreaks may be needed in more than one directon.

Windbreaks should have from 20-50% porosity, that is allow 20-50% of the wind to pass through them. If porosity is low, the windbreak will offer too much resistance and there is a good chance it will be blown over. The lower the porosity, the greater the reduction in erosion closest to the windbreak, but the shorter the distance over which this effect will be felt. Also, with lower porosity the speed and erosiveness of the wind as it passes around the ends of windbreaks will be increased, and these areas should be protected, for example, with perennial crops or stone mulch.

The effect of a windbreak is described in terms of distances equal to its height (H). Tall windbreaks of several rows of trees can produce a 50% reduction from 16H *leeward* (the direction the wind is blowing

Figure 9.24 Protecting Soil from Raindrops

toward, that is, downwind from the windbreak) to 2H *windward* (the direction the wind is coming from, that is, upwind from the windbreak).

Large windbreaks can protect the whole garden, a group of gardens, or adjacent farmland. Within the garden, smaller windbreaks are appropriate to protect single beds or even plants (Figure 10.12 in section 10.8.3). They can be made from living plants, mats, or stalks of bamboo, sorghum, or other grass, or walls of stone or mud brick. In Egypt, farmers and gardeners place lines of stalks in the ground to protect newly emerged seedlings. The Hopi in southwestern North America do the same, and also use empty tin cans to protect individual plants, and pile flat slabs of sandstone around the base of fruit trees to keep the wind from blowing the soil away from the roots. Windbreaks can also provide shade or serve as fences, protecting the garden from animals (section 13.3.3).

9.8 Building Garden Beds

Garden beds separated by walkways help contain and concentrate resources for plant growth. Using beds also makes it possible to reach and work on the garden without compacting the soil in the growing area by walking on it. Two types of garden beds are sunken and raised. Although raised beds have been promoted widely, they are not always the best design for dryland conditions.

9.8.1 Sunken Beds

Sunken beds are deep pockets of improved soil, and have a long tradition in drylands. In southwest North America, Zuni Native Americans traditionally used very small rectangular beds bordered with clay berms for growing crops such as melons, herbs, chilis and onions (see Figure 12.1 in section 12.3.2). These gardens, near the main villages and surrounded with juniper stick fences, were watered by hand from larger containers using small dippers.[45] The nearby Hopi people also use sunken beds in their terraced gardens.[46] Until the Aswan dam was built, Egyptians used basin beds for thousands of years along the Nile river to capture the annual floodwaters.

In drylands and during dry seasons, sinking the beds is better than raising them for several reasons:
- They are easier to water efficiently by flood irrigation.
- The berms give the moist soil and young seedlings and transplants some protection from drying winds and sun.
- Young plants can easily be protected by laying palm fronds or other material across the beds (Figure 9.25).
- When wet season rains are intense, the garden is not eroded and water is not wasted, since it is captured by the beds.

In drylands with heavy rainfall in the wet season, it may be better to switch to raised beds during this period. However, in drier areas this will not be necessary because not enough rain falls to cause waterlogging, even in the wet season. The choice of bed design will also depend to some extent on the crops grown and the soil and drainage in the garden. Figure 9.26 outlines the steps for making a sunken garden bed.

Figure 9.25 Sunken Beds Protect Young Seedlings

9.8.2 Raised Beds

Raised beds are common in humid areas of the world. They have also become part of the "alternative" gardening tradition in temperate industrial regions of North America and Europe and as such have been promoted by some development workers from those areas as part of an alternative to industrial gardens.[47] Raised beds are most commonly used in humid regions where they can improve drainage in areas where the soil may be seasonally flooded or waterlogged. Like sunken beds they also function to increase the depth of improved soil. However, under hot, dry conditions raised beds are difficult to water deeply. Raised

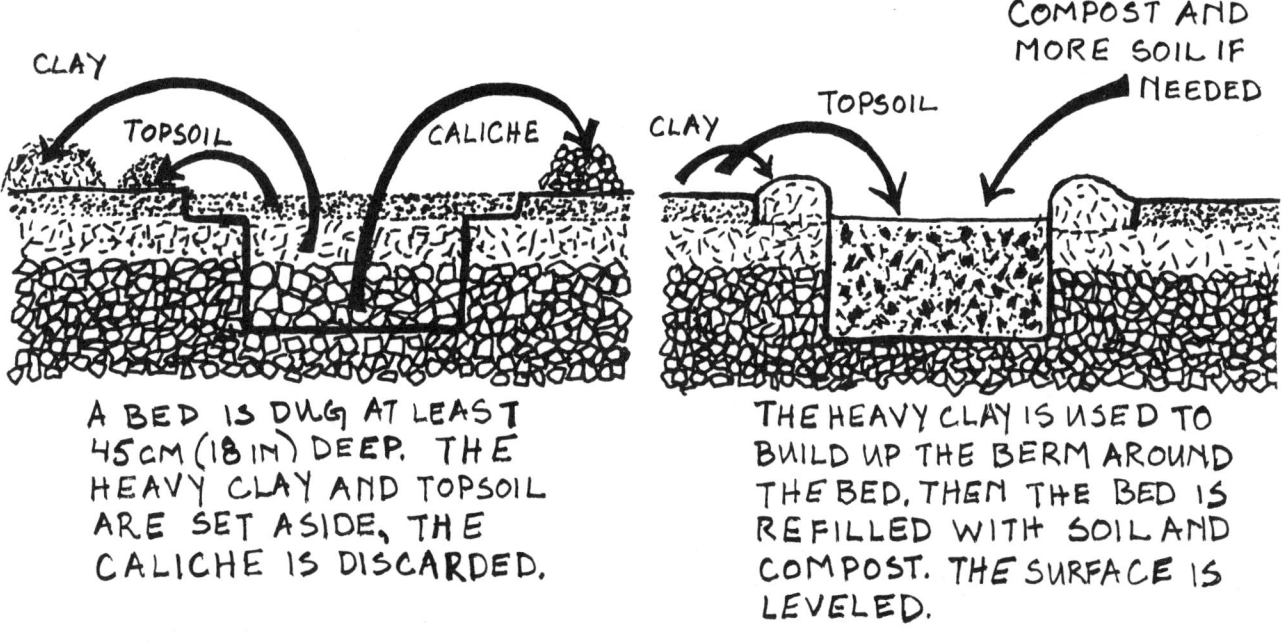

Figure 9.26 *Constructing a Sunken Garden Bed*

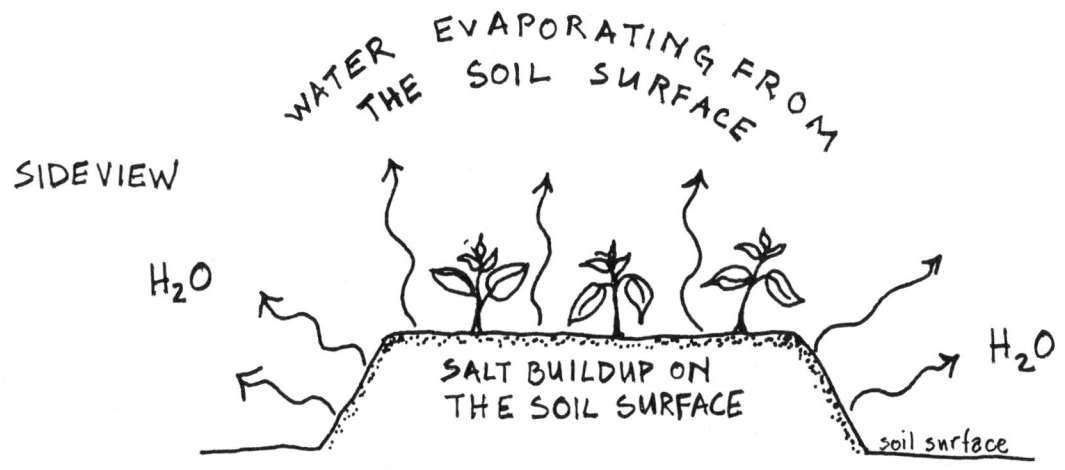

Figure 9.27 *Evaporation and Salt Buildup in a Raised Bed in Drylands*

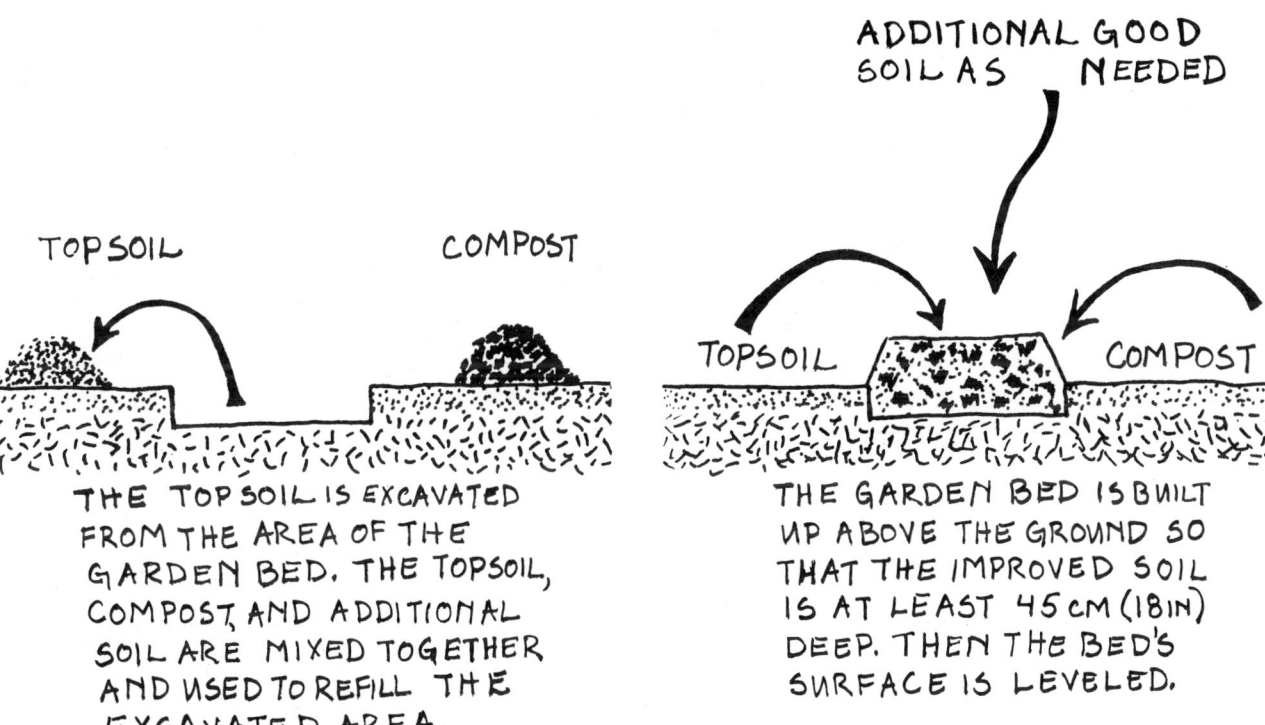

Figure 9.28 Constructing a Raised Garden Bed

beds also expose a lot of surface area which heats up soil temperatures, and from which moisture evaporates, leading to salt buildup in the growing area (Figure 9.27). Figure 9.28 outlines the steps for making a raised garden bed.

9.9 Resources

Blakie (1985) discusses the political and social basis of soil erosion and its control on both local and global levels. Good general introductions to applied soil science, both based on Africa, can be found in Ahn (1970) and Dupriez and De Leener (1983:Lessons 33-37, 40-43). We have used Donahue, et al. (1983), as a basic reference on Western soil science. A clear, simple description of how village people can build contour bunds and check-dams to reduce erosion is given in *Introduction to Soil and Water Conservation Practices* (1985), produced by World Neighbors and based on work in Indonesia.

References

[1] Blakie 1985.
[2] Chatelin 1979.
[3] Ahn 1970:202-213.
[4] Wilken 1987:28-36.
[5] Johnson 1974.
[6] Morgan 1974.
[7] USDA 1975.
[8] Donahue, et al. 1983:529.
[9] USDA 1975:155.
[10] Ahn 1970:101.
[11] Ahn 1970:18.
[12] Ahn 1970:19-21; Leonard 1980:11.
[13] Ahn 1970:25.
[14] Donahue, et al. 1983:62.
[15] Pacey and Cullis 1986:160-161.
[16] Donahue, et al. 1983:64-65.
[17] Birkeland 1984:138-146; Dregne 1976:168-173.
[18] Dupriez and De Leener 1983:93.
[19] Lal 1987:1070.

[20] Donahue, et al. 1983:90ff, 211-213, 267; Raven, et al. 1981:560, 563, 568ff.
[21] Agrios 1988:507-509.
[22] Ahn 1970:94; Donahue, et al. 1983:107-108; Dregne 1976:186.
[23] Lagemann 1977:38.
[24] Wilken 1987:57.
[25] Purseglove 1974:200.
[26] Ahn 1970:265.
[27] Ahn 1970:149-150.
[28] Donahue, et al. 1983:155.
[29] Donahue, et al. 1983:148-149; CFA 1980:127-128.
[30] Bassuk 1986.
[31] Bell, et al. 1987:50-52.
[32] Bell, et al. 1987:79.
[33] Hammond 1966:35,42.
[34] Delgado 1979:73-75.
[35] Lagemann 1977:39.
[36] DCFRN #9-7.
[37] UDS 1983.
[38] World Neighbors 1985.
[39] Wright 1984.
[40] Adapted from Wright 1984, cited in Pacey and Cullis 1986:171; see also Chleq and Dupriez 1984:43-46.
[41] Denevan 1980:222-229.
[42] Soleri 1989.
[43] Nabhan and Sheridan 1977.
[44] Troeh, et al. 1980:410-411.
[45] Ladd 1979:497-498.
[46] Soleri 1989.
[47] Cleveland and Soleri 1987.

9 Soils in the Garden

Figure 10.1 The Water Cycle

10
Water, Soils, and Plants

Water is essential for all life, but is often a scarce resource in drylands. Water carries plant nutrients from the roots upward, and food from the leaves downward. It is also a medium for chemical reactions, cools the plant by evaporating from the leaves, and plays an important role in photosynthesis. Water as a liquid and a gas is constantly being cycled among plants and animals, soil, air, and bodies of water such as streams, lakes, rivers, and oceans (Figure 10.1).[1]

The goal of water management in the garden is to provide adequate supplies of water to plant roots at a reasonable investment of time, money, and other resources, without creating salinity or waterlogging problems. Another important goal of garden water management is to ensure that all members of the community have access to good quality water, and that water use today does not jeopardize the quantity or quality of water in the future.

Many indigenous techniques for watering, cropping patterns, mulches, shades, and windbreaks appear to meet these goals but they are not well documented. They are based on the same principles of water, soil, and plant relations that Western science is based on. Knowledge of these basic principles will help readers understand indigenous practices and suggest improvements where needed.

10.1 Summary

This chapter discusses the movement of water in soils, and the relationship of soil water and garden yield as the basis for specific techniques to improve water management.

The storage and movement of water in the soil depends on the texture and structure of the soil and the amount of organic matter present. Water is removed from the root zone by gravity, evaporation, and plant roots. Successful garden harvests depend on maintaining an adequate amount of water in the root zone at a reasonable cost. This is done by applying water to the garden when the plants need it; by conserving water through preventing excess loss from evapotranspiration using mulches, windbreaks, and appropriate cropping patterns; and, by preventing excess drainage loss by controlling the amount of water applied.

10.2 Dryland Garden Water Management

Efficient use of water for gardens means using the smallest amount of water to produce the largest amount of harvest in ways that do not harm the environment, and that promote equitable local control of the source and distribution of water. Conserving scarce water resources by reducing unnecessary losses is a very important step in reaching this goal. Runoff, deep percolation beyond the root zone, evaporation, and excess transpiration are the four ways water is lost within the garden (Figure 10.2). Management strategies to prevent these losses are:

- Minimizing runoff and maximizing infiltration of rain or irrigation water into the garden soil by improving soil structure and by using terraces and vertical mulch.
- Minimizing loss of water below the root zone by not overwatering, and by increasing the soil's water-holding ability with organic matter.
- Minimizing evaporation and excess transpiration in the garden by mulching, close spacing of plants, multiple plant levels, shading, windbreaks, careful selection of planting times, and use of heat-tolerant and drought-adapted crops.

Ways to conserve water before it reaches the garden

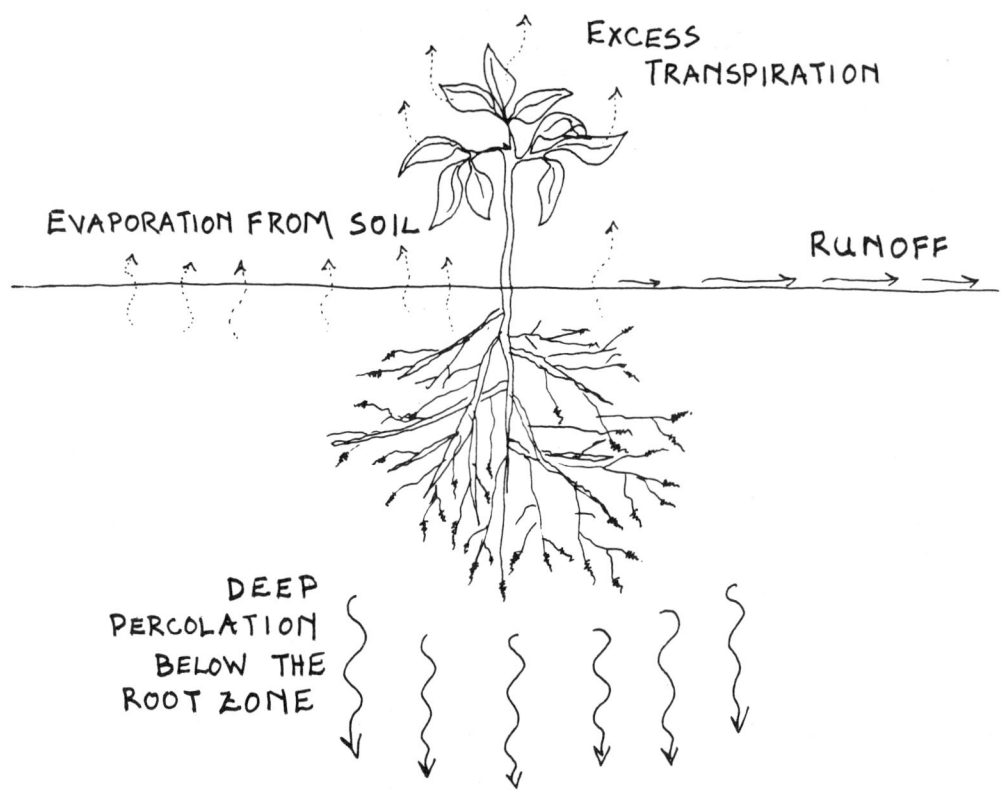

Figure 10.2 How Water is Lost in a Dryland Garden

are discussed in Chapters 11 and 12 and include:
- Improving and maintaining the quality of water used in the garden by minimizing salts, poisons, and organisms that cause plant and human diseases.
- Making use of local water resources such as rainfall, streams, floodwater, and shallow groundwater aquifers for the garden.
- Minimizing the loss of irrigation water from storage in reservoirs, tanks, pots, or other containers by covering or shading the surface to reduce evaporation, and by stopping any leaks in the container.
- Minimizing the loss of water while raising it from wells or rivers, and while bringing it to the garden in hoses, buckets, irrigation canals, or rainwater catchment plots by reducing the number of leaks or low spots and the amount of time the water is exposed to evaporation and to infiltration before reaching the garden.

10.3 Water, Soils, and Plants

Plants obtain the water they need to grow through their roots. Many characteristics of the soil affect the availability of water to plants, and in turn help to determine how much water the garden will need.[2]

10.3.1 Water Storage in the Soil

Water is stored in the soil pores, the spaces between particles of soil (section 9.3.2). Soil scientists have developed a system for describing water in the soil in relation to plant growth. If water is added to a soil until all the pores in the soil are filled and there is no room for any more, it becomes *saturated* (Figure 10.3). If a saturated soil is allowed to stand with no additional water being added, the *free water* or *gravitational water* will drain down and out of the root zone. Since most soils drain gravitational water from the root zone rapidly, it is generally not available to plants. When the point is reached where no more water drains out by gravity, the soil is at *field capacity* (FC).

Fine, clayey soils can hold more water than coarse, sandy soils. This is because clayey soils have many small pores that add up to much more total pore space than the fewer but larger pores of sandy soils. Clayey soils also hold water more tightly than sandy soils because the smaller pores provide more soil surface area per volume of water (Box 10.1).

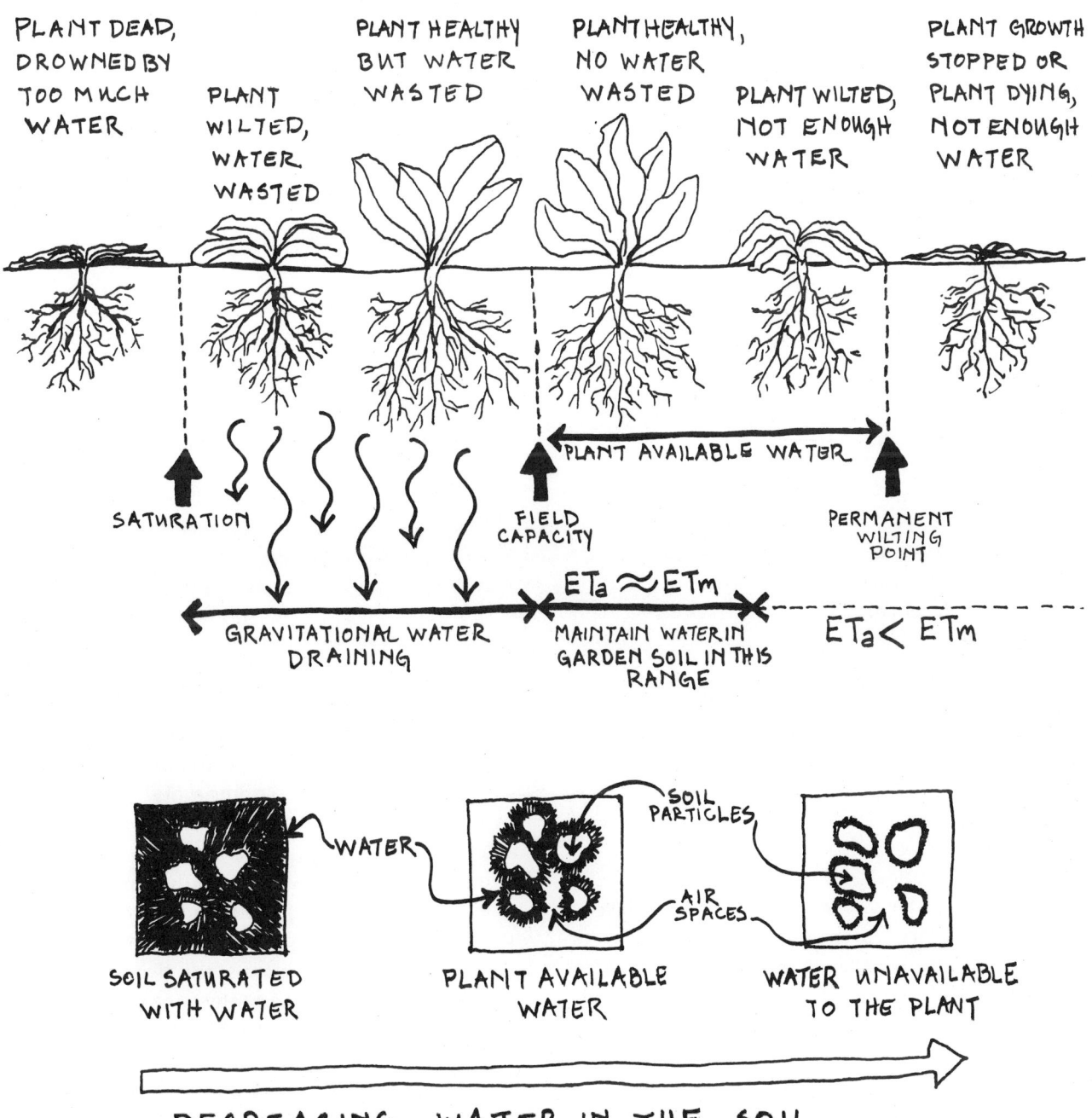

Figure 10.3 Soil Water and Plant Growth
(After Donahue, et al. 1983:171; and Doneen and Wescot 1984:6)

> **Box 10.1**
> *How Water is Held in the Soil*
>
> Water in the soil at or below field capacity is held by electromagnetic forces. One of these forces is cohesion, the tendency of molecules of a substance to stick to each other, as when water molecules are attracted to each other (section 5.2). Adhesion is the attraction of molecules of different substances to each other, as in the attraction between water and soil molecules. A combination of cohesive and adhesive forces holds layers of water molecules on the surface of soil particles.
>
> The water closest to the surface of the soil particles is held most tightly, so as water is taken up by the plant roots, the remaining water becomes more and more difficult to absorb. The energy needed to remove water from the soil, called the *soil water potential*, is most commonly measured in units called *bars*. The greater the negative value the more tightly the water is held, and the more energy is required to remove it. For example, the water in garden soil with a water potential of -0.5 bars is more easily available to plants than water in soil with a water potential of -0.7 bars. Field capacity is often considered to be -0.3 bars and permanent wilting point -15 bars.[3] In practice, however, the permanent wilting point depends on the crop and variety, and on the soil's properties including texture, structure, and organic matter.

When the water in the soil is reduced to the point where the plant cannot absorb it fast enough to grow or even to stay alive, soil water is at the *permanent wilting point* (PWP). The water held in the soil between field capacity and the permanent wilting point is water that can be absorbed fast enough by plants to grow and produce. This water is called *plant available soil water* (AW):

AW = FC - PWP.

Most garden crops should be watered when 50-75% of plant available water has been depleted (section 10.7). Even though a plant will be able to survive on the remaining 25-50%, it may develop a water deficit. This means that water will be lost through transpiration faster than it can be absorbed, reducing production in many crops (section 10.4). However, an experiment with tomatoes, sweet peppers, and cantaloupe melons in Tucson, Arizona, USA, showed that some crops or crop varieties may be able to survive without growing for several weeks when soil water is well below the wilting point, and then recover and produce a harvest when water content rises above the wilting point.[4]

Too much water can also harm or even kill plants. If a soil remains saturated because water is being added faster than it can drain, or because there is a high water table, the soil becomes *waterlogged*. Most plants cannot grow in waterlogged soils because there is no air to provide oxygen, since it has been squeezed out by the water (Figure 10.3).

10.3.2 Water Movement in the Soil

The way water infiltrates and moves within the soil influences plant growth.[5] The rate of infiltration is influenced by soil permeability (section 9.3.2) and slope (section 9.7.1). Holes in the soil surface and channels beneath the surface of the soil made by plant roots, earthworms, termites, moles, and other animals increase infiltration. Cultivating the surface improves infiltration, but it also increases evaporation in the cultivated layer (section 10.3.3). Because of smaller pore size, clayey soils have slower infiltration rates than sandy soils, which can mean higher runoff and evaporation from the soil surface. Adding organic matter and using vertical mulches both increase the infiltration rate of clayey soils.

The rate and pattern of water movement below the surface is not obvious from simply observing the pattern of wetting on the surface. It depends on soil texture, structure, depth, and the organic matter content. Most water movement in the soil after a good rain or irrigation is due to the downward force of gravity. However, because of capillary action, some movement takes place in all directions toward areas with less water. *Capillary action* is the movement of water through very small spaces (capillary spaces) from wetter areas to drier areas. In clayey soils with many fine pore spaces, water moves further horizontally by capillary action than in sandy soils with much larger particles and fewer and larger pores. Therefore, an equal amount of water wets a larger surface area of clayey soils than of sandy soils (Figure 10.4 a, b).

Soils with distinct layers of different textures in the root zone, for example, of sand or of clay in a loam soil (Figure 10.4 c), hold more water than uniform soils because the layers slow the downward movement of water, increasing the field capacity. An abrupt change from sandy to clayey texture will slow the movement of water, as will a transition from clayey to sandy texture (Figure 10.4 d, e). A dense layer of caliche, ironstone, or rock will practically stop water flow, causing the soil above it to become saturated (Figure 10.4 f).

A separate layer of organic matter will also slow the downward movement of water. However, when mixed in with the soil in the root zone, organic matter speeds downward movement (Figure 10.4 g). Organic matter improves soil structure as it breaks down to form humus, creating soil aggregates and larger pore spaces that water can penetrate faster and deeper. Organic surface mulches protect the soil surface from compaction by raindrops, and vertical mulches, which are open to the surface, allow water to penetrate quickly to the root zone (Figure 10.4 h; sections 10.8.1 and 10.8.2).

10.3.3 Evaporation

Evaporation is the change of water from a liquid to a gas form. When water evaporates from the soil into the air, plants can no longer use it. In drylands, evaporation from the upper 10 cm (4 in) of wet soil is very rapid because temperatures are high and capillary action moves water upward quickly in this layer. Below this level, soil water is protected from evaporation by the layer of soil above it which acts like a surface mulch (section 10.8.1). When soil is at field capacity, most of the water 10 cm (4 in) below the surface will be removed from the soil by plant roots, not evaporation, capillary action, or gravity.[6]

Shading crops from the sun, protecting them from the wind, and mulching the soil surface are all ways to reduce evaporation. Vertical mulches speed infiltration to the root zone (section 10.8.2). Maintaining a high organic matter content which promotes good soil structure (section 9.3.1), and efficient irrigation methods (section 12.2) also reduce evaporation.

Sometimes cultivation to break up the soil surface, which interrupts capillary action and so reduces evaporation, is recommended.[7] This method may be useful when it is difficult to find mulching materials, and when plants are young and do not cover the soil surface. In this situation cultivation can also help to discourage weed growth, and may avoid problems with pests hiding in the mulch and eating young plants. However, cultivation only slows down evaporation in the soil below the cultivated layer, while increasing evaporation in the cultivated layer itself. This technique requires continual work since cultivation has to be repeated after each irrigation or heavy rain. In addition, if cultivation is used over a wide area it can lead to soil erosion by wind and water.[8] For these reasons we recommend using mulch rather than cultivation, whenever possible, for reducing evaporation in gardens.

10.3.4 Water Uptake and Transport by Plants

Plants need a much greater proportion of water than animals of the same size and weight. This is because most water is recycled internally in animals, whereas in plants water is continually being lost through transpiration when stomata are open to obtain the CO_2 required for photosynthesis (sections 5.3 and 5.4). Water and dissolved nutrients move from the roots to the stems and leaves through the vascular system as described in section 5.2. When this water is lost through transpiration it must be replaced with more water absorbed from the soil by the roots in a continual cycle.

The rate of transpiration from garden crops is affected by many factors, such as the garden microclimate and the types of crops being grown. Transpiration is commonly measured in combination with evaporation, which is referred to as *evapotranspiration* (ET). *Maximum evapotranspiration* (ETm) is the rate of evapotranspiration that occurs when the crop requirements for water are fully met. Garden water management tries to maintain ETm by keeping water in the root zone at no less than 25-50% of field capacity.

A major goal of dryland garden management is to minimize excess evapotranspiration due to stressful environmental conditions such as high temperature, low humidity, wind, or poor infiltration. Shading plants and protecting them from drying winds, along with the measures to decrease evaporation from the soil given in the previous section, will help to reduce excess evapotranspiration.

The relationship between transpiration, soil water, and garden yield is discussed in the next section, along with more suggestions for increasing garden production.

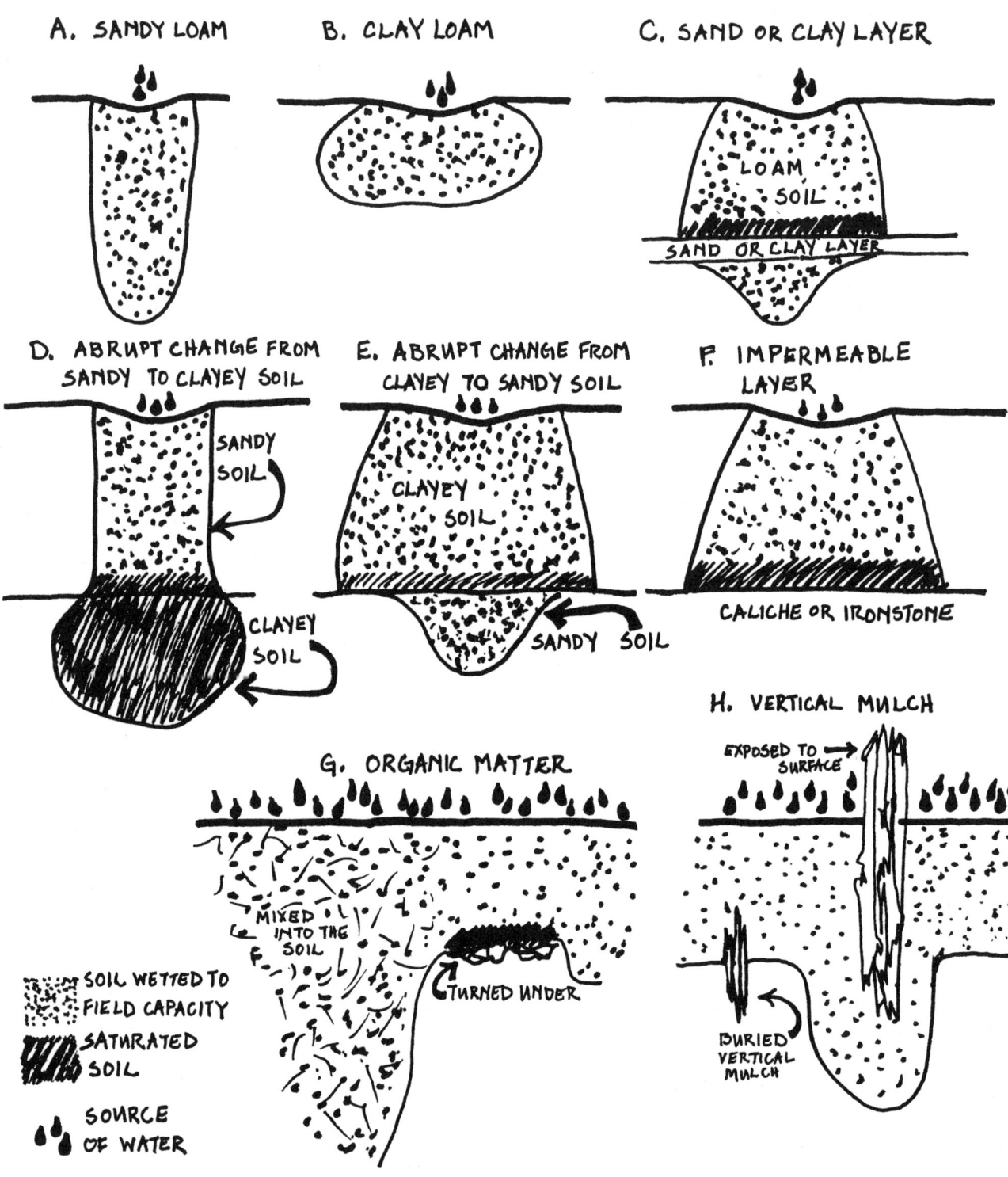

Figure 10.4 Water Movement in the Soil
(After Doneen and Wescot 1984:22-27; and Gardener 1979)

10.4 Soil Water and Garden Yield

A plant experiences water stress or water deficit when *actual evapotranspiration* (*ETa*) falls below ETm, that is, its requirements for water are not met (section 5.5). This is often the result of insufficient water in the soil, but may also be caused by disease or pest damage which hinders plants' ability to take water from the soil and use it.

When a crop experiences water deficit, the *yield*, or the amount that is harvested, is often reduced. The amount of reduction depends on the stage of the life cycle in which the water stress occurs, the crop and crop variety, and the amount of water deficit.

As crops develop through vegetative, flowering, yield formation (fruit, seed, tuber), and ripening stages, their sensitivity to drought changes. In general, crops are most sensitive to drought during flowering, followed by yield formation, with early vegetative growth and ripening being the least sensitive[9] (Figure 10.5). Therefore it is most important that garden crops have enough water during the flowering and fruiting stages.

The specific crop and variety also affect the reduction in yield as drought increases. The more drought-resistant a variety is the less its yield will be reduced by drought, and the more stable its production will be over time in a marginal environment.[10] Figure 10.6 shows that for more drought-resistant plants, drought results in a relatively smaller yield reduction (B) than with less drought-resistant (drought-sensitive) plants (A). New varieties bred specifically for high production may be more sensitive to drought than locally adapted, indigenous varieties. They may have higher yields (M) than drought-resistant indigenous varieties when there is no drought, but also have a greater reduction in yield with drought (D_1) (compare A with B

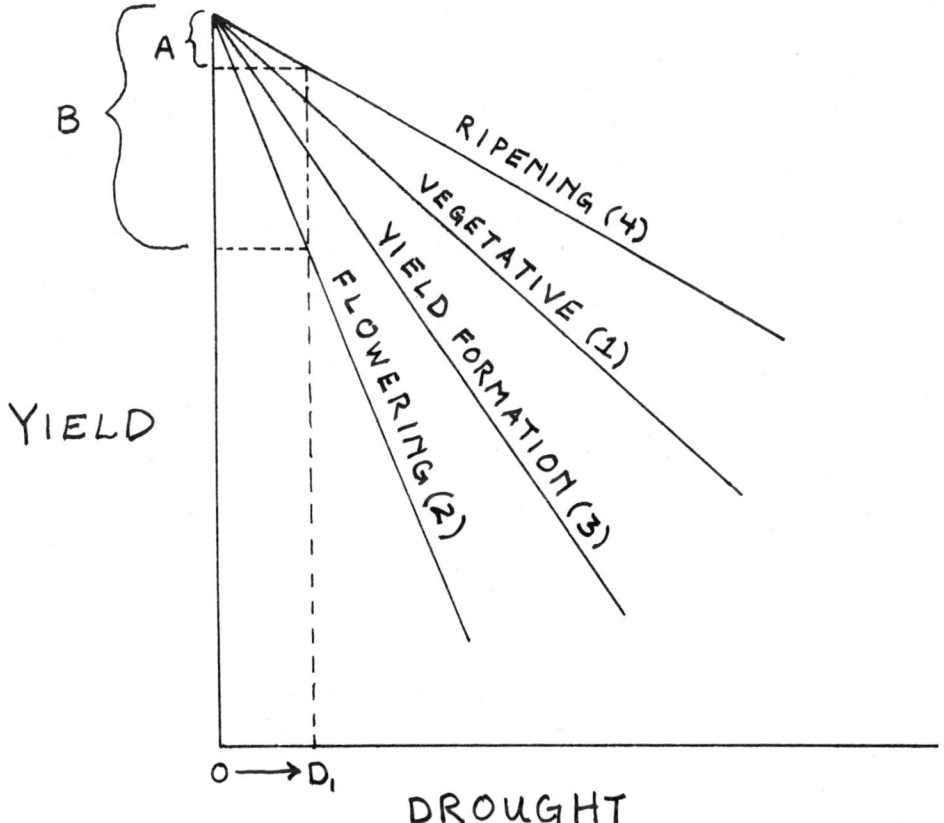

A = THE DROP IN YIELD FROM DROUGHT D_1 OCCURRING DURING RIPENING.
B = THE DROP IN YIELD FROM DROUGHT D_1 OCCURRING DURING FLOWERING.
$D_1 = ET_a < ET_m$

Figure 10.5 Water Requirements, Yield, and Crop Life Cycle
(After Doorenbos, et al. 1979:38)

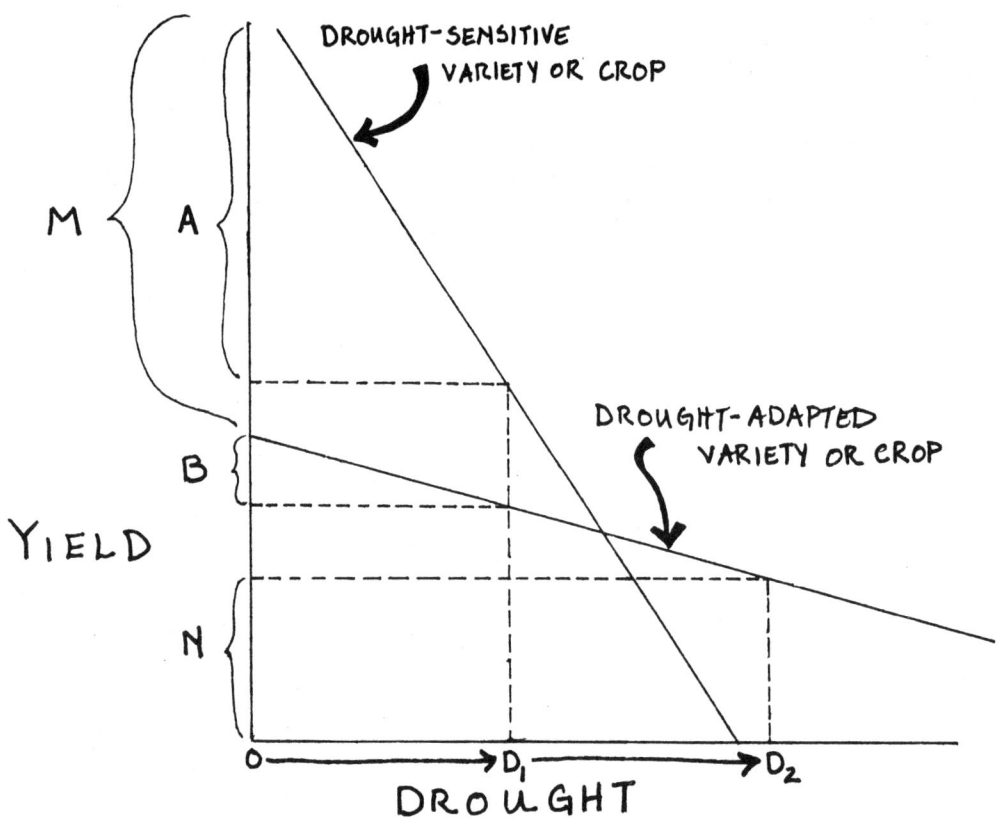

Figure 10.6 Yield Response to Water (After Doorenboos, et al. 1979:38)

in Figure 10.6). As drought worsens (D_2), yield is reduced below that of indigenous varieties (N). Thus, where drought is likely because of lack of rain or irrigation water during the garden growing season, drought adaptation is an important criterion for choosing crops (section 5.5). This means that growing a high-yielding new variety unadapted to drought is more risky for gardeners in areas subject to drought.

Thus, there are three main approaches to increasing garden yield by minimizing drought:
- Changing planting times to improve the fit between soil water availability and crop demand.
- Using more drought-adapted crops, crop varieties, and crop mixtures (section 5.5).
- Increasing available soil water by applying more water; by using mulch, shade, and windbreaks to decrease water lost to evapotranspiration; and by decreasing water lost to runoff and excess deep percolation (sections 10.3 and 10.8).

Trying any of these approaches to increase yield often means experimenting with new methods, and taking a certain amount of risk. Increased investments of labor and other resources such as organic matter, shading materials, or water may also be required. Each gardener will have to decide if these increased investments are worthwhile. The greater yields that may result from such increased investments must be considered in light of other benefits that could be obtained by investing the time and resources in different ways.

10.5 How Much Water?

Gardeners learn to judge when and how much to water their gardens by observing variation in plant growth and yields in relation to different watering methods. Most successful gardeners never make numerical calculations about water requirements. (See Box 10.2 for ideas about estimating water requirements for large projects.) To estimate water requirements for individual household gardens it is probably best to observe the amount of water being used by other gardeners or farmers in the area and use this as a basis for experimenting.

There is very little information on the water requirements of dryland mixed gardens. In one case study, an annual average of 48 liters/m^2/week (1.2 gal/ft^2/week) was applied to two urban gardens in the Sonoran Desert of North America.[11] The gardens were mixtures of annual vegetables, producing food throughout the year. The amount of irrigation water needed differed greatly during the year depending on temperature, rainfall, and humidity. For example, even with a moderate amount of shading and mulching the amount of irrigation water applied in one garden during the hot, dry season (April-June) (averaged for 2 years) was 67 liters/m^2/week (1.7 gal/ft^2/week), with rainfall supplying another 3.6 liters/m^2/week (0.09 gal/ft^2/week), or only about 5% of the total (Figure 10.7). In contrast, during the cold, wet season (January-March), only 7.7 liters/m^2/week (0.19 gal/ft^2/week) of irrigation water was applied, with rainfall supplying another 8.6 liters/m^2/week (0.2 gal/ft^2/week), or about 53% of the total.

The objective in watering is to wet the soil where the roots are growing and a short distance beyond (Figure 10.8 and section 5.2.1). Too little water can restrict root growth, increase salt in the root zone, and a larger proportion of water will be lost to surface evaporation. Too much water means a waste of resources and can

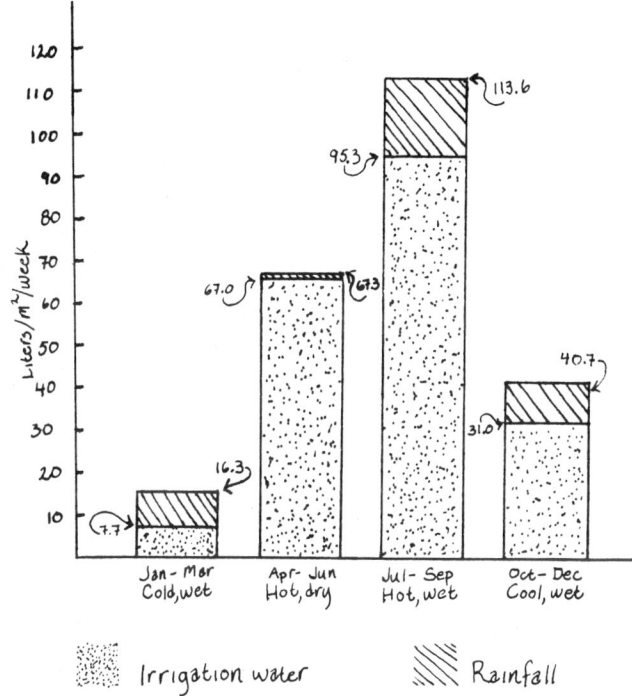

Figure 10.7 Seasonal Variation in Water Use in a Desert Garden
(After Cleveland and Soleri n.d.a)

cause waterlogging (section 12.6.1). When there is a problem with salty soil or water, more water will be needed to wash any accumulated salts below the root zone (section 12.6.2).

A sharp stick can be pushed more easily into wet than dry soil, and so can be used to check depth of watering. Depth of watering can also be determined by digging down through the root zone in several places after watering and checking soil moisture (section 10.7).

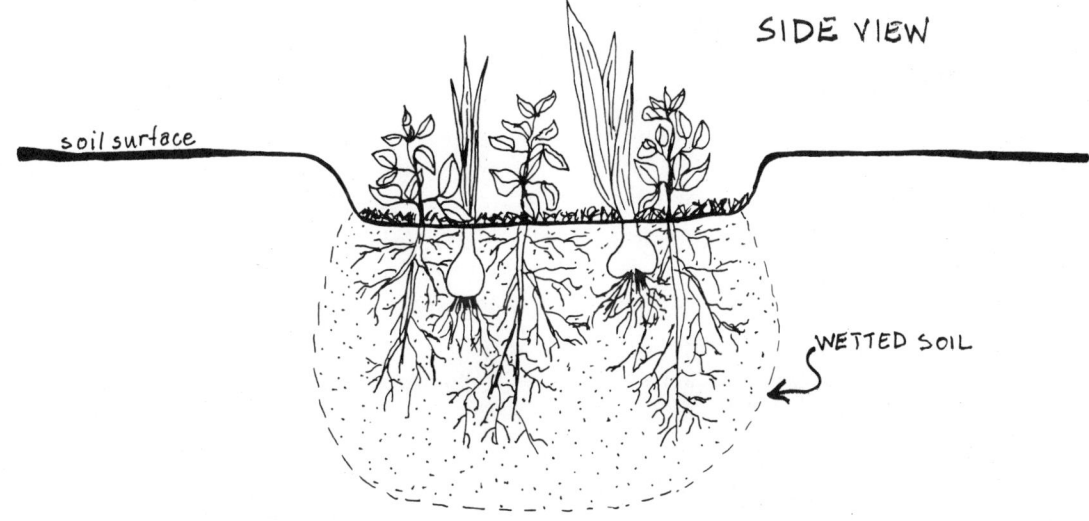

Figure 10.8 Wetting the Soil Beyond the Root Zone

> **Box 10.2**
> *Estimating Water Requirements for Large Garden Projects*
>
> For large-scale, monoculture production it is possible to make many sophisticated measurements which are then used to estimate watering frequency and the amount of water needed. For household gardens such measurements are usually unnecessary as well as impractical because of the equipment, experience, time, and expense involved, and because of the much greater complexity of mixed gardens compared with large fields planted with only one crop. However, in drylands where water is often scarce it may be possible to improve the design or actual water management of larger garden projects and anticipate their approximate water needs. This can be done with an understanding of how garden water requirements can be estimated using simple calculations.
>
> When designing a garden project involving a large number of gardens, or in situations where there is no local information on water consumption, as may be the case in refugee camps or new urban squatter settlements, estimates of the amount and frequency of watering can be useful. Making monthly estimates of water requirements involves knowing the water-holding capacity of the soil, the approximate rooting depth of the garden crops, the water requirement of the crops due to evapotranspiration, and how much water will be lost through drainage below the root zone.[12] Variation between seasons, or even within seasons, is often very large due to changes in rainfall, temperature, and type and growth stage of garden crops. After initial estimates have been made, the next step is to adjust them by measuring the amount of water applied, and monitoring its effects in the garden.
>
> The garden should be watered when about 50-75% of the plant available water in the principal root zone has been used up, that is, when the soil is at 25-50% of field capacity.[13] We will use 50% as a conservative estimate. If the soil water is allowed to drop below this level, reduction in yield may result, since many plants will not be able to maintain the transpiration rates required for maximum production.
>
> The amount of water needed at the time when the crop would otherwise begin suffering from drought can be calculated using the following formula:
>
> $W = (pr)(AW)(d)/E_a$,
>
> where
>
> W = amount of water needed to bring the soil in the root zone to field capacity (liters/m^2 or mm of depth; see Chapter 17 to convert English measures to metric);
>
> pr = the proportion of plant available soil water which a particular crop can actually use without ETa becoming less than ETm, that is without experiencing water deficit; estimated at 0.5;
>
> AW = plant available water (mm/m), the difference between field capacity and permanent wilting point (section 10.3.1);
>
> d = root depth of the crop (m);
>
> E_a = water application efficiency, the ratio of the water in the root zone to the total amount of water applied (section 12.2).
>
> Rooting depths of crops are usually given for soil

10.6 Measuring Water Applied to the Garden

If the amount of water needed by the garden is to be calculated, some method of measuring the water is required. Even if the water applied to gardens does not need to be calculated in advance, measuring that water may be an important way to estimate water needed when garden programs are expanded to other households, communities or seasons, or for estimating costs where gardeners will have to pay for water.

If water is being applied in containers, such as buckets or calabashes, either by hand or with a *shaduf* (section 12.7.1) or other mechanism, the volume of the container can be determined by pouring water from it into another container of known volume. Once the volume of the watering container is known, the number of containers of water used in the garden is counted and multiplied by this volume to find the total amount applied.

When water is delivered from a pipe or hose, the time taken for water to fill a container of known volume can be determined by using a watch that shows seconds. The number of seconds required to fill the container is then divided into the total time taken to irrigate. It is best to measure the rate of flow several

Box 10.2, continued

conditions which do not limit root growth. Potential rooting depth is not as important as depth of maximum water absorption by roots, which usually occurs at shallower levels. For example, for shallow-rooted garden crops such as onions and spinach whose potential rooting depth is up to 0.6 m (2 ft), depth of maximum water absorption is only 10-15 cm (4-6 in); crops with medium potential rooting depth (up to 1.2 m or 4 ft) such as chilis, beans, and squash may draw most of their water from the area 25-40 cm (10-16 in) below the surface; and for deep-rooted crops such as grapes, olives, almonds, and watermelons with a potential rooting depth of 2.0 m (7 ft), the depth of maximum water absorption may be 1.2-1.4 m (4-6 ft).[14] If root growth or water penetration is limited by caliche, ironstone, rock, or other layers in the soil, then rooting depth will be reduced. Root growth also responds to the amount of water applied to the garden. Small applications of water that wet only the top 5-8 cm (2-3 in) of the soil will limit root growth to this depth, and can cause a buildup of salt in the soil.

As an example, let us consider an urban neighborhood group that wants to start a community garden on vacant land. The group must estimate the amount of water needed in order to negotiate with the city government to obtain an adequate supply of piped water. Let us assume that the garden soil is sandy loam with plant available water storage capacity of about 120 mm/m, or a deficit of approximately 60 mm/m when 50% of the plant available water in the root zone has been used up. The garden is laid out in 100 1.5 m x 1.5 m garden beds with each participating household cultivating one or more beds and irrigating with a hose. They plan to grow annual vegetables for which a rooting depth of 0.45 m (18 in) is quite adequate. To irrigate the beds to field capacity when 50% of plant available water in the root zone has been depleted, and assuming a water application efficiency of 80% (allowing 10% of the irrigation water to leach salts below the root zone, and another 10% to be lost to evaporation before infiltration), the amount of water needed would be:

$W = (pr)(AW)(d)/Ea = (0.5)(120 mm/m)(0.45 m)/0.80 = 34$ mm of irrigation water.

Since 1 mm of water on 1 m^2 = 1 liter, and the area of the beds = 1.5 m x 1.5 m x 100 = 225 m^2, the gardeners will apply (34 liters/m^2)(225 m^2) = 7,650 liters of water.

If the soil were a clay loam with plant available water capacity of 200 mm/m, then they would need:

$W = (0.5)(200 mm/m)(0.45)/0.80 = 56$ mm of irrigation water,

= (56 liters/m^2)(225 m^2) = 12,656 liters of water.

However, they would not need to water as often, since there would be more water in the root zone:

200 mm/m - 120 mm/m = 80 mm/m, or
(80 mm/m)/(120 mm/m) = 67% more.

To calculate the total water requirements for the growing season, the gardeners would also need to know how often this amount of water would have to be applied. (Box 10.3 in section 10.7 discusses calculating frequency of irrigation.)

---

times and take an average, especially if the water pressure is subject to change. For example, if the valve is opened to maximum flow and it takes 1.75, 1.50, and 2.00 minutes to fill a 20-liter (5.3-gal) container, then the average is (1.75 + 1.50 + 2.00)/3 = 1.75 minutes. If it takes 25 minutes to water the garden then the total water used is approximately (25/1.75) x 20 = 286 liters (75.6 gal).

Water applied by a sprinkler can be measured the same way as rainfall (section 11.4.2), preferably by averaging the water collected at several places within the garden.

There are many indigenous methods for measuring irrigation water delivered in a canal. In Yemen, irrigation water taken from a cistern is measured in a number of ways: by the movement of shadows cast by fixed indicators, such as trees, along the cistern's edge; by a stick or stone submerged in the water which is gradually exposed as the water level drops; or by a "water clock," a copper bowl with a hole in the bottom which is filled with water—the time it takes to drain out is the unit of measurement used.[15] These are all good ideas for reliable, inexpensive measuring devises. For any system of measurement to work it is essential that all those using it understand it and agree to its legitimacy.

Water delivered in a canal can also be measured

with devices called *weirs* which require precision construction, calibration, and measuring.[16] Less accurate, but also inexpensive and easier, is the *float method*, in which the area of cross section of the canal (A in m^2) is multiplied by the velocity of water (V in m/sec) to obtain flow (Q in liters/sec):[17]

$$Q = (A)(V)$$

This requires measurement of the cross section of the canal at several places over a relatively straight 20-30 m (65-100 ft) length (L) to obtain an average (Figure 10.9).

If the average width of a 25-m (82-ft) section of a semicircular canal is 1 m, and the average depth is 0.5 m (1.64 ft), then the radius (r) is 0.5 m (1.64 ft), and the cross section (A) is found by using the formula for the area of a circle divided in half:

$$A = [(\pi)(r^2)]/2 = [(3.14)(0.5^2)]/2 = 0.39 \text{ m}^2.$$

A watch that shows seconds is then used to measure the time (t) it takes for an object to float between a string stretched across either end of the measured section of canal. A bottle half-filled with water or soil will be carried along below the surface, and a stick can be placed in the top to make it easy to follow. This should be done several times to obtain an average. The result is multiplied by 0.8 to adjust for the fact that surface flow is faster than average flow. If the bottle took an average of 40 seconds to move the 25 m (L) then the velocity (V) of water in the canal is:

$$V = (L/t)(0.8) = (25 \text{ m}/40 \text{ sec})(0.8) = 0.5 \text{ m/sec}.$$

Flow (Q) is then

$$Q = (A)(V) = (0.39 \text{ m}^2)(0.5 \text{ m/sec}) = 0.2 \text{ m}^3/\text{sec},$$
or 200 liters/sec.

10.7 When to Water

For individual gardens, the time to water can be judged by observing both plants and soil.[18] Perhaps the most obvious sign is wilting. However, a few plants such as cucurbits can wilt during hot afternoons before they need to be watered. If they recover quickly as the temperature drops in the evening, then they probably do not need watering, but may benefit from shading. Wilting even when the soil is wet is probably caused by disease or pest damage, for example, beetle larvae feeding on roots, or fungi blocking the movement of water in the plant's vascular system (Chapter 13). Plants should be watered before or when they begin to show signs of water stress, since prolonged wilting can cause a permanent reduction in yields.

Other signs that plants may need watering are leaves that feel very warm during midday, leaves turning darker green, and a lack of new growth. Dieback of growing tips is often a sign of severe water stress.

Table 10.1 Soil Texture and Water Deficit [a]

The amount of water in mm/m is the depth of water (equal to liters/m^2) that would have to be applied to bring the soil to field capacity. A permanent wilting point of -15 bars is assumed. See section 9.3.1 for determining soil texture.

Texture	Field capacity (0% deficiency)	Time to water (50-75% deficiency)	Permanent wilting point (100% deficiency)
Sand (coarse)	No free water when squeezed, will not form ball, but wet outline left on hand	Appears dry, will not form ball with pressure, but still some clumping (40-65 mm/m)	Dry, loose, single grains flow through fingers (85 mm/m)
Sandy loam (moderately coarse)	No free water when squeezed, makes weak ball, outline left on hand	Appears dry, will not form a ball, sticks together slightly (65-100 mm/m)	Dry, loose, flows through fingers (125 mm/m)
Loam (medium)	No free water when squeezed, can form cylinder, wet outline left on hand	Crumbly but makes weak to good ball when squeezed (85-125 mm/m)	Powdery, dry, small clods easily broken into powder (170 mm/m)
Clay loam (fine)	No free water when squeezed, can form ring, wet outline left on hand	Pliable, forms ball but not cylinder (100-160 mm/m)	Hard, cracked (200 mm/m)

[a] Based on Doneen and Wescot 1984:9, 11; Merriam, et al. 1980:759; and Stegman, et al. 1981:798.

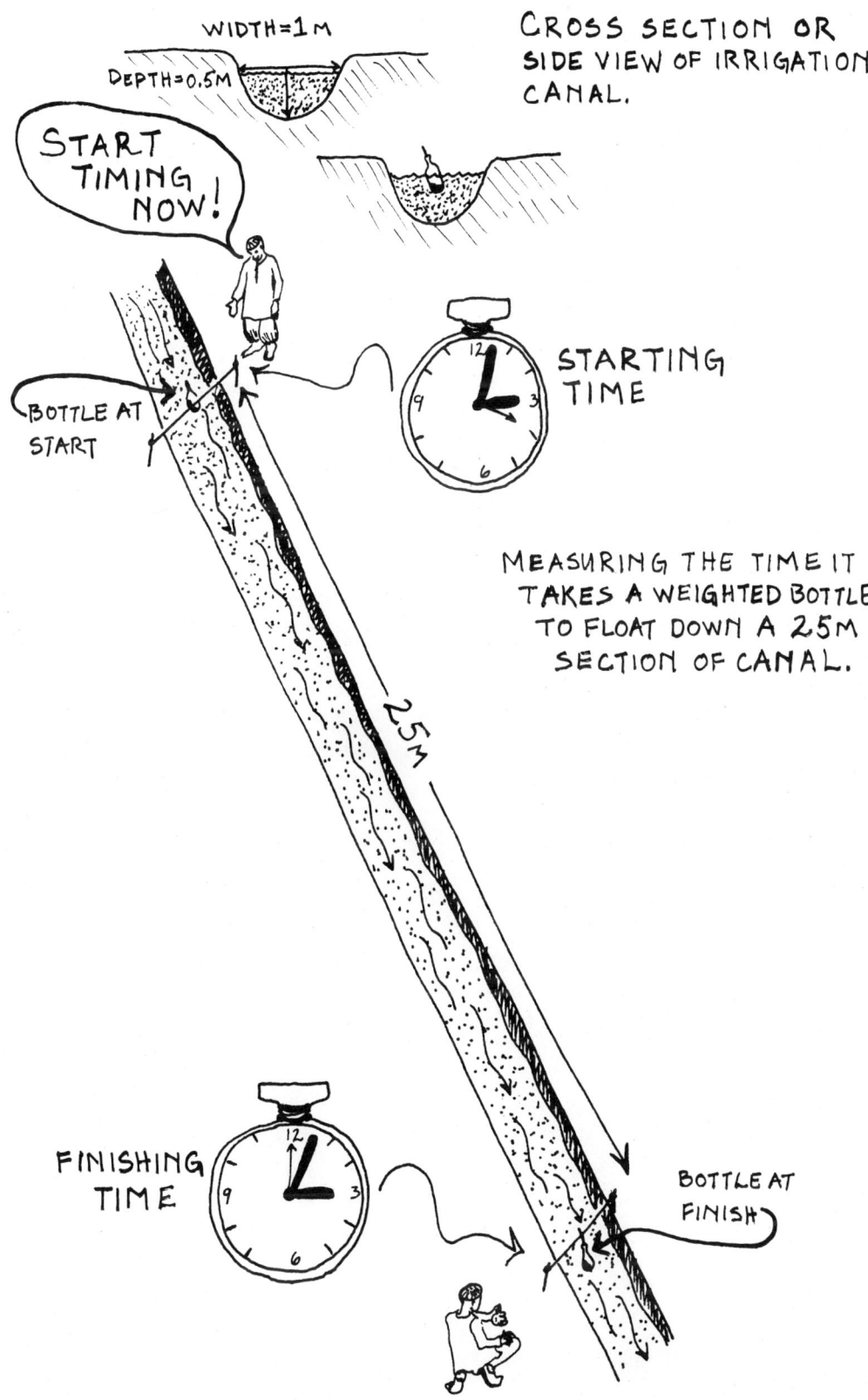

Figure 10.9 Measuring Water Flow in a Canal

Garden soil can also be read for signs that it is time to water (Table 10.1). Most crops should be watered when 50-75% of the plant available water has been used up. The soil in the root zone where most of the roots are growing should be sampled by digging down in several places. The descriptions in Table 10.1 are a guide for making rough estimations of soil water deficit based on handling of soil samples. Experimenting with crops and crop mixes, time of planting, time of day water is applied, mulching, and shading, can increase the length of time between waterings, reducing water use in the garden. In Box 10.3 we discuss a method that can be used to calculate irrigation frequency for large garden projects.

10.8 Mulches, Shades, and Windbreaks

Reducing evapotranspiration in the garden reduces the gardener's time and the amount of water that must be invested. Increasing infiltration and storage of water in the root zone, decreasing temperatures, and increasing humidity are the basic methods for minimizing evapotranspiration. This is done by using mulches, windbreaks, shades, and cropping patterns. Irrigation methods that decrease water use are discussed in Chapter 12.

Mulches are materials applied to the garden soil to modify temperature, air movement, water infiltration, and weed growth. There are two basic kinds, surface

---

Box 10.3
Calculating Irrigation Frequency for Large Garden Projects

Box 10.2 described how to calculate the amount of water needed for a single watering of a large garden project. While details are beyond the scope of this book, we describe here the basic principles of how to estimate the frequency of watering, in order to obtain a complete estimate of water requirements.

The frequency of irrigation required to maintain maximum yield can be calculated as:[19]

$i = W/ETm$,

where

i = irrigation interval in days;

W = amount of water needed to bring soil in the root zone to field capacity when 50% of the plant available water is depleted;

ETm = maximum evapotranspiration per day of the garden crop or crop mixture growing under optimal conditions including adequate water.

ETm can be estimated by observing local gardens, or calculated by

$ETm = (kc)(ETo)$,

where

ETo = theoretical evapotranspiration;

kc = crop coefficient for the garden crop or crop mixture.

ETo is a standard measure of the effect of climate on evapotranspiration, and has different values under different climatic conditions. It is "the rate of evapotranspiration from an extensive surface of 8-15 cm (3-6 in) tall, green grass cover of uniform height, actively growing, completely shading the ground and not short of water."[20] This is calculated by one of several methods using climatic data for the specific location including temperature, humidity, wind, and sunshine. The pan method uses actual measurements of evaporation from a pan of water which are then converted to ETo. Values of ETo and ETm are usually expressed in mm/day and calculated for 10 or 30 day periods. The crop coefficient (kc) varies with the crop, stage of growth, and environmental conditions. Values of kc or graphs from which they can be read are available, to be used with specific calculation methods.[21]

To calculate irrigation frequency for the example in Box 10.2 we assume that ETm has been estimated at 4 mm/day. For the sandy loam soil that needs 34 mm of water to bring soil in the root zone to field capacity when 50% of plant available water has been used, watering frequency is

$i = W/ETm = (34 \text{ mm})/(4 \text{ mm/day}) = 8.5$ days, that is, every 8-9 days.

For the clay loam soil with a higher water-holding capacity, and W = 56 watering frequency is

$i = (56 \text{ mm})/(4 \text{ mm/day}) = $ every 14 days.

mulches and vertical mulches. Mulches made of organic materials improve soil structure and fertility as they are broken down by soil microorganisms. Woody mulches with a high C:N may tie up soil nitrogen temporarily (section 9.5.2).

10.8.1 Surface Mulches

Surface mulches cover the soil surface, shading and cooling it, and so helping to reduce evaporation (Figure 10.10). Mulch protects the soil from the impact of raindrops, which compact the soil surface, thus decreasing infiltration and increasing runoff. A thick surface mulch also discourages growth of unwanted weeds which compete with crops for water (section 8.6). Plant debris such as leaves, straw, and weeds can be used as a mulch, as can sand or stones. Plants like purslane, mat bean, melons, and squash provide a living mulch by spreading out horizontally to cover the soil surface. Since much of the water held in the mulch itself after irrigating will be lost to evaporation, watering under the mulch or combining surface mulch and root zone irrigation—for example, by using a vertical mulch—are good strategies.

In one experiment in semiarid India, a grass mulch on three different annual pulses significantly improved production.[22] The crops received 160 and 230 mm (5 and 7 in) of rainfall in each of two 3-month growing seasons. The mulch reduced the soil temperature at a depth of 10 cm (4 in) resulting in increased root length and weight. It increased height and weight of the plant at the active growing stage, and resulted in less weed growth, and an increase in total plant yield, grain yield, and water-use efficiency in comparison with controls which had no mulch. Of the different rates of mulch application tested (3, 6, 9, and 12 T/ha), the most effective application rate was 9 T/ha. On a scale more appropriate for gardens this is equal to 0.9 kg/m^2 (0.2 lb/ft^2).

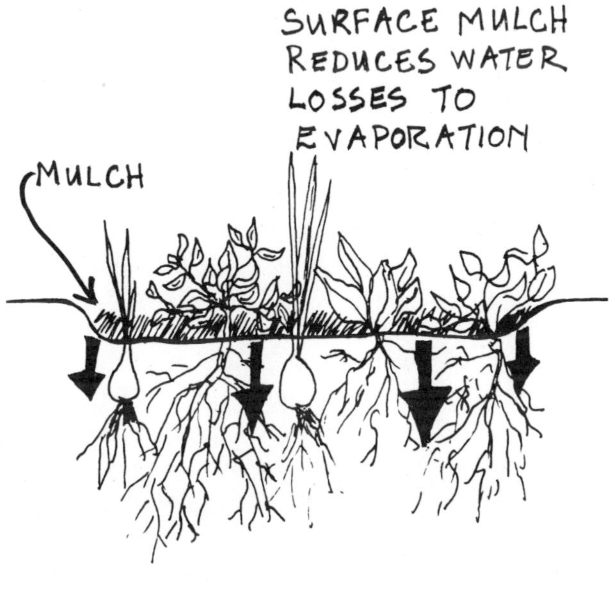

Special care should be taken when mulching around newly planted seeds (section 6.5.2). Sometimes mulches are hiding places for garden pests such as cutworms or sow bugs which can quickly destroy young seedlings (section 13.4). If the seedlings are being damaged, mulch should be removed from the planting furrow or area immediately around the seedlings.

Stones and rocks can also be used for surface mulching. In the process of cooling during the night, the moisture that evaporates from the soil condenses on the lower surface of the stones and rocks instead of being lost to the air. Because of the difficulty of planting and cultivating areas mulched with stone or rock, and the fact that they can reflect sunlight and hold and radiate heat, we suggest that these mulches only be used on perennial crops, especially trees. Lower-lying, more tender garden plants can suffer from the heat absorbed and radiated by the rocks. With annual crops gravel and sand can be used as a mulch. In arid northwest China, a 5-16 cm (2-6 in) thick gravel, pebble, and sand mulch has been used on melons and other annual vegetable crops for the past 300 years.[23]

10.8.2 Vertical Mulches

Vertical mulches provide water a pathway to the root zone, and can reduce the area of soil surface that is wet. Narrow trenches in garden beds can be filled with stalks like those of millet, maize, sunflower, Jerusalem artichoke, or amaranth. The air spaces created by these stalks conduct the water quickly down to the plant root zone where it is less likely to evaporate than when it is spread on the soil surface and infiltrates slowly (Figure 10.11). The principle is the same as root zone irrigation (section 12.4). In addition, these organic mulches will improve soil water-holding capacity as they decompose.

The beneficial effect of vertical mulch has been demonstrated in a laboratory experiment where 5.5 cm (2 in) of water was applied to a furrow, and the same amount poured into a vertical mulch of barley straw 7 cm (3 in) wide and 15 cm (6 in) deep.[24] After 17 days

Figure 10.10 Surface Mulch

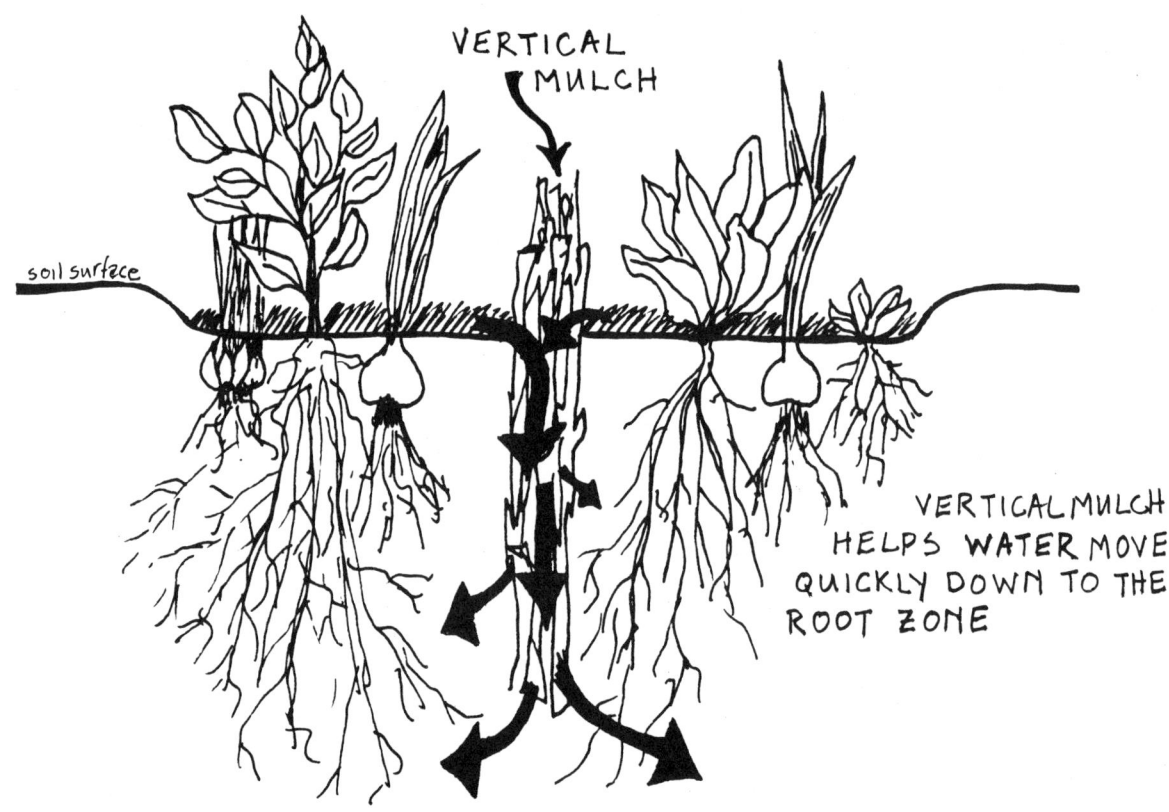

Figure 10.11 Vertical Mulch

of evaporation, 42% of the water applied in the furrow remained in the soil, while 72% of that applied in the vertical mulch remained. When water was applied rapidly, so that it overflowed the vertical mulch and wet the surrounding soil surface, only 60% of the water remained after 17 days of evaporation. Subsequent field experiments with sorghum showed that a similar mulch increased yields 20-40% over plots that were identical but with no mulch.[25]

Experiments with vertical mulches in gently sloping Vertisols in semiarid India have shown that sorghum stubble packed in trenches 15 cm (6 in) wide improved water infiltration and produced larger yields of grain and straw of an "improved variety" of sorghum, especially in years of drought.[26] Vertical mulch in trenches 30 cm (12 in) deep and 2 m (6.5 ft) apart gave 25 times more grain yield and 2 times more straw yield than the areas without vertical mulches in a very dry year. The beneficial effects appeared to last 4 years, probably a result of improved soil structure as the straw partially decomposed after the second year. While sorghum stalks are used to feed cattle in this area, the bottom third of the stalks is not well liked by the cattle. The experimenters therefore felt that using the stalks for vertical mulching would not be a great loss for the cattle, especially if the mulching increased total straw production. However, alternative uses of the bottom third of the stalks, such as for fuel, were not considered.

10.8.3 Windbreaks, Shades, and Cropping Patterns

Wind blows away the protective layer of moist air near the surface of leaves and soil, and direct sunlight raises leaf and soil temperatures. Thus wind and sun can greatly increase rates of evapotranspiration in the garden leading to greater water requirements. This means that more water must be applied to the garden or yields will be reduced. Windbreaks, shades, and cropping patterns that protect the garden from harsh winds and sun can reduce water requirements and increase yields (Figure 10.12). Young seedlings or new transplants often require extra protection from sun and wind, especially in the hot, dry season. (For more on windbreaks see section 9.7.4, and on orientation toward the sun, section 8.3.)

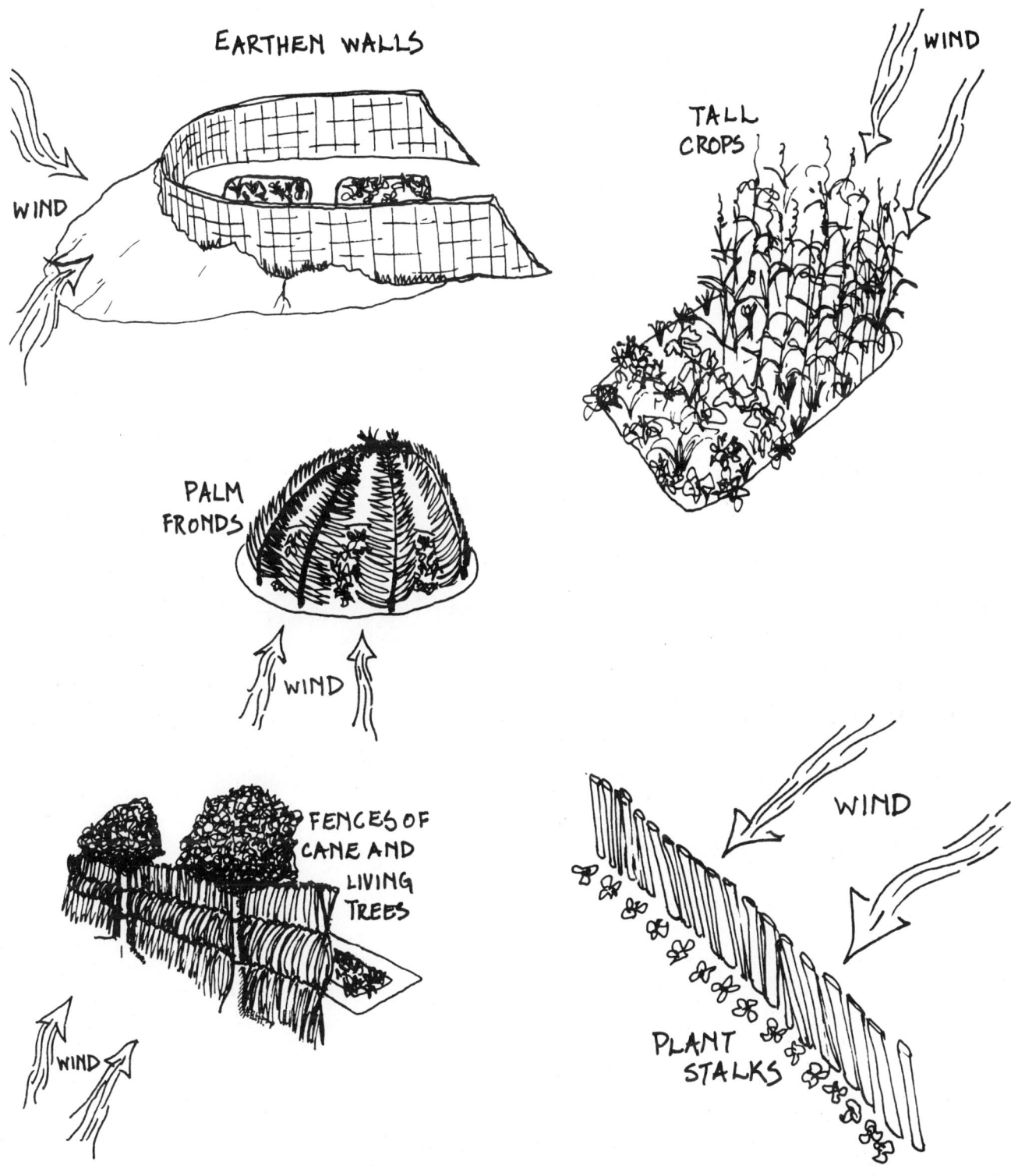

Figure 10.12 Windbreaks Protect Garden Plants from Wind and Sun

Mixed cropping in the garden, and dense planting of crops with different aboveground heights, creates shade for the soil and for shorter plants (section 8.5.1). In addition, different rooting depths help to make the best use of water in the soil.[27] In cold winter areas, trees such as peach, pomegranate, or fig that lose their leaves to let in the winter sunlight, but leaf out in the summer to provide shade and wind protection for other garden crops, are especially good (Figure 8.8 in section 8.3).

Branches, palm fronds, stalks, and other plant parts can be pushed into the soil or tied to frames of branches or bamboo. On the Mediterranean island of Pantelleria, Sicily, Italy, gardeners remove large pads from a prickly pear cactus and prop these up to shade their tomato plants.[28] Mats or cloth can also be tied to frames, and leaves, bark, straw, or old baskets can be used to shade small plants. All of these materials should allow some wind to pass through, and be anchored securely in the ground, or they will be blown over.

Houses, walls, trellises, and fences, especially tall, living fences can give protection from the wind and sun (section 8.8).

10.9 Resources

Technical details of crop, soil, and water relationships in production are presented in Doorenbos and Kassam (1979) and calculations of crop water requirements are shown in Doorenbos and Pruitt (1977). Ahn (1970:127-147) gives an excellent, easy-to-read description of soil and water relations with reference to West Africa.

References

[1] See also Dupriez and De Leener 1983:55-58.
[2] See Ahn 1970:127-147.
[3] Donahue, et al. 1983:166-170.
[4] Delgado 1984:57-58.
[5] Donahue, et al. 1983:185-191; Doneen and Wescot 1984:21-27; Gardner 1979.
[6] See Ahn 1970:133,137-139; Doneen and Wescot 1984: 27-28; Henderson 1979.
[7] E.g., Dupriez and De Leener 1983:107.
[8] Troeh, et al. 1980:475.
[9] Doorenbos and Pruitt 1977:61-63; Dupriez and De Leener 1983:134-140.
[10] Flinn and Garrity 1986.
[11] Cleveland and Soleri n.d.a; Cleveland, et al. 1985.
[12] Stern 1979:69-74; Doorenbos and Pruitt 1977:68-82.
[13] Halderman 1977.
[14] Doneen and Wescot 1984:33-35.
[15] Varisco 1983.
[16] Merriam, et al. 1980:749-758; Stern 1979:141-147.
[17] Stern 1979:141.
[18] Halderman 1977.
[19] See Doorenbos and Kassam 1979:26-27.
[20] Doorenbos and Pruitt 1977:1.
[21] E.g., Doorenbos and Pruitt 1977:38-41.
[22] Gupta and Gupta 1983.
[23] Ming and Yun-wei 1986.
[24] Fairbourn and Gardner 1972.
[25] Fairbourn and Gardner 1974.
[26] Rama Mohan Rao, et al. 1977.
[27] See Dupriez and De Leener 1983:115-123.
[28] Galt and Galt 1978.

11
Sources of Water for the Garden

Water for gardens can come from many different places: rainfall, rivers, streams, wells, and even water left from washing, bathing, or spilled at the pump (Figure 11.1). The cost of water, the amount available, its quality, and whether gardeners can depend on a steady supply, are often determined by the source of water.

11.1 Summary

The most important sources of water in drylands are:
- Rain, either falling directly on the garden or harvested from adjacent areas.
- Surface water from rivers, streams, or lakes, which may first be stored in reservoirs behind small dams, and can be transported to the garden through irrigation canals or by natural or managed flooding.
- Groundwater which may be obtained through shallow, hand-dug wells, or deeper wells and bore holes in the case of a community water supply.
- Piped water, originally from various sources, which is especially important in cities.
- Recycled water from household washing and food preparation, or from water spilled at a pump or stand pipe.

For later use during dry periods, water can be stored in the soil, behind berms or dams, or in cisterns, pots, and jars.

Figure 11.1 Using Water Spilled at the Pump

> ### Box 11.1
> ### Measuring Water Salinity
>
> The level of water salinity is usually expressed in terms of electrical conductivity of the water (EC_w). Conductivity is measured by special instruments in units of millimhos/centimeter (mmhos/cm), or the more recently adopted standard of Siemens/meter (S/m) or deciSiemens/meter (dS/m), where
>
> $$1 \text{ S/m} = 10 \text{ dS/m} = 1{,}000 \text{ mmhos/m} = 10 \text{ mmhos/cm}$$
>
> The better water conducts electricity the saltier it is, and the more production will be reduced. Water with a conductivity of 0.7 dS/m or less is good for all crops, 0.7 to 3.0 dS/m is harmful to sensitive crops, and greater than 3.0 dS/m is harmful to most crops.[3]
>
> Salt content can also be expressed in mg of dissolved solids/liter of water. Salinity can be determined in these terms if there is access to a scale that is accurate to 100 mg (0.1 gm), perhaps in a laboratory at a secondary school, university, or agricultural experiment station. A liter of water is completely evaporated in a pot protected from dust. This may be done slowly by leaving the pot to stand in the sun, or more quickly by boiling it over a heat source. The remaining solids are then weighed. The *total dissolved solids* (mg/liter) indicate saltiness: 0 to 500, good for all crops; 500 to 2,000, harmful to sensitive crops; and over 2,000, harmful to most crops.[4]
>
> Over time the amount of salts leaving the root zone through drainage should equal the amount brought in through watering and rainfall. To prevent buildup of salts it is very important that there be enough extra water applied to the garden at regular intervals to wash out the salt, and adequate drainage to carry that water away from the root zone.
>
> If natural drainage does not exist, then drainage should be provided (section 12.6.1). The saltier the irrigation water, the larger the amount of water that must drain below the root zone. The amount of water needed to *leach* or wash down salts carried in by irrigation water is expressed as the *leaching requirement* (LR), and can be calculated if the salinity of the water is known (section 12.6.2).
>
> Ion toxicity problems often accompany salinity problems. Ions are absorbed by plants along with water, they accumulate in the leaves as a result of transpiration, and can poison plants. Chloride is the most common cause of ion toxicity. It shows up as drying ("burning") of leaves beginning at the tips and moving along the edges. Leaves may drop prematurely. Leaf burn due to sodium toxicity begins at the edge of older leaves and moves between the veins toward the center.
>
> When crops continue to have problems after soil salinity has been reduced, then it may be worth having a water sample checked in a laboratory. If tests show that ions such as sodium, chloride, boron, and selenium are present in harmful levels in water used in the garden,[5] an alternative source of water should be found.

11.2 Water Quality for Plants

All water, even rainwater, contains some impurities. Impurities may include salts, toxic ions, manufactured chemicals such as pesticides, disease-causing microorganisms, and small animals such as nematodes.[1]

Salty or *saline* water is a common problem in drylands (Box 11.1). Surface and groundwater contain naturally occurring salts which can build up in the soil to levels that reduce garden yields. Salty water is more difficult for plants to absorb, and causes symptoms similar to those produced by water stress; leaf edges often turn brown, white or yellow.[2] Salty soils make the problem worse). However, sometimes salt-tolerant crops can be grown successfully (section 5.6).

11.3 Water Quality for People

If water is used only for the garden and not for human or animal consumption, the danger of disease is reduced, but not eliminated. In Zimbabwe, for example, wells located in gardens are used for irrigating and clothes washing but not for drinking water.[6] Wells for drinking are located outside the gardens and close to houses. There are a number of diseases that can be spread through contact of the skin with water, or by biting insects that breed in water. Making a great effort to improve household nutrition and income through increasing garden production is not worthwhile if these benefits are undone by an increase in infectious disease.

While methods to improve water quality may seem relatively simple to the outsider, there are many political, economic, social, and cultural reasons why people may find it difficult to change established patterns of behavior that affect their health. This is true even when they understand the negative effects of these behaviors on their health. This difficulty is as common in industrialized countries where the level of formal education is very high, as it is in Third World countries where the level of formal education is very low. For example, in the industrialized countries of North America and Europe the role of tobacco and alcohol in causing sickness, accidents, and death are well known. Yet well-educated people in those countries spend billions of dollars on these products every year, encouraged by manufacturers' advertisements. Tobacco producers in the United States even receive government subsidies.

In Box 11.2 we briefly review prevention of some of the most important infectious diseases that can be spread by contact with water in the garden. It may be very difficult to prevent children from drinking water meant only for irrigating, especially during the dry season when sources are limited, but every effort should be made to do so. Discussion of the diseases spread by drinking contaminated water can be found in community health-care books.[7]

Box 11.2
Human Disease and Water Quality [8]

Hookworm disease is caused by several species of microscopic hookworms which hatch in damp soil and enter the human body through bare feet. They then attach to the walls of the intestine and can cause loss of blood leading to weakness and severe anemia. Hookworm eggs leave the body in feces.

Strongyloidiasis is another disease spread through human feces. The worm enters moist soil and feeds there before penetrating the skin of another person. Strongyloides and hookworms are *nematodes*, a group of small roundworms, some members of which also infect garden crops (section 13.3.2). Wearing shoes and defecating well away from the garden and other places where people walk help to prevent the spread of these disease-causing nematodes.

Leptospirosis is a disease with a variety of symptoms lasting up to 3 weeks. Fatality is low but increases with age. Domestic animals like goats spread the disease-causing organism in their urine. From water or moist soil it enters the skin, especially the mucous membranes (moist areas such as the nose, mouth, or eyes), or where there is a cut or scrape. People should avoid walking barefoot where animals urinate or defecate.

Schistosomiasis (also known as *bilharzia*) is one of the most prevalent diseases spread by the use of irrigation water, especially slowly moving water in canals, lakes, and rivers. The disease is caused by tiny blood flukes, worms whose eggs are present in the urine and feces of infected people. These worms live part of their lives in a certain kind of small, freshwater snail, then leave the snail's body and enter directly through the skin of a person drinking, washing with, or standing in the water. Inside the body the worms cause bleeding of the bladder or intestine, leading to loss of blood in urine or feces. The best method of control is sanitation systems that eliminate urination and defecation in or near canals, ponds, lakes, or other sources of garden irrigation water. Building access areas so that gardeners do not have to stand in the water to irrigate, and removing from canals and other water sources the vegetation snails need for food are also important.

Mosquitoes can breed in even tiny amounts of water, such as in the ends of bamboo poles or pieces of a broken pot. *Malaria* is one of the most devastating diseases of the drylands and is spread by mosquitoes. Malaria organisms live in the blood and other organs where they can cause serious illness and death. A large number of other diseases are also spread by mosquitoes, including yellow fever and other virus diseases, as well as *filariasis* (*elephantiasis*), caused by a nematode. During the rainy season when mosquitoes are breeding, the incidence of these diseases increases dramatically. Tin cans, calabashes, pots, or other containers used for watering the garden should be emptied of water after use and stored upside-down, to prevent collection of rainwater. Water storage containers should be covered to prevent mosquitoes from laying eggs. Areas around wells, *shadufs* (section 12.7.1), and other water-lifting devices should be drained to prevent puddles from forming. One way to do this is to direct this water into a garden bed as shown in Figure 11.1.

11.4 Rain

Seasonal patterns as well as amounts of rainfall vary greatly in the drylands, and help to determine what kinds of gardens are grown. If rain is adequate, gardens can be grown in the rainy season with little or no additional water. This is especially true when the rainy season is also the cool season, as in North Africa, the Middle East, and the west coast of South America. Rainfall is also the source of water for streams, rivers, and lakes and for many underground aquifers which can be tapped with wells. The need to understand rainfall patterns is greatest when gardeners depend on rain falling directly on the garden, or harvested from nearby catchments.

Gardeners and farmers in drylands have methods of deciding exactly when to plant during the beginning of the rainy season. This is very important, since variability of rainfall is greatest at this time (and at the end of the rainy season). This is also a time when germinating seeds and young plants are very sensitive to drought, and planting materials and irrigation water are limited. Often there are rains before the true beginning of the rainy season. Planting after such rains could result in seeds or cuttings failing to grow if additional rain does not follow soon enough. On the other hand, if planting is too late, the garden crops' maximum need for water will occur when the peak rainfall has passed.

To determine the start of the rainy season so that they can plan the planting of crops, Nigerian farmers use many different signs: astronomical (stars, moon), social (elders, festivals), weather (clouds, heat, rainfall, humidity), and other signs in their environment (bird and insect behavior, vegetation changes).[9] In Nigeria's western savannas for example, the Yoruba use the following as indicators that it is time to begin planting: leafing out of two trees, iroko (*Chlorophora excelsa*) and baobab, sky signs such as changes in color and cloud formations, the cessation of singing of a certain bird that sings only in the dry season, and shifts in wind direction. While these farmers, like most farmers everywhere, may not understand the physical causes of rainfall patterns, they are good observers of these patterns and their relationship to rainfall and other environmental events.

11.4.1 Rainfall Records

Field workers can easily measure and record rainfall[10] and doing so increases their understanding of the relationship between water and garden management.

For planning large-scale projects, especially those involving water harvesting, records for the project site or for the closest weather station should be used, if possible. Even records for 1 year are better than nothing at all. Several publications describe regional climates and provide monthly and yearly weather data for representative weather stations. One is the *World Survey of Climatology* series.[11]

Rainfall records for a number of years can provide the average, maximum, and minimum rainfall, and number of rainy days over the period of record keeping, by the year, month, or week. They may also show the average, earliest, and latest dates for the beginning and end of the rains, or for the last and first freezing nights. For designing water-harvesting plots (section 11.5) it is also useful to know the amount of maximum daily rainfall and intensity of rainfall measured in mm/hr.

Decisions such as when to plant and what crop mixtures to grow often depend on amount and timing of rainfall. Such decisions are based on estimates of the probability of receiving a certain amount of rain during a given period. Probability means the chance that a certain event will occur based on the information available (Box 4.2). The greater the probability of adequate rainfall, the less the risk to the farmer or gardener. The traditional methods of judging the beginning of the rainy season described above are one way of calculating probabilities. Box 11.3 gives an example of another way to calculate rainfall probabilities.

When gardens depend on rainfall, there will always be a certain probability of failure due to lack of rains. The amount of failure acceptable to the gardeners will depend on the risks they take in terms of the labor, water, and other resources they invest, and on how important the produce from the garden is. Western agricultural scientists often use an 80% success rate as a standard on which to base management practices.[12] This means that, in terms of water, gardening systems would be designed so that, on average, they would produce a successful harvest on rainfall alone 8 out of every 10 years.

11.4.2 Measuring Rainfall

Simple plastic or glass rain gauges can be purchased from scientific, agricultural, or garden supply houses. One of good quality may cost around US $10.00 or more. For drylands the tapered type is far better than the flat-bottomed kind, because it makes reading small amounts of rainfall much easier (Figure 11.2).

Box 11.3
Calculating Rainfall Probabilities from Records

Probabilites can be calculated from rainfall records. For example, in the desert areas of southwestern North America, the beginning of the summer rains in July is an important time for planting summer crops including corn, beans, and squash. We can calculate the probability that the coming year (or any year) will have enough rainfall for July planting. Given data on the amount of rainfall in the month of July for the past 30 years, the amounts of rainfall from most to least are first listed (Table 11.1). Using a simple formula, the probability of any amount can be estimated, as shown in the table. For example, if a successful harvest from a July planting depends on at least 45 mm of July rainfall, then it will be successful in only 68%, or less than 7 out of every 10 years on the average. This means that, on the average, during 32% of the years, planting will have to be done again in August, or supplemental irrigation will have to be provided. It does not mean that in any given 10-year period there will be 7 years of successful harvests. Probability is a calculation based on long-term, average observations; it is a useful guideline but should never be taken as an absolute predictor of growing conditions and resulting harvests.

This method assumes that there are no trends through time toward more or less rainfall. The greater number of years the calculation is based on the greater the reliability. In calculating rainfall probabilities for water-harvesting design, it has been suggested that a minimum of 10 years of records be used, eliminating the 2 wettest years if they are extreme.[13]

Table 11.1 Calculating Rainfall Probabilities [a]
An example of July rainfall for Tucson, Arizona, USA, 1957-1986

m	Year	Rainfall (mm)	m	Year	Rainfall (mm)	m	Year	Rainfall (mm)
1	1981	156.7	11	1970	64.3	21	1980	45.2
2	1958	132.1	12	1975	60.5	22	1973	44.2
3	1964	122.4	13	1971	55.4	23	1963	42.2
4	1974	112.8	14	1982	54.1	24	1969	38.4
5	1959	99.6	15	1965	54.1	25	1962	35.1
6	1972	88.6	16	1979	51.8	26	1957	31.8
7	1985	79.8	17	1983	50.3	27	1976	30.0
8	1984	74.2	18	1968	50.0	28	1978	19.8
9	1967	69.1	19	1986	46.2	29	1977	19.3
10	1966	65.3	20	1961	46.0	30	1960	18.5

$p = m/(n + 1)$
p = estimated probability of equal or greater rainfall
m = position in the series, ranked from highest to lowest rainfall
n = number of years of measurements = 30

Example #1: To find the probability of a July with over 45 mm of rainfall, find 45.2 mm in the table above and note that its position is m=21, then
$p = 21/(30 + 1) = 0.68 = 68\%$
This means that there is a 68% chance (about 2 out of 3) of having a July with more than 45 mm of rain and a 32% chance of a July with less than 45 mm of rain.

Example #2: To find the amount of rainfall that can be expected with a 60% probability ($p_{0.60}$)
$m = p(n + 1)$
$m = 0.06(30 + 1) = 18.6$
Rounding down to 18 to be conservative we look for m=18 in the table and find that for m=18 rainfall was 50 mm. Therefore, 60% of the years the rainfall in July will be 50 mm or more.

[a]Data from NOAA (1987), procedure based on WMO (1983).

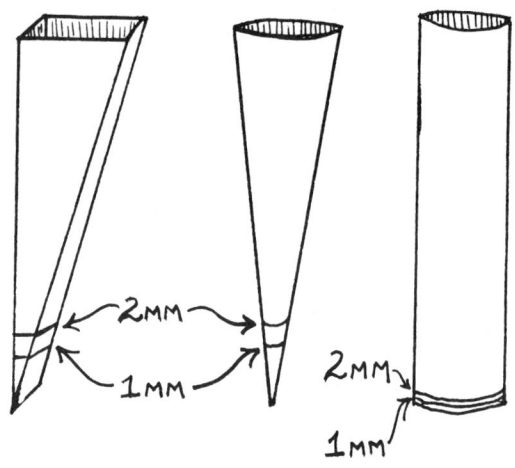

Figure 11.2 Measuring Rainfall with Rain Gauges

Rainfall can also be quite easily and accurately measured without a rain gauge. All that is needed is a large collecting container, such as a metal or plastic bucket, and a small calibrated container, such as a measuring cup used for cooking. Rainwater collected in the large container is poured into the smaller container and the volume is measured. The volume in cm^3 is then divided by the area of the collection container opening in cm^2. This could also be done using in^3 for volume and in^2 for area, although it may be more difficult with small quantities of water.

For example, a bucket with an opening that has a diameter of 26 cm (10 in) (Figure 11.3) is set out to collect rainwater. The collected water is then poured into a measuring cup and comes up to the 6.5 oz (191 ml) mark.

$\quad$ A (area of bucket opening) = πr^2 = (3.14)(13 cm)^2
$\quad\quad$ = 530.7 cm^2
$\quad$ Vol (volume of water) = (6.5 oz)(29.6 cm^3 per ounce) = 192.4 cm^3
$\quad$ P (Precipitation) = Vol/A = 192.4 cm^3/530.7 cm^2
$\quad\quad$ = 0.36 cm = 3.6 mm of rainfall.

11.5 Harvesting Rainwater for Dryland Gardens

Gardening with rainwater runoff can be a simple way to increase the effectiveness of natural rainfall without the expense or trouble of importing water. Water harvesting is often combined with measures to reduce soil erosion, such as terraces and contour bunds (section 9.7.1). It involves harvesting rainwater from one area (*catchment*) and applying it to the garden (a plot or a single tree or other plant), to add to the amount of water available from direct rainfall alone. Where there is abundant, reliable, and inexpensive irrigation water, rainwater harvesting may not be worthwhile. (Harvesting floodwater from streams, rivers, and lakes is covered in section 11.6).

Rainwater harvesting is not limited to the Third World. It is increasingly popular among industrial world farmers, for example in the western United States where the cost of groundwater for irrigation has risen due to falling groundwater tables and the growing cost of energy to pump it.

In this section we discuss local systems of water harvesting, some basic concepts of runoff, and how to measure runoff and estimate the ratio of catchment to growing area in designing runoff gardens. We emphasize microcatchments, which are much more efficient at harvesting rainwater than larger catchments.[14] There is no need to transport the harvested water to growing areas, and because the water flows fairly slowly, only simple structures are needed to control it. Few if any new engineering and construction skills are needed for design and construction of small microcatchments. However, even the smallest system, if improperly designed, may be destroyed during the first rainstorm.

It may well be a better investment in all but the largest projects to encourage local gardeners to experiment rather than hiring "experts" to come up with sophisticated calculations for design.[15] Even for large projects, it will be necessary to check calculations

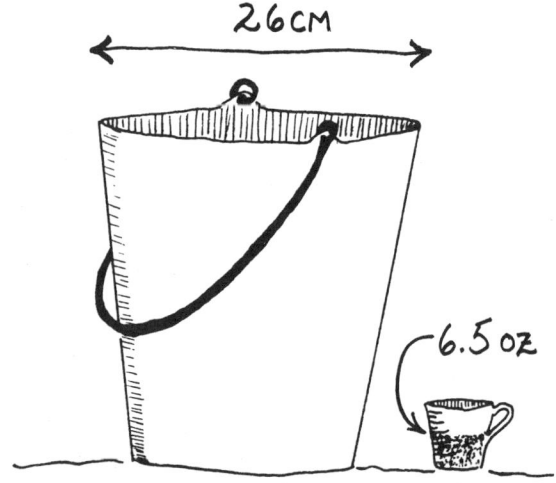

Figure 11.3 Measuring Rainfall with a Bucket and Measuring Cup

against experience in the garden. The simplified outline of runoff garden design given here along with local experience and knowledge, can provide a starting point for experimentation without the need for an expensive design process or detailed rainfall data. We emphasize that even this design process is not necessary to employ water harvesting for gardens. Experimentation beginning with small microcatchments and building on the knowledge gained may be the best choice in most situations.

11.5.1 Patterns of Water Harvesting

Rainwater harvesting gives gardeners more options in terms of the seasonal timing of gardening:

- During dry periods in the rainy season, in unusually dry years, or in areas with inadequate average seasonal rainfall, the rain that does fall can be harvested to supplement that falling on the garden.
- Gardens started at the end of the rainy season or the beginning of the dry season may be successfully grown on rainwater harvested and stored in the soil.
- Gardens grown mostly in the dry season can be irrigated with water harvested in the rainy season and stored in a container or the soil.

While many people in drylands probably practice some form of water harvesting, not much data has been gathered on existing traditions.[16]

In areas such as North Africa and the Middle East the rainy season occurs in the cool winter when evapotranspiration rates are low. Food can be produced there during the rainy season on much less rain than in most of the drylands of sub-Saharan Africa, India, and much of Mexico where the rainy season occurs in the summer, accompanied by high evapotranspiration rates. Small-scale water harvesting in areas of the Negev Desert, which receive only 100 mm (4 in) of rainfall during the cooler winter season, has been successful with some crops, especially drought-resistant fruit trees.[17] Olive trees have been grown in southern Tunisia using this method for centuries.[18] However, in summer rainfall areas, especially in the tropics, more total water is needed to compensate for losses to high evapotranspiration. Water harvesting here may only be appropriate where there is at least 300-600 mm (12-24 in) of rainfall.[19]

Even without any water control structures such as berms or terraces, cultivation methods are often adjusted to take advantage of rainwater which collects naturally in certain areas within a garden or field. An example is the planting depressions made by Mossi farmers in Burkina Faso described in section 9.3.2.

In Jaisalmer District in the desert area of Rajasthan, India, the technique of *khadins* was first employed over 500 years ago.[20] Sheet and gully runoff from low hills and ridges is collected behind berms in the hot, rainy season. Much of this water infiltrates into the soil, and any remaining water is drained off for planting in November, and production during the cooler dry season. Crop roots follow the dropping water table, which by the end of the dry season is 2 m (6.5 ft) below ground level. Such large-scale water harvesting systems are widespread in India.[21] Although normally planted primarily with major cash or staple crops, gardens are sometimes a part of these systems, as distinct plots in the field, or as a few vegetable crops interspersed among the main crops.

Many experiments with runoff farming have also been carried out in the Third World. In India, microcatchments for growing jujubes were evaluated over a 5-year experimental period in which mean annual rainfall was 558 mm (22 in).[22] It was found that a catchment area with a 10% slope and only 87% as big as the growing area, produced yields of jujube fruit 2.3 times that in the control with no microcatchments. Another example comes from Kenya where contour ridges 3 m (10 ft) apart and semicircular berms were compared with the traditional deep digging receiving no runoff.[23] It was found that during a rainy season with 539 mm (21 in) of rainfall, the plots with ridges yielded 2.3 times as much sorghum grain, and 7.8 times as many cowpeas as the control, while those with semicircular berms yielded 3.4 and 6.5 times more, respectively, than the control.

An experiment in the drylands of the western United States measured the difference between planting sorghum on a flat soil surface and planting sorghum in 50-cm (20-in) wide furrows with 75-cm (30-in), steep (33% slope) catchments on both sides (CGAR = 3:1, see section 11.5.4).[24] In a 14-week growing season with 180 mm (7 in) of rainfall, the rainwater catchment plots yielded more than twice as much sorghum.

11.5.2 Building on Local Knowledge

As with all aspects of garden design, the most important foundation for equitable improvements is what the local people already know and practice.[25]

An example has been recorded in northern Kenya where the Turkana traditionally grow sorghum in floodwater gardens[26] (section 11.6.2). The Israeli model of microcatchments for growing trees was the inspiration for outsiders to design a project to build contour ridges 10 m (330 ft) apart for annual crops in the area. However, the rainfall was far too erratic, both in timing and location, and none of the crops produced a harvest. This is an example of inappropriate technology transfer without consideration of environmental conditions or of the Turkana's own knowledge based on their use of seasonal stream and river flow to grow gardens (section 11.6.2). As a result, after several hundred Turkana had been paid with food to clear and contour 200 ha (494 a), the project was converted to an irrigation scheme using more reliable river water. However, working on their own, some observant Turkana were fairly successful in using water harvesting to extend a garden into a natural depression, similar to those they used traditionally.

Local knowledge and experience can provide the foundation for the evolution of a successful rainwater harvesting project. In Yatenga, northern Burkina Faso, project workers introduced the idea of microcatchments for growing trees for erosion control.[27] However, based on local knowledge and needs, the project was gradually changed to focus on rainwater harvesting for food production. The revised project included construction of stone contour berms modeled on the stone berms traditionally used in the area. These allowed the water to slowly infiltrate into the soil; any excess passed through the stone berms. Contours were defined with a tube level (Box 9.11 in section 9.7.1 and Figure 11.8 in section 11.5.4) which the farmers were trained to use, and no hydrological calculations were made, for example, to determine spacing of berms. Rather, the farmers started with one or more simple, small berms, and expanded them or added spillways based on their experience from season to season.

11.5.3 Catchments and Runoff

Estimating the amount of rainwater that can be harvested from a catchment is the key factor in designing and using rainwater runoff. Runoff can be calculated by placing values from field measurements or published tables into equations. This approach may sometimes be useful in project planning. For most garden applications it is more practical to actually measure the runoff in sample garden plots. No matter which method is used, experiments by gardeners will always be necessary to check the results.

Water can be successfully harvested from roofs of thatch, tile, packed soil, or metal and directed into garden plots. This water can also be stored in containers such as large clay pots next to the house for later irrigations when there is no rain. Roadways or pathways sloping toward the garden can also be modified to harvest rainwater. Special care must be taken to ensure that catchments are not polluted with poisons that could be carried into the garden. Small areas can be shaped into catchment systems, called *microcatchments*, up to about 0.1 ha (1,000 m^2, or 10,763 ft^2) (Figure 11.4). Microcatchments direct rainfall runoff into adjacent growing areas. Microcatchments and growing areas can take many shapes, but for simplicity and efficiency they should be next to each other. In general, the best collection and growing areas have quite different characteristics (Table 11.2).

The amount of rainfall that runs off the catchment depends not only on the amount of rainfall itself, but on the characteristics of the catchment area. Some of the rainfall is intercepted by vegetation and evaporates before it reaches the ground. Some of the rain that does reach the ground infiltrates into the soil. Infiltration is increased by anything that slows the water flow—such as vegetation, organic matter, rocks, or gravel—or that increases porosity—such as cracks caused by shrinking of soils containing montmorillonite clay, coarse (sandy) texture, and looseness (lack of compaction), or holes made by animals or plant roots. Also, depressions in the soil surface capture water in puddles allowing time for the water to infiltrate.

Soil texture has a strong influence on the amount of runoff.[28] Sandy soils should be avoided for water harvesting catchments because they are so permeable. Exceptions are some sandy clay loams and sandy clays, which are excellent. Soils high in silt and fine sand tend to crust easily and so make good catchment surfaces. Loam soils are not too good but can work. The clays can be excellent or very poor depending on clay type and soil structure. Clays high in montmorillonite tend to crack, creating a broken-up surface with high permeability, therefore, soils containing much montmorillonite, such as Vertisols, should be avoided. Oxisols with sesquioxide clays also tend to be highly permeable. In contrast, clay soils high in kaolinite crack very little, are very impermeable, and make good catchment areas.

The *threshold rainfall* is the amount that falls before water begins running off the catchment. The more rain that is trapped by vegetation and in puddles, or infiltrates into the soil, the higher the threshold rainfall will be. *Runoff percent* (R%), also called *runoff ratio*, is

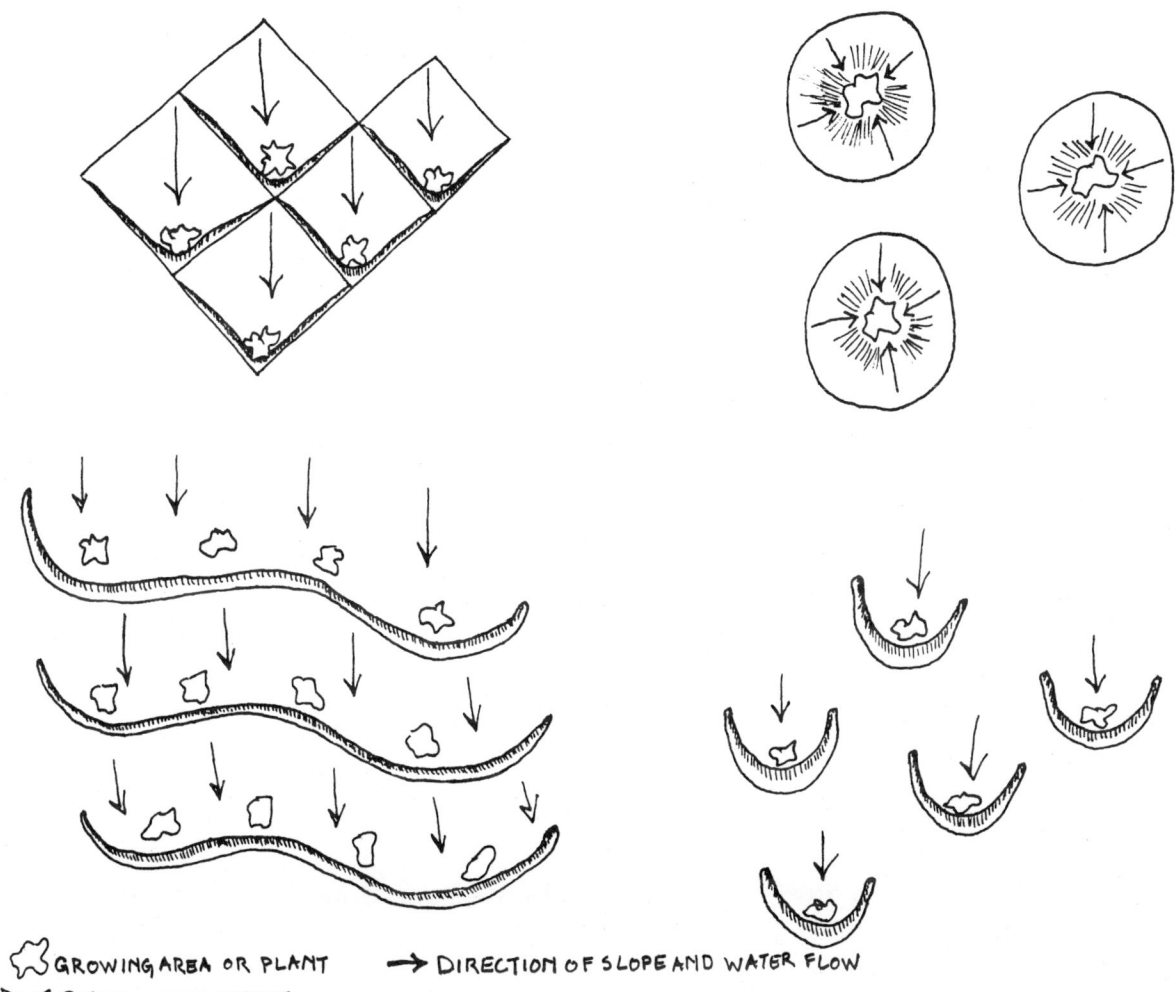

Figure 11.4 Microcatchments

Table 11.2 Comparison of Desirable Characteristics for Catchment versus Growing Areas for Rainwater Harvesting

Characteristic	Catchment area	Growing area
Runoff	maximum	minimum
Permeability	minimum	maximum
Slope	3 to 7%	0%
Vegetation	none	dense
Soil surface	smooth, no openings, tendency to crust	rough, many openings, no crusting
Mulch	none	yes
Organic matter	none	yes
Salts	some help crusting	none

the percentage of rain falling on a given catchment that runs off the catchment:

$R\% = R/P \times 100$

where

R = runoff

P = precipitation.

A large range of R% values can be expected even from soils of similar texture. Although estimates can be made using published values as a guide, it is better to make actual measurements of the proportion of rainfall that runs off.

Field measurements of R% can be made quite simply. If the catchment area is bordered by low dikes to prevent water from running onto or off of the plot, a container (such as a cleaned 208-liter [55-gal] steel drum) can be be sunk in the low corner to collect the runoff. The amount of rainfall and runoff should be measured every day on which there is rainfall. Even a few weeks of such measurement during a rainy season will give a far better estimate than any calculations can.[29] If long-term rainfall records are available, they can be used to extrapolate and gain an idea of how big a catchment must be to reliably supply the water needed by a garden.

Runoff percent is calculated by dividing the total amount of rain falling on the catchment area (P) into the volume collected in the container (R) (Figure 11.5). For example, if it rains 12 mm on a 10 m^2 catchment area, this is 120 liters of water; if 45 liters are collected from the runoff plot, then the runoff percent is

$R\% = (R/P) \times 100 = (45/120) \times 100 = 38\%$.

Runoff percent from small, smooth, impermeable surfaces like plastic sheeting or sheet metal roofs is almost 100%. In contrast, runoff from larger, irregular, permeable, vegetated soil surfaces may be very low. Runoff from soil surfaces can be increased by a number of treatments such as asphalt or wax, but these are expensive and can be harmful to the environment.

Treatment with common salt (NaCl) or other sodium compounds disperses the soil particles to seal the surface, and has been used successfully in Arizona, USA, and other areas.[30] However, the limited increase in R%, the cost of salt, the high salt content in runoff

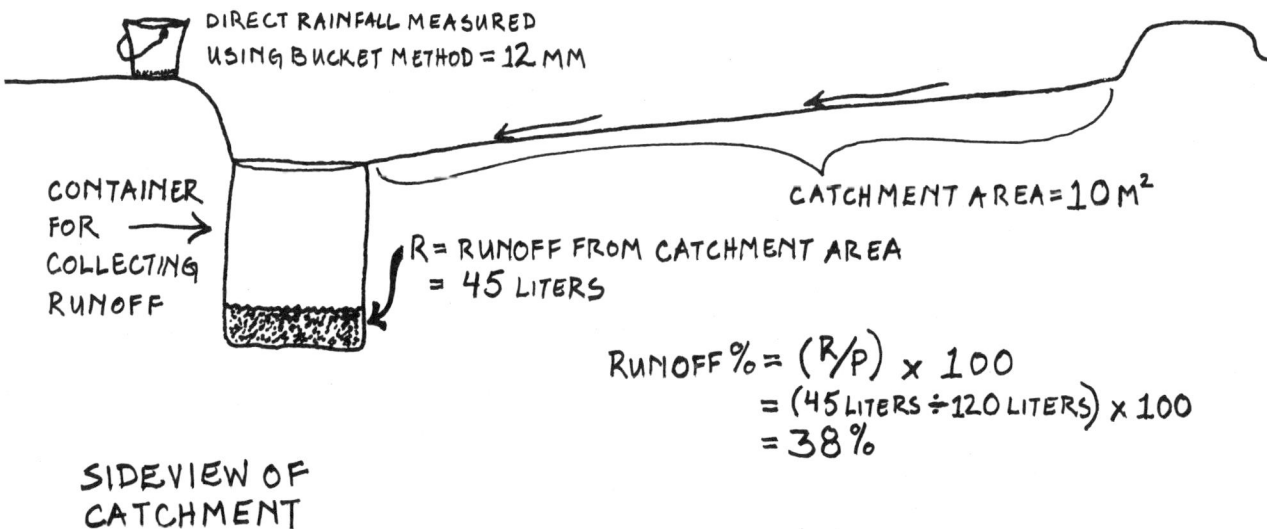

Figure 11.5 *Calculating Runoff Percent*

water during the first year, and the inability to use the catchment area for cultivation in the future limits the usefulness of this technique, especially for Third World gardens.

Microcatchments for harvesting rainwater for gardens are best treated only by clearing vegetation and large stones, smoothing, and perhaps compacting. A number of studies have shown that on catchments 10 m (33 ft) long or less, with slopes 7.5% or less and on sandy loam to clay loam soils, R% is about 25-30% with only clearing and smoothing. The surface of the soil can be smoothed when wet with a flat rock or shovel. When the soil is also compacted, R% can increase up to 70%.[31] Treatment of these microcatchments with sodium compounds increased R% only a small amount to 30-75%.

Most of the characteristics of the catchment area that increase runoff also increase soil erosion, which is undesirable. Therefore, a balance must be reached between maximizing runoff and minimizing erosion. On collection areas longer than 10 m (33 ft), slopes greater than 7% may lead to severe erosion. (Measuring a slope is discussed in Box 11.4.) On very short collection areas, slopes may be up to 10% without causing serious erosion. A 3-7% slope is good for most microcatchments. However, if the soil surface has been smoothed, any increase in catchment slope above 3% will do little to decrease the threshold rainfall and will increase erosion instead.[32] While it is generally best to keep the runoff areas weed free, weeds such as amaranth or purslane, or drought-resistant perennials like prickly pear cactus, may sometimes provide enough food to make them tolerable, and may help to prevent erosion.

11.5.4 Estimating the Catchment to Garden Area Ratio

The *catchment to garden area ratio* (CGAR) is the ratio of the area of the rainwater catchment to the garden area where the water is used. If the ratio is too small then there will not be enough water for the garden, and if it is too big, there may be so much water that part of the catchment or the garden itself is washed away. The highest R% will be from the shortest slopes. This is because there is less distance for the water to flow over and infiltrate into. Also, the slope can be steeper than on longer catchments without causing soil erosion. This is illustrated by a 5-year experiment in India growing jujube trees on loamy sand, with average annual rainfall of 558 mm (22 in).[33] A 5.1-m (17-ft) long catchment with a 10% slope stored 472 mm (19 in) of rainwater in a 3-m (10-ft) soil profile in the growing area, compared with 313 mm (12 in) for a 14.5-m (58-ft) catchment with a 0.5% slope, and 380 mm (15 in) for a 10.7-m (35-ft) catchment with a 5% slope.

We can use the following equation to estimate the CGAR:

CGAR = [(ET_m/E_a) - P]/R, where

ET_m = maximum evapotranspiration for the garden crops (section 10.3.4),

E_a = efficiency of water application (section 12.2),

P = precipitation (rainfall) (section 11.4), and

R = runoff (section 11.5.3).

ET_m can most practically be estimated by talking to local gardeners and observing the amounts of water needed to grow gardens in the area.

When the harvested water is applied directly to the growing area, as with most microcatchments, then Ea

---

Box 11.4
Measuring a Slope

Slope is usually measured as percent of change in vertical distance for a given horizontal distance (Figure 11.6). Thus a drop of 1 m (3.3 ft) over a horizontal distance of 3 m (10 ft) is a 33% slope (1/3 x 100). A drop of 0.5 m (1.5 ft) over a horizontal distance of 10 m (30 ft) is a 5% (0.5/10 x 100) slope. When the horizontal distance is an even number, calculations are easier. If possible the slope should be measured over the entire distance under consideration for a garden or catchment area, to eliminate the effect of irregularities.

If a line level or carpenter's level and a long piece of string are available it is quite easy to measure the slope (Figure 11.7).[34] It is important that the string be stretched tightly and that the level be placed exactly in the center of the string. When the bubble is centered between the two lines, the vertical distance or drop is then measured and divided by the horizontal distance.

The tube level described in Box 9.11 can also be used to measure slope (Figure 11.8). The vertical distance is the difference between the water levels.

can be assumed to equal 1.0 and will drop out of the equation. When the harvested water is led through canals, used in sprinklers, or other irrigation systems, or when the garden plot being irrigated is large, then an efficiency rate will have to be entered (section 12.2).

It is more realistic, although more complicated, to use estimates of ETm, P, and R for each week of the garden growing season so that CGAR can be calculated for the week where (ETm - P) is greatest, that is, when the difference between water needed and water available is greatest. In areas where P is high early in the growing season then it may be possible to store some of the runoff in a reservoir for use during periods when (ETm - P) is greatest, thus reducing the CGAR.

Box 11.5 gives an example of microcatchment design using the CGAR.

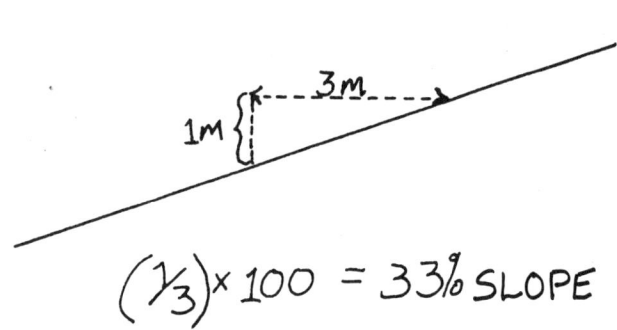

Figure 11.6 Slope

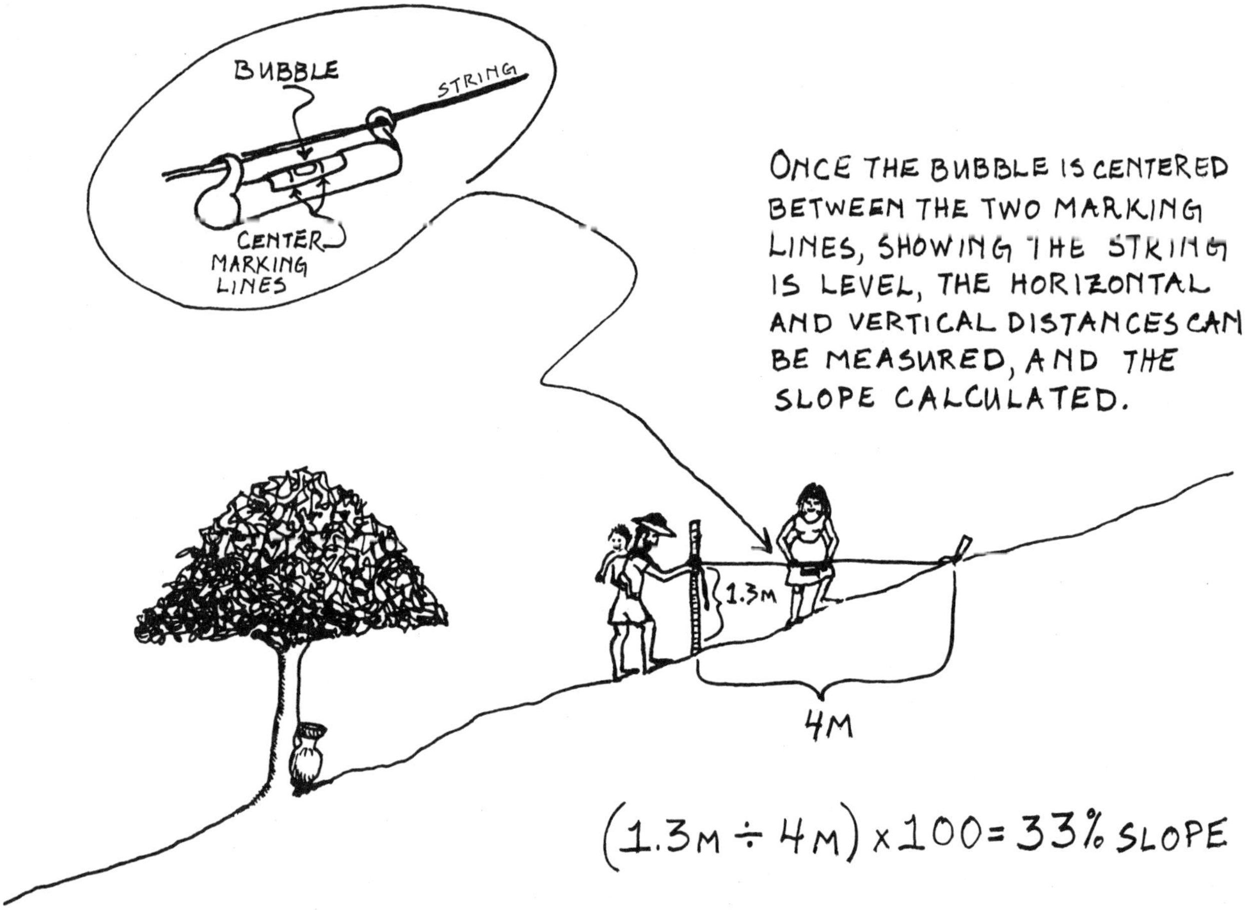

Figure 11.7 Using a Line Level to Measure Slope

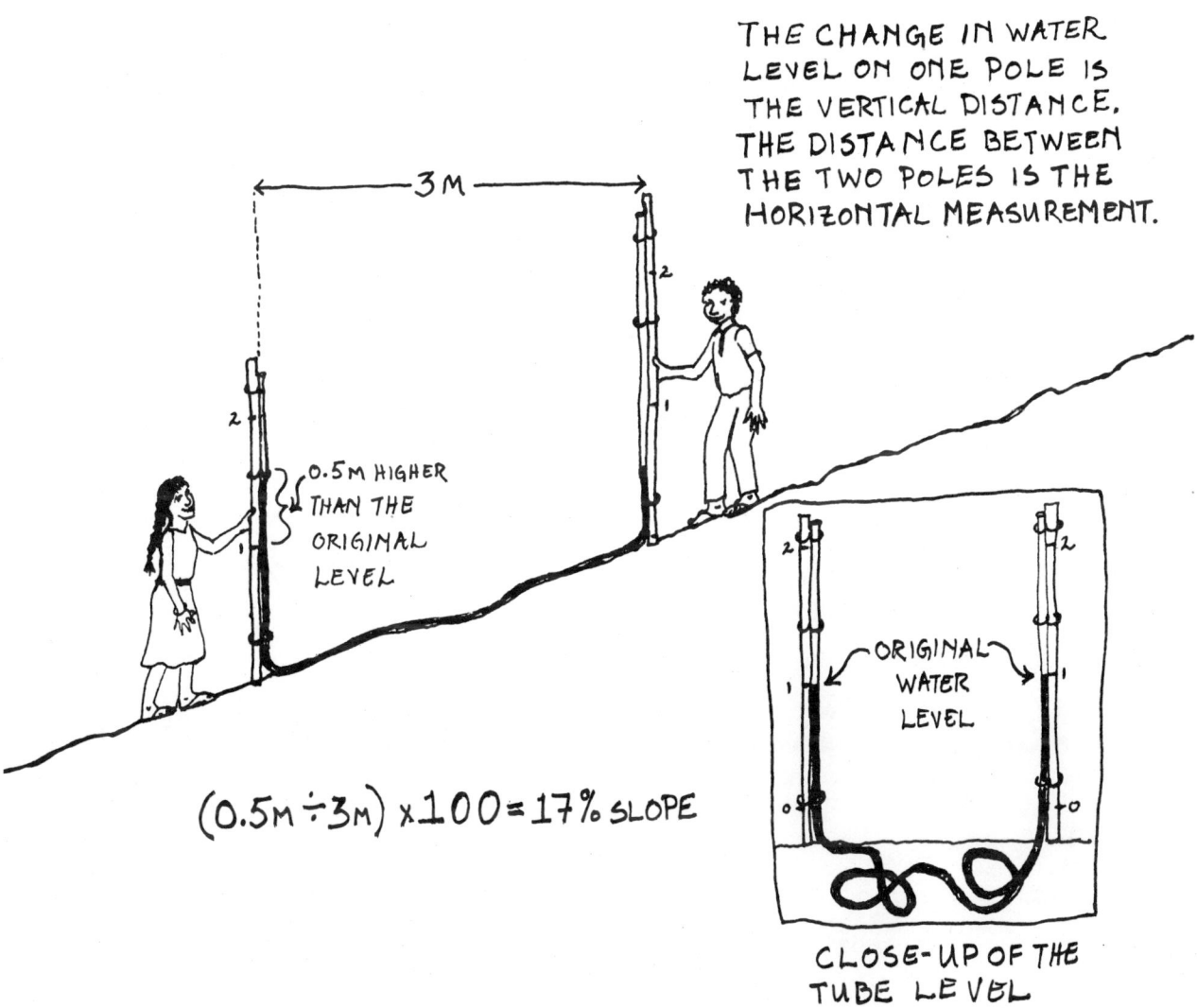

Figure 11.8 Using a Tube Level to Measure Slope

11.6 Harvesting Stream Flow and Floodwater

Rainwater harvesting captures water flowing in sheets or rivulets, but water flowing in streams, rivers, or lakes can also be captured for use in the garden (section 12.3).

Floodwater gardening makes use of seasonal water flows in streams and rivers to produce crops. Compared with other sources of water, floodwater is not as easy for the gardener to control, yet it often requires much less investment of time and resources. In addition, alluvial deposits of clay, silt, and organic matter, made by flowing water as it slows down, create soils that are deeper, more fertile, and have higher water-holding capacity than adjacent upland soils. Many people in drylands use some form of floodwater gardening, and understanding and building on local knowledge is important in the improvement of existing systems or the design of new ones.

At least two different methods are used for floodwater gardens: water spreading at the base of seasonal streams, and flood recession cultivation on seasonally flooded river and lake terraces.

> **Box 11.5**
> *Designing a Microcatchment Garden Plot*
>
> In this example, traditional varieties of beans, corn, squash, and greens are to be grown in the summer rainy season in the Sonoran Desert of southwestern North America.
>
> - Growing area: Sunken beds 1 m² will be built on a site with a slope of 5%.
> - Rainfall: It is decided that failure 2 years out of 10 is acceptable. Since there is some production even when available water is less than optimum (ETm), a rainfall probability of 60% is chosen ($P_{0.6}$). Monthly rainfall data are available for the last 30 years. The corn, beans, and squash will be planted in early July and will flower and set fruit in late August. Much less water will be needed during ripening in September, which has a lower rainfall. Therefore, CGAR will be calculated based on July rainfall, which is very similar to that of August (see example #2 in Table 11.1). $P_{0.6}$ for July is 50 mm.
> - Runoff: Runoff percent (R%) is first calculated from trial catchments of the compacted and smoothed clayey soil and found to be 50%. Runoff is then calculated as R = (R%)($P_{0.6}$) = (0.50)(50 mm) = 25 mm.
> - Catchment to garden area ratio: Mixed gardens in the area growing a variety of different vegetables use 113.6 mm/wk or (52 weeks/12 months = 4.33 wk/month, and so 113.6 mm/wk x 4.33 =) 492 mm/month during this period. CGAR = (ETm - P)/R = (492 - 50)/25 = 17.
> - Catchment area: Area x CGAR = 1 m² x 17 = 17 m².
> - Depth of planting basin: Since the depth of the root zone (d) is limited by a caliche layer at 0.8 m which allows only very slow drainage, and available water in the soil is 60 mm/m, storage in the root zone is: (AW)(d) = (60 mm/m)(0.8 m) = 48 mm. To avoid waterlogging during intense rainstorms, we limit the depth of the planting basin to 60 mm with excess flowing around the upslope end of the sides of the basin. The berm on all but this overflow area is raised another 60 mm to 120 mm to create a *freeboard*, a margin of safety protecting it from erosion due to rapidly moving overflow.

11.6.1 Water Spreading

Water spreading is a technique of diverting seasonal stream flow to flood low-lying adjacent areas for irrigating crops. It can be practiced with a fairly elaborate system of water control using dikes, berms, terraces, and even canals and storage tanks, or with a very simple system requiring little construction and using the natural pattern of water spreading and storage in the soil.

Water spreading was commonly used by natives of arid southwestern North America including the Tohono O'Odham, although only a few practice it today.[35] Stream channels (known by the Spanish term *arroyos*) are dry most of the year, but will flood after an intense summer rainstorm. At the mouth of the arroyo where the slope changes abruptly, the waters naturally slow and spread, depositing their load of sediment and organic matter in an alluvial fan. The soil here is deeper and more fertile than in surrounding areas. To increase production the Tohono O'Odham build dikes of soil, rocks, or brush to further slow and spread the water. A number of domestic garden crops such as melon, watermelon, and squash are interplanted with grain and dry beans. Many wild species, some used as leaf vegetables, are also encouraged.

Floodwater can also be diverted artificially from streams. In the Oaxacan valley of southern Mexico, for example, for extra water to supplement direct rainfall, many fields depend entirely on floodwater which flows along dirt roads and is diverted by 20-30 cm (8-12 in) high berms.[36] Where the level of the road is up to 1 m (3.3 ft) below that of the field, small rock dams are built to raise the water to the level of the field.

Forms of water spreading are common from the northern savannas of West Africa eastward to Sudan and Somalia.[37] This technique may be especially appropriate for pastoralists in arid and semiarid regions who rely primarily on animals or trade, and for whom any big investment in irrigation may not be economical.

11.6.2 Flood Recession Gardening

Flood recession gardening is growing fruits and vegetables in areas that are flooded in most years. In northwestern Nigeria 100-500 m² (1,076-5,380 ft²) dry-season gardens are cultivated in the flood plain of the Sokoto valley by 80% of the farmers.[38] A variety of annual vegetables are grown primarily on moisture left in the soil by the flooding river, although hand-dug wells provide supplemental water.

Traditionally the Tonga of Zambia and Zimbabwe in central Africa depended primarily on sorghum and millet cultivated in regularly fallowed, rain-fed fields, and secondarily on maize, cucurbits, groundnuts, sweet potatoes, and tobacco (*Nicotiana* spp) cultivated in both wet and dry seasons on Zambezi River terraces.[39] The potential ETm of 2,296 mm (90 in) is 3.5 times higher than the yearly rainfall of 651 mm (26 in), which falls in a single rainy season. The Tonga use soil moisture from flooding in three distinct locations as they are exposed by dropping floodwaters: *kalonga*, small pockets of moist tributary streambeds; *kuti*, on the first river terrace; and *jelele*, on the riverbank itself.

In northern Kenya where the annual rainfall is less than 200 mm (8 in), Turkana pastoralists practice a combination of water spreading and floodwater cultivation.[40] They grow gardens of about 1,000 m^2 (10,763 ft^2), consisting primarily of quick-maturing (65 days) sorghum. When the rains fail, there is no harvest. We consider these to be gardens because sorghum is a minor crop, yielding only about 128 kg per 1,000 m^2 plot (282 lb/10,763 ft^2), but providing a valued supplement to the main source of food from herding goats.

The Turkana plant at the beginning of the rainy season at three different sites:
- Where minor tributaries enter the Kerio, a major seasonal river, and most of the water soaks quickly into the alluvial soil.
- In depressions or former meanders of the river where there are rich soil deposits occasionally flooded by the river, and groundwater is sometimes tapped by hand-dug wells.
- In depressions subject to flooding in the delta of the Kerio River where it enters Lake Turkana, and where the water table is higher because of the lake.

No berms or any other water-control devices are used. The gardens are surrounded by thorn brush fences to protect them from livestock and the area is divided into individual women's plots. Platforms from which young girls scare birds are also built in the gardens. Soil fertility is provided by silt brought in by the floodwaters, and by the cattle and goats that graze the plots during the rest of the year.

11.7 Groundwater and Wells

Groundwater may be the only source of water for dry-season gardens in areas where streams or rivers dry up when the rains stop.

11.7.1 Groundwater

Rainwater infiltrates down into the soil until it meets an impermeable layer in the soil profile, and then saturates the layer above it (Figure 11.9). This underground water is known as *groundwater*. If water can flow in this saturated layer, it is known as an *aquifer* and its upper surface as the *water table*. Aquifers are usually made of sand, gravel, or porous rock. The water may flow through the aquifer until it emerges as a *spring*, or until it flows into rivers, lakes, or oceans.

Wells are holes dug down into the aquifer to obtain groundwater. When there is no impermeable layer between the aquifer and the surface it is called an *open aquifer*, which is the most common type of aquifer tapped by hand-dug wells. When the aquifer is overlain by an impermeable layer and the water is under pressure that forces it up to the surface through any openings, it is an *artesian aquifer*. Wells that tap artesian aquifers are *artesian wells*. Water flows up and out of artesian wells naturally because water enters the aquifer at a higher level creating pressure that forces the water up the well.

The water table of an open aquifer, and thus the supply of water in a well, will vary depending on rainfall and amount of rain that percolates down to recharge the aquifer. The amount of percolation can be increased by slowing runoff so that the water has enough time to infiltrate into the soil. This can be done by using terraces, contour berms, garden basins (sunken beds), or dams built across seasonal streams. Some of the same terms used to describe soil (section 9.3.2) are also used when referring to aquifers. *Porosity* measures the amount of water an aquifer can hold, and *permeability* refers to the ease with which water flows through the aquifer.

11.7.2 Locating a Well

The best method for deciding where to locate a well is to survey existing wells in the area and to talk with people who have built wells. In many drylands there are experts in indigenous methods of digging wells; they will have many valuable ideas about well location using observations of the environment. For example,

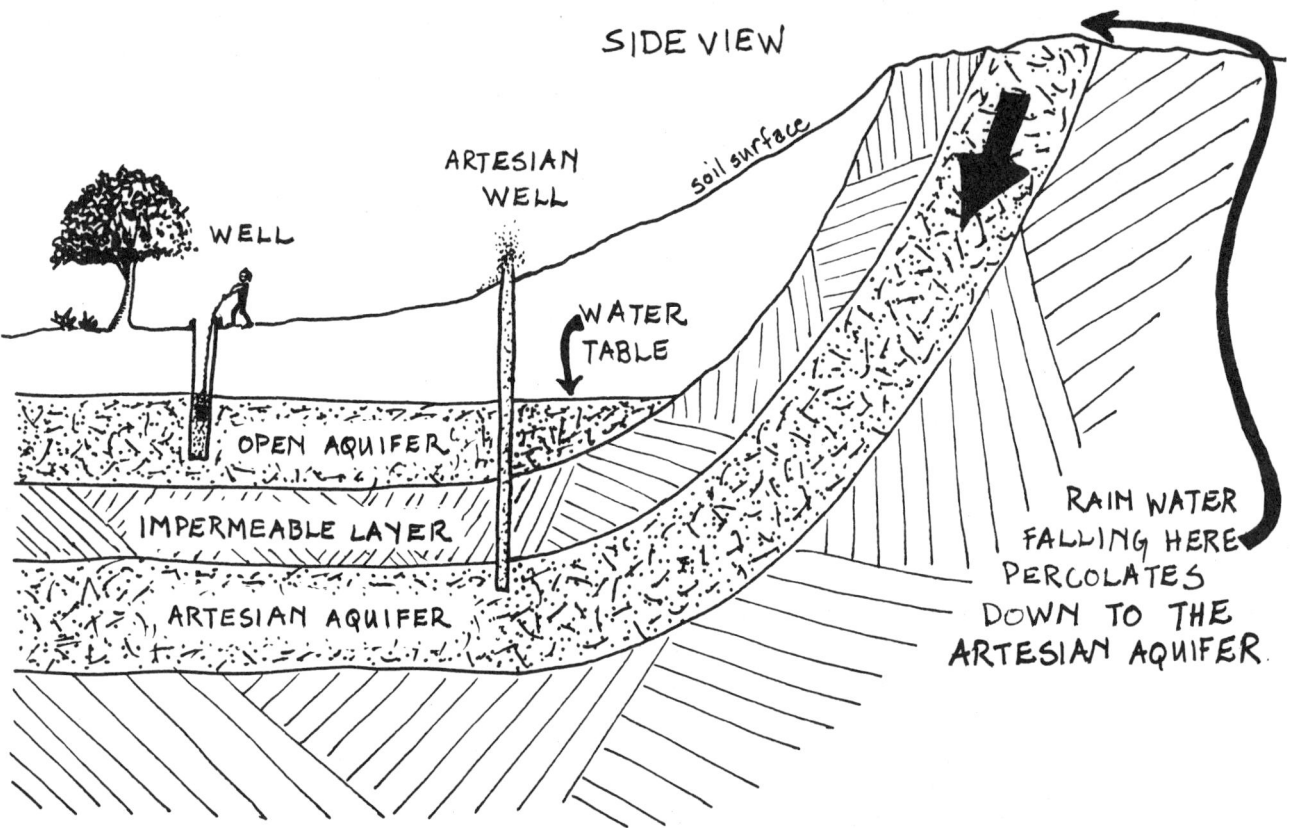

Figure 11.9 Wells to Bring Groundwater to the Garden

vegetation that stays green into the dry season may be tapping a groundwater aquifer with its roots. Geologists, hydrologists, botanists, and others from universities, government departments, and development projects may also have information about local groundwater conditions.

11.7.3 Hand-Dug Wells

Hand-dug wells are also known as *large-diameter wells* because they are 1 m (3.3 ft) or more in diameter to allow workers room to excavate them. Many people have traditionally dug wells by hand to reach groundwater. Very-large-diameter wells that are enlarged below the surface, can also serve for collecting and storing surface runoff during the rainy season, as well as for groundwater during the dry season. These are called *cistern* wells and water from them should not be used for drinking. In northern Ghana for example, there are many of these ancient cistern wells consisting of a 0.4- to 2.0-m (1.3- to 6.6-ft) diameter hole through a cap of plinthite (Box 9.4) up to 2.0 m (6.6 ft) thick, with a cavern 0.6-1.5 m (2-5 ft) high and 6 m (20 ft) or more across hollowed out of the clay and shale below.[41] The market town of Salaga in the north of that country is known as the city of 1,000 wells, many of which have linings of plinthite blocks and wooden lids with locks.

In sandy soils hand-dug wells over 1 m (3.3 ft) may have to be lined to prevent the sides from collapsing, while in clayey soils lining may not be required until after 5 m (16 ft). It is best to ask for advice from local people with experience. The lining can be of brick or stone, but reinforced concrete is much stronger and is preferred if the cement and the steel reinforcing rods are available and affordable. A field-tested method for constructing one type of hand-dug well has been described.[42] Since the cost of lined wells increases in direct proportion to the diameter, it is best to keep the diameter to a maximum of 1.3 m (4 ft), which is enough space for two workers. These wells consist of three units (Figure 11.10):

- The shaft to the water table, which is constructed by digging down in sections of up to 5 m (16 ft), depending on the soil type. In sandy soils the sections should be much shorter for safety reasons. After each section is dug a reinforced concrete liner is poured.
- Precast reinforced concrete caissoning rings within

the aquifer, which allow water to pass through and into the well.
- A well head, also of reinforced concrete, which protects the top of the shaft from crumbling and from water washing in soil, debris, and microorganisms which spread human or plant diseases. It also helps to protect people and animals from falling in.

Water tables are usually higher in the area of streams, rivers, and lakes, even in the dry season. People often dig shallow wells in these locations to water dry-season gardens, as do the Turkana in their floodwater gardens (section 11.6.2) or the people of Dar Masalit in Sudan (section 12.3.2). In northern Ghana, wells dug in dry streambeds to water gardens are severely damaged in the rainy season when flowing water collapses and fills them in; they have to be redug annually.[43] One drawback of streambed wells is that the clayey alluvial soil in which they are often located is low in permeability, so water moves slowly through them and therefore the wells take a long time to fill. This can be remedied to some extent by increasing the depth, and in the case of very shallow wells, the diameter. In permanent wells the investment of making lateral extensions at the bottom of the well to increase inflow may be worthwhile.

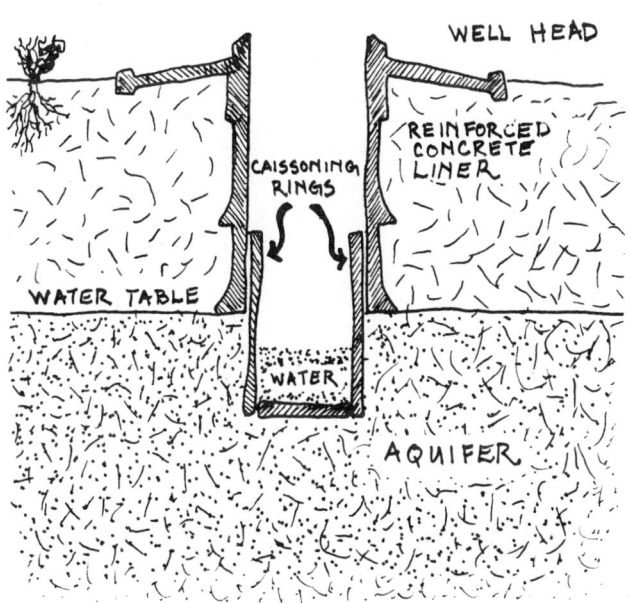

Figure 11.10 Cross Section of a Lined, Hand-Dug Well
(After Watt and Wood 1979:27)

Qanats (a Persian word) consist of a slightly inclined tunnel bringing water from an aquifer to the surface with no need of water-lifting. They are excavated by a series of vertical wells joining the tunnel sections. Qanats have been used in Iran and other areas of the Middle East and North Africa for over 2,000 years, and are still being used in some places. They are called *foqjara* in northwest Africa and *aflaj* on the Arabian Peninsula.

11.7.4 Small-Diameter Wells

Small-diameter wells are constructed using a variety of methods and without the need for underground workers.[44] Thus they are much safer to construct than hand-dug wells. In addition, they have potentially greater discharge because it is easier to extend the well deep into the aquifer. They are also easier to keep clean. The major disadvantages of small-diameter wells are that specialized equipment is needed for their construction, and a pump or special small-diameter buckets are needed to raise the water. Also, because they are narrow they cannot store as much water as larger-diameter hand-dug wells, which limits their usefulness in low-permeability aquifers where water flow is very slow.

11.8 Water Storage

The best place to store water is in the root zone of garden plants. There the water is readily available to the growing plants when they need it, there is no water surface exposed to evaporation, and no special storage containers are necessary. Water from intermittent streams may be stored in the alluvium deposited behind small dams.[45] Such dams are very popular in Mexico where they have been used for thousands of years.[46]

However, water storage in the soil is not always possible because sometimes too little water will be available from rainfall, the soils will not be able to hold enough water, or drainage will be insufficient, leading to too much water in the root zone (sections 10.3.1 and 10.3.2).

There are a variety of different containers that can be used for water storage. Large, locally made ceramic vessels are often easily available. Other popular containers are the steel drums used for packaging substances such as fuel, pesticides, soap, and foods. Only those used for soap and foods are acceptable for water storage. Determining the size of the container needed depends upon the amount of water available to store and the most practical and affordable container.[47] For example, for storing rainwater runoff from small roofs, start with the maximum runoff available:

(horizontal area of roof used for collecting [Figure 11.11]) x (runoff percentage) x (monthly rainfall).

The runoff available is then divided by the water requirement per m^2 (or per ft^2) of garden for the needed growing season to see what size garden could be maintained going into the dry season.

Storing water from the peak of the rainy season could allow extending the growing season, on a small plot, or even growing a second crop in a dry-season garden. However, storing water for gardens should be carefully considered because for many households the cost of constructing or purchasing a container will be too great to justify its use for watering a garden.

Figure 11.11 The Horizontal Area of a Roof

11.9 Resources

As mentioned in this Chapter, local experts such as well diggers, and gardeners and farmers practicing water harvesting, are the best resources for information and ideas about local water resources. For a general review of water harvesting and floodwater farming see Pacey and Cullis (1986) and UNEP (1983). Watt and Wood (1979) give detailed instructions for hand-dug wells. Chleq and Dupriez (1984:Chapters 7,8,9) and Koegel (1977) give a brief review of both hand-dug and small-diameter wells. Watt (1978) describes construction of steel-reinforced concrete storage tanks. Nissen-Petersen (1982) has technical information on building various size water containers, mostly larger than household size. For a discussion for field workers of the treatment of human diseases carried by water we recommend *Where There is No Doctor* (Werner 1977). Chapter 12 in that book describes the prevention, signs, and treatment for some of those diseases.

References

[1] Ayers and Wescot 1985; CFA 1980:23ff.; Cox and Atkins 1979:300-304; Stern 1979.
[2] Ayers and Wescot 1985:15-21.
[3] Ayers and Wescot 1985:8.
[4] Ayers and Wescot 1985:8; Stern 1979:75.
[5] Ayers and Wescot 1985:77-83.
[6] Bell, et al. 1987:39-40.
[7] See for example, Werner 1977:Chapter 13.
[8] Based on Beneson 1985; and Werner 1977.
[9] Oguntoyinbo and Richards 1978:181-187.
[10] Dupriez and De Leener 1983:125-133.
[11] Bryson and Hare 1974 for North America; Griffiths 1972 for Africa; Schwerdtfeger 1976 for Central and South America; Takahashi and Arakawa 1981 for Southern and Western Asia.
[12] Dancette and Hall 1979:110.
[13] Fraiser and Myers 1983:7.
[14] Shanan and Tadmor 1979:6.
[15] Pacey and Cullis 1986:170-172.
[16] Dupriez and De Leener 1983:100-108; Pacey and Cullis 1986:127.
[17] Evenari, et al. 1982.
[18] El Amami 1979.
[19] Pacey and Cullis 1986:130.
[20] Kolarkar, Murthy, and Singh 1983.
[21] Pacey and Cullis 1986:135-141.
[22] Sharma, et al. 1982.
[23] Smith and Critchley 1985.
[24] Fairbourn and Gardner 1974.
[25] Pacey and Cullis 1986:154-155.
[26] Van Doorne 1985:15-16,47,54.
[27] Pacey and Cullis 1986:165-173; Thomson 1980.
[28] Evett 1985b.
[29] Evett 1985b.
[30] Dutt 1981.
[31] Evett 1983:97.
[32] Evett 1983.
[33] Sharma, et al. 1982.
[34] Leonard 1980:21-22.
[35] Nabhan 1979.
[36] Kirkby 1973:40.

37 Pacey and Cullis 1986:154.
38 Adams 1986.
39 Scudder 1962, 1982.
40 Morgan 1974.
41 Cleveland 1980:76.
42 Watt and Wood 1979; see also Chleq and Dupriez 1984:Chapters 7, 8, 9.
43 Cleveland 1980:76.
44 Koegel 1977.
45 Chleq and Dupriez 1984:114-119.
46 UNEP 1983:23-24, 127-133.
47 Pacey and Cullis 1986:59-62.

12
Irrigation and Water-Lifting

In drylands, rain falling directly on the garden is often inadequate, and additional sources of water must be found. *Irrigation* is the conveyance of this water from its source and its application within the garden. Regardless of the source of water, the principles of irrigation are the same.

Small-scale, locally controlled systems of water management are often environmentally sustainable. One reason for this is that they can respond and adapt to changes in local needs and conditions. Their replacement by large-scale, centrally controlled systems has frequently increased production, but in many cases it has also led to increasing inequity and loss of water and soil resources.[1] This has happened in ancient systems as well as in modern ones. Falling water tables, soil erosion, waterlogging, salinization, groundwater pollution, and increasing poverty are common problems with large-scale water control systems. While small-scale systems are not trouble free, these problems are more common and harder to prevent in large-scale systems. As a result, there is growing interest in small-scale water management and irrigation, including household gardens.[2] For example, in Zimbabwe, indigenous irrigated gardens in valleys existed before the Europeans arrived, and are today being recognized as an important national and household resource, after years of being ignored by irrigation development projects and discouraged by governments.[3]

In areas of irrigated agriculture, gardens often depend on water from the same irrigation system that delivers water to the fields. Gardens here may be small plots in the fields or near the house, or narrow strips along canals or roads. These gardens have been neglected and undervalued, but they can help improve household income and nutrition even though significant improvement in equity and environmental sustainability will require major changes in the values and structure of the irrigation system.[4]

There are many examples of indigenous irrigation systems that function successfully without imported or capital-intensive technology. Some of these systems and the principles on which they are based are discussed in this chapter. Irrigation systems for large garden projects, especially those involving heavy construction and imported designs and equipment, require consulting with experienced local people or professional engineers.[5]

12.1 Summary

Water can be applied to the garden by surface, subsurface or overhead irrigation. Surface irrigation is the application of water using basin, furrow, or trickle systems. Water can also be applied directly to the root zone using pots, vertical mulch, or sand and gravel columns. Overhead or "sprinkler" irrigation in gardens is usually done by hand. Salinity and waterlogging are often the result of irrigation in drylands, but can frequently be avoided in small-scale systems with proper design and management. If the water supply is at a lower level than the garden, water can be lifted by hand, or by using the power of animals, water, wind, solar energy, or fossil fuels.

12.2 Irrigation Efficiency

Irrigation efficiency is a measure of how much water is actually placed in the root zone of growing crops, compared with the amount of water that is extracted from a well, stream, or other sources and put into the irrigation system. Understanding irrigation efficiency is important for saving water, a scarce resource in drylands; but first we need to discuss extraction, conveyance, and application.[6] The quantity of water taken from a particular source such as a well is the amount *extracted*. What happens to the water during *conveyance*, that is, from the time of extraction until it reaches the garden, will affect the amount of water the garden receives. If all of the water extracted reaches the garden then the *conveyance efficiency*, that is, the amount of water applied to the garden (Id), as a percentage of the amount of water extracted, is 100%. However, losses of water by evaporation, leaking through holes and cracks in hoses and canals, and percolation into the soil of earthen canals and ditches, mean that conveyance efficiencies are usually less than 100%, often much less.

Application efficiency (Ea) is the water applied to the garden (Id) divided into that portion needed to bring the soil in the root zone to field capacity (W): (Ea = W/Id x 100). (Multiplying by 100 gives the efficiency as a percentage). Application efficiencies are less than 100% because of evaporation from the surface, runoff of excess water out of the garden, and deep percolation below the root zone.

Irrigation efficiency equals application efficiency multiplied by conveyance efficiency. An irrigation efficiency of 50% means that 50% of the water extracted from the source infiltrated to the root zone and stayed there. It also means that twice the amount of water needed to bring soil in the root zone of the garden to field capacity will have to be extracted, since 50% of it is lost in conveyance or application in the garden.

An irrigation efficiency of 100% means that all of the water extracted from the source reaches and stays in the root zone, and is equal to the amount needed to bring the soil in the root zone to field capacity (W)[7]. While no irrigation system can achieve 100% irrigation efficiency, good management can keep efficiency relatively high, that is over 50%. In addition to simply wasting water, low irrigation efficiency reduces production because it leaches nutrients below the root zone, and over the long run in some areas it can raise the water table into the root zone, causing waterlogging and an increase in salt content of the soil.

Shading and mulching the soil surface and adjusting the timing of irrigation reduce evaporation. Gardeners also try to increase efficiency by ensuring equal time for water infiltration in all areas of the garden being irrigated, with no water lost in runoff, or deep percolation below the root zone, except as required to leach out salt (section 12.6.2). The smaller the garden and the closer the water supply, the easier it is to achieve irrigation efficiencies well above 50%.

In large garden projects there are several steps that can be taken to make irrigation efficiency as high as possible:
a) Estimate the amount of water needed (W) to bring garden soil to field capacity (Box 10.2, section 10.5).
b) Apply water until all soil in the root zone is at field capacity; determine this by checking soil water in the root zone at various sites within each bed, furrow, or other irrigation unit of the garden (section 10.7).
c) Measure the water actually delivered (Id) to the garden (section 10.6).
d) Determine application efficiency (Ea) by calculating W/Id x 100, and adjust the amount of water and method of application accordingly.

12.3 Surface Irrigation

Basin, furrow, and trickle irrigation are the methods of surface irrigation most often used in gardens. In some cases, water may first have to be transported to the garden from some distance.

12.3.1 Transporting Water to the Garden

Water can be delivered to the garden directly from a rainwater catchment area, floodwater diversion, well, stream, or standpipe, but is sometimes carried in a bucket, pot, hose, or canal. Delivery systems do not have to be elaborate or costly; they can be built from inexpensive, readily available local materials. An additional advantage of such systems is that they are easy to adjust and repair. Large-scale irrigation systems are usually accompanied by an elaborate social organization for regulating distribution to all those with rights to cultivate.

Indigenous irrigation systems using perennial and nearly perennial stream flow for food production are fairly common, for example, in hilly areas of east Africa. In Tanzania the Sonjo use water from streams and springs to irrigate heavy bottomland in the dry season, and supplement rainy-season production on sandy upland fields.[8] *Flumes* (elevated canals) made of hollowed logs are used to bridge low spots. Beans and two to three kinds of cowpeas are interspersed with the main crops of sweet potatoes, millet, and sorghum in these plots.

Another example is found among the Taita living in the hills of Kenya.[9] Water from the Mwatate River is obtained by intakes built of sticks, rocks, and soil which are located in small pools along the steep gorge through which the river flows. The intakes divert water from the river into canals that carry it to the fields. Water from the canals is also used for drinking and washing, livestock, construction, and a school brick-making project. The heaviest rains wash out the intakes, but this protects the canals and fields below from too much water flow which could wash them out also. The earthen canals are partly dug into the upper slope and partly built up from the lower slope. In some places flumes made of hollowed banana stems are used to bridge short gaps, and in other places canals are dug into the hillside to avoid tree roots or rocks.

Traditionally these canals have been used to minimize risk from variable rainfall at the end of both the short and long rainy seasons, and if rains are good in a particular season, the canal system may not be repaired and used. With gravity delivery and plentiful supply, the benefits of such a low-cost, low-risk system may indeed outweigh the advantages of maximizing conveyance efficiency. In other cases high pumping costs may justify creating delivery systems with greater conveyance efficiency.

The Taita grow their staple crops of maize and beans in most of the irrigated fields, but minor crops such as bananas and sugarcane are also grown. Farmers with fields close to the main road can produce vegetables for market, some even year-round, and many fields have been completely given over to commercial vegetable production. The system is maintained and the water distribution managed quite successfully by user groups based on kinship. This is in contrast to user group organizations created by some new irrigation schemes, without regard for existing social institutions.

12.3.2 Basin Irrigation

We discussed building raised and sunken beds in section 9.8. Sunken beds are also referred to as *basin beds* because the berms form a basin that holds water. Basins can also be formed on raised beds or on level ground by simply hoeing up soil into ridges to enclose areas of various sizes. Terrace basins on a slope should be smaller than those on flat ground to improve application efficiency, to minimize the amount of soil moved, and to avoid rocky or infertile horizons or layers (section 9.4). Basins can be filled by hose, with water from an adjacent irrigation ditch, or by pouring water from a container (Figure 12.1).

An example of basin irrigation using canals comes from Dar Masalit in east central Sudan, where people cultivate both rainy-season and dry-season gardens in the clay soils of dry streambeds, called *wadis* in Arabic.[10] Groups of people usually cooperate to dig a well and fence in an area against animals, then divide up the land inside the fence. Each gardener makes about 10-50 basins 0.5-1.0 m^2 (1.6-3.3 ft^2) and a network of small canals to water each basin in succession. Every 3-4 days they spend the morning drawing water from the well and pouring it into canals. For young people this is mostly a pleasant way to pass time with friends, but in large gardens the work can be tiresome.

In basin irrigation the goal is to have level basins and to fill them as quickly as possible. They should be filled with enough water so that soil in all parts of the basin is wetted throughout the root zone with minimal loss of water below the root zone, except for deep irrigations to flush out salts (section 12.6.2). Efficiency is decreased by low spots that get too much water. It is also decreased when there are different types of soil within the bed, so that some areas, for example, of sandy soil, reach field capacity sooner than other areas, for example, of clayey soil. In this case much water could be lost to deep percolation in the sandy soil by the time the clayey soil reaches field capacity. In general, because of deep percolation, sandy soils have lower application efficiencies than do clayey soils.

Given the goal of efficient application, the size of the basin will be determined by a combination of crop water needs, soil texture and structure, and water quality and speed of delivery. The depth of the basin, that is, the distance between the top of the berm and the soil surface within the basin, depends on the amount of water needed, which in turn depends on soil texture, rooting depth, and the crops grown (sections 10.5 and 5.2.1). The deeper the rooting depth of the crop, the deeper the basin needs to be.

The faster the rate of infiltration, the smaller the surface area of the basin should be so that water is not lost to deep percolation in one area of the basin before other areas are wet. Porous soils with sandy texture or strong structure have higher rates of infiltration than soils with clayey texture and weak structure. Porous soils hold less water, and the depth and surface area of the basin should be smaller than in clayey soils. This is because water will sink into the soil quickly so that the basin does not need to hold as much on its surface; water will not spread evenly over a large surface area from a single point of application. Infiltration is slowed when large quantities of suspended matter in the irrigation water clog the pores in the

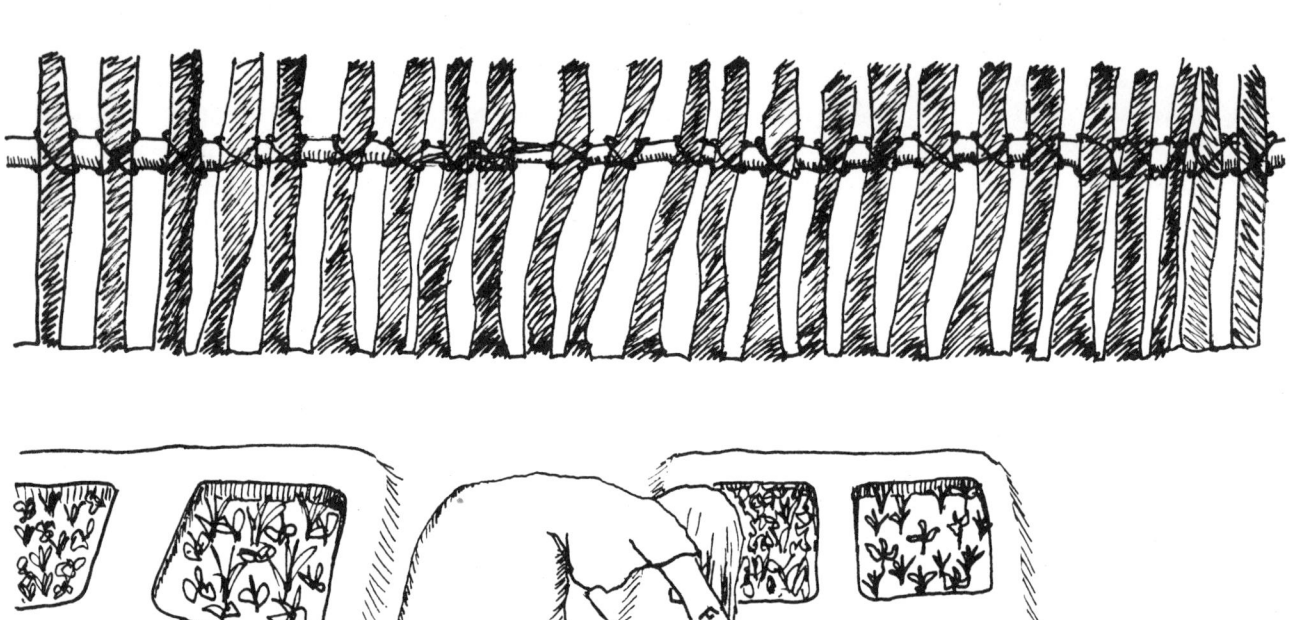

Figure 12.1 Basin Irrigation of Zuni Sunken Garden Beds (After a photograph in Ladd 1979:497)

surface layer of the soil.

The faster water can be applied, the larger the basin; and the slower the application, the smaller the basin. If the basin is too big, or the water is applied too slowly, too much water will have infiltrated at the point of application by the time all soil is at field capacity. In this case the basins will have to be watered at several different places, or made smaller.

12.3.3 Furrow Irrigation

In **furrow irrigation** water is delivered to the garden through a network of furrows (Figure 12.2). For example, among the Taita living in the hills of Kenya, a farmer cuts away the side of the canal at the upper end of a field and puts a rock in the canal to divert the water; a network of earthen furrows distributes the water within the field.[11] In small gardens water can be guided carefully to each plant by tiny earthen dams and furrows.

Furrow irrigation can be used where water is available at the high point of the garden, and the garden has a fairly uniform gentle slope. If the slope is greater than about 2%, then the furrows should not run straight up and down the slope, but at an angle to it, in order to maintain the furrows at a 2% slope.[12] This allows the water to flow down the furrow, but prevents the water from flowing too quickly and eroding the furrows.

The goal of furrow irrigation is an even flow along the length of the furrow so that by the time water has reached the low end of the furrow, an adequate amount of water will have infiltrated the soil at the high end[13] (Figure 12.3a). This will depend primarily on the rate of application, length of furrow, degree of slope, and soil texture. The length of furrow is often the easiest variable to adjust. Gardeners can experiment. Efficiency is decreased when furrows are too long, or water is applied too slowly, so that by the time the low end is adequately wetted, the high end has excess water (Figure 12.3b). If the water is applied too fast, there will be excess water at the low end of the furrow by the time the middle part is at field capacity (Figure 12.3c).

12.3.4 Trickle Irrigation

Also known as drip or localized irrigation, *trickle irrigation* is increasingly popular in industrialized countries because of its high irrigation efficiency. It is especially appropriate for sandy soils with low water-holding capacity. In the most commonly promoted system, water is delivered to each plant or group of several plants through *emitters* (small holes or valves) on a small-diameter, flexible plastic line. Water must be filtered to prevent clogging of the lines and emitters, and pressure must be regulated to ensure even distribution. For most limited-resource households in the Third World, this type of trickle irrigation is inappropriate because of the expense of purchasing the system and replacement parts, and the difficulty of maintaining it.[14]

An alternative system of trickle irrigation using

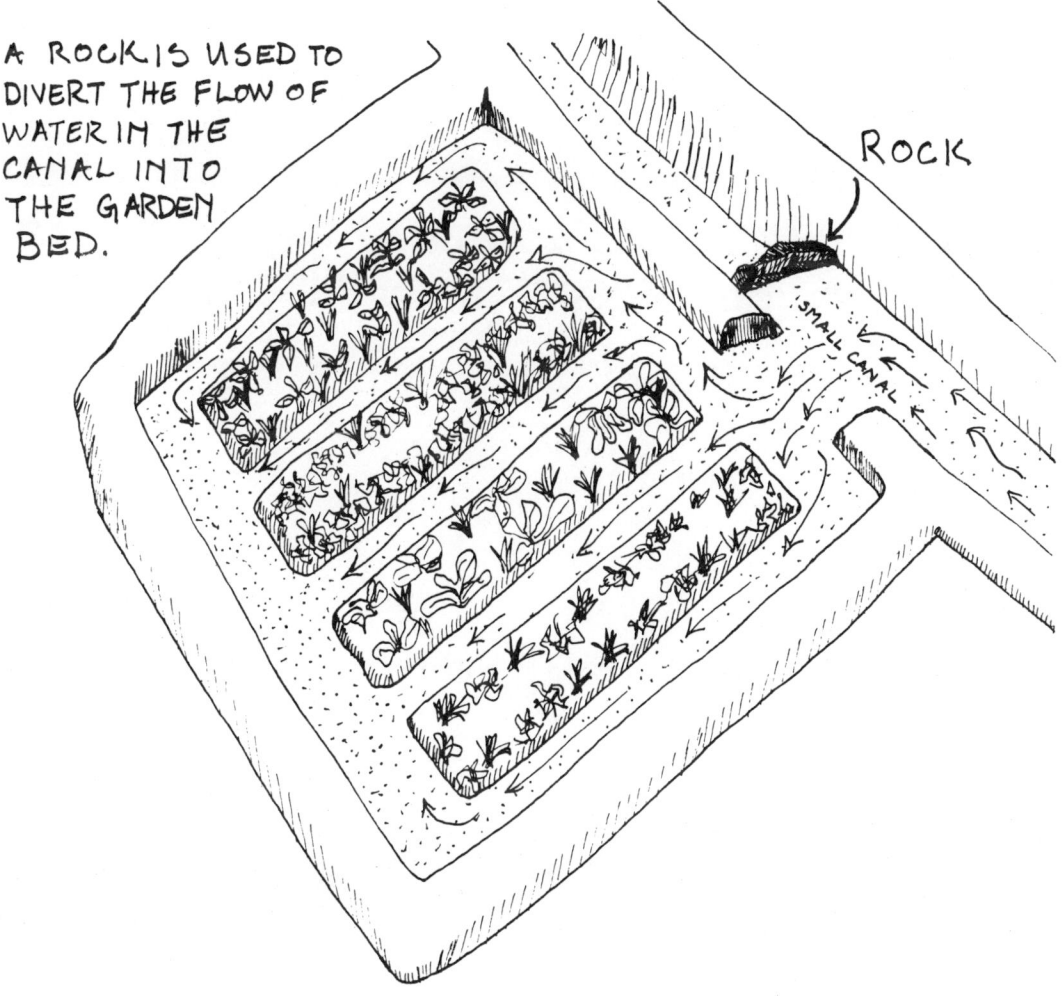

Figure 12.2 Furrow Irrigation

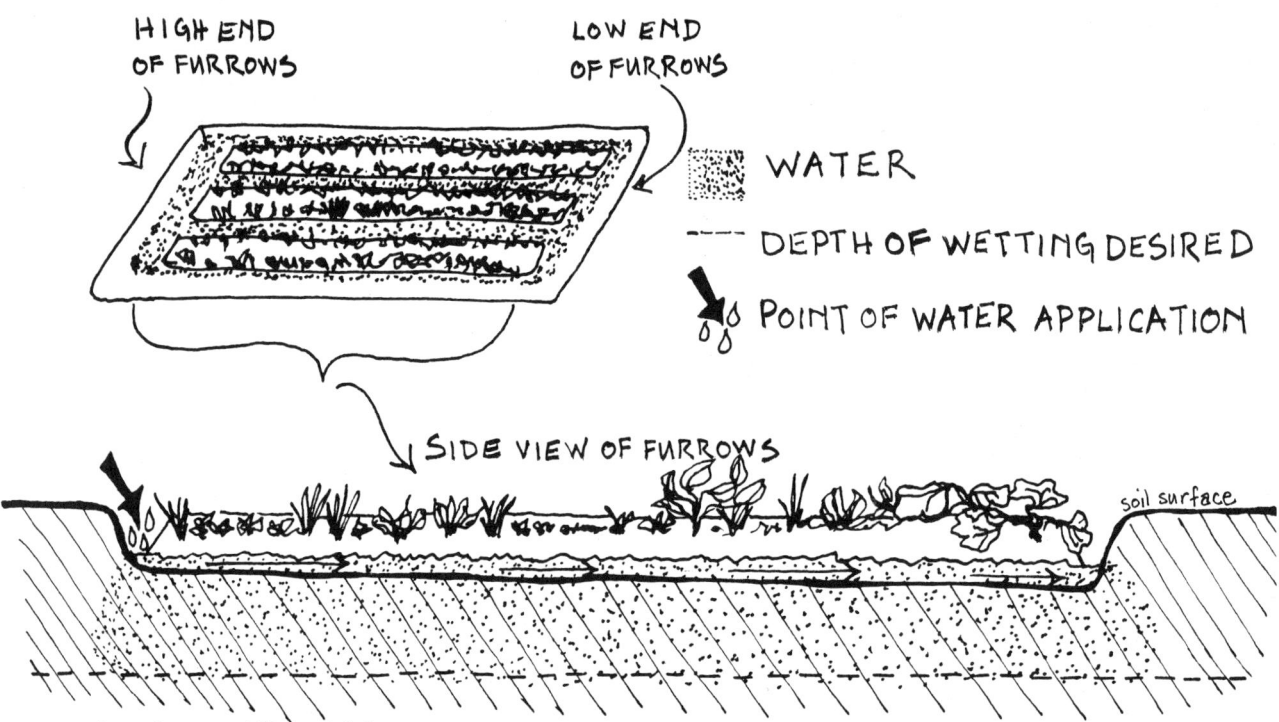
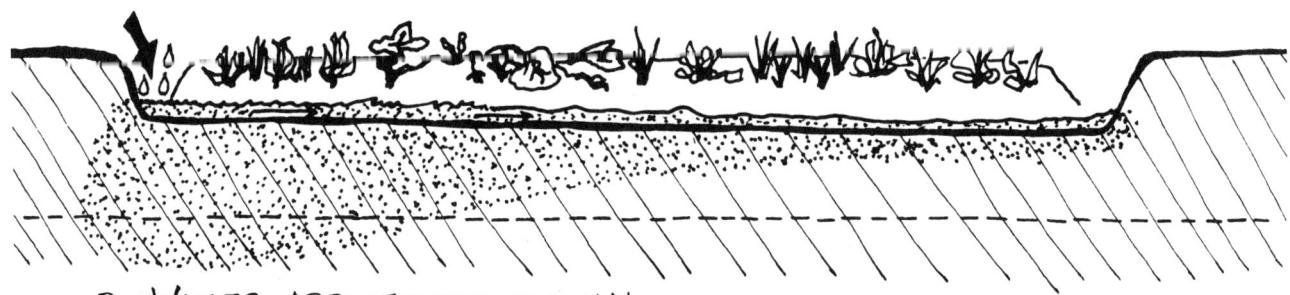

Figure 12.3 *The Effect of Water Application Rate on Efficiency of Furrow Irrigation*

plastic infusion or drip sets discarded by hospitals has been experimented with in desert areas of western Rajasthan, India, for growing cauliflower.[15] The area has loamy sand soil, average annual precipitation of 360 mm (14 in) and potential ETm of 2,063 mm (81 in). Application of 315 mm (12.5 in) of water supplied 100% of ET requirements for cauliflower during the 88-day growing season with yields of 2.3 kg/m^2 (0.5 lb/ft^2). Results of experiments showed yields comparable with those using conventional trickle irrigation under identical circumstances, yet this system cost nothing, and any clogging was easily remedied by squeezing the suspended water bag or removing the emitter tube from the ground. (CAUTION: If discarded infusion sets are tried in the garden, all needles and syringes should first be removed and destroyed and medical workers should certify that there is no danger of spreading disease.)

12.4 Root Zone Irrigation

Root zone irrigation is the delivery of water below the surface of the soil, directly to the root zone. In drylands it has the advantage of minimizing evaporation of water from wet soil surfaces and upper soil layers. One technique is vertical mulching discussed in section 10.8.2. Trenches or holes in the garden bed, which are filled with sand and topped with gravel to slow evaporation, also lead water to the root zone. If trickle irrigation emitters are buried, they too can be used to create a root zone irrigation system.

12.4.1 Pitcher Irrigation

Pitcher irrigation is watering by filling a buried, unglazed ceramic pitcher or pot through which water slowly seeps into the root zone. Covering the top of the pitcher reduces evaporation. An extra advantage of the pitcher method is that clay pitchers are readily available because they are commonly used for storing and cooling water in many drylands. However, they still cost money and this may be prohibitive for poor households.

A problem that may occur is salt accumulation in the soil between pots, since not enough water is applied to flush the salt below the root zone. In addition, clean water must be used because any clay, silt, or organic material in the water can clog the pores of the pitcher and prevent water from seeping out. Leaving water to stand in a container for a while allows some of these particles to settle out before pouring the water into the pots. Salty water will also clog the pores.

Some informal experiments with pitcher irrigation in India used growing holes 90 cm (24 in) in diameter and 60 cm (35 in) deep filled with manured soil[16] (Figure 12.4). Unglazed ceramic pitchers were placed in the center of the holes and filled with water 2-3 days before seeds were planted. Germinated seeds were planted 2-3 cm (0.8-1.2 in) from the outside of the pitcher in soil that had been wetted by adding a small amount of water directly. About 3 liters (0.8 gal) of water was added to the pitcher daily. Using one pitcher per 12.5 m^2 (135 ft^2) several species of cucurbits were grown with yields ranging between 0.16 kg/m^2 (0.5 oz/ft^2) on 19.8 mm (0.8 in) of water for pumpkin to 2.56 kg/m^2 (8.4 oz/ft^2) on 18.6 mm (0.7 in) of water for watermelon. Bottle gourd used only 12.3 mm (0.5 in) of water, but produced only 0.53 kg/m^2 (1.7 oz/ft^2).

12.4.2 Water Table Irrigation

Water table irrigation is the method used in flood recession gardening. Plant roots follow dropping water tables as the floodwaters recede during the dry season, as in Tonga cultivation along the Zambezi River in Africa, or they obtain water from a high water table during the rainy season, as in Turkana cultivation on the Kerio River delta at Lake Turkana (section 11.6.2). Gardens located below earthen dams benefit from the raised water table due to the seepage of water. In Sri Lanka, for example, this allows the cultivation of fruit trees such as plantain, which normally only grow in more humid areas.[17] The Hopi Native Americans of North America plant peach and apricot trees and melons on sandy hillsides where the heavier subsoil is kept moist by subsurface flow of water from nearby seeps and springs.[18]

12.5 Sprinkler Irrigation

Sprinkler irrigation is similar to rainfall. The easiest method is to sprinkle water from a small, hand-held container. This is appropriate for small seedlings or transplants or a very small garden (section 6.5.1). Watering cans can be used, but are not necessary, and are often expensive. Larger systems with a hose and a nozzle or a mechanical sprinkler can also be used, but require a delivery network of pipes or hoses and water

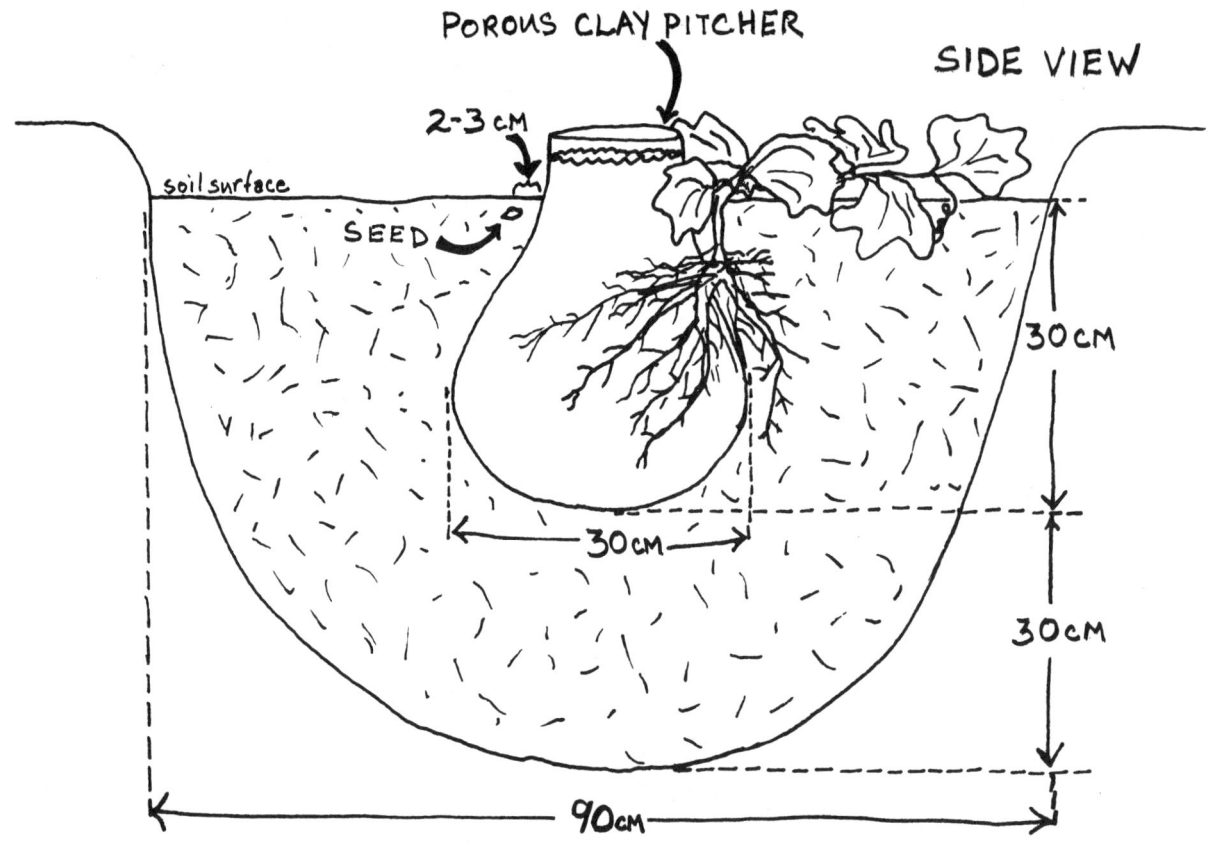

Figure 12.4 Pitcher Irrigation (*After Mondal 1974*)

under pressure produced by a pump or an elevated tank or reservoir. Sprinkler irrigation is not efficient in areas with high winds which distort the watering pattern and increase evaporation. Overall, sprinkling systems are expensive and require constant maintenance and the availability of spare parts. In many situations where piped water is available it is good drinking water, and it may be more important to reserve it for domestic use than to use it to irrigate a garden.

A major advantage of sprinkler irrigation is that irregularly sloping areas can be watered with no need for leveling. In spite of the fact that more water may be lost to evaporation from the air and plant leaves, in some situations sprinkler irrigation can have a greater efficiency than surface irrigation. In sandy soils, especially, it allows more even distribution than furrow or basin irrigation. In clayey soils with slow infiltration rates, the rate of water application for sprinkler irrigation may have to be very slow to avoid surface runoff and soil erosion.

12.6 Irrigation Problems

Waterlogging and salinity are two of the major problems that occur with irrigation, and are common in drylands everywhere.

12.6.1 Waterlogging

Waterlogging occurs when the water table rises into the root zone of garden crops. Since the roots of almost all garden crops require oxygen, waterlogging causes reduced yields and eventually death because the water forces the air out of the soil. Waterlogging also causes nitrogen deficiency shown as a yellowing of the leaves, because anaerobic bacteria which convert nitrates to ammonia multiply under these conditions. Soil becomes waterlogged when there is lack of drainage because of a naturally occuring high water table or a soil layer or horizon that is much more impermeable than the layer above it (Figure 12.5). Because more water than is needed by crops must be applied to flush out salts, this water can eventually raise the water table when there is poor drainage. Low irrigation efficiency due to leaky canals, low spots in the garden, and excessive deep percolation make the problem worse.

To provide for drainage during especially heavy rains, cuts may have to be made in berms around basins to let out excess water, or drainage ditches may have to be dug at the end of furrows. Raised beds can help keep plant roots out of the waterlogged soil but if the water table has risen substantially the garden site may have to be abandoned.

12.6.2 Salinity

Salty soils occur naturally in arid areas where not enough rain falls to wash soluble salts down and out of the root zone. Irrigation makes the situation worse, since surface water and groundwater contain more salt than rainwater. Salt tends to build up in the soil as water is continually added through irrigation. As water is used by plants and evaporates from the soil surface, the salt in that water concentrates in the soil. The high temperatures and low humidity in drylands mean that salinization often accompanies irrigation.

It is important to water to the bottom of the root zone and slightly beyond, and to reduce excess evapotranspiration so that irrigations can be spaced as far apart as possible to minimize wetting of the upper layers of the soil from where most water is lost to evaporation.

Salts are redeposited at the point where the water movement stops or where the water evaporates. White salt deposits can be seen on high spots in the garden after rain or irrigation water has evaporated. Planting at low points in the furrow or in basins rather than at the highest point of the raised bed, mound, or ridge helps plant roots avoid those salt deposits (Figure 12.6). In areas where there is poor drainage and a high water table, water with its dissolved salts may be constantly drawn up into the root zone. Waterlogging makes this problem worse. Raising the root zone by building up mounds or beds for planting may be the only alternative to try before abandoning the site.

If soils are salty then extra care must be taken not to add salty kitchen waste to the compost. Plants growing in salty soils, like the salt bush (*Atriplex* spp.) in the deserts of southwestern North America, accumulate salt in their leaves, and should not be used for compost.

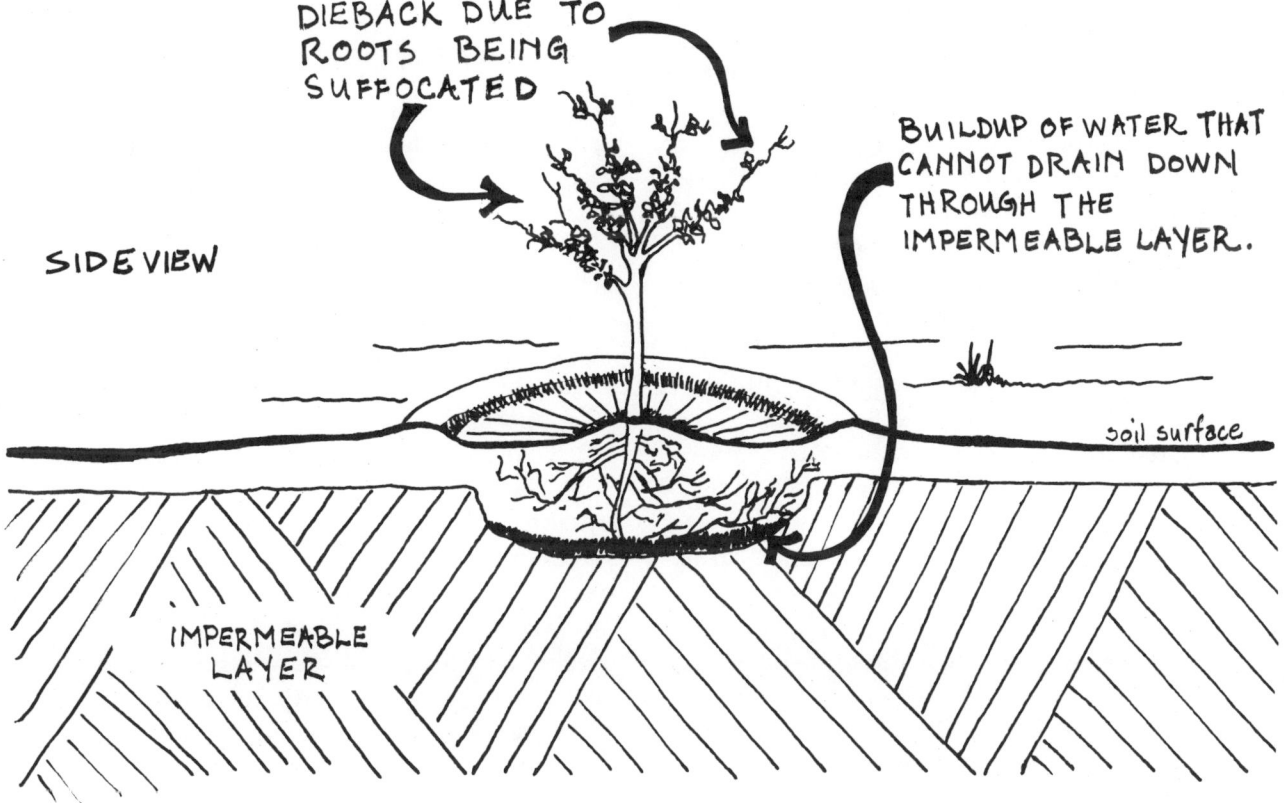

Figure 12.5 Waterlogging Due to an Impermeable Layer in the Soil

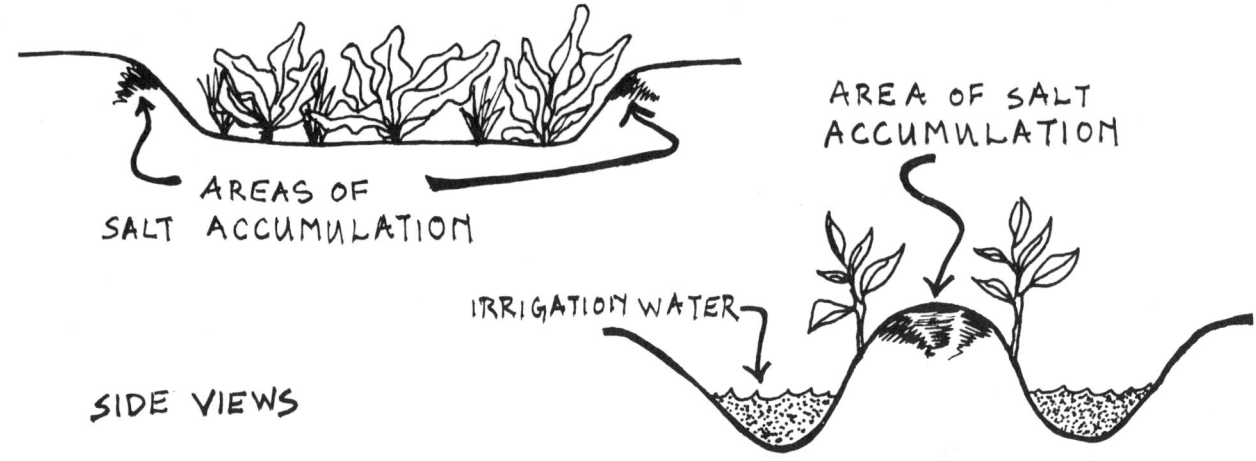

Figure 12.6 Planting to Avoid Areas of Salt Buildup

With large amounts of sodium (Na⁺), the pH (section 9.5.1) can reach 10 or higher and only a few salt-tolerant crops are able to grow. Obviously these are not good soils for gardens, and where they occur, the best solution may be to bring in better soil from elsewhere, plant in containers, or find another location.

We visited a squatter settlement located in a dried-up lake bed outside of Mexico City. Faced with the extremely poor, salty soil, many women had brought in soil and manure from other areas and were growing trees, herbs, vegetables, and flowers in boxes, used tires, cans, shampoo bottles, and other containers they had found on the streets (Figure 8.2 in section 8.2.2).

Where high water tables are not a problem, leaching can be used periodically to rid the root zone of salt. Leaching is washing salt from the root zone by adding excess irrigation water, and is commonly used to prevent salt buildup. The amount of water needed will depend on the salt sensitivity of the garden crops grown and the amount of reduction in yield that can be tolerated. For many commercially grown crops these values can be obtained from published tables.[19] The leaching requirement may be calculated as follows:[20]

$LR = EC_w/EC_c$

where

LR = leaching requirement, or fraction of applied water necessary to leach salts, beyond amount to meet ETm,

EC_w = salinity of water applied to the garden (section 11.2 and Box 11.1),

EC_c = salinity of the soil which allows an acceptable yield for the garden crops (section 5.6).

Calculating LR is a complex problem subject to great error as a result of slight differences in estimates used in the calculation.[21] It is also important to note that there are several different methods for calculating LR in use, based on different assumptions.[22] For example, the one recommended in a recent FAO publication[23] differs markedly from that given above for all but a narrow range of values. Whichever calculation is used, there is no substitution for starting small, experimenting, and understanding the local situation.

12.7 Water-Lifting

Water for irrigation from wells, cisterns, canals, rivers, streams, and lakes needs to be lifted whenever it occurs at a level lower than the garden.[24] The choice of a water-lifting method is determined by the gardener's needs and the power requirement. The power requirement is the product of the necessary flow rate and the lift, adjusted for efficiency losses in the lifting system. This requirement can be expressed as the *volume-head product* in units of m⁴/day (flow in m³/day x vertical lift in m). Human-powered systems can operate up to 100 m⁴/day, while a large animal can increase this output approximately 5 times. Most gardening requirements fall within this range. However, gardens will often be only one of many uses of water-lifting systems. Therefore, mechanical pumping systems with capabilities between 1,000 and 10,000 m⁴/day may be used for gardens. Detailed discussion of water-lifting devices is beyond the scope of this book, but we offer a brief review of some of the most common ones particularly appropriate for gardens.

Depending on the type of device used, the efficiency of water-lifting can vary from less than 5% to 75%.

Efficiency in this case is a measure of the amount of energy expended by the human, animal, fossil fuel, or other energy source in the process of lifting water, compared with how much water is actually lifted. If a water-lifting method is only 5% efficient, this means that 95% of the energy used in the operation did not result in water lifted to the garden. Most energy is lost due to water spilling out or leaking while being lifted, awkward and strenuous movements required by the person or animal, and water-lifting devices that are poorly designed or maintained.

12.7.1 Lifting with Human and Animal Power

Direct-lift methods are the most common for human- and animal-powered water-lifting systems. When the water only needs to be raised a small distance, a variety of hand-held containers can be used. In Oaxaca, Mexico, and other areas of southern Mexico and Central America, small-scale commercial vegetable producers irrigate small basins using a 10-14 liter (2.6-3.7 gal) clay or metal jar called a *cantaro* lowered into shallow wells and poured onto the garden[25] (Figure 12.7). These gardeners can irrigate at a rate of 10-20 liters/min (2.6-5.3 gal/min). At a rate of 15 liters/min (4 gal/min), a gardener could irrigate 15 m^2/hour (160 ft^2/hr) (calculated by the authors, assuming a need for 60 mm/m^2 [0.75 in/ft^2]).

The rope and bucket is one of the most available and easy-to-use methods for lifting water from deeper levels, although it is still hard work (Figure 12.8). The bucket can be made of leather, discarded rubber tire tubes, tin, or other material, and, like the rope, can be made locally.

Scoop irrigation is another traditional method used in Central America.[26] It involves lifting water a short distance from canals with a wooden or metal scoop, plastic bowl, or gourd and splashing it over the grow-

Figure 12.7 Water-Lifting and Watering by Hand Using a Ceramic Jar in Oaxaca, Mexico (After Wilken 1977)

Figure 12.8 Water-Lifting with a Bucket and Rope

ing area. The delivery rate is about 60 liters/min (16 gal/min). Scoops are more efficient if suspended, and can lift water up to 0.5 m (1.6 ft), at about 165-250 liters/min (66 gal/min) or 10-15 m³/hr (353-550 ft³/hr), although this is only about 25% efficient.[27]

The counterpoise lift, or *shaduf*, is a traditional method from the Middle East and East Asia consisting of a container (leather bag, metal bucket, lined basket) on one end of a pole with a counter weight on the other and a fulcrum in the middle. The counter weight lifts the water, and the operator's weight is used to return the bucket to the water source against the counterweight. It can lift water up to 3 m (10 ft) at 30 liters/min (8 gal/min). In northwestern Nigeria, for example, farmers use *shadufs* with buckets to raise water from hand-dug wells to water dry-season gardens,[28] and they are used extensively in Egypt to water small gardens of vegetables and fruit trees along the canals.

The *pi cottah*, used primarily in India, is similar to the *shaduf* but is operated by two people, one of whom acts as a moving counter weight to eliminate much of the strenuous work of returning the water container against a stationary counter weight.[29] Although it can lift water 5-8 m (16-26 ft), its output is small, and it is used primarily to water small vegetable plots.

Bucket systems may also be adapted to animal power to increase flow, such as with the *mohte*, or self-emptying bucket. This traditional device employs either a tipping action or simple flap valves in the bucket or bag to discharge the water at lifts of 5-10 m (16-33 ft). The system can be arranged for the animal to walk back and forth in a straight line or in a circle, thus requiring less supervision.

The low-lift *dhone*, or see-sawing gutters from Bangladesh, can deliver about 300 liters/min (80 gal/min) at a 1 m (3.3 ft) lift. This device uses flap valves and can be operated by a single person shifting the weight back and forth at the fulcrum.

The *Persian wheel* is a direct-lift device that uses an animal like a water buffalo to turn a wheel, around which is traditionally strung a rope with earthenware pots attached to it. By increasing the length of rope, while keeping the wheel diameter fixed, Persian wheels are capable of lifts up to an effective height of about 10 m (33 ft), where discharges may range on the order of 50 liters/min (13 gal/min), down to about 2 m (7 ft), where they can approach 600 liters/min (160 gal/min). Efficiency is high, approximately 50% or more.

Variants of the Persian wheel include the *zawaffa*, *sakia*, and *tablia*, some of which can attain 75% efficiency but are limited to a lift of 2 m (6.6 ft) or less by the diameter of the wheel. Discharges over 2,000 liters/minute (500 gal/min) can be achieved with the devices when driven by large animals, although they are normally operated at 50% to 75% of this flow rate. They can also be fitted with belt drives for engine-driven operation for even greater flow rates, as long as fuel purchase is possible.

Pumps that displace water by pushing or pulling are also suited to human- and animal-powered systems. The *Archimedean screw* is an example of a *displacement pump*. It is a long tube with a helical screw inside which is turned by hand to raise water a short distance, such as from a canal to the garden. It was supposedly invented by Archimedes around 2,000 years ago and today is used in India and Egypt.[30] The Archimedean screw is very efficient, up to 75% when working at full capacity, and about 50% overall.[31] It is limited to a maximum of 1 m (3.3 ft) lift and requires two operators. Archimedean screws of traditional construction have efficiencies in the range of 30% and provide discharges from about 300 to almost 1,000 liters/min (80-260 gal/min).

Another rotary displacement water lifter is the Chinese *chain and washer pump*, a method thought to have been in use for 2,000 years. Steady rotary power is applied to a shaft that drives a series of disks linked on a chain and pulled through a pipe, pulling water through at the same time. The energy source can be humans, animals, or engines. Chain and washer pumps are effective up to about a 15 m (50 ft) lift and are capable of discharges up to 300 liters/minute (80 gal/minute), with an efficiency usually less than 50%.

The mechanical hand pump does not have a good record in the Third World. Compared with other hand-powered lifting methods it has been expensive, subject to breakdown, and is often poorly maintained. One of the major problems with maintenance is the lack of control that local people have over water projects using mechanical hand pumps, including a lack of training in maintenance and repair. Some of these problems can be traced to poor project design and implementation. However, sturdier pumps are now being manufactured and may be a good choice for small-diameter wells. Many such pumps are designed primarily for the low flows needed for domestic water supply. There have recently been many evaluations of the wide variety of hand pumps, and they should be consulted before a choice is made.[32]

12.7.2 Lifting with Other Power Sources

Mechanical pumps powered by engines obtaining energy from gasoline or diesel fuel are becoming more and more popular in the Third World. When used in hand-dug wells care must be taken not to empty the well faster than water can infiltrate into it. If this does occur it can lead to sediment from the aquifer flowing into the well and causing sides of the well or the lining to collapse. Unless there is a reliable supply of fuel and spare parts, and maintenance is available at a reasonable cost, such pumps can make the risk of gardening unacceptable for poor households.

A market garden project in Lesotho used pumps to create pressure for irrigating sloping land with sprinklers.[33] Gardeners grew mostly cabbage on plots up to 500 m^2 (5,400 ft^2). Fuel costs (in 1985), at about US $0.50/liter ($1.67/gal), took up to 50% of potential profits. It would have been cheaper to locate gardens far enough down the slope from the dam to create pressure for sprinkling by gravity. The cost of installing a plastic line for the required distance would have been less than the cost of a pump, and would have eliminated the need for fuel. Alternatively, level beds could be created and irrigated by flooding, without the need for high pressures for sprinkling.

To avoid dependence on fossil fuels, many devices for lifting water using energy from the sun (photovoltaic cells), the wind (wind mills), or composted organic matter (biogas) have been promoted as "intermediate" technology for the Third World. However, most of these are intermediate only from the standpoint of the industrialized world, and human and animal power remain the most useful and reliable technology for most Third World rural areas,[34] and anywhere else that households wish to be able to buy and maintain the pumps themselves.

12.8 Resources

A brief review of small-scale irrigation techniques, though not generally applicable to garden-scale production, can be found in Stern (1979: Chapters 5, 6, and 7). Jensen (1980) contains chapters on technical aspects of larger-scale irrigation that may be consulted by those planning large irrigated garden projects. Kennedy and Rogers (1985) discuss basic principles of a wide variety of human- and animal-powered water-lifting devices, while Hofkes (1983) gives a brief review of a variety of water pumps, and Chleq and Dupriez (1984: Chapter 10) review water-lifting with reference to dryland West Africa.

Fraenkel (1986) gives a comprehensive discussion of water-lifting for irrigation and a basis for comparing human, animal, fossil fuel, and renewable energy sources in a handbook form. Arlosoroff, et al. (1987), is one of a series of World Bank publications documenting the progress of an extensive United Nations project conducting laboratory and field tests of a wide variety of hand pumps to assess their technical and economic performance in different Third World settings.

References

[1] Denevan 1980; Lawton and Wilke 1979; Manners 1980:51.
[2] Stern 1979:19-23.
[3] Bell, et al. 1987:4, 41.
[4] Chambers 1988.
[5] See Jensen 1980; Stern 1979:Chapters 12, 14.
[6] Doorenbos and Pruitt 1977:79-81.
[7] Doneen and Wescot 1984:54-59.
[8] Gray 1963:36-37, 55.
[9] Fleuret 1985.
[10] Tully 1988:128-130.
[11] Fleuret 1985.
[12] Stern 1979:43.
[13] Stern 1979:45.
[14] Doneen and Wescot 1984:52.

[15] Kolarkar, Singh, and Lahiri 1983.
[16] Mondal 1974.
[17] Leach 1961:17-18.
[18] Soleri and Cleveland 1989.
[19] See e.g., Ayers and Wescot 1985:31-35.
[20] Donahue, et al. 1983:380.
[21] Stegman, et al. 1980:801.
[22] Smith and Hancock 1986.
[23] Ayers and Wescot 1985:24.
[24] See Stern 1979:Chapter 15; Watt and Wood 1979:Chapter 22.
[25] Wilken 1977.
[26] Wilken 1977.
[27] Kennedy and Rogers 1985.
[28] Adams 1986.
[29] Kennedy and Rogers 1985:18, 20.
[30] Stern 1979:116.
[31] Kennedy and Rogers 1985:15-16.
[32] E.g., Arlosoroff, et al. 1987; Kennedy and Rogers 1985.
[33] Evett 1985a.
[34] Kennedy and Rogers 1985:1.

13
Pest and Disease Management

Gardens in drylands can be oases of green vegetation that attract insects, worms, birds, rodents, and other wild animals, and provide good growing conditions for weeds, fungi, viruses, and bacteria. Most of these do not affect garden production, and many improve it. Those that reduce production are called pests or pathogens. However, most problems in the garden are not just due to a pest or pathogen, but to temperatures that are too high or too low, too little sunlight, poor soil quality, too much or too little water, planting inappropriate crops, or planting at the wrong time of year.

A *pest* is any insect or other animal that has a negative effect on the garden. A *parasite* is an organism that lives on or in a *host* organism without benefiting it, and parasites that cause disease in their host are *pathogens*. Healthy plants will resist pests and pathogens much better than those that are unhealthy. In this book we emphasize reducing damage from pests and pathogens by enhancing natural controls through garden management, not by using commercial pesticides which kill both pests and beneficial organisms. (We use the word "pesticides" as a general term to include insecticides, fungicides, nematicides, herbicides, etc.)

In addition to general principles, we give examples of a few pests and pathogens common in dryland gardens and ways to manage them. This is not an exhaustive description of all dryland pests and diseases, rather it is meant to inspire creative thinking and experimentation by both gardeners and project workers when responding to pest and disease problems in local gardens.

13.1 Summary

The most important way to prevent pests and diseases from causing problems in the garden is good plant, soil, and water management as described in the preceding chapters of Part II: appropriate crop varieties; pest- and pathogen-free propagation materials; good soil texture, structure, and pH; adequate nutrients and moisture in the soil; and control of temperature, sunshine, and wind. A key ingredient is diversity; in the genetic makeup of crop varieties, and in the mixture of crops and crop varieties which encourage variety in the insects and microorganisms in the garden.

When pest and disease damage does decrease production enough to cause concern, there are diagnoses and responses that can limit the damage. The first step is careful observation. Insect damage includes chewing, cutting, and sucking of leaves, stems, and roots; transmitting disease; and eating or parasitizing beneficial insects. Infectious plant diseases are caused by fungi, bacteria, nematodes, protozoa, and viruses, which can attack any part of the plant. These small organisms cannot usually be seen with the naked eye, and so diagnosing disease relies even more on the plant's symptoms than is true with pests.

Looking at the garden in terms of yearly cycles, and comparing it to other gardens, fields, and natural plant communities in the area can help to diagnose a problem and find appropriate solutions. If the problem is severe and difficult to control, considering other crops or even garden sites may be an option.

In tables and accompanying figures in this chapter we summarize some examples of observation and management of pests and pathogens, including those that cause wilt, abnormal growth, and leaf and fruit problems. The goal of diagnosis is to understand the situation well enough to choose the most appropriate management strategy.

13.2 An Ecological Approach

Agriculture, including gardens, always involves the need for increased management and resources in return for increased production. The selection of plants for increased yield and better taste, for example, means reducing resistance to pests. While gardeners try keeping losses to pests at a minimum, some pest or disease damage is a natural part of a healthy garden, so it is important not to overreact to a few chewed, spotted, or dead leaves. Understanding the basic principles of ecological pest and disease management is more important than trying to remember all of the specific diseases and pests and the actions that can be taken to manage them. This is because many problems do not require specific identification and the same pest or pathogen acts differently and requires a different response depending on the crop, location, weather, time of year, gardener's time and expectations, and other factors. The goal of management is not having a perfect harvest from every crop in the garden, but having the largest harvest for the smallest amount of work and resources invested.

Most insects and microorganisms either cause no damage, or are beneficial to crop production. In nature many pests and pathogens do not grow to large enough numbers to cause significant damage because they are kept in check by natural controls. In general, as their numbers increase, so does the rate at which they die from lack of food, predators, disease, the weather or other causes.[1] We emphasize managing the garden to provide enough water and nutrients for healthy crops, and encouraging natural controls on pests and pathogens so that they do not cause enough damage to lower yields appreciably. In some cases pests and pathogens can be kept away from the crop or garden by adjusting planting times, or by making physical barriers, for example, fences to keep goats out of the garden, or ant barriers on trunks and stems. Another precaution is preventing the introduction of soil nematodes when transplanting seedlings into the garden.

Sometimes pest or pathogen populations do get large enough to cause extensive damage to many crops in the garden, or to one or two very important crops, such as a stand of sweet corn that is almost ripe, a prize mango tree, or a hedge of pomegranates. This is when specific management techniques are used to eliminate specific pests and pathogens that are threatening the garden. This approach is a short-term one compared with ongoing garden management and exclusion of pests or pathogens.

Pest and pathogen populations are kept under control naturally or through garden management in one of the following ways:

- The crop plant itself (e.g., timing of its life cycle, thorns, unpleasant chemicals).
- The physical environment (e.g., temperatures, moisture, separation in space or time, barriers).
- Other organisms (e.g., birds, snakes, spiders, insects, nematodes, fungi, bacteria, and viruses).
- The application of chemicals (e.g., botanical or manufactured chemicals on plants and soil).

13.2.1 Pest and Disease Management by the Crop Plant

Crop plants have evolved along with pests and pathogens, and have genetically adapted to living with them. Healthy crops can usually resist the buildup of pests or pathogens, and can tolerate even heavy damage when it occurs. Most plants produce far more leaves and flowers than required so that partial defoliation by insects or attack of buds, blossoms, or young fruit may not decrease yield significantly.[2] In a mixed garden if one or two crops do succumb, others can grow to take their place.

As with adaptation to drought (section 5.5), we can think of adaptation to pests and pathogens in terms of escape or resistance. Plants escape damage through life cycles in which growth and reproduction occur when there is little threat from pests or pathogens.

Plants may resist damage by avoiding it. Chemicals in the leaves may slow down eating by pests or even kill the pests. Some plants have chemical odors that repel insects, while others lack the color or odor that attracts certain pests. Plants can make it physically difficult for animals to eat them. Thorns may repel rabbits; a thick cutical prevents caterpillars from chewing; and silica (sand) particles that occur naturally in the leaves of many grasses like maize, sorghum, and rice, wear down the mouth parts of insects that chew the leaves.

Resistant plants that are eaten or are diseased may also tolerate attacks of pests or pathogens and suffer little loss in yield. Root or tuber crops, for example, may have many leaves removed or have extensive disease in their aboveground parts with little loss in harvest.[3] Aphid attacks on beans at the time fruit is setting may increase yields by diverting energy away from vegetative growth to seed formation.[4] Increased yield in cereals frequently results from tillering after the initial shoot of the seedling is eaten by a shootfly, stem borer, or cutworm.[5] When chewing removes the main growing tip in young plants of leaf crops like amaranth or basil, it can lead to bushier plants and increased leaf production, just as pruning would.

One of the most important ways of reducing pest and disease problems is to use varieties that produce good harvests without much care. We advocate using indigenous varieties for a number of reasons (section 14.2). Indigenous crops are adapted to the local soils and climate and therefore are likely to be more resistant to local pests and diseases than commercial varieties. Gardeners are familiar with their indigenous varieties: they know how and when to plant them, how they grow, what pest and disease problems those varieties have, and how they like to eat the harvest. All of this means less risk for the gardeners, an important consideration for low-resource households. Another advantage of using local seed and other planting material is that it eliminates worry about introducing exotic pests or pathogens that could be harmful to local varieties. Some of the varieties may only be known and grown in the immediate area, and talking to local gardeners is the best way to learn about them.

Obviously, knowing his crops and garden environment is a big advantage for the gardener. For example, some pests and diseases may be a problem in just one area of the garden and for a limited amount of time. Many crops can be planted and harvested during a range of times, and adjusting these times can help to escape pests and pathogens. Often if a crop fails miserably because the soil or air temperature was too hot or too cold, or because a pest or disease was especially abundant, it will prosper if planted at a different time or in a different location.

13.2.2 Environmental and Mechanical Management of Pests and Diseases

Managing soil, water, temperatures, and other aspects of the garden environment is critical for keeping pest and pathogen populations low. Creating a garden environment through soil and water management that encourages healthy plant growth is the best pest and disease control strategy, because healthy plants are better able to withstand attack. Anything in the plant's environment that decreases vigor makes it more susceptible to pests and pathogens. The plants most frequently attacked are very young or old; those that have too much or too little water or shade; and those growing in soil that is too low in organic matter or nutrients, or too high or too low in pH. However, some conditions that would otherwise be good for the plant may also encourage pests and disease. For example, mulch—which is so beneficial under hot, dry conditions—may harbor insects, fungi, and bacteria that can harm plants, especially young ones. This helps illustrate why absolute rules about management are not useful, because every situation is unique, requiring observation and experience to adjust management strategies.

Another way to manage the garden environment is with mechanical control, that is, physically protecting crops with *trap crops* (plants that pests will feed or live on instead of the garden crops) and barriers, and by picking off, washing off, or crushing pests. Compared with fields, gardens are smaller, intensively managed production systems. Because of this some management methods such as mechanical control, which are not always appropriate for fields, can be very effective in gardens. Picking pests such as beetles, caterpillars, and grasshoppers off garden crops by hand (Figure 13.1), or crushing them as with aphids, is practical and very effective especially if done before the populations get very large. Success depends on understanding the life cycle and habits of the pest. For example, watching for aphids and destroying them as soon as they are seen prevents growth in aphid population to

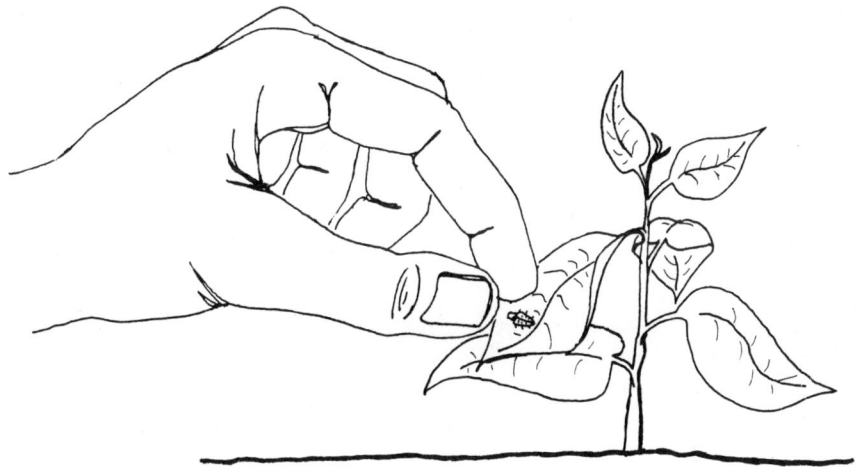

Figure 13.1 Hand-Picking is a Good Way to Control Garden Pests

the point that they start being born with wings. Early in an infestation most aphids are wingless, but as stress increases due to rising population density or disappearance of food sources, more and more of them are winged (Figure 13.2). Avoiding this winged stage keeps the aphids from spreading to new areas where their population would grow rapidly once again.

Small insects like mites, aphids, and thrips, can be washed from leaves with plain water or crushed with fingers. Removing diseased and insect-infested plants and burying them in the ground or compost pile also helps. Other means of mechanical control include making a barrier that prevents the pest from getting to the plant, such as collars against cutworms (Figure 13.9 in section 13.3.1). In northern Thailand, grass, raffia, paper, or plastic bags are placed over fruit like guava, citrus, pomegranate, and jackfruit to protect them from fruit flies.[6] The lasora tree (*Cordia myxa*), is a drought-resistant tree grown for its fruit and used as a living garden fence in arid northwestern India. Researchers at the Central Arid Zones Research Institute in Jodhpur, India, told us that the large, tough lasora leaves are used by local farmers to cover ripening pomegranates, protecting the fruit from pests.

Sometimes wild plants, weeds, or other crops can be secondary hosts for pests or disease (section 8.6.2). Solanaceous weeds are important alternative hosts for tomato, eggplant, and potato pests, and all important sweet potato pests feed on the widespread morning glory.[7] Careful observation and experimentation is the only way to find out if a weed or other plant is attracting pests that cause damage to crops, or is trapping those pests, thus keeping them from harming the crops. Once pests have settled on trap plants these plants can be destroyed by burning or burying them before the pests move to the crop plants.

If pest and disease problems increase with continued planting of the same crop or closely related crops in the same location, planting other crops in rotation, especially those from other botanical families, often helps. A clean fallow, one with no weeds, also helps to break the life cycles of pests and diseases like soil nematodes and some soil-borne pathogens by separating them in time from their host plants. However, many pests easily move such long distances that fallowing is not effective.[8]

13.2.3 Pest and Disease Management Using Other Organisms

Biological control is the action of parasites, predators, and pathogens in keeping another organism's population density lower than it would otherwise be.[9] Gardeners use biological control when they encourage organisms that help decrease pest or pathogen numbers on their crops. For example, encouraging predators such as spiders, toads, guinea fowl, geese, or chickens in the garden can help control insect pests (Figure 13.3).

Dryland gardeners have for a long time understood the ecology of the garden and have distinguished between similar animals whose habits are very different. Medieval date growers on the Arabian Peninsula were among the first practitioners of biological control. They transported predatory ants seasonally from nearby mountains to their oases to control another species of ants that attacked the date palms.[10]

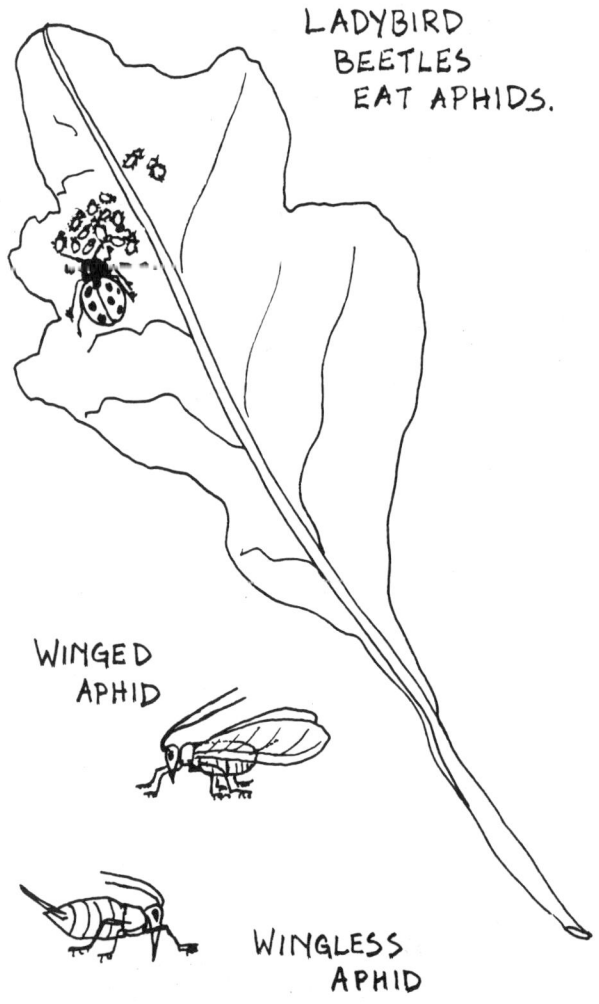

Figure 13.2 Aphids

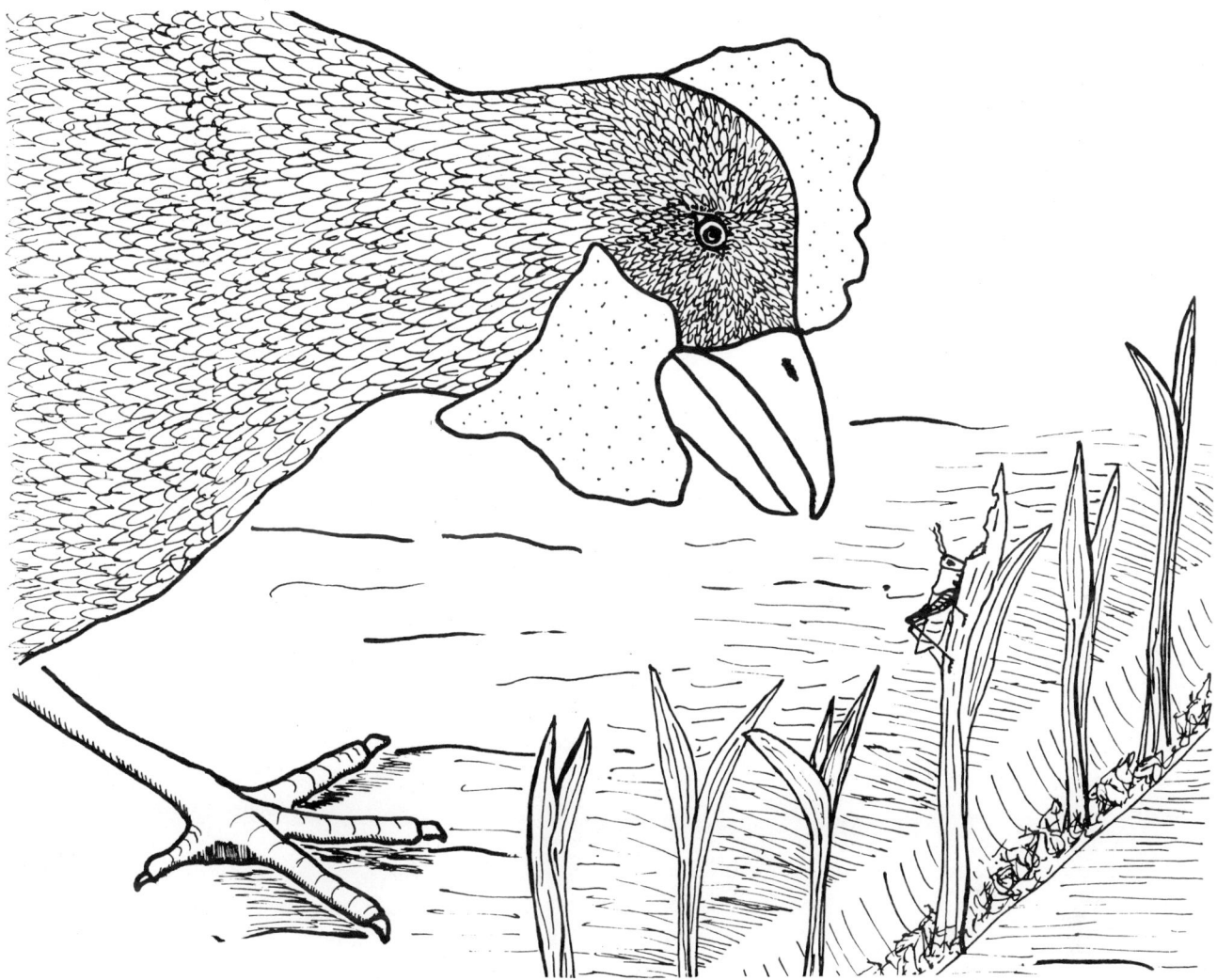

Figure 13.3 Chickens and Other Birds can Help Control Some Garden Pests

Most cases of successful modern biological control of insect pests involve bringing predators or parasites from the pest's area of origin.[11] Most insects now bred commercially for biological control are nonspecific pests or pathogens; that is, they attack more than one prey or host. Examples are lacewings, ladybird beetles, and Trichogramma wasps. Importation or large-scale breeding of predators of common garden pests are not practical or affordable for most dryland gardeners, who practice biological control through garden management for healthy plants and by maintaining a diverse garden environment.

Biological control is defined not only by the use of organisms, but also by the ways in which they are used to cope with pest and pathogen problems. Biological controls are meant to control pest or pathogen populations, keeping them at acceptable levels but not eliminating them.

There is concern among some researchers who believe that biological controls are currently being developed and used in the same way as chemical pesticides, that is, to completely kill off the target pest or pathogen, not to control the population.[12] Used in this way the "biological control" provides a strong selective pressure for genetic resistance among the targeted pest or pathogen, and eventually the controlling organism is no longer effective and other methods must quickly be found. This appears to be the case with *Bacillus thuringiensis* (Box 13.1).

13.2.4 Pest and Disease Management with Chemicals

Chemicals are nonliving substances, some of which are essential components of living cells. Certain quantities of some chemicals can repel, make sick, or kill pests and pathogens. As mentioned in section 13.2, the chemical composition of some crop plants deters pests

> **Box 13.1**
> **Biological Control Using Insect Pathogens**
>
> Since pathogens of plants and insects will not harm people or other large animals directly, some can be used for pest control. Probably the most widely known pathogen for control of pests is *Bacillus thuringiensis* (Bt), a bacterium that produces a toxin that kills certain kinds of caterpillars and fly larvae, but is not poisonous to other animals.[13] It is not an important control in nature, but it is cultured commercially in large-scale fermentation vats and sprayed onto crops, just as are synthetic insecticides, and so also kills many beneficial caterpillars (larvae of Lepidoptera). The Bt gene that manufactures the toxin that kills the caterpillars is also being genetically transferred into crop plants, with the likely result that with such wide use insect resistance to it will increase, thus decreasing its effectiveness.[14]
>
> Although little information is available on this subject, gardeners can experiment with making their own pest spray. Some researchers claim that it is possible to collect diseased pests and use them to make a spray that will spread the disease pathogen to other, similar pests.[15] When applied to plants or areas in the garden where that pest is a problem the spray infects and eventually kills those pests.
>
> Caterpillars with viral infections commonly hang by their back "legs" from stems or branches.[16] This behavior can be used to identify specimens to use in making the spray. In one example,[17] diseased cabbage looper caterpillars (*Trichopulsia ni* Hubner) were identified this way. Eight to 10 of these caterpillars were ground up and diluted with water to make enough spray to treat 0.5 ha (1.2 a) of crops. Three to 4 days after spraying, the caterpillars on those crops died. There are claims of positive results using this method of biological control with several other caterpillars including fall armyworms (*Spodoptera litura*) and corn earworms (*Heliothis zea*). The one thing that is certain is that this is an area needing much more research by gardeners and others interested in the use of insect pathogens.

or pathogens from damaging those plants. *Botanical* chemicals occur naturally in plant products, and some of them, for example, neem and citrus oils, garlic juice, and chili powders, are used to protect other crops (Box 13.2). Some botanical chemicals like nicotine from tobacco are very poisonous to people and animals and should be used with great care.

Synthetic chemical pesticides are manufactured chemical compounds, such as dichlorodiphenyltrichloroethane (DDT), malathion, or captan usually specifically created for pest or pathogen control. Synthetic chemical compounds manufactured for other purposes, like kerosene or soap (Box 13.2) are also used. Naturally occurring chemicals such as sulfur and bordeaux (copper sulphate and lime) have been used as fungicides; copper compounds have also been used to control bacteria.

The discovery of DDT during World War II ush-

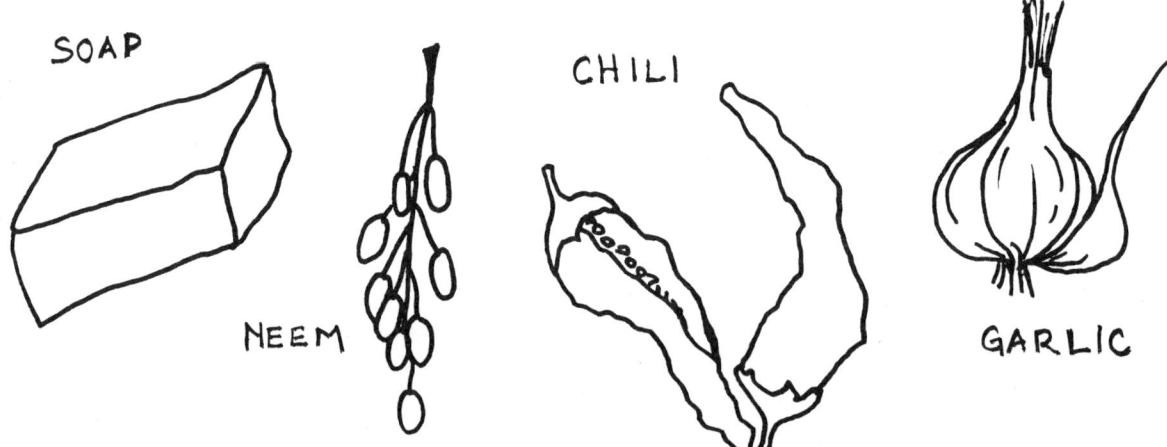

Figure 13.4 Ingredients for Safe Homemade Pesticides

> **Box 13.2**
> *Safe Homemade Pesticides*
>
> Some safe ingredients said to be useful for controlling garden pests are listed in this box (Figure 13.4). The "sprays" described can all be applied using bunches of grass or small twigs to splash and paint the mixtures on plants. While very little research has been done to determine if and how these ingredients work, many gardeners (including us) feel that they often do work. The best approach is to experiment, trying them first on small areas to determine if they have any positive or negative effects.
>
> *CHILI EXTRACT*[18] USE: Spray to repel insects or slow their feeding. RECIPE: Whole hot (spicy) chili peppers are ground, including the seeds which are a concentrated source of *capsaicin*, the fiery-hot active ingredient in chili sprays. The ground chilis are left to soak overnight in water. Soap can also be added to this mixture. The concentration of this spray is best determined by testing; if too weak it will be ineffective, if too strong it can burn leaves. Strength will depend on how spicy the chilis are. Care should be taken as this spray can burn the skin and eyes. REPORTED TO CONTROL: Aphids, caterpillars, beetles, and other insect pests.
>
> *NEEM SEED EXTRACT*[19] USE: Spray to repel insects or slow their feeding; kills pests when they eat it. RECIPE: Seeds of fruits fallen from the tree are cleaned, dried, and stored in a dry, ventilated place. When needed, seed hulls are removed, and seeds are finely ground and hung in a cloth sack in a container of water, using between 25-50 gm seed/liter of water (3-7 oz seed/gal of water). The ground seeds are soaked overnight in the water before using. This mixture should be made fresh for each use as it can lose its effectiveness over time and with exposure to sunlight. REPORTED TO CONTROL: Caterpillars, beetles, grasshoppers, and other garden pests.
>
> *SOAP*[20] USE: Spray to repel insects. RECIPE: 30 cm^3 (1 fl oz) of soap mixed into 5 liters (1.3 gal) of water. Some soaps contain harsh additives so it is best to test the mixture on a small area first to make sure the plants will not be harmed. REPORTED TO CONTROL: Piercing and sucking pests like aphids and thrips.
>
> *GARLIC EXTRACT*[21] USE: Spray to repel insects. RECIPE: 100 gm (3.5 oz) crushed or grated garlic soaked 24 hours in 2 teaspoons of oil, then 0.5 liters (17 fl oz) water and 10 gm (0.4 oz) of soap are added. When ready to use add about 20 liters (5 gal) of water to this mixture. REPORTED TO CONTROL: Aphids and some caterpillars and beetles.
>
> *BAIT FOR FRUIT FLY TRAPS*[22] We list here three recipes for fruit fly bait (Figure 13.11 in section 13.3.1) [CAUTION: Human and animal urine should be used with care because they can spread disease, see Box 11.2 in section 11.3]:
> - 1 liter (34 fl oz) water, 150 ml (5 fl oz) urine, 100 gm (3.5 oz) sugar, 1.5 small spoons (teaspoons) vanilla, 10 gm (0.4 oz) pyrethrum (Box 13.3 describes pyrethrum and the synthetic pesticides that have been modeled on it).
> - 1 small spoon pyrethrum, 300 ml (10 fl oz) honey, 1 small spoon vanilla, 300 ml (10 fl oz) fruit pulp (melon, etc.), 10 liters (2.6 gal) water.
> - Peel or pulp of 1 orange, 100 ml (3 fl oz) urine, 0.5 liters (17 fl oz) water. Mix and let stand overnight then dilute with 15 liters (4 gal) water before use.

ered in the modern era of synthetic organic pesticides and their use has been growing ever since. Presently more than 4.5 million metric tons of pesticides are used in world agriculture each year, 30% of this use is in the Third World.[23]

Integrated pest management (IPM) evolved in response to increasing awareness of the environmental and health costs of synthetic pesticide use. Originally IPM was meant to reduce the use of synthetic pesticides by integrating them with other control methods. In this way pesticide use could be limited to specific, critical times such as just prior to pest reproduction, instead of being continually applied as the sole form of control. However, IPM is frequently used by the pesticide industry to preserve or even intensify the commercialization of pest control.[24]

Although naturally occurring and botanical pesticides can be harmful, synthetic ones are especially dangerous because these pesticides are available in large amounts. Because they are created specifically to kill pests or diseases, synthetic pesticides contain high concentrations of poisons, which are toxic to many living things like beneficial insects and other animals, including people.

Synthetic pesticides usually kill both pest and predator, leading to a more severe problem. For example, some early olive orchards in southern California, USA, suffering from black scale insects were treated with a kerosene spray.[25] Comparisons between sprayed and unsprayed orchards found that sprayed orchards continued to have high populations of black scale, and no beneficial predators, while unsprayed ones had less black scale and much greater numbers of the black scale predatory beetle, *Rhizobius ventralis*. When infestations of scale insects in orchards are fought with DDT and dieldrin, chalcid and braconid wasps that parasitize scales are destroyed, and population outbreaks of scale result.[26] Another problem with synthetic chemical controls is that some can take a very long time to break down, remaining toxic for many years. As a result they spread through the food chain, increasing their damage to living organisms and the environment.

The persistance of toxic synthetic chemicals in the environment and throughout the food chain is one reason why we do not recommend the use of synthetic chemical controls in the garden. In addition, these pesticides are often expensive and hard to obtain for poor households; they undermine self-reliance, and lead to increased pest problems. However, because synthetic pesticides are being heavily promoted in the Third World, often in small-scale garden projects, it is important to have some knowledge of them (Box 13.3).

Pesticides are promoted worldwide by companies through advertising whose purpose is to increase sales. These advertisements contain information that is even less objective than the information provided by labels or extension agents.[31] Yet, even pesticide labels often fail to give information necessary for safety.

Unfortunately, social responsibility for safety often succumbs to the profit motive, and pesticides are presented as the answer to all problems. At the national level it is usually the poor who pay the highest costs of pesticide use and receive the smallest benefits.[32] The inability to read and understand labels, lack of awareness of the dangers, no access to medical care, and poor safety equipment make poor households and agricultural workers especially vulnerable to the dangers of synthetic pesticides. In fact, inadequate protection from those chemicals is a serious problem even for farm workers in rich countries like the United States.

Box 13.3
Synthetic Pesticides[27]

Following is a list of pesticides by broad chemical categories with some examples of each (some of their trade names are given in brackets). There are many more synthetic pesticides than those listed here, and new ones are being constantly introduced to the marketplace and the environment.

Contact poisons are absorbed through the body surface, stomach poisons have to be ingested by the insect or other animal, and systemic poisons are first absorbed into the plant tissues before being eaten by the pest or affecting the pathogen.

ORGANOCHLORINES (e.g., aldrin [Aldrex], BHC Lindane, Chlordane, DDT, dieldrin [Dieldrex], endrin [Hexadrin], heptahlor [Drinox], endosulfan [Cyclodan], toxaphene). Organochlorines are used primarily as insecticides, but some are also used as *herbicides* (weed killers) and *fungicides* (fungus killers). They are contact and stomach poisons affecting a wide range of animals. Organochlorines are soluble in water and fat, and persist in the environment. They are therefore effective pesticides because they can be dissolved in water to make a spray which is then absorbed through insects' exoskeletons, and remains active for a long time. These same characteristics make organochlorines extremely dangerous since they accumulate in the body fat of vertebrates (birds, amphibians, reptiles, and mammals, including humans). For example, even years after exposure, a lactating woman can have organochlorines in her breast milk, which are then passed on to the nursing child.[28] Immediate effects of poisoning in humans include skin rash, dizziness, and excitability, while long-term dangers include cancer and damage to the liver, brain, kidney, and reproductive system.

ORGANOPHOSPHATES (e.g., azodrin, diazinon [Dianon] [Neacide], malathion, methyparathion, para-

---

The danger of pesticide poisoning is probably worst in the Third World. For example, the government of Kwara State in West Central Nigeria has for the last 10 years mounted a strenuous campaign to increase synthetic pesticide use among farmers, and it is claimed

Box 13.3, continued

thion [Fusferno], tepp [Vapotone]). The first organophosphate compounds were developed during the Second World War by a German team researching nerve gases, and are some of the most toxic substances known to mammals and birds.[29] Used as contact and systemic insecticides and *acaricides* (materials toxic to mites), they have replaced organochlorines for many uses because they break down relatively quickly. Organophosphates interfere with nerve transmission, and early symptoms of poisoning in humans include headache, dizziness, and flu-like symptoms. Residues of these chemicals may be more dangerous than the original compound especially when temperatures are high, when there has been no recent rainfall, and when there is a large amount of dust in the air and on plants.[30]

N-METHYL CARBAMATES (e.g., aldicarb [Temik], carbaryl [Sevin], carbofuran [Furadan], methiocarb, propoxur [Baygon]). N-methyl carbamates are used as insecticides and *nematicides* (nematode poisons). Symptoms of poisoning are similar to those of organophosphates, but these pesticides are not as dangerous, except for aldicarb, which is extremely toxic.

DITHIOCARBAMATES (e.g., thiram, ziram, ferbam, vapam, maneb, zineb). Dithiocarbamates are used as fungicides. Exposure to large quantities of these chemicals causes skin rash and respiratory problems in humans, and birth defects and cancer in other mammals.

NITROPHENOLS (or substituted phenols, e.g., dinoseb, DNOC [Sinox], pentachlorophenol [Dowicide]). Nitrophenols are used primarily as herbicides and fungicides, but also as insecticides and acaricides. These chemicals are easily absorbed through the skin. Large exposures cause headache and overheating, and so they are very dangerous in hot weather.

DIPYRIDYLS (e.g., diquat, morfamquat, paraquat). Dipyridyls are used as herbicides. Contact with skin or breathing spray can cause inflammation of eyes, nose, mouth, and throat; nosebleeds; and coughing.

CHLOROPHENOXYS (e.g., 2,4-D, 2,4-DB, 2,4,5-T, dichlorprop, erbon, falone, MCPA, MCPB, MCPP, silvex). When fresh these herbicides are easily absorbed into the skin, although when dry they break down very quickly in the environment. Symptoms of poisoning in people are skin rash and eye, nose, and throat irritation. TCDD is a contaminant of 2,4,5-T which appears to cause liver and kidney damage, cancer, and birth defects in animals. While chlorphenoxys are quickly eliminated from the body, TCDD is not. Agent Orange, a combination of 2,4-D and 2,4,5-T, was widely used during the Vietnam War and led to many lawsuits against the US government by members of the US military who claim to suffer serious long-term health problems due to exposure to it.

PTHALAMIDES (e.g., captan, captafol, folpet). These fungicides are very irritating to the human skin and respiratory track, and cause birth defects in animals.

PYRETHROIDS (e.g., bioallethrin [D-Trans], cypermethrin [Cymbush, Ripcord], permethrin [Ambush, Kalfil], pyrethrins [Pyrethrum], resmethrin [Chryson, Synthrin]). Pyrethrins were originally extracted from pyrethrum flowers (*Chrysanthemum cinerariaefolium*) and are powerful contact insecticides that rapidly break down in sunlight. A number of synthetic pyrethrins have been developed that have higher toxicity and last longer. Prolonged skin contact can result in a rash; headaches and sickness can be caused by inhaling the dust or spray.

---

that many traditional pest control methods have been displaced by the use of synthetic chemicals.[33] The chemicals are subsidized by the government and sold at only one-third of their cost. While the state had formerly provided overalls, boots, gloves, and soap to those using dieldrin, DDT, and other very dangerous pesticides, this has been eliminated for financial reasons, and there has been an increase in serious health problems. We give some emergency first aid suggestions for pesticide poisoning in Box 13.4.

Box 13.4
Emergency First Aid for Pesticide Poisoning

Knowing what the pesticide is and having its chemical formula are very important for medical workers to decide the best treatment. Often however, this information is not provided by the manufacturer, distributor, or salesperson. In addition, trained medical workers are not always available. David Werner gives the following emergency treatment for pesticide poisoning in his book *Where There is No Doctor*.[34]

First remove all clothing that might be contaminated. Wash the body with soap and water, and if the eyes are affected, rinse them for 5 to 10 minutes with plenty of clean water. If the person swallowed pesticides try to make him vomit. Then get him to drink a lot of either flour mixed with water, or beaten eggs, or milk. Continue to alternate this with trying to induce vomiting. Seek medical care if at all possible.

13.3 Examples of Pest and Disease Management

The following are examples of the management of insect, nematode, and large animal pests, and crop diseases. These examples are intended to help readers understand the ecological approach to pest and disease management and inspire them to develop methods appropriate for their specific needs. This understanding will also help them assess the potential of introduced pest and disease management methods.

13.3.1 Insects

Arthropods are small animals with segmented bodies and hard outer skins (*exoskeletons*). An *insect* is an arthropod with a three-part body and six legs (Figure 13.5). Insects are the most abundant type of arthropods with several hundred thousand different species, most of which are neutral or beneficial for gardens and people. For example, insects are extremely important in pollination, and due to pesticides that have inadvertently killed many pollinators, numerous crops are

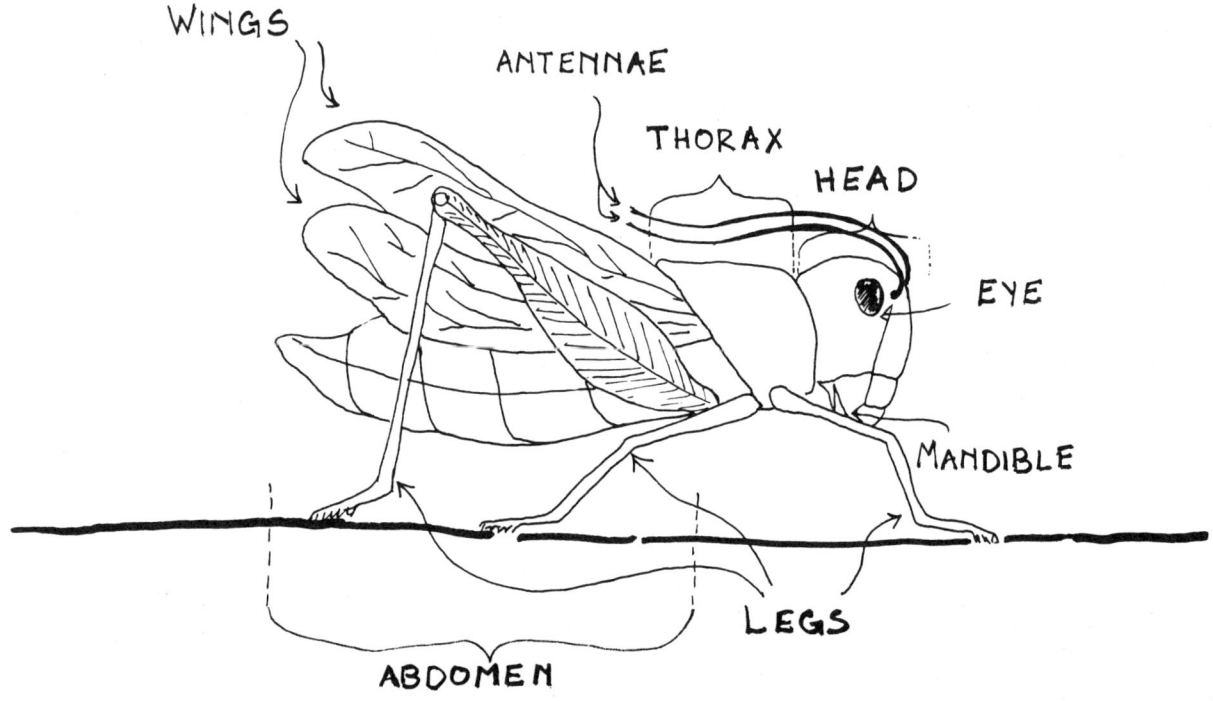

Figure 13.5 Insect Anatomy

Figure 13.6 Ants can Quickly Destroy Small Plants

now suffering from lack of pollination.[35] In temperate regions honeybees (*Apis mellifera*) are probably the most important pollinators along with bumblebees (*Bombus* spp). Leaf-cutting bees (*Megachile* spp.), mason bees (*Osmia* spp), honeybees and flies are the main pollinators in the tropics. Other minor pollinators include ants, beetles, thrips, and moths. The benefit for the gardener of having pollinators and pest predators is a strong argument against using synthetic chemicals that indiscriminately kill all insects.

Some noninsect arthropods that can be pests in the garden include sowbugs and mites; spiders, centipedes, and others are beneficial. Most arthropods in the garden will have no noticeable effect on production, and many are just visitors passing through.

CHEWING INSECTS Insects can cause damage in the garden by chewing leaves, stems, flowers, fruit, and roots, but as we mentioned in section 13.2.1, in some cases light insect "damage" to a plant may actually increase production. Ants are a common garden insect worldwide, and can cut down young seedlings, damage tree bark, or cut and carry away sections of leaves from mature plants (Figure 13.6). Gardeners may need to protect seeds and young seedlings from ants until they are big enough to survive an attack.

Ants are difficult to control because their nests can be several feet underground. Citrus oil is toxic to ants and fresh, mashed citrus peels may help repel them, as will garlic. Starting seeds in containers or easily protected nursery beds is a good way to avoid losing seedlings to ants (section 8.2). The gardener can also try to slow the ants down, discouraging them from doing any major damage while the plants are young and most vulnerable. Pouring boiling water down their holes is one way to do this. We have seen small-scale farmers in southern Mexico placing piles of Bermuda grass over ant holes to distract the ants from the crops. Once plants are mature, most ants are a much less serious problem. In addition, some ants help pollinate crops like squash.

Grasshoppers and the closely related locusts will feed on just about any plant, but prefer young shoots. Larger grasshoppers do the most damage and if there are not a lot of them, can be controlled by handpicking, especially in the early morning when cooler temperatures slow them down. Sprays made with hot chili peppers and soap (Box 13.2) may help repel them when there are large numbers.

In southern and central Nigeria variegated grasshoppers (*Zonocerus variegatus*) cause considerable damage to dry-season garden and field crops such as banana, citrus, cowpea, maize, okra, yam, and other vegetables.[36] Based on local farmers' careful observation of this grasshopper's life cycle (Figure 13.7) an experiment was conducted using a simple, cooperative control technique. In two 200,000 m² (49 a) test sites farmers dug up the egg pods on their land, killing them by exposing them to the sun. This control measure resulted in an 80% drop in variegated grasshopper populations on those sites. The key to this technique is cooperation because individuals acting alone cannot protect their gardens and fields from a large population of grasshoppers, which can move quickly and easily.

Caterpillars are the *larvae*, or immature form, of moths and butterflies, and many of them chew leaves, stems, and fruit (Figure 13.8). Some are general feeders while others are limited to a single crop. Many feed in the open on leaves and can be easily spotted and picked off, but others are well camouflaged. Chewing marks on the borders of leaves, small, often black particles of feces on the ground or on leaves below the damaged area are signs of caterpillars. Some caterpillars roll up leaves in webs and are easy to find and dispose of.

Like weeds, some insect pests including grasshoppers, termites, ants, and certain kinds of caterpillars and beetle larvae (*grubs*), can be food for humans or domestic animals. For example, each July in savanna West Africa, a certain caterpillar invades the shea butter tree and eats most of the leaves.[37] However, the

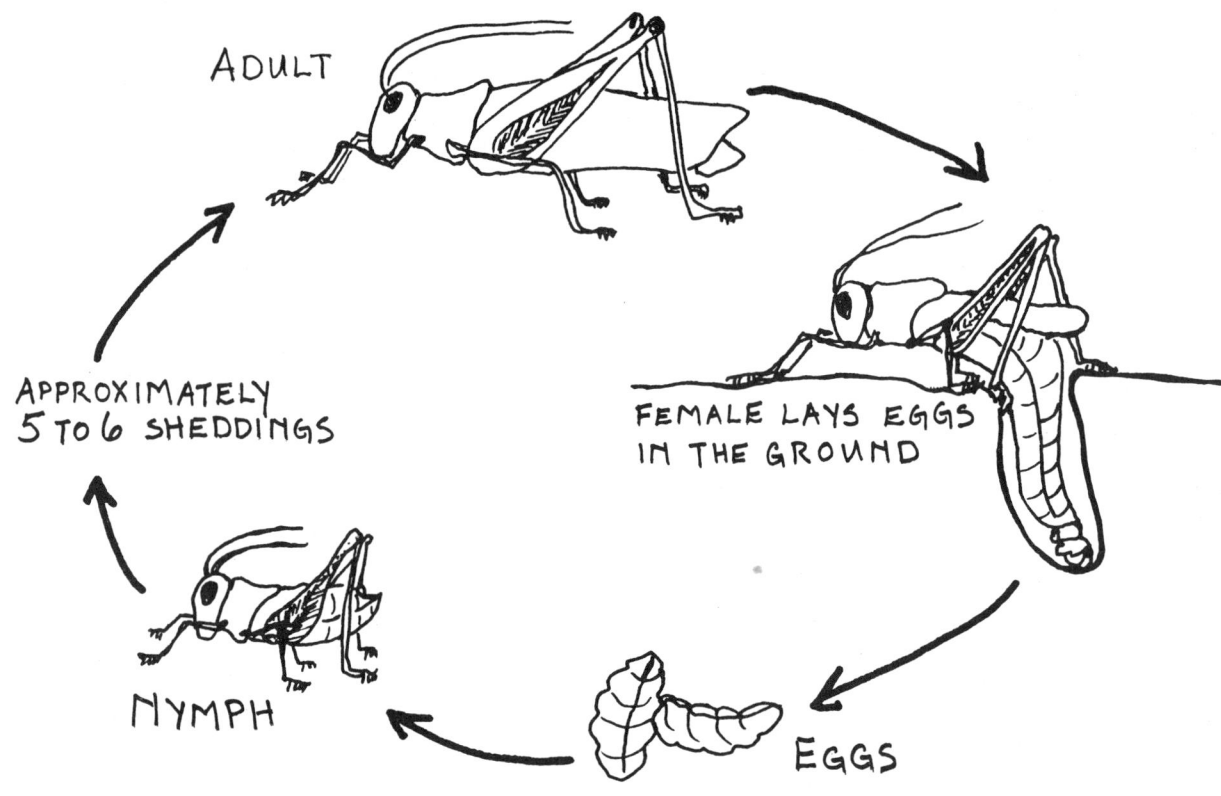

Figure 13.7 Life Cycle of Grasshoppers

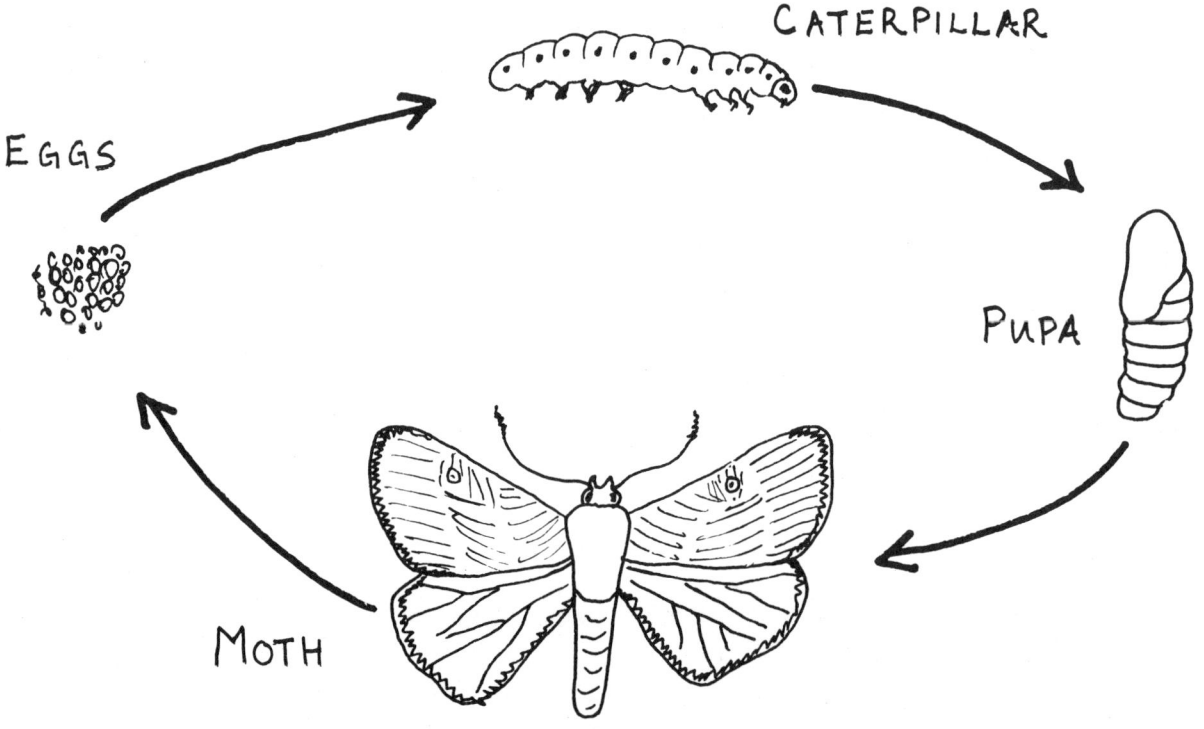

Figure 13.8 Life Cycle of Moths

caterpillars are eaten dried or roasted, or used directly in sauces, and they provide a significant source of income for those who harvest and sell them.

Cutworms are the caterpillars of various moths. These worms eat many crops and different species are common pests in drylands all over the world. In warmer areas there can be as many as four generations per year. The adults are up to 5 cm (2 in) long, and a dull pinkish grey color, sometimes with markings. Cutworms feed during the night, often attacking new seedlings and cutting them off at the ground. But they also feed on roots, tubers, stems, and leaves. For example, the common cutworm (*Agrotis segetum*), which occurs throughout dryland Asia, Africa, and Europe, cuts off seedlings of many crops at ground level and also feeds on roots, potatoes, carrots, and other crops. During the day cutworms rest curled up under debris on the soil surface or up to 10 cm (4 in) below it at the base of the stem, where they can be found and destroyed. Collars of cardbord or of metal cans which extend 5-13 cm (2-5 in) into the ground and about 10 cm (4 in) above the soil surface are an effective control (Figure 13.9). Thoroughly flooding the garden bed drowns the cutworms or at least forces them to the surface where they are easier to find and destroy.[38]

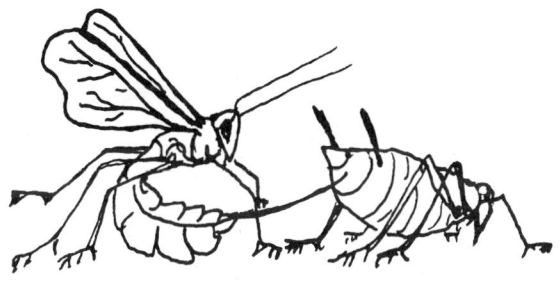

Figure 13.10 A Wasp Laying Eggs in an Aphid

SUCKING INSECTS Sucking insect pests have piercing mouth parts which they use to penetrate plant cells and suck out the contents. Sucking insects include whiteflies, aphids, scale insects, thrips, plant hoppers, and leaf or plant bugs. They often carry and spread bacteria and/or viruses that may cause disease. Some mites also damage plants this way. Many other sucking insects and predatory mites kill pests in the garden and are thus beneficial. Common signs of damage caused by these animals are leaves that are curled, twisted, or have dry spots, and abnormal-looking twigs. Thrips, for example, are common on onions; they eat the leaves from the inside which creates transparent, thin spots eventually causing the leaves to fall over.

Tapping the plant to dislodge mites and small insects onto a cloth or other clean surface makes them easier to see as they crawl for cover. If a hand lens or magnifying glass is available it can be used for identification. Larger ones can be controlled by hand-picking, smaller ones by crushing, washing them off the plants with water, or spraying with soap or other home mixtures.

Aphids are very common light green insects 1-2 mm (0.04-0.08 in) long (Figure 13.2 in section 13.2.2). Adults and nymphs suck plant juices from the undersides of leaves, causing wilting and distorted growth. Aphids also transmit viral diseases such as mosaic viruses in squash, melons, and lettuce.[39] Aphid "shells" are the remains of aphids that have been eaten by the larvae of parasitic wasps (Figure 13.10), an example of biological control. The wasps can be encouraged by growing plants like cilantro, fennel and dill, members of the umbellifer family, which the adult wasps feed on.

Figure 13.9 Collars as Protection from Cutworms

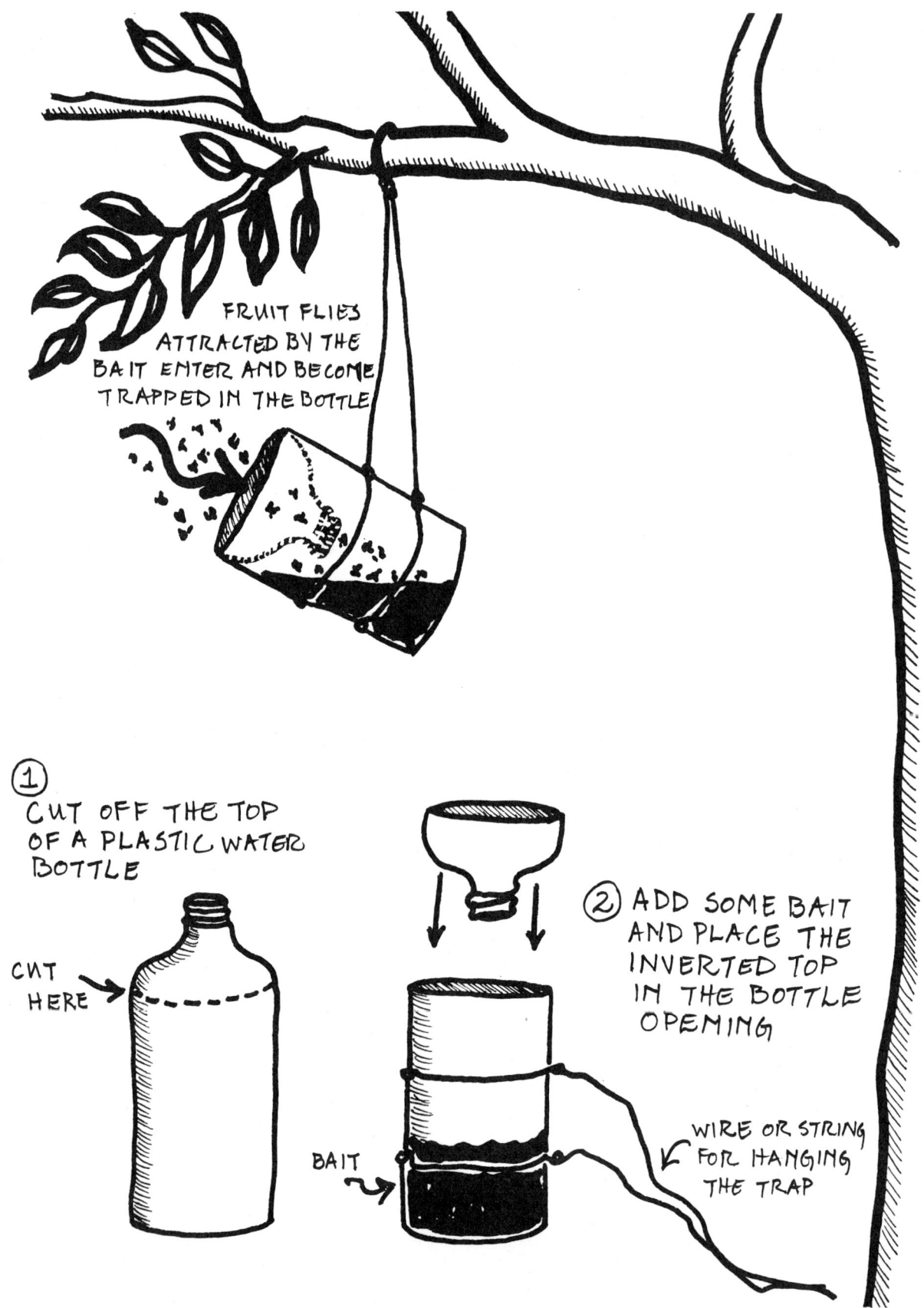

Figure 13.11 A Fruit Fly Trap Made from a Plastic Water Bottle

Aphids secrete a sugary, high-protein substance called *honeydew* which some kinds of ants harvest from them. This honeydew also encourages the growth of a black mold which can eventually cover the leaf surface, destroying its ability to photosynthesize.

Most aphid attacks can be controlled by hand crushing and washing, or removing heavily infested leaves and burying them in the compost pile or ground. Timing is important, so that the population is reduced before it has a chance to become large and spread (section 13.2.2). Aphids are attracted to yellow objects, and a bowl of soapy water with something yellow resting on the bottom is a trap which will attract and drown them. Soapy sprays or washes may be helpful if aphid populations are growing rapidly.

BORING INSECTS Boring insects make holes in stems, roots, and fruits and do most of their damage from the inside. Often a small hole and perhaps some *frass* (feeding debris and feces) will be the only evidence on the outside. Fly larvae (*maggots*), adult beetles and grubs, and caterpillars are among the most common boring insects. Spiking boring grubs that were preying on coffee bushes by pushing a bicycle spoke into their holes has been successful.[40]

There are many species of fruit flies (*Dacus* spp. and *Ceratitis* spp.) that together cause damage to a variety of host crops including mangoes, citrus, peaches, guavas, olives, cucurbits, and coffee.[41] Female fruit flies lay their eggs under the skin of ripening fruit, where the larvae feed on the fruit. Often the spoiling fruits drop from the tree and most species then *pupate* (undergo the transformation from larvae to more mature, winged forms) in the ground nearby. The adult flies emerge from the ground and continue this cycle.

One way of controlling fruit fly damage is covering the fruits to prevent egg-laying (section 13.2.2). Chickens, which eagerly eat fallen fruit and the larvae inside, control fruit flies biologically.[42] Another technique is to trap and eliminate the flies before they lay their eggs. According to some researchers, simple traps can be made using plastic water bottles (Figure 13.11) or jars containing bait and hung in the garden.[43] See Box 13.2 for ideas for bait recipes.

13.3.2 Nematodes

Nematodes are roundworms, of which there are several thousand species. Most live freely in the soil and feed on microscopic plants and animals. Some cause human disease (section 11.3), others parasitize insects and are an important natural control, while several hundred species feed on plants and cause disease. Some are widespread and serious pests of many annual and perennial garden crops in the drylands.

Most nematodes that attack plants are microscopic and live below ground where they feed on roots, although some feed on flowers, seeds, and leaves. They also spread viral diseases and interact with some disease-causing fungi and bacteria, making damage to the plants greater than the sum of each separately.[44] Identification of particular species requires microscopic examination by trained observers, but nematode problems can often be identified by gardeners, who can then take some simple actions to control them.

Some nematodes feed on plant roots from the outside, while others live inside of the roots, causing root galls, root lesions, excessive branching, injured root tips, or root rot. Aboveground symptoms are chlorosis (yellowing) of the whole leaf, wilt, failure to thrive, and poor yield. One of the most common and destructive nematodes, especially in gardens in warm or hot areas with mild winters, is the root knot nematode (*Meloidogyne* spp.). This nematode causes knots or galls on the root (Figure 13.12) which, unlike nodules caused by nitrogen-fixing bacteria (Box 9.7 in section 9.5.2), will not rub off without breaking the root apart. Also, unlike the nodules, a layer of soil often sticks closely to the knot galls. Sweet potatoes and other tubers have cracked skin when infested with nematodes.

In many areas where nematode pests are present, traditional gardening and farming practices, such as mulching, burning crop residues, rotation, fallowing, and making mounds or ridges for planting keep nematode numbers low.[45] Often in these farming systems some loss due to nematodes is acceptable, but this balance is upset by the introduction of new crops, varieties, pests, or management techniques. Nematode problems in the garden can be managed using a number of low-cost techniques that help keep nematodes out of uninfected areas, and when present, keep their numbers in both soil and plants low enough that a sufficient harvest is still produced.

EXCLUDING NEMATODES FROM CROPS AND GARDEN PLOTS Nematodes travel through the soil at a rate of only 1-3 m (3.3-10 ft) per year on their own, however, they can be transported much further if infested water, soil, or plants are moved from one area to another. They can also be transported by insects. This means preventing the movement of soil, transplants,

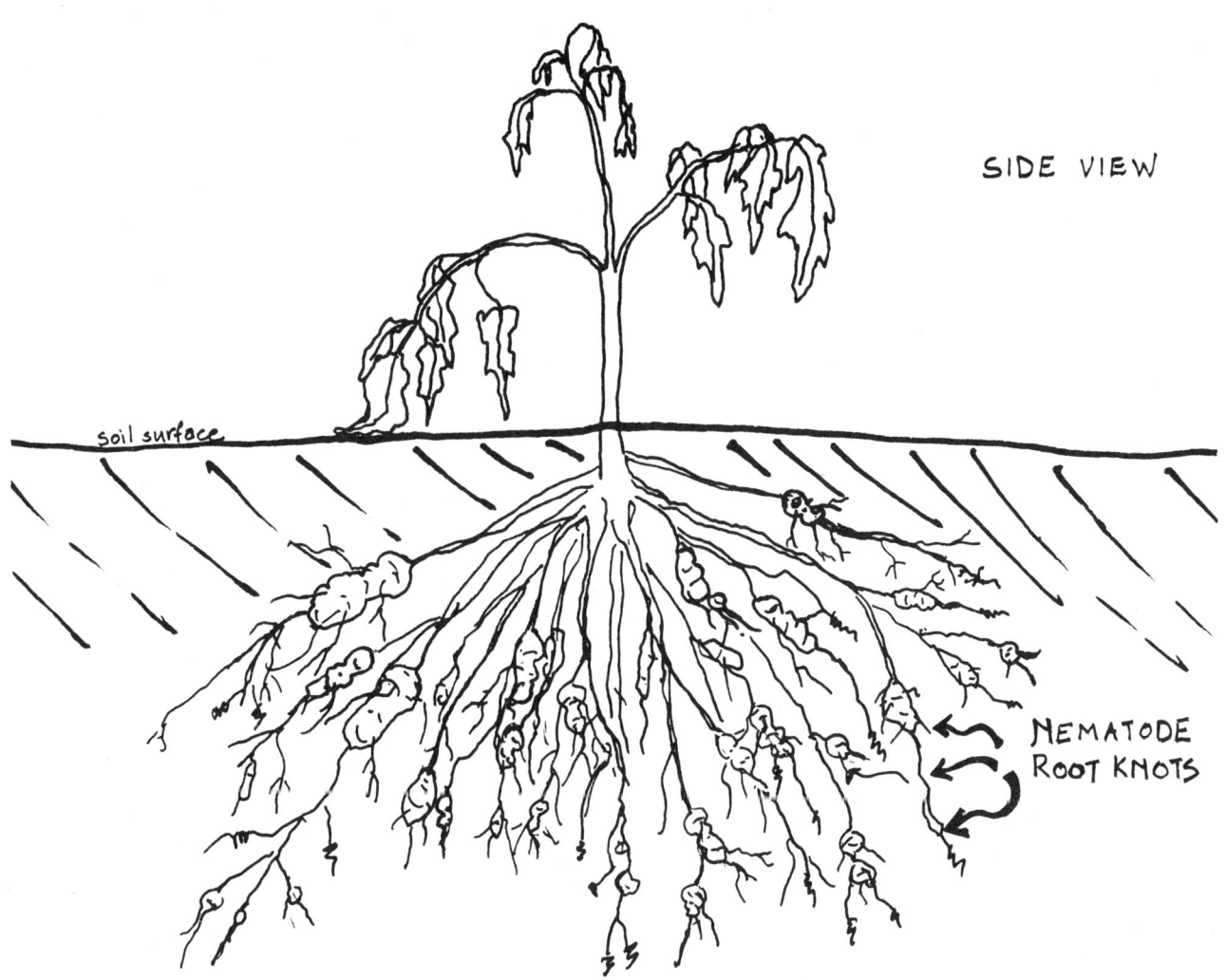

Figure 13.12 Tomato Plant with Root Knot Nematodes

and irrigation water from infested to noninfested areas, for example, by maintenance and design of irrigation systems, and by cleaning soil from tools and gardeners' hands and feet.

Planting material for vegetative propagation (Chapter 7) should also be from nematode-free plants. In some plants there is a choice of planting materials. For example, sweet potatoes can be propagated by stem cuttings which do not become infested with nematodes, rather than tuber cuttings which do. If planting materials are infected, or might be infected, and if no other materials are available, nematodes can be killed by placing the planting materials in hot water. Care should be taken that the water is not too hot, because it can kill the planting material. Some approximate guidelines for a few crops are: banana, citrus (bare root), and garlic cloves: 25 minutes at 55°C (131°F), and sweet potato: 65 minutes at 50°C (122°F).[46] A researcher in Nigeria found that putting infected yams in 50-55°C (122-131°F) water for 40 minutes was very effective for killing the nematodes.[47] This technique worked very well on yams that had been in storage for two to six months, however, treating freshly harvested yams with hot water caused them to rot.

KEEPING NEMATODE NUMBERS IN SOIL LOW
Once an area has become infected with nematodes, it will probably always be infected. Even highly toxic synthetic pesticides cannot completely eliminate nematodes, and most of these pesticides have now been

banned in some countries, such as the United States, because they are dangerous to human and environmental health.

Nematode pests must feed on plants to survive, so nematodes in infested beds or areas of the garden can be starved out by fallowing for a year or two. The fallowed area should be kept weed free, since weeds can also provide food for the nematodes, allowing them to survive and multiply. If nursery beds are kept in the same location for many years, populations of nematodes, especially root knot nematodes, can build up and infect transplants (section 8.2.1).

Nematodes are very sensitive to heat and drying. Therefore turning the soil regularly during the hot, sunny, dry season while the plot is being fallowed, helps reduce nematode populations even more. Wetting the soil occasionally will increase the effect, because it causes nematodes to change to more active states, making them more likely to be killed when the soil dries and heats again. Up to 80% of nematodes can be killed in less than 1 month using these combined methods.[48] Roots of plants can be exposed to the sun to kill internal nematodes like root knot.[49] Burning crop residues helps to heat the soil, and also kills varieties of nematodes that live in aboveground plant parts.

Some researchers advocate killing nematodes by first moistening the soil and then covering it with a thin, transparent polyethylene sheet. The sheet traps the heat and the water increases its movement. However, simply turning the soil may work as well, and does not require purchasing poly sheeting (made from nonrenewable resources) which litters the garden area as it is broken apart by sun and heat.

When there are crops growing in the garden, adding a lot of organic matter to the upper layers of the soil and mulching heavily helps to keep nematode populations low. This is probably because several varieties of fungi, insects, and other nematodes, whose reproduction is encouraged by organic matter, compete with or attack pest nematodes.

In Nigeria a researcher compared the effectiveness of several different treatments on yam production in soil infested with the yam nematode (*Scutellonema bradys*).[50] The treatments were as follows: 1 and 2) each a different synthetic chemical pesticide, 3) chemical fertilizer, 4) coating the yam setts with wood ash before planting, 5) mixing 1.5 kg (3.3 lb) of manure into each yam heap or planting site, and 6) a control plot, with nothing added. The results showed the highest yields from the manure treatment, which also suppressed the nematodes. The wood ash treatment showed the next highest yield, followed by the control. Yields from the plots treated with the chemical pesticides and fertilizer were lower than those from the control plot. Even though one of the pesticides did reduce nematode populations, under these three treatments yields suffered.

KEEPING NEMATODE POPULATIONS LOW IN PLANTS Because nematodes move so slowly, population densities can vary greatly over short distances. Therefore, replanting in another part of the garden may improve production, at least for that season. Nematodes multiply faster when the weather is warm, so nematode-sensitive crops grown in the cool season will probably do better than those grown in the warm season.

Observation in the garden will reveal that certain crops are more resistant to some nematodes than others. In addition, nematode resistance can often be found in local crop varieties.[51]

Some plants repel nematodes or are toxic to them. Crops like sesame, mustard, and asparagus are resistant.[52] Asparagus releases compounds from its roots that kill some nematodes in the soil. Neem trees (*Azadirachta indica*) also release substances from their roots that have been observed in India to reduce the numbers of at least six *genera* (the plural of genus) of nematodes parasitic on tomato, eggplant, cabbage, and cauliflower.[53] Azadirachtin, the compound found in the leaves and seeds of the neem tree, introduced from India into many dryland areas, is said to kill nematodes.[54] There is little information about what kinds of nematodes are affected and in what form the neem should be applied. Since the seed contains the highest concentrations of azadirachtin, gardeners can experiment with incorporating ground neem seeds into nematode-infested soils. Even if this treatment is not successful for reducing nematodes it will benefit the garden soil because neem seeds are high in nitrogen and other plant nutrients.[55] The castor bean (*Ricinus* spp.), a common plant in many drylands, appears to be effective in reducing populations of root knot and other nematodes when interplanted with tomatoes.[56]

Some varieties of domesticated marigolds may reduce populations of root knot nematodes. In India, where marigolds are traditionally planted among other crops, African marigolds (*Tagetes erecta*) intercropped with tomatoes or okra, were found to be effective for reducing six kinds of nematodes including the root knot nematode, and for increasing the quality and quantity of crop yields.[57] There appear to be several

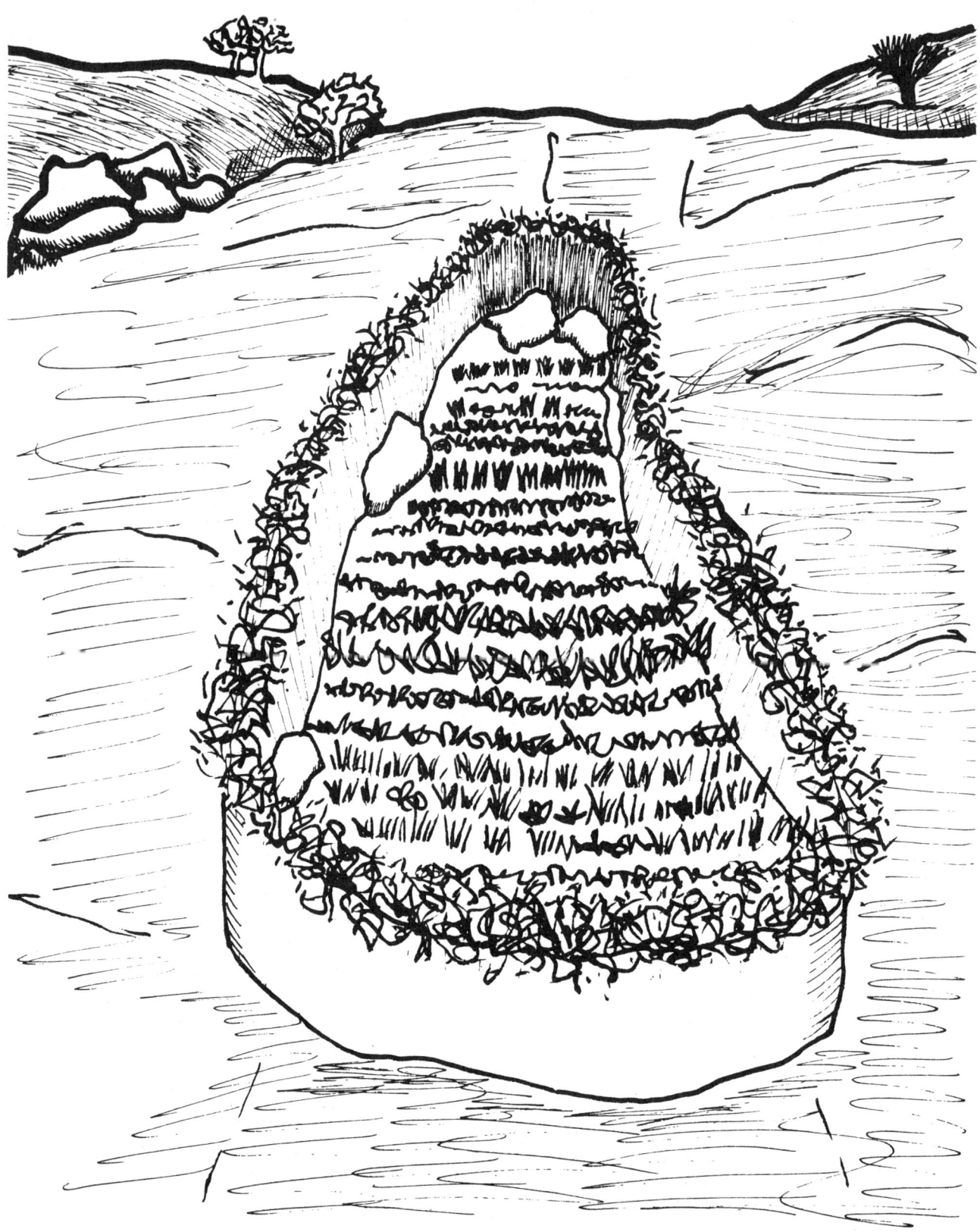

Figure 13.13 A Walled, Dry-Season Garden in Northeast Ghana

ways in which marigolds affect nematodes: substances exuded from marigold roots are nematocidal, the nematodes' life cycle is interrupted because nematode larvae have difficulty penetrating marigold roots, and nematodes are unable to develop once inside the roots. Mulches made from marigold leaves and stalks have also reduced nematode populations.

13.3.3 Large Animals as Pests

Because of their size, hungry rabbits, goats, pigs, or cattle, can quickly ruin a garden. Not only do they eat garden plants and fruits but some large animals can trample a whole garden, destroying everything they do not eat, including perennials. Some birds can also be bad garden pests, eating seeds, seedlings, leaves, and fruit.

REPELLENTS One way to prevent large animals from eating garden plants is to cover the plants with a repellent. In many places gardeners and farmers use mixtures containing feces as repellents, with good success. Hopi farmers in the United States told us that they mix dog feces and water to make a foul-smelling liquid which they spray or splash on tender bean plants to keep rabbits from eating them. Some farmers add other ingredients to their mixtures, like rabbit intestines or garlic.

In Ghana, farmers and gardeners use the feces of the pests to make a repellent for protecting young trees.[58] For example, if goats are the problem then goat feces are used. The fresh feces are mixed with enough water to make a soupy liquid, which is left standing to ferment for about 3 days. The gardeners then paint the mixture all over the tree. These repellent mixtures will wash off with rain or may eventually be blown off in a dry, windy area, and will need to be reapplied.

BARRIERS TO ANIMALS If domestic animals cannot be tethered or confined, or if there are wild animals that eat garden crops, then fences or other barriers will have to be built around the garden or gardens to protect them. In northern Ghana, for example, traditional law calls for animals to be controlled by their owners in the rainy season to protect the field crops, but the animals roam freely in the dry season, so that anyone with a dry-season garden must build a stout wall of sticks and earth topped with thorn branches (Figure 13.13). Even then, animals may break in if the garden is not guarded.

Fences can be made from a variety of different materials, depending on what works best for the gardener (Figure 13.14). In the Sonoran Desert of northern Mexico farmers have been planting living fences in the beds of rivers for at least 100 years, where they grow many beans, squash, and other vegetables.[59] The fences keep out cattle in addition to protecting the fields from erosion by the river, and capturing sediment to enrich

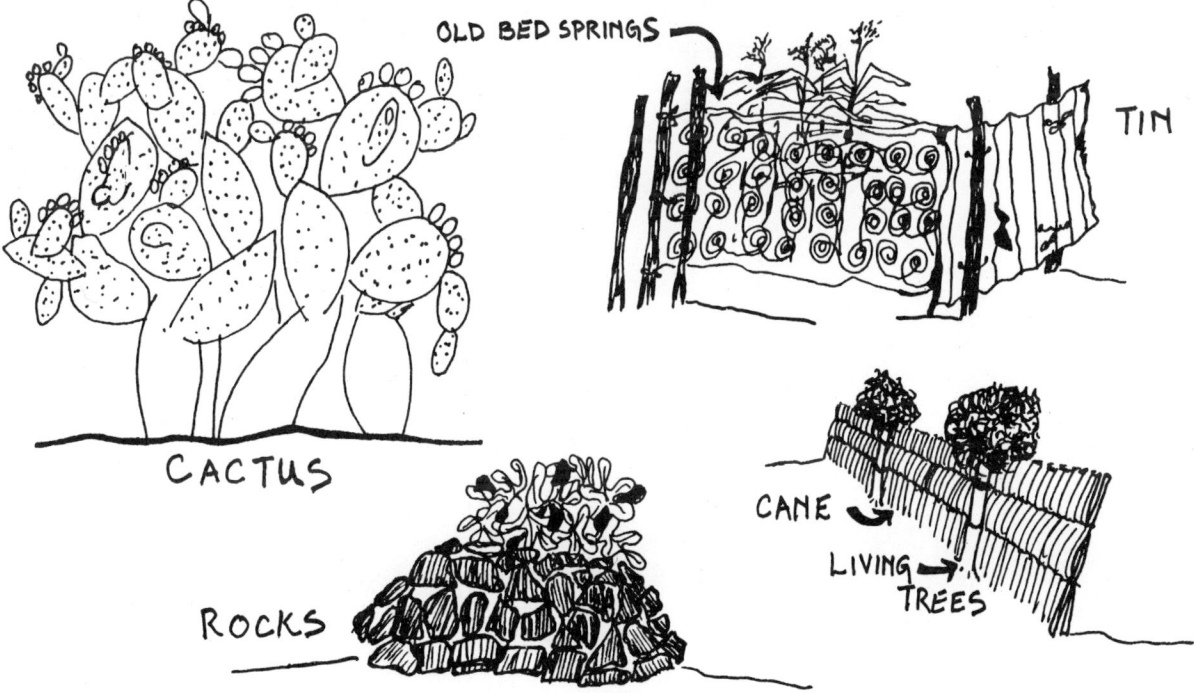

Figure 13.14 Gardeners Make Fences out of Many Different Locally Available Materials

Figure 13.15 A Young Mango Tree is Carefully Protected from Animals in Northern Ghana

and enlarge the fields. The fences are made mostly of willows (*Salix goudingii*) and some cottonwoods (*Populus fremontii*). Cuttings are taken from trees in existing fences, side shoots and leaves are pruned off, and the cuttings are planted in trenches. Cuttings from trees and shrubs are then interwoven to provide a solid barrier to keep out cattle. Farmers say that birds living in the trees eat insect pests. The trees also provide fuel and cuttings for new fences.

In Burkina Faso market gardeners have traditionally used millet stalk fences to protect their gardens from cattle. However, these fences must be repaired or replaced because of termite damage, and many millet stalks are required. These stalks have other uses as fuel or organic matter for the soil. A recent Forestry Department project is working with gardeners on experiments with living fences made from the local tree *Acacia nilotica*.[60] Once established, the living fences require little maintenance, they act as windbreaks, their roots hold soil and control erosion, and their leaf litter contributes organic matter to the garden soil.

Gardeners in Zimbabwe have used sisal (*Agave sisalana*) for fences.[61] Sisal is a fiber plant related to agave with sharp, pointed leaves. The gardeners find that the sisal plants are very effective against goats, although goats are still the main reason for the small size of gardens in some areas.

Barriers can be used around individual plants such as important trees in northern Ghana (Figure 13.15), or even fruits or seed heads. In southern Iran women make loosely woven baskets from fan palm fronds and use them to cover and protect ripening dates from birds.[62] The baskets are carefully designed so they will not impede air circulation, which would cause the fruit to rot. We have had success using cloth and paper bags to protect seed heads of sunflower, amaranth, and sorghum from birds in our garden.

FRIGHTENING AWAY Fences are useless for protection against large wild animals like elephants or hippopotami, such as along the Zambezi River in northern Zimbabwe and southern Zambia. Gardens must be guarded by household members at certain seasons of the year to scare off the intruders. When the garden is close to the house this is much easier.

Birds can also cause major damage to many garden crops such as sunflowers, figs, peaches, dates, maize, and young seedlings, especially in the early morning and late evening. Scaring birds away is easier when the garden is near the house and is the center of many other activities. Scaring them before they have discovered there is something good to eat, and have encouraged other birds to join in is best. Anything that will flap or make noises in the wind like strips of cloth, clusters of sticks (Figure 13.16), gourds, or the tape from spoiled tape cassettes can be fastened to poles or tree branches in the garden to help scare off birds.

Figure 13.16 Bird Scarers

In Sierra Leone some gardeners and farmers scare birds away from crops using tin cans with pebbles inside.[63] Tall, flexible poles are driven into the ground in the garden and rope is tightly strung in a web from one pole to another. Many short (20-30 cm, 8-12 in) pieces of string are hung from this web, and old tin cans with a few pebbles in them are tied onto their ends. A long cord is tied to the web so that someone can jerk it when birds are in the garden, shaking the poles and cans and making a lot of noise that scares the birds away.

13.3.4 Diseases

In plants *disease* is abnormal growth or functioning that harms the plant. Diseases can be infectious or noninfectious. Examples of noninfectious diseases are nutritional deficiencies, sunburn, and severe drought stress. This section is about infectious plant diseases, those caused by pathogens. While some crop diseases are fairly easy to identify, there are many that can only be identified by *plant pathologists* (specialists in the study of plant diseases) using specialized equipment and processes, and even they are not always successful. In the tables at the end of this chapter we list a few dryland garden crop diseases that are relatively easy to identify and for which there are specific responses. However, in most cases precise identification is not necessary because it will not help gardeners respond to the disease. Instead, we provide basic guidelines for coping with diseases identified by symptoms.

Fungal, bacterial, and viral pathogens causing disease in plants are different from those causing disease in animals and people and therefore most diseased plants and food harvested from them will not harm people or animals. Important exceptions to this generalization are some fungi that grow on plants and produce toxins that are poisonous to animals. *Aspergillus flavus*, for example, infects grain and legume seeds while still in the field and can produce levels of the poison aflatoxin that cause severe illness in humans and other animals (section 2.10).[64]

Fungi (the singular is *fungus*) are nonphotosynthesizing organisms that obtain their food through the decomposition of organic matter or from living organisms. Fungi secrete a sticky substance that helps form water-holding aggregates in the soil. They are also important in the garden soil because they break down organic matter into forms that are easier for plants to use. Some also attack and devour nematodes, including harmful ones, while others help to control disease-causing fungi by competing with them. Mycorrhizae are an extremely important group of fungi that grow in and around the roots of most flowering plants including all garden crops and help them to take up nutrients (Box 9.5 in section 9.5).

Some fungi, however, are parasites that attack garden crops, causing diseases. In fact, most plant diseases are caused by fungal pathogens. In the dry season, most fungal pathogens occur below ground in the root system, but are expressed above ground as wilting or chlorosis. Fungal diseases can be spread by wind, water and by contact from insects and other vectors. Generally they are encouraged by warm, moist conditions which in dryland gardens occur near the soil surface and in the middle of lush garden beds.

Bacteria are simple, one-celled organisms. They are the most abundant organisms in the world, and play an important role in gardens. Bacteria are very small (0.001-0.003 mm, 0.00004-0.00012 in), about the size of clay particles. Most bacteria found in the garden are beneficial, helping decompose organic matter in the soil, and making nutrients more available to crops (section 9.6 and Box 9.7 in section 9.5.2). Some make their own food from mineral nutrients in the soil, often transforming them to forms that are more available to plants. There are some, however, that harm the growth and productivity of dryland garden crops. Some of these bacteria live only as parasites in plants, while others can also survive for a period in the soil, living off dead organic matter. In warm, moist environments bacteria can multiply extremely rapidly and are easily spread by insects, people, and water.

Fungal and bacterial diseases often appear first on lower stems, trunks, fruit, and older leaves near the soil surface. On leaves they show up as specks or spots that are water-soaked, dark green, or brown. Lesions may look like a target, having alternating dark and light concentric circles, or may have a furry, moldy appearance. The fruit may feel like a bag of water, look rotten and moldy, or have a scablike wound, depending on the disease.

A *virus* is a parasite that can only reproduce by invading and taking over cells of other organisms. In plants viruses are spread mainly by infected seeds, cuttings, grafts; and by sucking insects or people touching infected plants and then uninfected plants. Viruses cannot live without a host nor can they lie dormant in fallowed soil. Viruses harm their hosts by diverting the resources and processes of those cells into the production of more viruses. In humans viruses cause many different diseases like AIDS (acquired immune deficiency syndrome), influenza, and harmless skin warts. In plants there are viral diseases that affect only specific crops and others with a wide range of hosts. Often viral symptoms in plants are most obvious on new growth as deformities, dieback, and discoloration. Most importantly, plant diseases caused by viruses are almost always systemic.

Distinguishing between systemic and localized plant diseases can help gardeners plan their disease control strategies. *Localized* diseases are only active in certain parts of the plant such as the roots or leaves and fruits. *Systemic* diseases, including vascular wilts and almost all viral diseases, are spread throughout the plant, for example through the vascular system (section 5.2). Some localized diseases or pest problems, especially in the roots, may first become evident to the gardener in symptoms that look systemic, like wilting. Checking the roots of garden plants periodically helps the gardener detect these problems early, before they cause too much damage. Checking roots for signs of disease is discussed in section 13.4.

If caught early in their development, localized fungal and bacterial diseases can be controlled by removing the infected parts and using other management strategies. For example, if a fungal leaf spot is found on a few tomato leaves in the garden, these leaves can be removed, the plants pruned and staked, and organic matter added to the soil. However, if leaves on the growing tip of a squash plant appear mottled and deformed, signs of a systemic viral disease, removing those leaves will not provide control because the entire plant is infected.

Management strategies for systemic disease will vary depending on the crop and severity of the disease. Many systemic plant diseases can be transmitted by insects and humans to other plants and this should be considered when deciding whether or not to keep a diseased plant alive. If other, healthy plants of the same crop are growing in the garden, removing the diseased plants helps control further spread of the disease.

Some crop varieties may be able to overcome or tolerate systemic diseases, especially if the infection is not severe. We have seen both bean and squash plants infected with viral diseases survive to produce good harvests. In general, it is a good idea to remove from the garden young plants with signs of systemic disease infection. It is often not worth the effort to keep sick young plants alive until they become productive, and removing them prevents the disease from spreading to other young plants. The gardener must decide if the plant is capable of surviving, if it is worth any extra work to save it, and if leaving it will lead to further spread of the disease.

It is also useful to check the area surrounding the garden for weeds that may serve as alternate hosts for systemic plant diseases. Experienced local gardeners may know of weeds associated with plant disease problems in their area. Removing those weeds before planting can help prevent or control disease.

Using locally adapted, disease-resistant crop varieties is one of the best ways to control disease problems. Local varieties may have been selected for resistance to diseases as an adaptation to local growing conditions. Familiarity with local crops and their growing habits helps the gardener take advantage of a crop's ability to resist disease and the environmental conditions that make this resistance most successful.

Some diseases only affect certain crop families, for example tobacco mosaic virus (TMV) only occurs in solanaceous crops (tomato, pepper, eggplant, potato, tomatillo), and the fungus causing clubroot only on crucifers (mustards, cabbage, cauliflower, broccoli). Planting crops from different crop families is importantl for controlling these disease pathogens. In the case of TMV, nonsolanaceous crops can replace those that have succumbed to virus, although solanaceous crops may be tried again the following season because the virus does not remain in the soil. However, in the case of a fungi such as clubroot, which can live on in the soil, it is more appropriate to plant a rotation of noncrucifer crops for several seasons. In fact, if the pathogens are well established, planting nonsusceptible crops may be the only control method available short of abandoning the site.

Interesting new work is being done on using watery compost extract (WCE) as a treatment to encourage crop plant resistance to some fungal diseases.[65] Watery compost extract is a solution made by soaking one part well-rotted compost in six parts water for about 1 week. The compost must include some animal manure, although the best kind of manure and its proportion in the compost have not been explored. Spraying WCE on leaves of garden crops like fava beans, tomatoes, and grapes has helped control fungal disease in those plants. One of the pathogens controlled is *Uncinula necator*, powdery mildew of grapes, a common dryland fungus. In addition, researchers suggest soaking seeds overnight in WCE before planting to prevent damping-off fungus (section 13.4). Since WCE itself has no fungicidal properties, it is thought that it stimulates the plant to produce fungicidal substances.[66]

Fungal root rots can occur when crops are grown out of season. For example, cool soil temperatures favor the fungi more than the roots of cold-sensitive squash varieties. Maize should be planted in warm soils (20-25°C, 68-77°F) and wheat in cool soils (15-20°C, 59-68°F) to control seedling blight caused by the same strain of a fungal pathogen.[67] This is because the different soil temperatures favor seedling growth of the two crops differently, and vigorous seedlings resist penetration by the pathogen. So, if grown under unfavorable soil temperatures, seedling growth is very slow, and the plants succumb to fungal disease.

Mechanical and environmental controls of plant diseases focus on preventing the spread of those diseases. Removing and destroying diseased plants or plant parts is advisable to prevent the spread of disease by vectors like wind and insects. Some insects that commonly spread plant pathogens are aphids, leafhoppers, cucumber beetles, and whiteflies. If the presence of any of these insects coincides with the onset of disease in the garden it is likely that they are the source, and control measures should be taken. Familiarity with life cycles of insects that are disease vectors helps in timing plantings to avoid the insects' active periods.

While not all pathogens can be spread from infected plants or plant parts, many can. Burning these plants or burying them away from the garden is recommended. With systemic diseases the entire infected plant should always be destroyed when it is removed from the garden. Although only some diseases are seed-borne, it is wise not to use seeds from diseased plants or fruit, especially in the case of systemic diseases. People can also spread plant diseases from infected to uninfected soil or plants. This can be prevented by carefully washing hands and feet with water and soap, if possible, and cleaning tools with bleach or soapy water after contact with a source of disease and before touching healthy plants or entering an uninfected area.

Many disease pathogens thrive under moist, warm conditions, so staking or trellising plants and pruning lower leaves helps because it allows better air circulation, reducing some of the moisture. Placing dry mulch or sticks under fruit in the garden keeps the fruit from lying on the moist soil surface. We sometimes place sticks or a flat stone under ripening melons in our garden to keep them from rotting due to fungal or bacterial infections (Figure 13.17). Immediately removing and destroying infected fruit and leaves prevents spread of the pathogen by insects.

Traps are mechanical control measures that help prevent the spread of pathogens by controlling their insect vectors. The yellow aphid traps described in section 13.3.1 attract and drown aphids. When approaching a garden or field, aphids will stop first to feed on a tall trap crop such as corn planted around peppers, beans, or squash. In doing so, many of the aphids lose any pepper-, bean-, or squash-infecting viruses they may have been carrying, thereby greatly reducing the amount of virus reaching those crops.[68]

Figure 13.17 A Flat Stone Keeps Ripening Fruit Off Moist Soil and Prevents Rotting

Figure 13.18 Careful Observation is the Best Tool for Diagnosing Garden Problems

Suppressive soils are those that inhibit the growth of pathogens because of the microorganisms contained in the soil.[69] Often it is the diverse mixture of soil microorganisms that seems to suppress growth of pathogens. For this reason, some fungal and bacterial diseases can be controlled biologically by adding lots of organic matter to the garden soil. This encourages growth of fungi and bacteria that compete with and thus control some disease pathogens. For example, studies in Mexico of soil in indigenous fields showed that those soils suppressed damping-off disease caused by fungi significantly more than did soils from industrially farmed fields.[70]

Another way to biologically control diseases is by using predators to control those insects and other animals that spread the disease pathogens. An example of this is the parasitic wasps that prey on aphids (section 13.3.1).

13.4 Diagnosing Pest and Disease Problems

The key to diagnosing problems in the garden is careful observation that leads to an understanding of the ways in which plants, animals, insects, microorganisms, and their environment interact. Traditional gardeners and other small-scale food producers are usually excellent observers of the world around them, and are continually testing, evaluating, and redesigning their management strategies in response to garden conditions. Always look first to local gardeners for help with diagnosing, understanding, and managing problems in the garden.

When trying to diagnose problems, it is important to look at the garden as a whole. Is there a pattern among the plants that are not doing well? Are there pests or eggs on the undersides of leaves, in the soil and mulch around the plants (Figure 13.18)? Night is an active time for many insects and other animals in hot, dry areas and is a good time to check the garden for them. Although not necessary, a small hand lens or magnifying glass can sometimes help, for example, in determining if leaf spots have small insect holes in the middle. Are the affected plants young or old, scattered or concentrated, of one kind or different kinds? How does the distribution pattern of plants with the problem correspond to that of water drainage, soil, shade, wind? What is the history of the garden site? What was grown there before? Was anything buried there? Have chemicals been applied nearby? Has there been standing water?

It is a good idea to plant a few "extra" plants of each crop in the garden for digging up occasionally to check on progress or to diagnose problems below the soil surface. Many problems—especially soil conditions, fungal and bacterial diseases, and some insects—cause damage below ground in the root system. These problems are then expressed above ground in the stems, leaves, and fruit. (See Chapter 5 for definitions of plant anatomy terms used in this Chapter.)

Learn what a healthy root looks like by digging up a healthy plant and washing the roots carefully in water. A rotted root system may have fewer lateral roots, be off-color (tan or brown), and collapse when squeezed between a finger and thumb (Figure 13.19). *Cortical sloughing* is the loss of cortex (cortical tissue) on the roots and occurs in several plant diseases such as fungal root rot. If a plant is suffering from a disease that causes cortical sloughing, but that tissue has not yet been shed, it can be easily tested by pinching the root lightly and pulling toward its tip. This will pull off a "sleeve" of cortical tissue, revealing vascular tissue underneath (Figure 13.20).

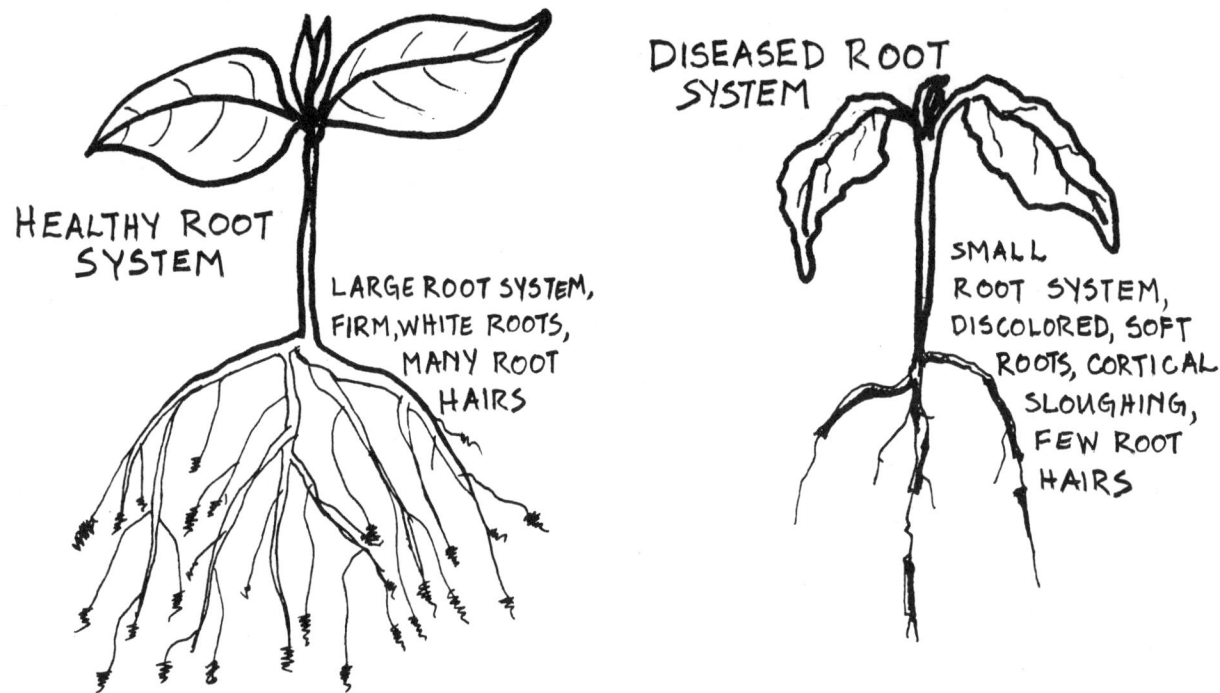

Figure 13.19 Comparing Healthy and Diseased Root Systems

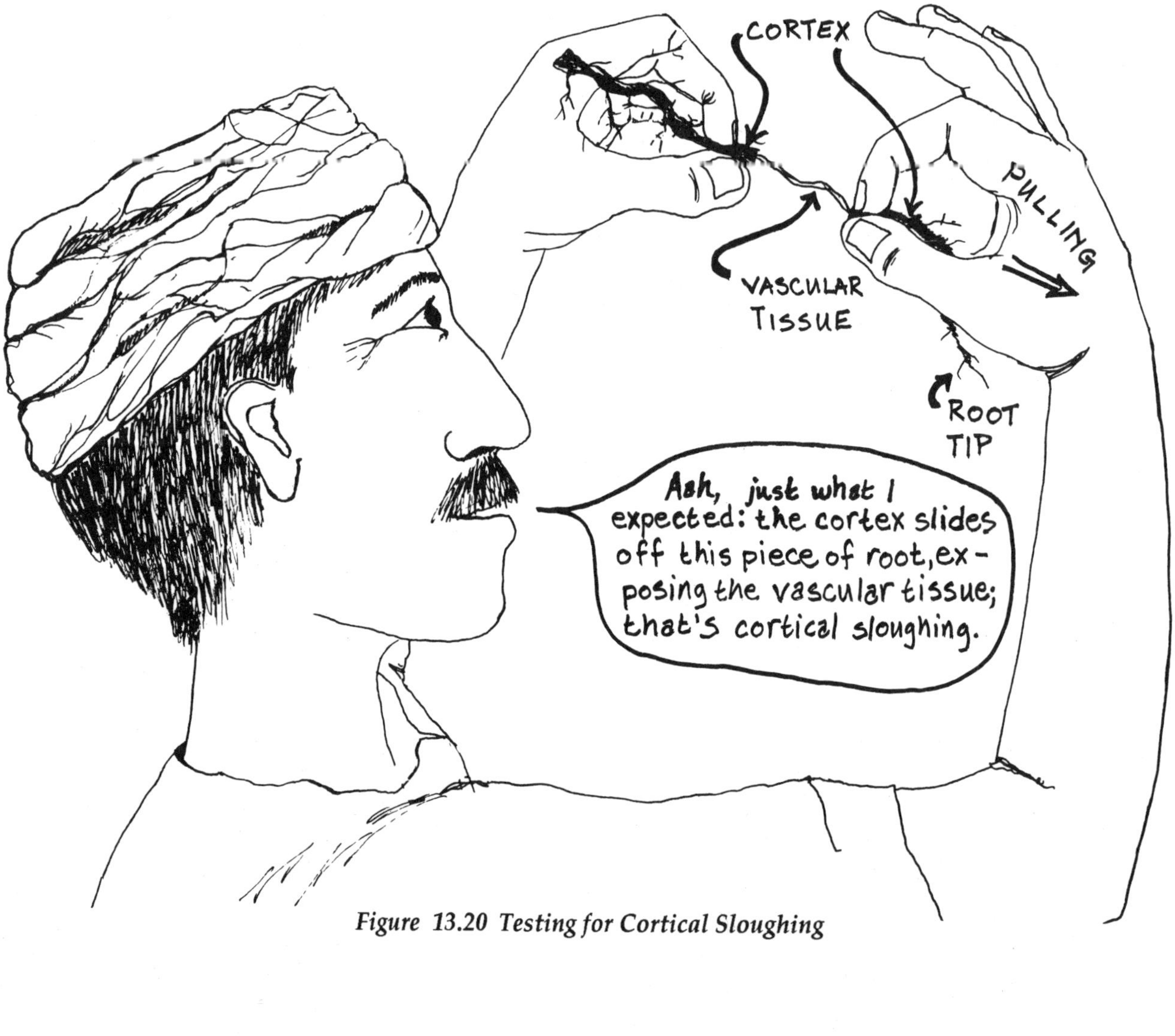

Figure 13.20 Testing for Cortical Sloughing

It is not necessary to dig up a whole tree to examine the roots (Figure 13.21). At a point at or just inside the *drip line* (the line indicating the maximum spread of aboveground growth from the central stem or trunk), a 20- to 50-cm (8– to 20-in) deep hole can be dug and roots about the diameter of an adult's small finger exposed. These can be checked for cortical sloughing, discoloration, and deformities.

Many agricultural and other chemicals, including fertilizers, can cause damage to plants, and should be considered a possible cause of problems. Poisoning by herbicides such as 2,4-D (Box 13.3 in section 13.2.4) can cause symptoms that look like a viral infection. Symptoms include abnormal stem and leaf growth, and leaf chlorosis, curl, and mottling. Check for evidence of nearby spraying, or contamination by tools, containers, soil, or clothes brought into the garden. Be sure a poisonous chemical was not accidentally put in a bucket, sprayer, or on other tools used in the garden, and find out the chemical history of the garden plot. This is especially important in urban areas where empty lots are often used as waste dumps.

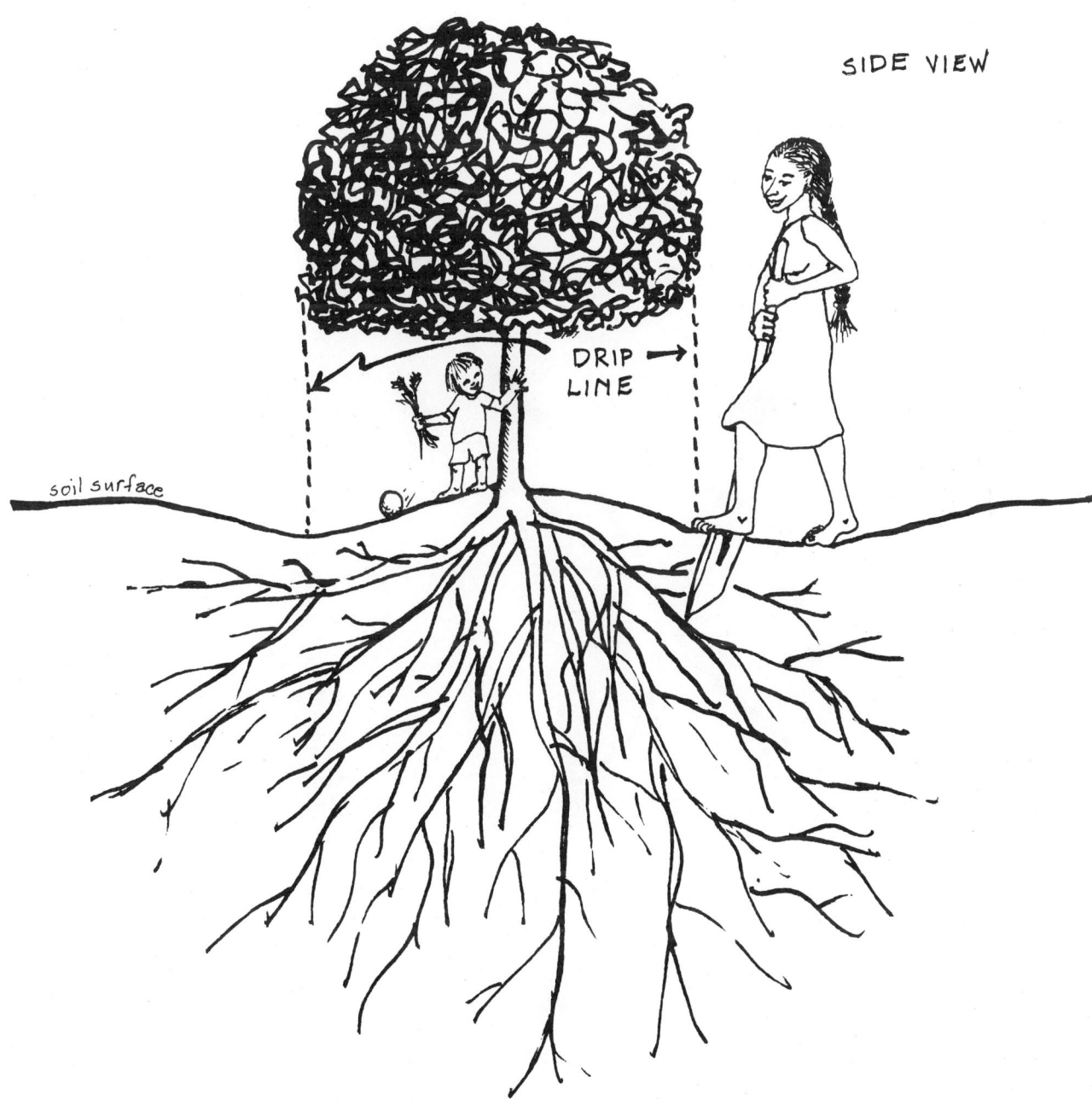

Figure 13.21 Digging for a Root Sample from a Tree

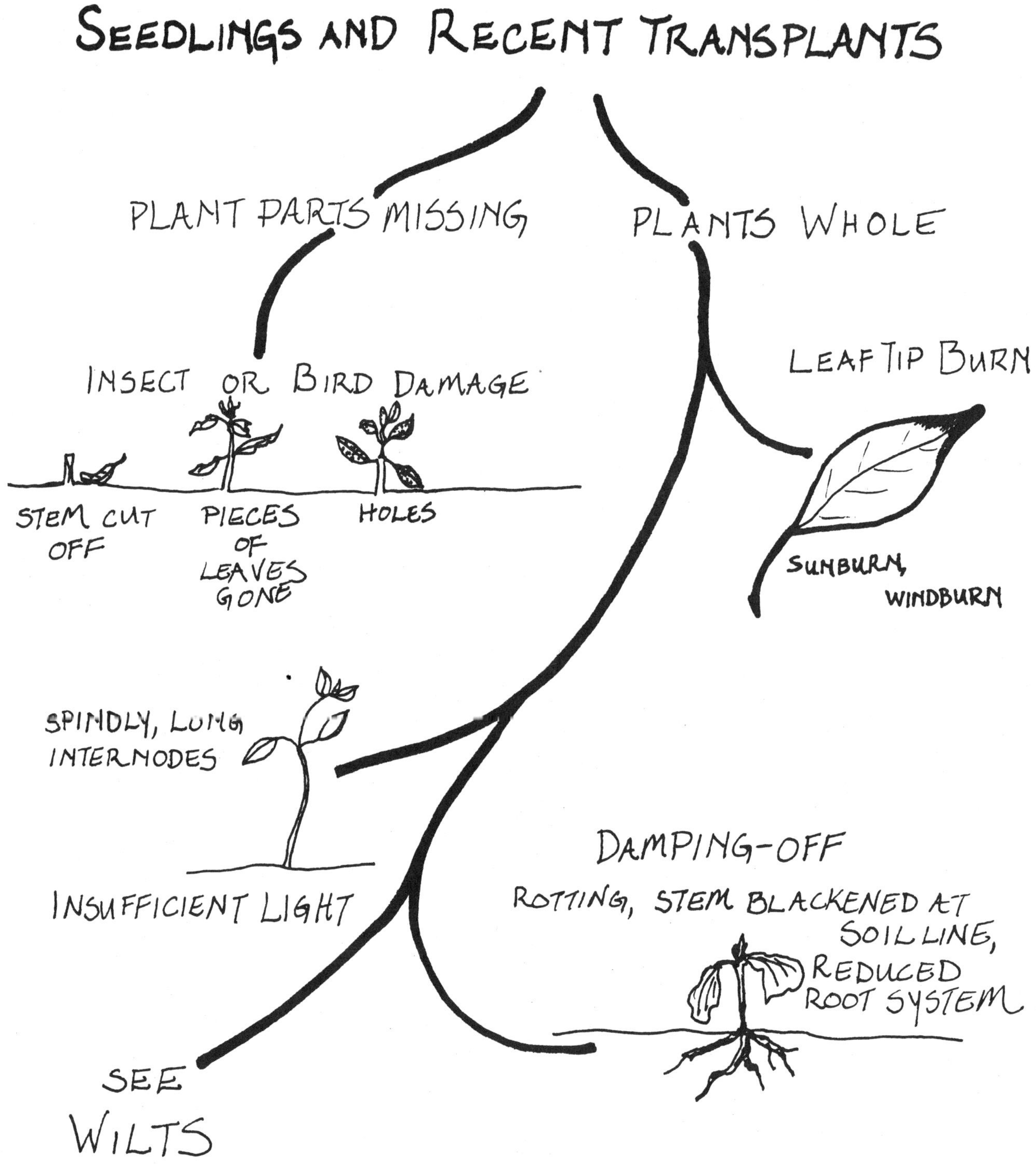

Seedlings and Recent Transplants

Plant Parts Missing

Insect or Bird Damage
- Stem cut off
- Pieces of leaves gone
- Holes

Plants Whole

Leaf Tip Burn — Sunburn, Windburn

Spindly, Long Internodes — Insufficient Light

Damping-off — Rotting, stem blackened at soil line, reduced root system

See Wilts

Figure 13.22 Diagnosing Problems with Seedlings and Recent Transplants

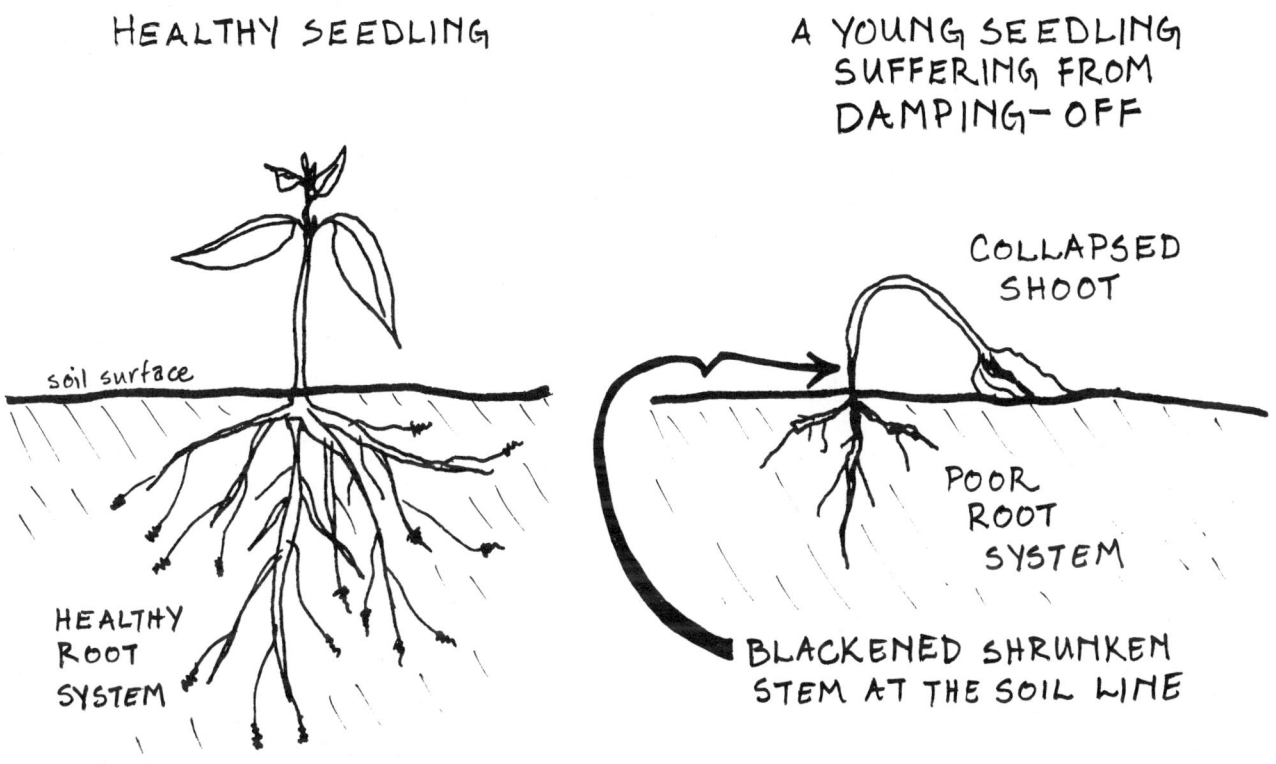

Figure 13.23 Signs of Damping-Off

SEEDLINGS AND RECENT TRANSPLANTS (Figure 13.22) The younger and more tender its tissues the more susceptible a plant is to pests, diseases, and difficult growing conditions. As described earlier, some garden crops contain chemicals that repel or are toxic to pests, but only start producing these chemicals as they grow and mature. In addition, young plants with small root and shoot systems are much more likely to be killed by pest or disease damage than more established plants. This makes careful observation and quick diagnosis and action especially important when caring for young plants.

The most common disease problem in seedlings is damping-off, caused by several soil fungi and some bacteria that rot the plant roots or stems at the soil line (Figure 13.23). Infected seedlings wilt, the tops dry out, they fail to grow, or even fall over. Damping-off is diagnosed by digging out the seedling, washing it carefully in a container of water, and checking the roots and stem for soft brown areas and cortical sloughing. In advanced cases, the entire root system will be rotten and will disintegrate during washing.

There are three common causes of damping-off:
- An unusually large population of the fungi or bacteria that cause the disease due to insufficient competition from other microorganisms in the soil. Adding compost can reduce damping-off problems because compost contains many active soil organisms, some of which compete with and reduce the damping-off fungi.
- Waterlogged soil due to overwatering or poor drainage causes damping-off. This can be remedied by watering only as much as is necessary, and by improving soil drainage, for example, by adding sand to the soil.
- Soil temperatures too hot or too cold for the crop variety result in lack of vigor and susceptibility to disease. This can be avoided by modifying soil temperatures (use a covering of surface mulch to cool soil, remove it to allow soil warming), or replanting at a better time of year. Soaking seeds overnight in watery compost extract (WCE) before planting may be helpful for controlling damping-off fungus (section 13.3.4).

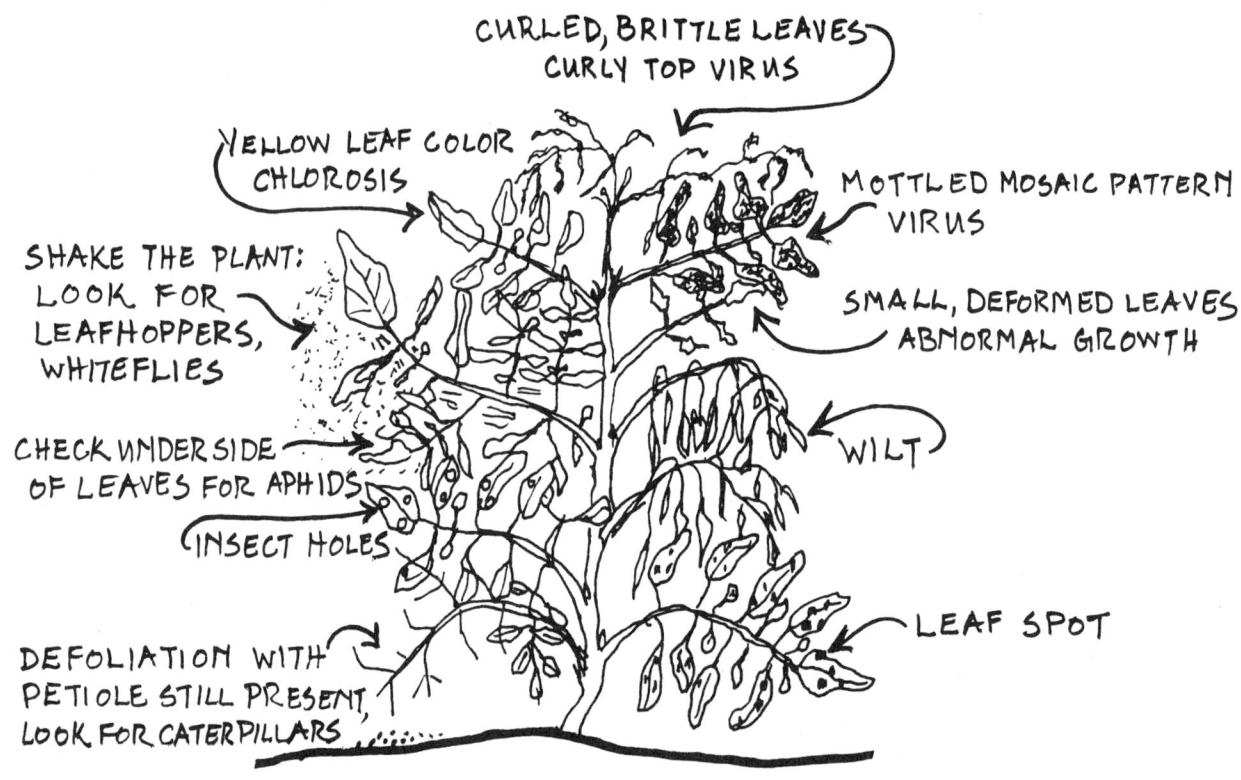

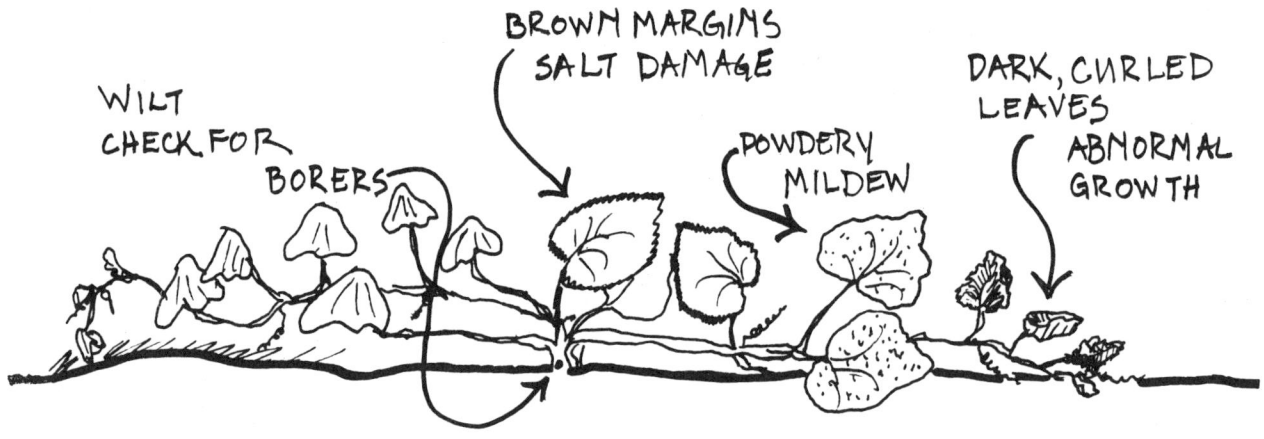

Figure 13.24 Problems with Established Garden Plants

Problems with transplants are often caused by transplant stress or exposure to pests or diseases in the planting site. Transplant stress weakens the plant making it vulnerable to pests or pathogens which were not a problem before transplanting. Bruised and broken tissues from handling during transplanting create openings where pests and pathogens can enter the plant.

One of the best ways to overcome problems with seedlings and transplants is to just keep trying. For example, try replanting or transplanting when the soil is a little warmer or cooler, wait until insect populations have died down, or plant where there is more or less sunlight. Often these experiments lead to an understanding of how to work with the garden environment instead of fighting it, with better long-term results using fewer resources.

ESTABLISHED PLANTS Most of the symptoms of disease in established herbaceous annuals, such as wilt and chlorosis, will be expressed above ground (Figure 13.24). However, the majority of these symptoms, except those due to insects and viruses, are caused by problems below ground. Digging up an extra plant to look at its roots helps in diagnosing problems. Problems with trees and other woody perennials also frequently originate in the root zone, and checking the roots is important for diagnosing these symptoms.

If a tree seedling or cutting grown in a container was root-bound at the time of transplanting, the roots can twist around each other and in 3 to 4 years choke each other off. Above ground the leaves may become chlorotic, turn orange, or wilt. Spreading the roots at the time of transplanting prevents this problem (section 8.4.4). *Verticillium* wilt and a variety of nematodes can also cause problems in woody perennials. Impervious soil layers can slow or stop growth of perennials and can also contribute to waterlogging and salt buildup. Above ground the most common problems of woody perennials are insects (twig damage, leaf eating, or causing abnormal growth), powdery mildew, birds and other large animals, and wind damage.

FAILURE TO THRIVE All gardeners at one time or another have plants that just do not grow. They may look perfectly fine or show a variety of difficult-to-classify symptoms. Failure to thrive and grow may be due to a single factor, but just as likely it is the result of a combination of interacting factors. Planting again at different times of the year, in different soil conditions, checking the roots of a few plants to look for clues, and talking with other gardeners to see what is happening in their gardens all help. If other gardens have similar problems, then a common climate (section 5.7), water quality, or seed source may be the cause.

The following four sections include figures and tables to help in diagnosing garden problems according to types of symptoms. Figure 13.25 is a guide to these sections.

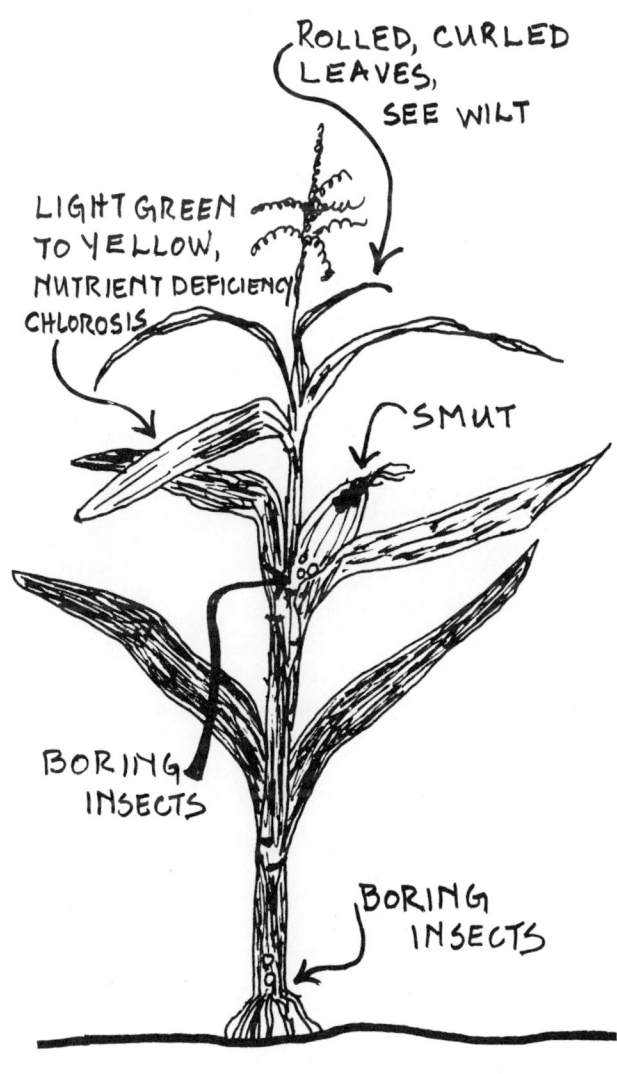

Figure 13.24, continued

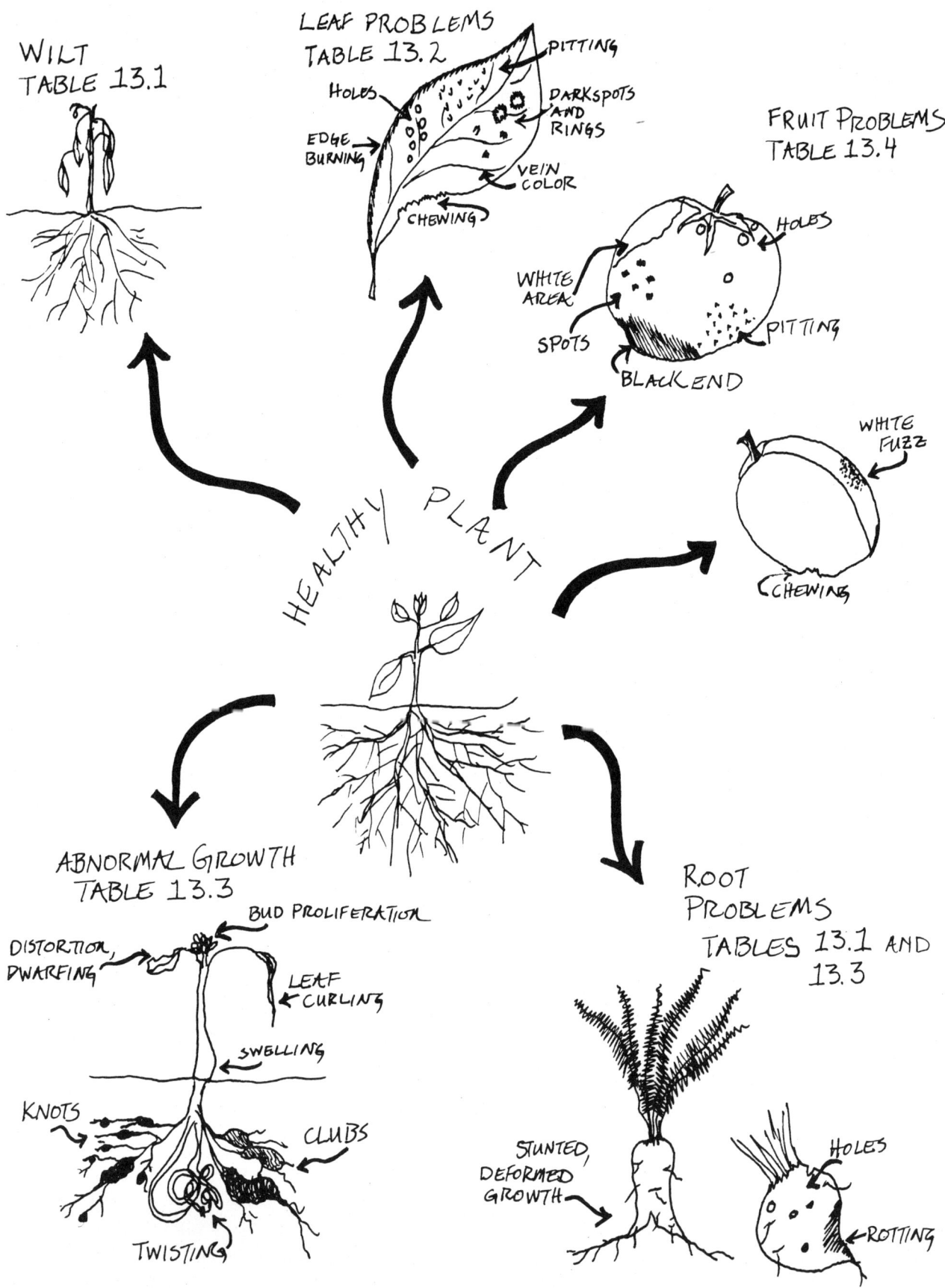

Figure 13.25 Diagnosing Problems in the Garden

13.4.1 Wilts *(Table 13.1 and Figure 13.26)*

When a plant wilts it loses its rigidity, its leaves droop, and its branches bend. In drylands the most common cause of wilting is lack of water in the root zone (section 10.3.1). On very hot days, with temperatures higher than 38°C (100°C), some plants like cucurbits, will wilt even with sufficient soil moisture, but will recover in the cool of the evening. Check the plants in the early morning during the hot season to see if they have recovered. Shading plants and providing windbreaks also helps. Plants may also wilt while being hardened off for transplanting. This is to be expected but should not be so severe that the plant's growing tips die back or that it does not recover in the evening (section 8.4.4). Other important causes of wilting in drylands include fungal and bacterial wilts, insects, and nematodes.

In general we recommend adding organic matter to the soil when fungal diseases are a problem. Although this is not a cure for those diseases, in some cases the fungi present in the organic matter will compete with, and therefore help control, the disease-causing fungi in the future (section 13.3.4).

Larvae or grubs in the soil, which eat root hairs and even larger roots, will cause the plant to wilt, especially under hot, dry conditions. When these pests attack the storage roots of crops like carrots and Jerusalem artichokes, the plant may not wilt. However, those wounds are sites for diseases to become established which can destroy the root in the ground, or later in storage. Clusters of frass along a plant's stem are a common sign of stem boring caterpillars.

13.4.2 Leaf Problems *(Table 13.2 and Figure 13.27)*

Many soil nutrient deficiencies (section 9.5) show up as discoloration of the leaves. A common sign of many problems is chlorosis, a light green or yellow color which contrasts with the normal, darker green leaf color. The pattern of the chlorosis is often a good clue to the cause of the problem. Frequently leaf chlorosis or other discolorations due to nutritional deficiencies are *symmetrical*, (the same on both halves of the leaf) and follow leaf vein patterns, but leaf discoloration caused by diseases is often *asymmetrical* (the opposite of symmetrical). Mottling refers to mixed patches of light green or yellow and normal green leaf color caused by viral diseases.

Leaves that are yellow or have brown edges, and poor plant growth may indicate damaging amounts of salt in the soil or water, which is common in arid areas (section 11.2). Blowing sand, also common in some drylands, can tear and pit leaves with tiny scars. Fungi and bacteria affecting primarily the leaves are less common in drylands because of the lower humidity, and under dry conditions most of these diseases are primarily focused in the root system. An exception is a fungal disease called powdery mildew, which thrives in dry conditions. However, during the rainy season, or with watering methods such as sprinkling, fungi and bacteria, can be a problem especially on lower leaves. Insect damage of leaves can be severe, but can often be controlled by hand picking, washing the leaves, or replanting. Some of the sprays described in Box 13.2 in section 13.2.4 are also effective until they are washed off.

13.4.3 Abnormal Growth *(Table 13.3 and Figure 13.28)*

Abnormal growth means that the roots, stems, leaves, or flowers are not developing normally and are misshapen in some way. This growth is the plant's response to stresses including pests, disease, the physical environment, management practices, or even genetic abnormalities.

13.4.4 Fruit Problems *(Table 13.4 and Figure 13.29)*

Fruit problems can be especially disheartening for the gardener because the harvest seems so near. Here we cover fruit problems of some herbaceous annual crops, as well as some perennial tree crops. Fruit can show abnormal growth and the mottling symptoms of virus infections. Fruit symptoms are usually associated with leaf symptoms (section 13.4.2). Some of these symptoms may also develop in harvested, stored fruits.

13.5 Resources

College and high school textbooks about *entomology*, the study of insects, are good sources for learning more about insects, their life cycles and habits. In addition, for understanding basic principles of some control methods, textbooks on diseases of plants, such as *Plant Pathology* (Agrios 1988) are useful. Unfortunately, many such texts most often recommend using synthetic chemicals as solutions to pest and disease problems.

Agricultural Insect Pests of the Tropics and Their Control (Hill 1983) seems to be one of the most thorough books on this subject in the English language. The dis-

Table 13.1 Wilts

Observations	Possible causes	Suggested actions
SOIL AND TEMPERATURE		
Heat stress	High temperatures and transpiration rates	Shade, mulch, check soil moisture, consider adjusting planting time
Soil in root zone dry	Underwatering	Irrigate, mulch, and shade
Water-saturated soil	Overwatering and/or high water table	Reduce watering, try raised beds for high water table, add organic matter for more open texture
Chemical injury	History of herbicide use in area	Move garden site or use containers and soil from other source
LEAVES		
Many tiny insects on underside of leaves	Aphids, mites, thrips	Wash off, crush, spray, use a yellow trap, trap crops
Mottled, curled, misshapen, dwarfed	Viral disease	Burn or discard plant, clean hands and tools
Young tomato leaves stunted, curled inward, bumpy on lower surface	Curly top virus spread by insects (leafhoppers)	Shade tomato plants to discourage insects
Dwarfed, curled, misshapen	Chemical injury	Move garden site or use containers and soil from other source
STEMS		
Insect holes on main stem	Stem- and vine-boring caterpillars	Remove borer, cover stem with soil; on squash stems encourage rooting elsewhere
Ring of vascular browning seen in stem cross section	Fungal vascular wilt	Plant resistant varieties, add organic matter, do not overwater, vary planting times
White ooze from cut stem, discolored tissue	Bacterial wilt	Rotate crop families, plant local resistant varieties, 1 year dry fallow
Chew marks on base of stem	Caterpillars, sow bugs	Hand pick, especially at night, remove mulch immediately around stem, use stem collars
ROOTS		
Roots brown and soft, cortical sloughing	Fungal root rot	Add organic matter, do not overwater
Round knots, lesions or swellings	Nematodes	Rotate crops, add organic matter, use noninfested planting material, heat and dry soil
Club-shaped swellings (galls) on crucifers	Clubroot fungi	Rotate noncrucifers, resistant varieties, raise pH above 7.2
Small root system, few root hairs	Larvae or grubs eating roots	Dig around plants or turn soil to expose and kill pests, allow domesticated or wild birds to forage for larvae or grubs

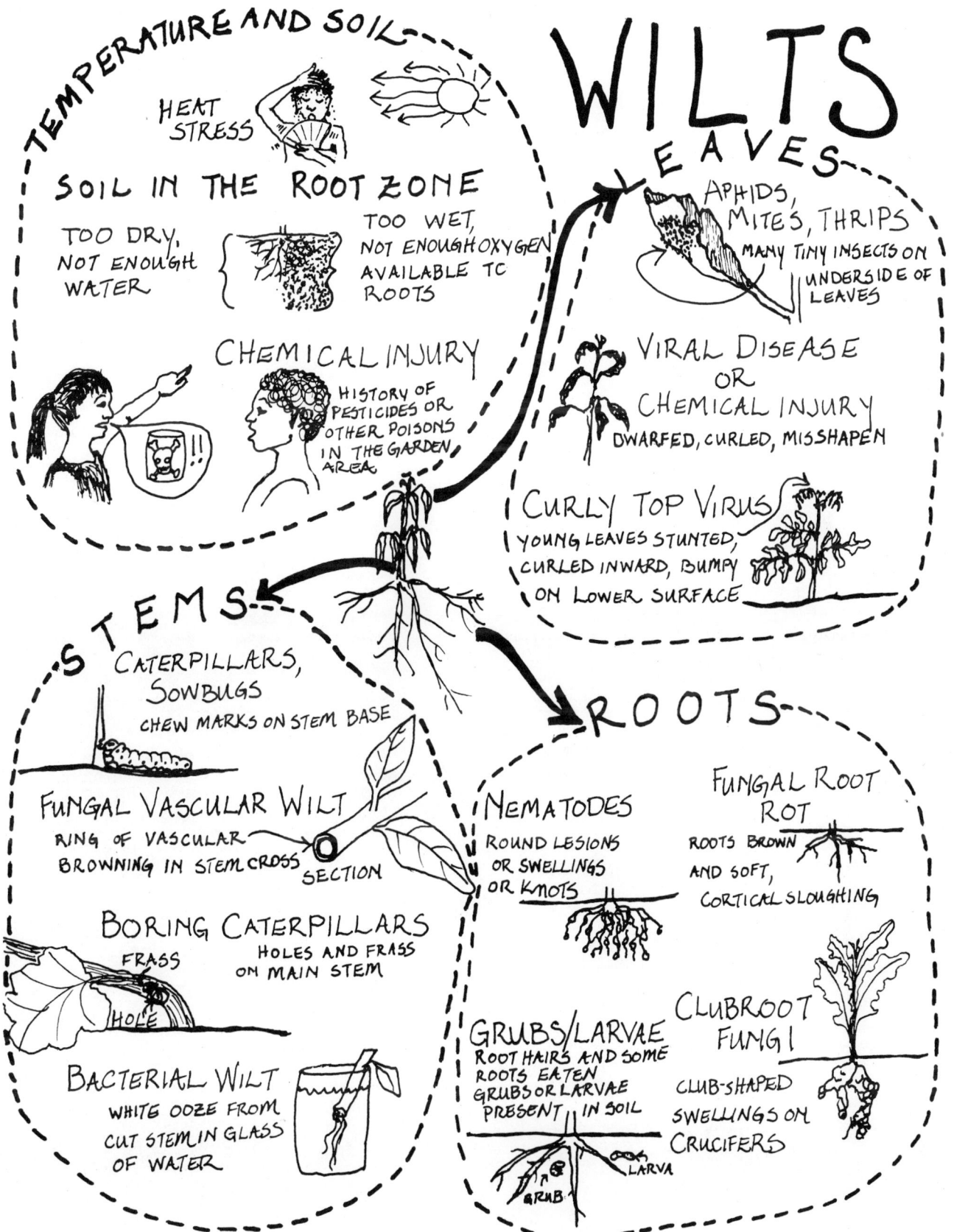

Figure 13.26 Wilts

Table 13.2 Leaf Problems

Observation	Possible causes	Suggested actions
CHLOROSIS (leaf yellowing)		
Whole leaf (old leaves)	Nitrogen deficiency	Add high N organic matter
Whole leaf	Nematodes	Check roots for swellings or knots, rotate crops, add organic matter, fallow and/or heat soil
Whole leaf	Fungal root rots	Check roots for browning and soft tissue, add organic matter, do not overwater
Whole leaf	Root-bound	Check roots for tangled, twisted growth, untangle roots, transplant to larger container or into the ground
Between veins of new leaves	Iron or zinc deficiency	Lower soil pH
Veins in old leaves	Viral disease or herbicide damage	Check for herbicide damage, burn or discard plant if young, clean tools and hands, control insect vectors, rotate in different crop family
Half of leaf and vascular browning	Fungal vascular wilts	Plant resistant varieties, rotate crop families, add organic matter
Mottled or mosaic pattern	Viral disease	If plant is young burn or discard it; clean tools and hands, control insect vectors, rotate in different crop family
OTHER		
Leaves unusually purple	Phosphorus deficiency	Add high phosphorus organic matter
Black spots or rings	Fungal leaf spot	Remove and discard affected leaves, clean hands and tools, add organic matter to soil, rotate in different crop family
Salt burn: edges brown, white, or yellow	High salt concentration in soil or water	Water deeply, flush soil, check water quality
Dry, brown patches, holes visible with lens	Sucking insects	Pick off or crush, wash or spray with water or a repellent, use traps
Chew marks on edges, small holes in center	Chewing insects	Pick off, wash or spray with water or a repellent
Pitting and tearing	Windblown sand	Windbreaks
White, powdery spots	Powdery mildew fungi	Remove and destroy infected plants or parts, add organic matter to soil, plant resistant varieties
Small (5-mm, 0.02-in diameter) spots, dark with yellow margins	Bacterial spot	Remove and destroy affected leaves and fruit, stake plants, wash hands and tools
Rusty yellow, brown, or white spots, also on stems, can form galls	Fungal rusts	Remove and destroy affected leaves and fruit, stake plants, wash hands and tools, add organic matter to soil

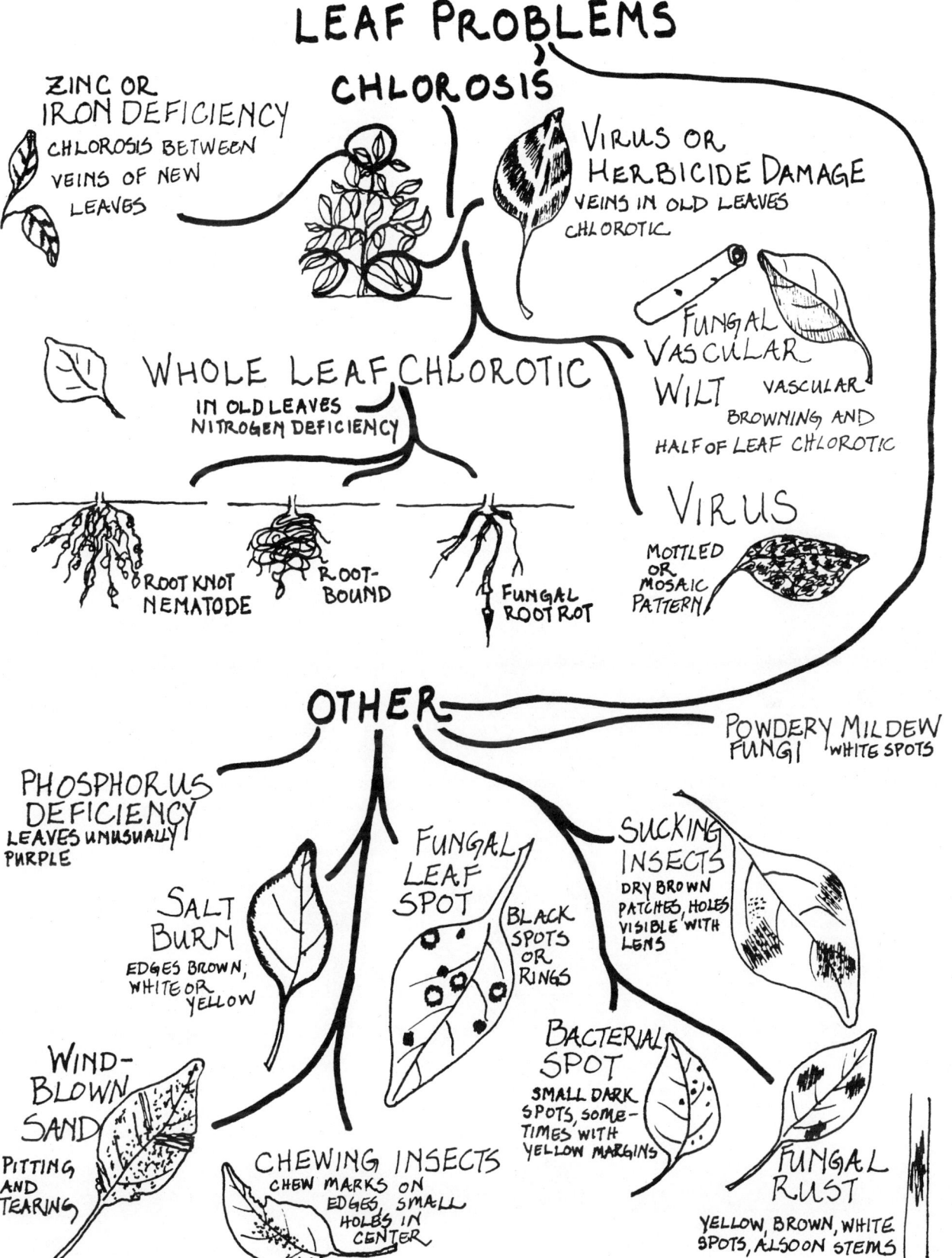

Figure 13.27 Leaf Problems

Table 13.3. Abnormal Growth

Observation	Possible causes	Suggested actions
ROOTS		
Round swellings or knots	Root knot nematodes	Rotate crops, add organic matter, solarize soil, turn soil regularly during hot season, fallow
Spindle- or club-shaped swellings on crucifers	Clubroot fungi	Rotate in noncruciferous crops, raise pH above 7.2
Plant in container with misshapen, twisted, forked roots	Root-bound	Untangle roots, transplant into larger container or into ground
Plant in ground with stunted, twisted, forked roots	Impermeable layer in soil e.g., caliche	Dig out deep planting hole, build up topsoil
Misshapen, twisted, forked roots	Transplanting inappropriate crop e.g., carrot	Plant seeds directly instead of transplanting
STEMS		
Galls or swellings on crown, or higher, especially on stone fruits and grapes	Crown gall bacteria	Destroy infested young perennials, tools used near or on diseased plants should be thoroughly cleaned
Swellings or holes higher on stem, dieback above these	Insect eggs	Allow beneficials to hatch, remove others
Many buds	Damage from thrips, mites, and other insects, may carry viral diseases	Remove and destroy affected parts, eliminate insect vectors, clean hands and tools
Pale, internodes long and spindly	Insufficient sunlight	Gradually reduce shade and/or remove mulch
Large, dark, sunken spot at soil line	Collar rot fungi	Discard or burn plant, clean hands and tools, add organic matter to soil, do not overwater, plant resistant varieties
Swollen growths with black dust inside, at the joints, in maize and teosinte	Maize smut fungi	Remove affected stalks before growths open, discard fungus
LEAVES		
Misshapen, curled	Mites, thrips	Remove and destroy affected parts, crush, spray, trap pests, encourage predators
Misshapen, curled, sometimes sticky	Aphids, other sucking insects	Wash off, crush, spray, use yellow trap
Dwarfed, misshapen, chlorosis	Chemical damage	Move garden or use containers with soil from another source
Misshapen, dwarfed mottled, especially the new growth	Viral disease	If symptoms severe or plant is young burn or discard it, clean hands and tools, use disease-free seeds/stock
Curled and brittle, thickened midribs, leaf hoppers active	Curly top virus	If symptoms severe, burn or discard plant, shade plants, plant resistant varieties
Misshapen, dwarfed with wilt or unilateral chlorosis, vascular browning	Fungal vascular wilts	Add organic matter, do not overwater
FLOWERS		
In maize, sugarcane, sorghum: swollen black growths with dust inside	Smut fungi	Remove affected fruit before growths open, discard fungus

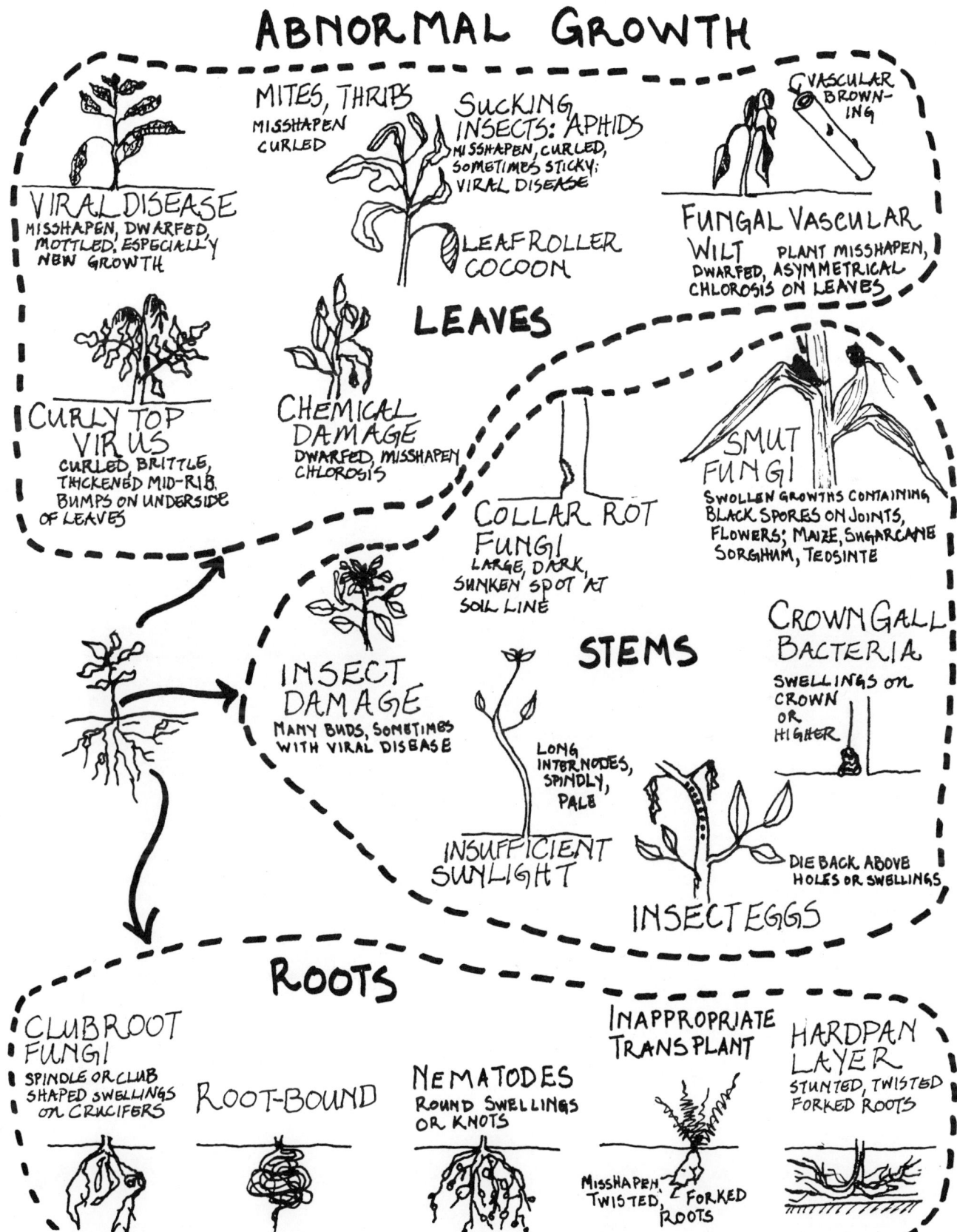

Figure 13.28 Abnormal Growth

Table 13.4 Fruit Problems

Observation	Possible causes	Suggested actions
FRUIT HARD		
Mottled yellow/green, faint yellow rings	Viral disease	If severe, remove and destroy fruit and plant, control insect vectors, do not plant seeds produced in these fruits
Tomatoes: black/tan hard spot on end (blossom end rot)	Calcium deficiency	Water more regularly, mulch to maintain even soil moisture
Hard white lesion or scar on exposed part of fruit	Sunburn	Shade plant
Yellow spots with holes in middle	Sucking insects	Remove, trap or crush insects wash or spray fruit with water
Pitting with no holes in middle	Windblown sand	Windbreaks, harvest early if danger of rotting, replant if possible
Bruised or scarred, chunks of skin missing	Hail	Harvest early if danger of rotting, replant if possible
Small, scablike dark spots with yellow margins	Bacterial spot	Remove and destroy affected fruit and leaves, prune and stake plant, add organic matter to soil
Pomegranates: fruits crack open	Uneven or inadequate water supply	Establish regular, deep watering schedule beginning just after flowering through fruit ripening
FRUIT SOFT		
Soft spots where fruit touches the ground, e.g., melons	Fungal soft rots	Elevate fruit above moist ground on stones, sticks, mulch, or trellis
Holes eaten in fruit, e.g., figs, dates, peaches	Fruit beetles or other boring insects	Harvest early, trap and remove beetles in morning, remove ripe, rotting fruit
Small holes on skin, inside eaten, rotten	Fruit fly larvae	Cover fruit to protect from egg laying, use homemade sprays, traps
Holes eaten in fruit	Birds	Frighten off, harvest early, cover fruit or whole plant
Fuzzy brown growth on stone fruits, also occurs in storage	Fungal brown rot	Remove and compost fruit before spores spread
Fruit feels like a bag of water	Bacterial soft rot	Pick and discard affected fruit to control fruit flies and other insect vectors, prune and stake plant to allow air circulation

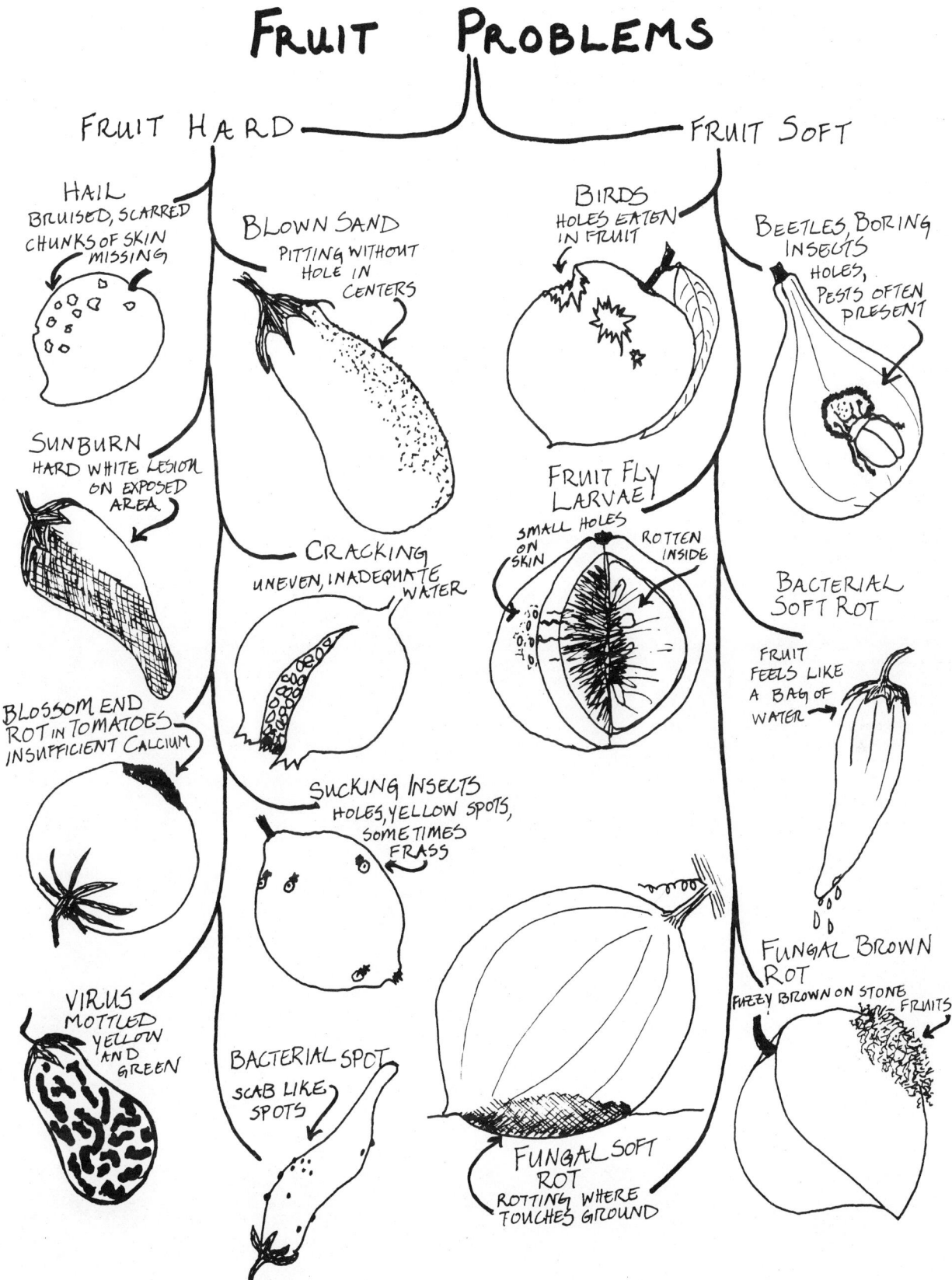

Figure 13.29 Fruit Problems

cussions of principles and methods of pest control are useful, although the emphasis is on large-scale agriculture and using manufactured pesticides. The bulk of this book is a catalog of major pests organized according to order and accompanied by distribution maps and excellent drawings or photographs.

Pests and Diseases of Tropical Crops: Volume 1, Principles and Methods of Control (Hill and Waller 1982) is a useful, easy-to-read handbook. Again, the emphasis is on large-scale production and using manufactured chemicals. See Onwueme (1978) for tuber crops.

Natural Crop Protection (Stoll 1987) focuses on solutions using local resources available to farmers and gardeners in the Third World tropics and subtropics. It is divided into sections for specific crops, and for specific methods in the field and in storage. *Natural Crop Protection* is really an annotated bibliography on the subject and contains some gaps and discrepancies. However, because this kind of research has been ignored for so long, these problems are to be expected. The author has tried to carefully document the techniques described and she invites readers to help improve future editions. We feel this book is a good first step in a new direction and includes helpful ideas and interesting references on the subject.

Developing Countries Farm Radio Network (DCFRN) has some program scripts that are useful for pest and disease control including: "Insect control" (#1-1), "Fruit and vegetable soft rot" (#8-2), "Pest life cycles" (#10-8), and "Preventing bird damage" (#12-10).

References

[1] van den Bosch, et al. 1982:16.
[2] Hill 1983:63.
[3] Hill 1983:242.
[4] Cammell and Way 1987:4.
[5] Hill 1983:49.
[6] Hill 1983:39.
[7] Hill 1983:42.
[8] Hill 1983:43-44.
[9] van den Bosch, et al. 1982:2.
[10] van den Bosch, et al. 1982:22-23.
[11] van den Bosch, et al. 1982:24-27.
[12] Garcia, et al. 1988.
[13] van den Bosch, et al. 1982:63-64.
[14] Gould 1988.
[15] Stoll 1987:138-139.
[16] van den Bosch, et al. 1982:63.
[17] Stoll 1987:138.
[18] Based on Stoll 1987:84-86.
[19] Based on Stoll 1987:96.
[20] Based on Stoll 1987:140.
[21] Based on Stoll 1987:91.
[22] From Stoll 1987:130.
[23] Pimentel 1988.
[24] Levins 1986.
[25] van den Bosch, et al. 1982:28.
[26] Hill 1983:80.
[27] Primarily based on IOCU 1984 and Hill 1983.
[28] IOCU 1984:101.
[29] Hill 1983:124.

[30] IOCU 1984:105.
[31] Abrahamse and Brunt 1984; Bull 1982:87-123.
[32] Bull 1982:76ff.
[33] Atteh 1987.
[34] Werner 1977:103.
[35] Hill 1983:16.
[36] Page and Richards 1977.
[37] Dupriez and De Leener 1983:66.
[38] DCFRN #7-9B.
[39] MacNab, et al. 1983:23, 29.
[40] Hill 1983:38.
[41] Hill 1983:382-393.
[42] DCFRN #4-9B.
[43] Stoll 1987:129-132.
[44] Agrios 1988:713-714.
[45] Bridge 1987.
[46] Radewald 1977.
[47] Adeniji 1977.
[48] Maas 1987.
[49] Bridge 1987.
[50] Adeniji 1977.
[51] Bridge 1987.
[52] Bridge 1987.
[53] Rice 1983:59.
[54] NAS 1980b:114.
[55] Radwanski and Wickens 1981.
[56] Rice 1983:58.
[57] Rice 1983:51-56.
[58] DCFRN #10-6.
[59] Nabhan and Sheridan 1977.
[60] Ouangraoua 1988.
[61] Bell, et al.1987:38, 39.
[62] FAO 1982c:225.
[63] DCFRN #9-1A.
[64] Agrios 1988:445-447.
[65] ACPP n.d.; Weltzien and Ketterer 1986.
[66] Weltzien and Ketterer 1986.
[67] Roberts and Boothroyd 1984:204,230.
[68] Agrios 1988:192.
[69] Baker 1987.
[70] Lumsden, et al. 1990.

Part III
Garden Harvest

In the Chapters of Part I we discussed the nutritional, economic, environmental, and social contributions of gardens to sustainable development. Part II presented the principles of plant, soil, and water management to meet these goals and ideas for applying these principles in ways consistant with the criteria for sustainable development. But the story does not end with the harvest. Many of the benefits of gardens depend on what happens to seeds and food after they are harvested. The goals of local self-reliance and control, and an approach to gardens that builds on local resources and knowledge, will help insure that the garden harvest promotes equity as well as social and environmental sustainability, and that the benefits will endure well beyond the life of the project.

14
Saving Seeds for Planting

Gardeners and farmers all over the world have been selecting and saving seeds and other plant propagation materials since the beginning of agriculture and plant domestication over 10,000 years ago. Almost all of the crops grown today are products of this selection and new varieties created by scientists are based on the work of these past generations. Plant selection by gardeners and farmers continues to be vital for conserving genetic resources and for producing crop varieties best suited to local needs. Years of experience and observation give people an understanding of desirable crop characteristics. This long and successful history is the reason why existing seed selection and saving techniques must be understood before suggestions for improvement are made.

Even when appropriate seeds are available for purchase, the advantages of saving her own seeds often make it worthwhile for the gardener. Saving seeds reduces the costs of gardening and takes advantage of locally adapted varieties. If the gardener does not save her own seeds, a good alternative is some other source of locally selected and grown seeds, such as other gardeners, local seed co-ops, or regional seed houses. (Suggestions for selecting materials for vegetative propagation are given in Chapter 7).

Introducing new crops and crop varieties has a long history and many now "traditional" or indigenous local crops or crop varieties, such as watermelon in southwestern North America and chilis in West Africa, were introduced from other continents or regions. (See section 1.1 for a definition of the word "indigenous" as it is used in this book). However, the most common obstacles to productive gardens for improving household nutrition or economic conditions—such as lack of control over resources, poor water quality, or marketplace competition from agribusiness—will not be resolved by introducing a new crop. For this reason we believe that nonlocal, new crops should not be the focus of garden projects, and if used, should only be introduced as part of an experiment, and never as a replacement for local garden crops.

The implications of saving seeds, especially of local varieties, go far beyond the effects on the gardener and her garden. Genetic resources for the future world food supply are becoming scarcer with the loss of indigenous varieties and the diverse local agricultural systems that produce and maintain them. At the same time, new and expensive plant-breeding technologies are being used by commercial, multinational seed companies. These changes raise new issues about the control and economics of seed production, and the genetic diversity upon which all food systems rely. These are issues that affect the gardener and that she addresses when saving her own seeds.

In this book the terms *folk variety* or *indigenous variety* refer to crop varieties that have been selected and managed by local people and the local growing environment. In the past these varieties have been called "landraces" or "primitive varieties." We use the term folk varieties in support of the efforts of Third World countries and the United Nations Education, Scientific and Cultural Organization (UNESCO) to have these varieties recognized as a part of the "folk" heritage of indigenous communities, and thus give them control over and compensation for the use of these genetic resources.[1]

14.1 Summary

Saving garden seeds for planting is important because it reduces investments and risks while promoting self-reliance. When local seeds are saved, genetically diverse folk varieties are conserved on-site for future generations.

Diversity in indigenous gardens results from planting genetically diverse varieties, several varieties of some crops, and many different crops, with combinations changing over the seasons and years. Commercial or industrial crop varieties are usually bred for industrial agriculture, have more genetic uniformity among individual plants of the same variety, and fewer varieties are grown in the cropping system. Bred for production under optimal conditions, industrial varieties generally require more purchased inputs and are a greater risk for the gardener in marginal lands. Although they may be appropriate in some situations, such as intensely managed gardens with abundant water and nutrients, they should not be promoted as a replacement for folk varieties.

Saving seeds can be done easily along with other work in the garden. Seeds should be harvested after they are mature, or they may not grow. Techniques for cleaning, drying, and storing seeds are easy to learn and are vital for maintaining seed stocks for next season's garden and for future generations. If not properly cared for, stored seeds can spoil or be damaged by pests. However, simple methods, many of them locally developed, can prevent or greatly reduce these problems.

14.2 Seeds, Gardens, and Diversity

The indigenous crops, gardens, and fields of the world contain great biological diversity. This diversity occurs on several levels:
- Folk crop varieties contain a diversity of genetic information.
- Often there are many different varieties of each crop.
- Gardens and fields frequently contain mixtures of many different crops and varieties.
- Gardens and fields in different ecological or cultural settings have different crops and crop mixtures.

The biological and ecological diversity of indigenous systems is good for gardens and other small-scale food production because it increases yield stability and local self-reliance. *Yield stability* is a measure of the variation in the amount of usable harvest from year to year. When exposed to stresses such as drought, flooding, pests, diseases, and high temperatures, total crop yields will decrease relatively little in mixed plantings of diverse indigenous varieties compared to monocultures of genetically uniform industrial varieties.[2] In other words, diversity reduces the risk of having nothing to harvest from the garden. Indigenous agriculture also reduces the risk of going into debt and losing land, because it is more self-reliant, and does not depend on obtaining credit to purchase expensive seeds, fertilizer, pesticides, and irrigation pumps.

Industrial agriculture is characterized by varieties that respond to favorable growing conditions created by increased inputs of irrigation water, chemical fertilizers and pesticides, and often mechanization. Compared with traditional agriculture, industrial agriculture often increases yields per unit of land and labor, and has contributed a great deal to increasing the world food supply, not only through production in industrialized countries, but as the green revolution in the Third World. This industrial approach has also been applied to gardens.[3]

In contrast to indigenous varieties, very few varieties of industrial crops are bred and released to the public. Within each industrial crop variety there is less diversity than in folk varieties, and the industrial varieties are developed for, and planted over, much larger areas. Table 14.1 is a simplified comparison of the differences in diversity at several levels between industrial and indigenous agriculture.

Choosing between indigenous and industrial varieties and gardens involves a trade-off. Even though the individual plants lack diversity, industrial varieties can sometimes be desirable for dryland gardeners. Also, not all new varieties are genetically uniform or dependent on expensive inputs. To minimize risk to the gardener, some diversity should be maintained at all levels (individual plants, crop varieties, and crops in the garden). This can be done by encouraging gardeners to continue growing their local folk varieties even when they are also trying new, industrial varieties, and to plant their gardens with a mixture of different indigenous and new crops and crop varieties.

There is some evidence that small farmers who do grow industrial varieties often continue to grow folk varieties for a number of reasons.[4] For example, Hopi Native American farmers in dryland southwestern United States continue to maintain their own Hopi sweet corn variety even when cultivating an industrial sweet corn variety.[5] A common reason given by farmers for doing this is that the industrial varieties produce large sweet ears, but the Hopi variety is better adapted to the local environment. If the growing conditions are especially dry the Hopi variety will produce a harvest but the industrial varieties will fail.

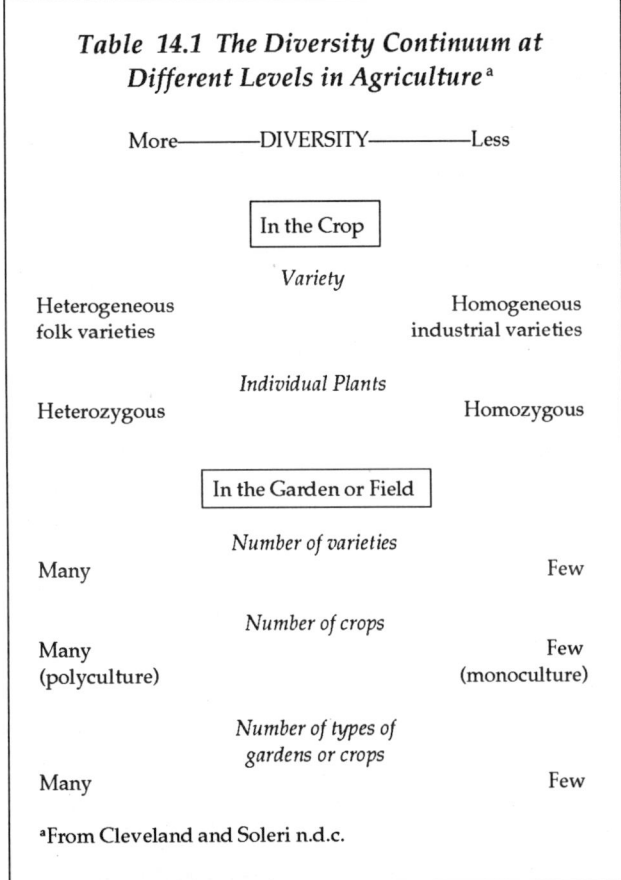

Table 14.1 The Diversity Continuum at Different Levels in Agriculture[a]

[a]From Cleveland and Soleri n.d.c.

14.2.1 Diversity in the Seed

Folk varieties have been selected for their adaptation to local growing conditions and local people's needs. Along with related wild species, they are the world's major store of genetic diversity for crop plants (Box 14.1). The number and genetic diversity of folk varieties is maintained, or even increased, by gardeners and farmers over generations in the following ways:

- Domestication of new wild plants.
- Introduction of new genetic information by crossing between folk varieties, and between crops and their wild and weedy relatives.
- Introduction of new crops and varieties from other villages or regions
- Selection of advantageous spontaneous mutations.[6]

In these ways genetic diversity is conserved even though some folk varieties are always being lost, either abruptly through replacement by other varieties, or slowly through selection by gardeners, farmers, and the environment.

While indigenous crop selection has not been well documented, some examples do exist, such as Richard's description of rice variety selection be Mende farmers in West Africa.[7] The farmers keep not only their existing 70 rice varieties pure, but are constantly searching for and experimenting with new varieties, a process they enjoy greatly. When potential new varieties appear in their fields as a result of mutations, cross pollination, or accidental mixing, the farmers carefully remove plants with undesirable variations and save those with desirable traits. Farmers test these new varieties, or ones aquired from neighbors or markets, in fertile, moist patches of soil near their houses. If the tests appear successful they will multiply the seed for full-scale planting. Farmers also sort a variety into separate lines when it becomes highly variable, harvesting seeds with different characteristics separately.

Many crops grown in both indigenous and industrial gardens are introductions from other regions. For example, in southwestern North America beans, maize, chili peppers, and other crops came from Central America, while watermelon, wheat, fava beans, garbanzos, and peaches were introduced from the Mediterranean and Africa by Spanish colonialists only a few centuries ago. Over time both native and introduced crops have diversified into many folk varieties which are important in local food systems today.

Folk varieties are constantly evolving in response to changes in the environment and in gardeners' needs. The gardener who selects and saves her own seeds may be particularly interested in a few characteristics, such as timing of production or flavor. But at the same time, the plants she chooses are being selected for their ability to grow and produce in the garden environment. In choosing the most vigorous and desirable plants, the gardener is selecting for a complex combination of characteristics particularly suited to her resources and needs as well as the garden environment.

The increasing production and promotion of industrial seeds, many of which are hybrids (Box 14.2), is radically changing this traditional system of conserving genetic diversity in folk varieties.[8] When folk varieties with their adaptive characteristics selected for by generations of gardeners and farmers are replaced by industrial varieties, the result is usually a reduction in diversity, both within the variety and often in the cropping system itself. When this happens, gardeners and farmers must rely on seeds that they may not be able to produce themselves and that may not meet their needs as well as the abandoned folk varieties. The result may be to increase the risk of survival for the food system as well as the social and cultural foundation of the community.

> **Box 14.1**
> *Genetic Diversity in the Seed*
>
> Genetic structures and processes are complex, but the following simplified explanation will help readers understand how the issue of diversity is affected by the kind of seeds used in the garden.
>
> A plant's *genotype* is the genetic information that it inherited. A genotype is composed of all the genes located on the chromosomes in the nucleus of a cell (Figure 14.1). Each cell in that plant contains this same genotype (except the gametes, as explained below). The genes contain all of the information the plant needs to grow and reproduce. In all plant cells except gametes, chromosomes come in pairs. Onions, for example, have 16 pairs, and tomatoes 12 pairs of chromosomes. However, there is only one of each chromosome in sex cells or gametes, the pollen from the male parent and the ovule in the female parent. That is, 16 single chromosomes in onion gametes and 12 single chromosomes in tomato gametes. When the male and female gametes are united at fertilization, they form an embryo which again has two of each chromosome. In each chromosome pair in the embryo one chromosome is from the male parent and the other from the female parent. This embryo will grow into a seed.
>
> *Alleles* are the two or more alternative forms of one gene and are used as a gauge of diversity in individual plants and varieties. Because chromosomes usually occur in pairs there are two locations for each gene in each cell. Therefore an individual plant can have a maximum of two different alleles for each gene. However, within all the plants of a crop variety there can be many alleles for one gene, with different combinations of these alleles being expressed in individual plants in the population.
>
> That is, a single plant or a crop variety can be either genetically diverse or uniform (Figure 14.1). *Homozygous* plants are those with only one allele for a gene. A *homogeneous* variety is one with many plants which are homozygous for one or more genes, that is they all contain the same allele for that gene. Individual *heterozygous* plants contain two different alleles for a gene and *heterogeneous* varieties include plants heterozygous for many genes, having two or more alleles for those genes.

Many industrial seeds have been selected for high production under conditions where there is pest protection with chemicals, plant nutrients supplied with chemical fertilizers, good soil, plenty of water, a single large harvest, and no competition from other crops or weeds. Maximum yield, disease and pest resistance, adaptedness to mechanized cultivation and harvesting, and fruit size and appearance are frequently the kinds of traits selected for. Other characteristics like drought hardiness, tolerance of marginal soils, continuous harvesting, adaptation to mixed cropping, taste, grinding texture, and "by-products" like bean leaves as a green vegetable, have not been considered relevant. This means that industrial varieties will often not be adapted to the low-input, and less-than-optimal conditions of most dryland gardens and fields.[9] Seed produced this way, frequently far from the environment where it will be grown and the people who will grow it, fails to take advantage of local expertise and the genetic resources available from folk varieties. Industrial varieties usually cannot fulfill the combination of gardener's needs and demands of the local growing conditions.

It is now recognized that the costs of controlled, optimum-environment agriculture are prohibitive for gardeners and farmers in many parts of the world. This is especially true for low-resource gardeners and farmers in the Third World. As a result, some breeding programs at international agricultural research centers have been changing. For example, the International Institute for Tropical Agriculture (IITA) is working to develop cowpea varieties that are drought-resistant and others that can be grown for both food and fodder, and the International Crop Research Institute for the Semi-Arid Tropics's (ICRISAT's) recent work with sorghum and millet focuses on drought adaptation.[10] ICRISAT is also working on mixed cropping of cereals and legumes such as millet and cowpeas. The dominance of production as a breeding criteria has lessened somewhat, and some crop breeders are focusing their work on developing varieties adapted to local growing conditions, sometimes using local folk varieties as raw breeding material.[11] However, there is still a long way to go in terms of breeding programs recognizing the value of folk varieties, the expertise of local gardeners and farmers, and the

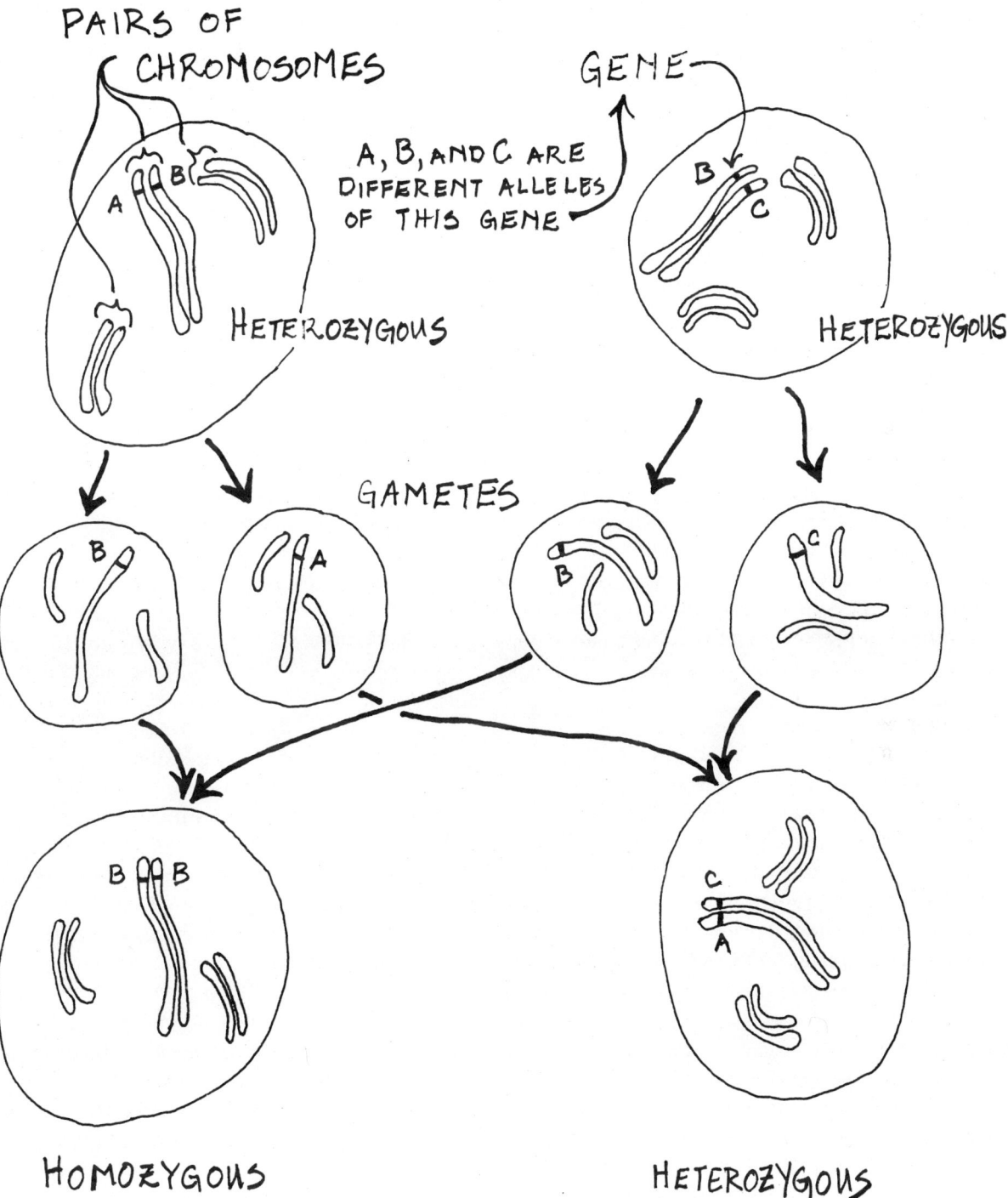

Figure 14.1 Chromosomes, Genes, Alleles, and Their Role in Genetic Diversity

> **Box 14.2**
> *Hybrid Seeds*
>
> A *hybrid* is the product of the cross between any two genetically different individuals. However, in popular usage the word has come to refer to a particular kind of hybrid developed for propagating agricultural crops.[12] This more common definition is the one we use here. *Hybrid seeds* are the product of a cross between homogeneous inbred lines.
>
> To help understand the process we give a simplified description of how single-cross hybrid maize seeds are produced (Figure 14.2). First, different plants from the same species (*Zea mays* in this case) are *inbred*, or self-pollinated, for many generations, with offspring selected for particular desirable traits. Eventually this inbreeding results in very homogeneous populations of plants whose offspring therefore, always have the desired characteristics. These breeding populations are referred to as *inbred lines*, for example, lines A and B.
>
> While inbreeding fixes desirable traits, it also fixes undesirable traits and can produce weak, unhealthy plants. However, these undesirable characteristics are unimportant to the breeder as long as they do not interfere with the characteristics being selected for.
>
> In the second step, selected inbred lines are paired. To ensure that all of the seed saved is the result of cross-pollination, one line is identified as the female parent and the other as the male. The female line is rendered incapable of self-pollinating either by cutting off the male flowers (tassels) of these plants in the field, or through introducing male sterility genes when creating that line. Thus, all of the seeds borne by the plants of the female parent line are a cross of lines A and B (A x B).
>
> Because all A plants are genetically identical to each other, and all B plants are genetically identical to each other, the result of the A x B cross is predictable. The seeds produced by this cross are F_1 generation hybrids, which are sold to farmers. These seeds produce strong, healthy F_1 plants because of the mixing of genetic material. The cross-pollination of these identical F_1 plants (F_1 x F_1) or [(A x B) x (A x B)], produces seed meant for consumption. However, if this seed is saved and planted it grows into F_2 plants, different both from their parents and from each other. Some of these plants are weak and lack vigor and many of them will show the undesirable traits of the original inbred parent lines, A and B. They may have some desirable traits but overall are unpredictable, unhealthy, and inadequate for food production. Therefore new hybrid seed must be purchased by the gardener or farmer every year.
>
> Commercial seed production of some other crop hybrids also depends on male sterility in one parent. For example, if seeds saved from fruit of most hybrid watermelon varieties are planted, there will be a complete crop failure, because there will be no pollen to fertilize female flowers. In the 1970s the genetic material widely used to create male sterility in maize in the United States resulted in widespread genetic uniformity and vulnerability to disease (Box 14.3).
>
> Today many varieties of commercially produced garden vegetable seeds are hybrids. This includes tomatoes, peppers, squash, melons, cabbage, lettuce, onions, and other popular garden crops.

adaptive complexity of the varietal selection that occurs in indigenous gardening and farming.[13]

The major influence on crop breeding in the decades to come will be the new agricultural biotechnologies, especially genetic engineering, which makes possible the direct transfer of genes even from distantly related organisms to crop plants. While this technology has the potential to help gardeners and small-scale farmers, this help is not inevitable. Agricultural biotechnology is dominated by private industry, and their profitmaking goals will undoubtedly have an influence on the varieties produced.[14] For example, many of the world's largest seed companies are now owned by private, multinational corporations which specialize in chemical manufacturing.[15] A major objective of this new corporate strategy is developing varieties that will increase sales of the company's other products. Worldwide, over 27 corporations have begun research on creating crops that are tolerant to herbicides, thus encouraging increased herbicide use.[16] An example is Monsanto's new soybean (*Glycine max*) which was bred to tolerate large quantities of that company's herbicide glyphosate, known commercially as Roundup.

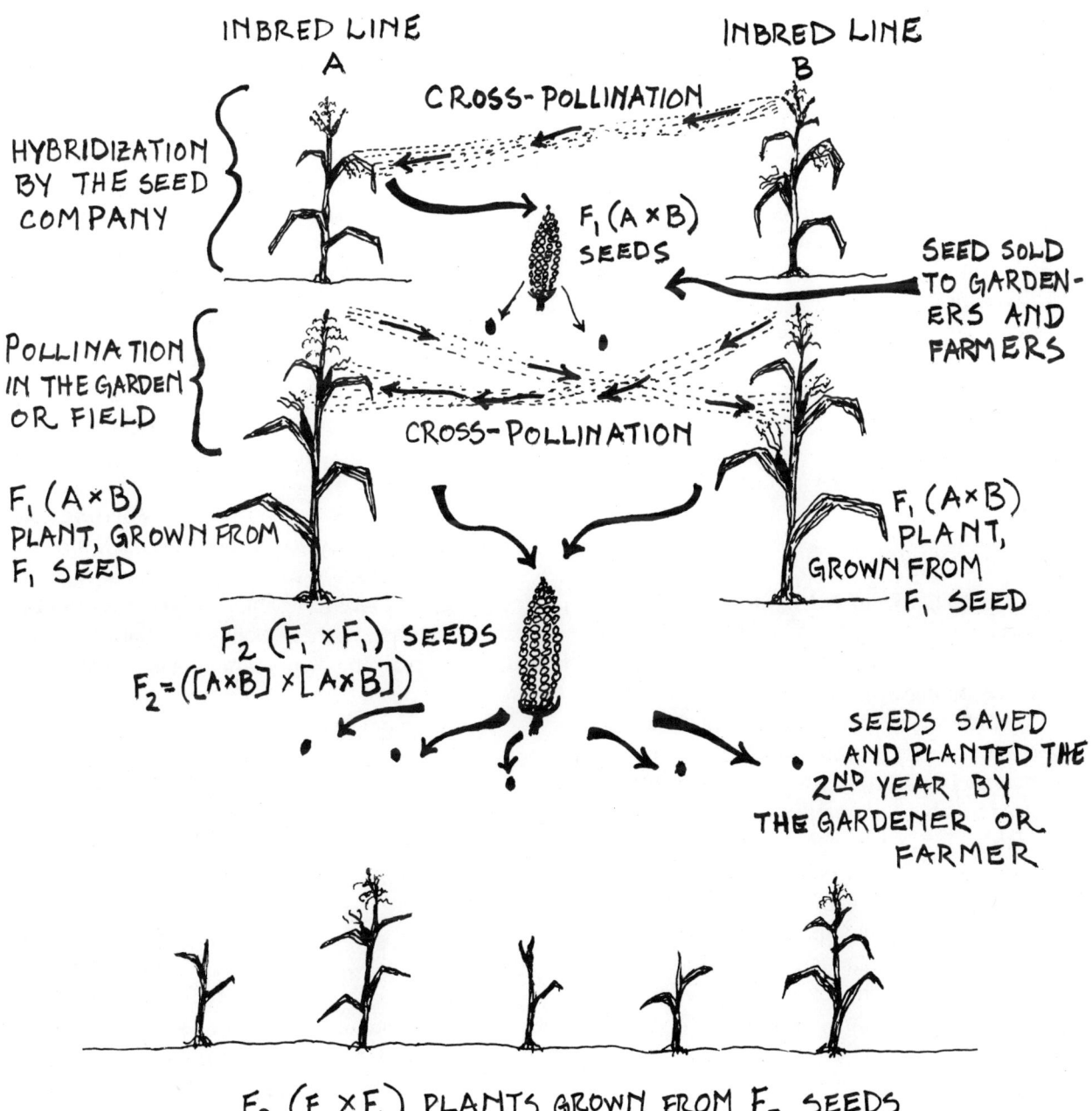

Figure 14.2 An Example of Hybrid Seed Production in Maize

> **Box 14.3**
> **The Southern Corn Leaf Blight**
>
> A well-known recent example of the risks of genetic uniformity is the southern leaf blight epidemic which swept maize (corn) plantings in the United States in 1970.[17] At the time, approximately 75% of the maize grown in that country shared identical genetic material for male sterility used in hybrid seed production. However, this was linked genetically to susceptibility to a fungus causing leaf blight. The blight spread rapidly through the country and losses were estimated at 710,000,000 bushels of maize, worth US $1 billion in 1970. New genetic material was essential in subsequent breeding programs seeking resistance to the fungus and other problems. This disaster drew attention to the trend in industrial agriculture away from biological diversity, and the risks of this trend.

14.2.2 Diversity in the Garden

Indigenously based mixed gardens are living gene banks that conserve genetic diversity while serving gardeners' needs. Gardens change through time as the household changes in size and needs, as perennial crops grow bigger, or die and are cut down, and as new crops and varieties are introduced or old ones are abandoned. In addition, gardens differ from household to household, community to community, and region to region, with changing climate, soils, diets, and history.

Gardens often contain many crops that serve different purposes. Some may provide fruits or vegetables, some medicine or craft materials, others may be grown for the beauty of their flowers, and all may also be grown for market. An irrigated garden in arid northern Pakistan may include vegetables such as eggplants, tomatoes, bitter gourds, amaranth, portulaca, and chilis; perennial fruit producers like grapes and mulberries; chinaberry trees (*Melia azedarach*), whose wood is used for construction; and a species of jute for making rope.[18] In northern Mexico, household gardens produce fruits, vegetables, flowers, and chicken eggs for sale and household consumption, and medicinal herbs for home use.[19] A survey of 145 gardens of Tswana agropastoralists in Botswana revealed 45 domestic species and 8 wild species which had been transplanted or were sprouting spontaneously.[20]

Different varieties of the same crop may be grown in gardens for several reasons,[21] such as their adaptedness to environmental conditions at a specific growing time. Data on different varieties is scarce for gardens, but there are examples for field crops. For example, Mende farmers in Sierra Leone (section 14.2.1), recognize 70 different rice (*Orriza* spp) varieties by name and sight, and each household grows an average of 4 to 8 varieties.[22] These varieties are suited to different growing conditions and have diverse growing characteristics, especially time from sowing to harvest. In one short-season variety, some farmers are selecting for tough outer *glumes* (the papery coat or bract around the seed) and long *awns* (the hair-like bristle growing out from the glume) which help protect the grains from birds, a major pest of early rice.

In the small country of Malawi in southern Africa there are many varieties of common bean. Even though only 4 varieties make up the majority of area planted, farmers grow an average of 13 varieties per household.[23] Local farmers, primarily women, explain that they grow so many varieties because of differences in flavor, cooking quality, market demand, time to maturity, digestibility, and ability to cope with pests, disease, and environmental stress like drought. For example, most farmers plant early-, middle- and late-maturing varieties to maintain a constant food supply. Other varieties are planted for good leaf and pod production or fast cooking time. Similarly, in Rwanda 18 bean varieties are grown.[24] One of these varieties is almost completely destroyed by pests under normal conditions. However, when drought occurs every 5 to 15 years, this variety produces a harvest when the others do not, and it is for this reason that farmers continue to grow it.

Another advantage of diversity in the garden is that it can provide a number of crops that meet the same needs. This adds variety to the diet, increasing production stability and diminishing the risk of reduced food or income should one crop fail. For example, a West African garden can provide dark green leaves from amaranths, jute, baobab, or cowpea, and oil-rich nuts and seeds from groundnuts, sesame, or egusi melon seeds. An Egyptian oasis garden produces many popular tree fruits such as guavas, pomegranates, apricots, and oranges which can be sold in local markets or consumed at home as delicious, high-vitamin treats.

In southern Mexico new housing projects do not provide as much garden area as local people are accustomed to having. A nutritionist working with children in the area found that these new gardens contained significantly less crop diversity than more traditional village gardens in the same area.[25] She also found that as garden crop diversity decreased, so did household vitamin C intake. Research on homegardens in Java found that the greater the cropping diversity the higher the overall nutrient production/m^2, and the higher the production of vitamins and minerals, in particular.[26]

14.2.3 Conserving and Using Genetic Diversity: How and for Whom?

Whether the goal of feeding the world population in the future is pursued primarily through industrializing world agriculture or improving on indigenous agriculture, genetic diversity will be essential.[27] By contributing resistance to pests, disease, drought, and poor soil, and to improved processing, cooking, and nutritional quality, the genetic diversity contained in folk varieties and wild crop relatives is a valuable resource both for gardeners and farmers, as well as for government and private commercial plant breeders.

Folk varieties and wild crop relatives are essential as sources of resistance when industrial varieties succumb to diseases and pests. This happens regularly and leads to what is called the "breeding treadmill" or "varietal relay race" as breeders rush to replace varieties that have "broken down" with new ones containing a new source of resistance.[28]

But the need for crop genetic diversity is even more critical for low resource gardeners and small-scale farmers in marginal areas. Diversity is one of their most important resources, giving them the flexibility to adapt and survive, minimizing the risk of working in an unpredictable, harsh environment. When gardeners and farmers are less vulnerable to environmental stress and input shortages, they can be more self-reliant and have more control over their own food system.

One of the most important sources of genetic material is the crop's center of diversity. The **center of diversity** is the area in which the species has its greatest genetic variety, in number of varieties of both that species and of related ones. The center of diversity can be, but is not necessarily, the area where the species originated or was domesticated, and where diversity is maintained and increased by crossing with related wild and weedy species. Varieties from areas outside the center of diversity may also possess valuable characteristics because they have adapted to different conditions.[29]

The growing destruction of both the natural environment and indigenous farms and gardens means that genetic diversity is rapidly disappearing. Loss of genetic diversity is also occurring for most major commercial vegetables in the industrial world, where diverse, open-pollinated local varieties are being replaced by a few hybrids marketed by multinational corporations.[30] Once all plants of a species or crop variety are gone, the information they carried in their genes can never be recovered.

There are two major ways of saving crop genetic diversity. *In situ* (or on-site) conservation is maintaining genetic diversity in gardens and fields, or in wild natural areas. *Ex situ* conservation is collecting and preserving plant genes in seed or gene banks, away from the environment where they are growing. More and more of our genetic diversity is being stored in freezers in gene banks for plant breeders creating industrial varieties. For those who assume that progress means eliminating indigenous agriculture, there is little need for alternative approaches to crop genetic conservation.

However, along with an increasing number of people, we believe that indigenous gardening and farming have important cultural and biological value and in many situations offer advantages over industrial food production (Part II).[31] According to this perspective, in situ conservation of crop genetic diversity is essential and offers some of the following advantages:

- Genetic diversity is widespread and available to local gardeners and farmers; it is not controlled by a distant bureaucracy unfamiliar with local needs whose main purpose is to serve the needs of plant breeders.
- Genetic diversity is not isolated from its growing environment and subject to deterioration and accidental loss.
- In situ conservation is not dependent on expensive equipment which may break down, or on government or international financing which can fail.
- At the same time that in situ conservation maintains crop genetic diversity, it maintains local knowledge of plants and their uses, and diversity in the food system; it also helps support the communities that depend on it.
- Growing folk varieties and encouraging wild and weedy crop relatives not only conserves genetic diversity, but supports socially and environmentally sustainable food production.

This is not to say there is no use for ex situ conservation. Ex situ conservation can still play an important role as a complement to in situ efforts. Small-scale, low-technology, locally controlled community or regional seed banks can avoid many of the shortcomings of larger, more expensive ones, and provide a valuable back-up in cases of emergency. For example, in Ethiopia researchers from that country's Plant Genetics Resources Centre are working with farmers to identify important folk varieties. They are establishing regional seed banks to maintain reserves of these varieties and make them available to people who have lost their own seed stocks due to war and famine.[32]

No matter where genetic diversity is preserved, very controversial questions remain about the control and use of this diversity. In the capitalist world economic system, genetic resources have become a commodity; they are being turned into property with a monetary value which can be owned and sold. There is now a heated debate about new laws created in the industrialized countries enabling plant breeders to patent crop varieties.[33] A *patent* is a legal contract which, for a designated time, gives an individual or organization the sole right to produce and sell a particular commodity such as seeds for a new crop variety.

Such laws favor the industrialized nations and their commercial plant breeders over the Third World which has few commercial breeders or the facilities they require. The cost of large-scale commercial breeding programs is prohibitive for Third World countries, therefore most of the patents and profits go to industrialized nations.

Another reason for the current controversy is that much of the genetic material used in modern breeding programs for the world's major food crops comes from centers of crop genetic diversity in the Third World. For example, although the United States and Canada make up one of the world's major food-producing regions, all of their 20 major food crops (measured in quantity produced) originated in other areas, mostly in the Third World.[34] Over 40% came from Latin America (e.g., maize, tomatoes, and potatoes), 36% from the Middle East (e.g., wheat, grapes, and apples), and 4% from Africa (e.g., sorghum and millet). Of the remaining 20%, 16% are from China and Japan (e.g., rice, soybeans, and oranges).

As the center of diversity of most major food crops, the Third World would be the source of the so-called raw materials used to develop many patented crops. Until now these materials have been seen as the common heritage of all people and collected free of charge. However, when genetic material is used in the breeding program of a commercial seed company it is that company, not the Third World, that will earn the profits from it. There is even a chance the product may end up being sold to the Third World. Third World gardeners and farmers would then purchase a new crop variety created with genetic material from folk varieties which they and their families have been developing for generations.

This aspect of the debate is an excellent example of how values and assumptions have a major effect on scientific activity and national and international policy. Current laws and procedures for control of plant genetic resources and compensation for them is based on the assumption that the work of a few individual breeders in laboratories and test plots over several years is more worthy of recognition and compensation than the work of a farming community over many generations. Factors contributing to this value-based assumption include the disregard of the skills and knowledge of indigenous farmers and gardeners, and a world economic system that favors the rich and powerful.

Since 1975, coordinating the collection and seed banking of crop genetic diversity worldwide has been the responsibility of the International Board for Plant Genetic Resources (IBPGR), a part of the Consultative Group on International Agricultural Research (CGIAR). CGIAR is controlled by private donors, primarily from the industrial nations, causing many Third World countries to fear that they may lose control over their own genetic resources.[35] In fact, most major collections of genetic material are in seed banks in industrial nations or at CGIAR member organizations.[36] Exceptions to this are the growing number of small, independent seed banks being established by regional conservation groups.

Concern about the pattern of genetic resource control and use resulted in demands by some members of the FAO that the world community recognize the contributions of Third World ecosystems, gardeners, and farmers to the world's genetic resources. Two suggestions of how to do this have been: 1) extending the principle of free exchange to include not only gardeners' and farmers' folk varieties and wild plants in Third World countries, but also the plant breeders' varieties and elite breeding lines which are currently considered private property, or 2) recognizing "farmers' rights" and offering compensation to Third World countries for their plant genetic resources, as is now given to commercial plant breeders and seed companies.[37] More recently discussion of this last idea was continued at an international conference on plant genetic resources in Madras, India.[38] The consensus statement issued by the group of participants from both industrial and Third World countries calls for a US $500 million fund to compensate Third World countries for the use of their genetic resources and to assist their efforts to conserve them. The fund, representing only 3% of the global seed industry's annual sales, would be maintained through subscription fees paid by industrial countries based on the size of their commercial seed industry and their use of Third World genetic resources. Many questions, including how and to whom such a fund would be distributed remain to be resolved. Meanwhile the larger debate about control and ownership of genetic resources continues.

14.3 Seed Saving

Saving of seeds by gardeners and farmers is the only way to preserve the full diversity of locally adapted folk varietes. Seeds from many garden crops can simply be collected in the garden and stored, while others need to be processed and dried for best results.

Seed saving is selection in action. Gardeners look for seed produced by plants with desirable characteristics such as drought and heat adaptation or pest and disease resistance. In addition to characteristics affecting production, the flavor, texture, size, color, and cooking and storage qualities of the food the plants produce are also important. For example, in Togo, local maize varieties are preferred over high-yielding new hybrids because the tight husks of the local ones significantly reduce damage by grain beetles during storage.[39]

Still other characteristics are sought in garden plants used for medicine, crafts, and other purposes. By selecting seeds of devil's claw plants (*Proboscidea parviflora*) with dark, long fibers, the most desirable characteristics for their craft, Tohono O'Odham Native American basket weavers in the southwestern United States and northern Mexico have created several new folk varieties.[40]

14.3.1 Seed Harvest and Processing

Saving seed from several of the best and healthiest plants in the garden maintains genetic diversity and reduces the risk of having poor seeds. Seeds that are saved for planting must be viable, that is, capable of growing into a healthy plant (section 6.2.4). If seeds are harvested before they are mature, development of the embryo and seed coat will be interrupted and the seed will not be viable. Seeds that are an abnormal shape, very small, or damaged in some way should not be saved. Larger seeds are best because they contain more food to support the seed embryo before and during germination.

The length of time seed can be stored depends on the type of seed, its quality, and the storage conditions. As seeds get older they become less viable, because the embryos weaken and die. Therefore the longer they are stored, the lower their germination percentage becomes (section 6.6.1). If stored for a long time, some seeds such as beans become so hard and dry that water cannot penetrate them to swell and break the seed coat. Growing out saved seed after 2 to 3 years helps avoid this problem and ensures fresh, viable seed stock. To allow for losses during storage, germination, and early growth, about 50% more seed than needed for planting should be saved.

Harvesting and processing seeds from different types of plants is described below and summarized in Table 14.2. (Section 14.4 discusses tree seeds.)

POD-BEARING Seeds of plants like fenugreek, sesame, beans, peas, and arugola are mature when the pods have dried to a light brown color and are starting to become brittle. At this time the pods are ready to *dehisce*, or pop open to distribute the seeds. Pods of some crops such as peas, do not easily dehisce, but others, such as arugola, do so at the slightest touch, throwing seeds quite a distance (Figure 14.3). If dehiscing is a problem, seed stalks can be harvested while the pods are turning brown but are not brittle, and left to finish drying, out of the sun in a bag or on a piece of cloth, where the seeds can be easily collected.

Figure 14.3 Arugola Pods Dehisce Readily

Table 14.2 Seed Harvest and Processing

Crop	When to harvest seeds	Processing
Pod-bearing: e.g., pulses, okra, crucifers, sesame, peas	Just before pod dries, while it is turning brown but still pliable	Dry pods on cloth, paper, mat, so that seeds can be collected when pods open
Cucurbits: e.g., squash, gourds watermelons, melons	Squash and gourds: at least 6 weeks after fruit is considered ripe; watermelon: when fruit is eaten; other melons: when fruit is overripe	Wash seeds and dry slowly in shade
Peppers	When fruit is ripe; remove seeds as fruit is eaten, either fresh or dried	Dry seeds
Soft, small-seeded fruit: e.g., tomatoes, tomatillos, eggplant	When fruit is ripe to overripe	Fermentation (see description in text); for eggplant no fermentation, wash seeds and dry
Seed-bearing flower heads: e.g., cilantro, Niger seed, sunflowers, amaranth	Just before seed head becomes completely dry and brittle	Cut seed heads, lay on cloth or in bag, when dry remove seeds by rubbing
Maize	Past milk stage, when color has developed, may be left on plant until dry if no pest or mold problems	Dry seeds on cob, husk may be left on or removed
Fruit trees: e.g., mango, cashew dates, citrus, papaya, jujube, stone fruits	When fruit is completely ripe	Remove fruit flesh and clean seed; plant fresh, dry, and/or stratify, depending on crop

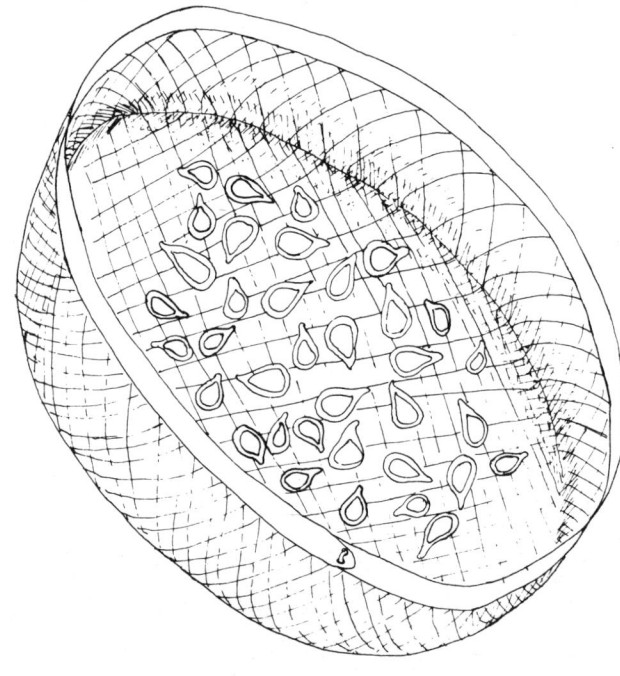

Figure 14.4 Squash Seeds can be Dried in a Basket

CUCURBITS The seeds of most squashes, pumpkins, and gourds continue to mature even after the fruit has reached its full size and is ripe for eating. Keeping mature fruit from which seed is to be saved in cool, dry storage for 6 weeks or longer, known as *after-ripening*, ensures time for seed development.[42] When slightly immature fruit is picked, for example after an early frost, viable seeds can sometimes be saved if the fruit is allowed to after-ripen. When the seeds are removed from the fruit they should be rinsed and separated from the pulp. Any small, flat seeds that float while being washed can be composted, as they are hollow and not viable. Seeds should be dried in a well-ventilated place like a basket (Figure 14.4).

Watermelon seeds are mature when the ripe fruit is eaten, and they can be rinsed and dried immediately. Leaving other melons to continue ripening for a few days after they are first ready to eat allows the pulp and seeds to separate more easily. After cutting the melon open, the seeds are rinsed, and all remaining pulp is removed before they are laid out to dry.

PEPPERS Seeds from sweet peppers and hot chilis are obtained from fruit that has matured on the plant to a red, orange, or black color, depending on the variety. The fruit can be either fresh or dried (Figure 14.5). Seeds taken from fresh fruit may need rinsing before drying.

SOFT SMALL-SEEDED FRUIT Tomatoes and tomatillos contain many small seeds, so only a spoonful of pulp provides more than enough seeds for most gardens. The mashed pulp of mature fruits from several selected plants is placed in an uncovered container such as a bowl or jar. The pulp is left to ferment (section 15.6) with occasional stirring. Fermentation takes 3 to 7 days depending upon the air temperature:[41] the warmer the air temperature, the less time required. A sour smell and bubbles on the pulp's surface are signs that fermentation is occurring.

Fermentation destroys microorganisms on the seeds which can cause some diseases, and it thins out the pulp, allowing the heavier seeds to separate and sink to the bottom (Figure 14.6). Fermentation also removes the gelatinous coating on these seeds, changing their texture from slippery, to rough and nonslippery. This can be felt by rubbing them between two fingers. This stage of seed processing is complete when the seeds

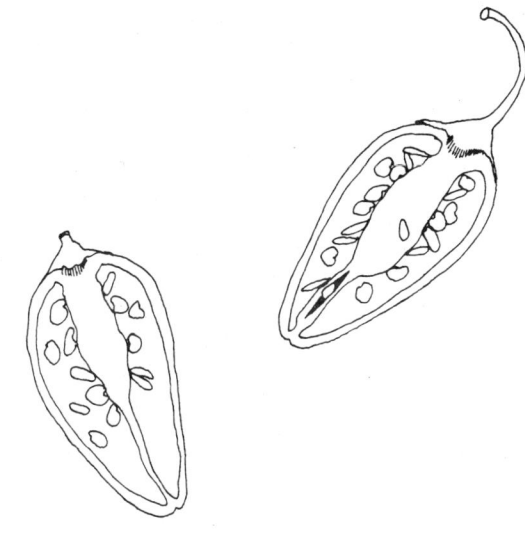

Figure 14.5 It is Easy to Collect Seeds from Chilis

and pulp have separated. Any seeds floating near the surface are hollow and not viable. These and the pulp can be skimmed off the surface and composted. The viable seeds are rinsed with water and laid on a cloth, piece of screen, or similar material, to dry in a place protected from the wind and direct sun.

Eggplant or garden egg seeds do not need to be fermented. The pulp of soft, overripe fruits from several selected plants is mashed. The seeds are separated by rinsing them with water, after which they are laid out to dry.

SEED-BEARING FLOWER HEADS Cutting off the seed heads of plants like sunflower, onion, carrot, amaranth, cilantro, and chia just before they have completely dried is a great help in collecting their many seeds. The seed heads are placed in bags or on cloth in

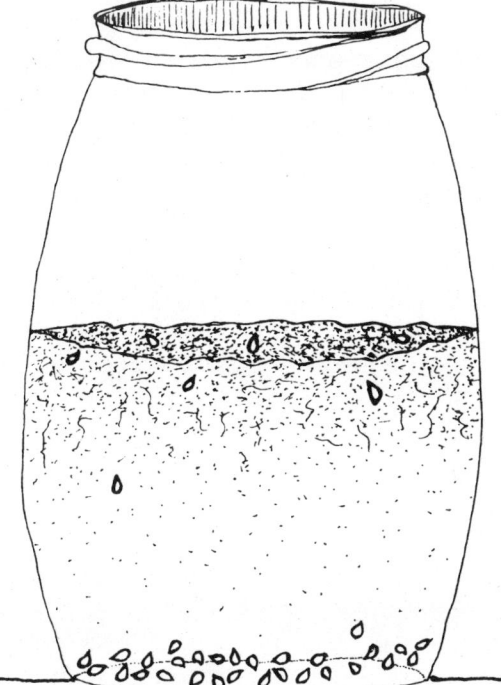

Figure 14.6 Fermentation and Separation of Tomato Pulp and Viable Seeds

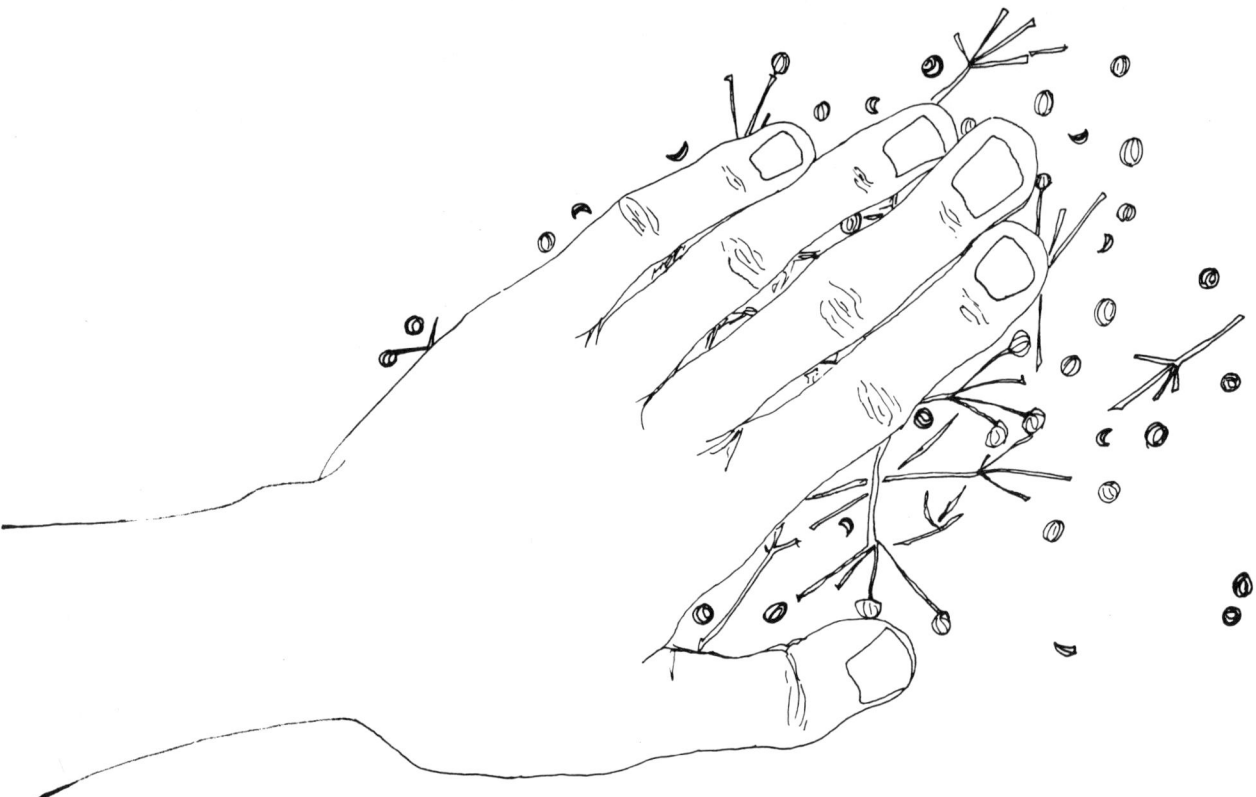

Figure 14.7 Gently Rubbing Dried Cilantro Seed Heads on a Firm Surface Easily Removes the Seeds

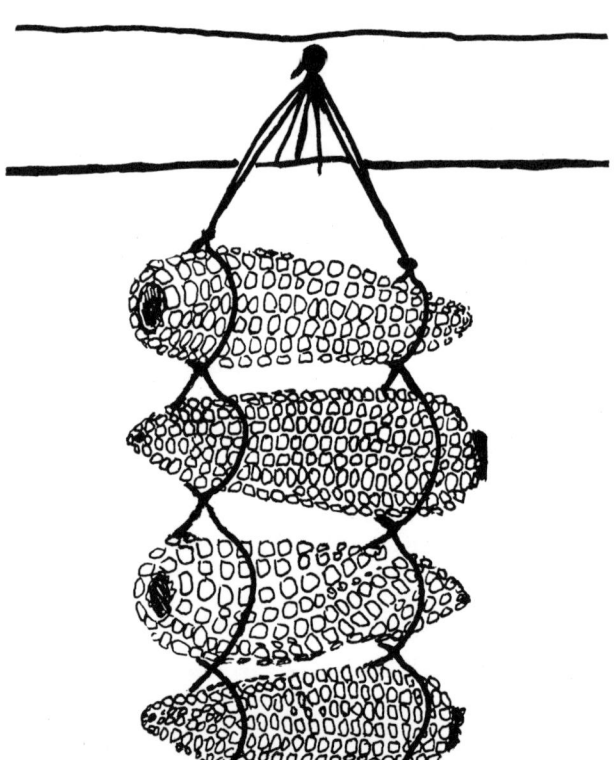

Figure 14.8 Maize Hung to Dry for Planting Seed in Northern Mexico

the shade to finish drying. When brittle and dry, seeds are easy to remove by rubbing or shaking the seed heads (Figure 14.7).

MAIZE A good indicator of seed maturity in maize is the browning of the *husk* or leaves around the ear. If birds and other pests are not a problem the ears do not need to be harvested until the whole plant has died. However, in very hot areas the sugars in the kernels may start to ferment inside tight husks, especially in the sweet varieties. The fermentation causes the kernels to explode, destroying them as seeds. To avoid this the husks can be opened slightly although this may allow pests inside. If the kernels have passed the stage when the juice inside is "milky," and have developed their mature markings or color, they can be harvested and allowed to dry in the shade with their husks open or removed.

Maize is frequently stored on the cob which may reduce some pest damage to the softer part of the seed where it was attached to the cob. There are many traditional ways to store maize seed, either with or without the cob or husks: in containers, hung in bunches, or strung up in hanging racks as is done in parts of northern Mexico (Figure 14.8).

14.3.2 Seed Drying

All seeds must be dry before storage. Small hard seeds harvested in the dry season can be stored immediately with no further drying. Larger, moister seeds usually require extra drying after harvest. When seeds are spread out to dry, turning or mixing them several times a day speeds drying and helps prevent mold. In dryland areas with high daytime temperatures (greater than 35°C or 95°F), it is best to dry seeds in the shade to avoid the danger of overheating and overdrying, which can damage the seed coat and embryo. Sometimes if drying is too rapid, case hardening can occur in larger, moist seeds such as those of squash and melons. **Case hardening** is the drying of the outside surface while the inside is still moist. The moisture trapped inside the seed encourages the growth of fungi and bacteria,[43] and can also attract insects. However, some people such as the Tohono O'Odham of the southwestern United States and northern Mexico have a long tradition of drying seed in the sun with good results.[44] The exposure to sun and heat may also help rid the seeds of some insect pests. If a successful tradition like this does exist, understanding how it works and supporting its continuation is the best approach.

Baskets, pieces of cloth or mats, calabashes, and pots can all be used for drying seed. Seeds may need protection to prevent birds and insects from eating them and to keep insects from laying eggs which may later hatch in storage. An upside-down basket (Figure 14.9), or a piece of cloth stretched over the container, protects the seeds while still allowing air circulation.

Rubbing dry seeds between hands or on a piece of

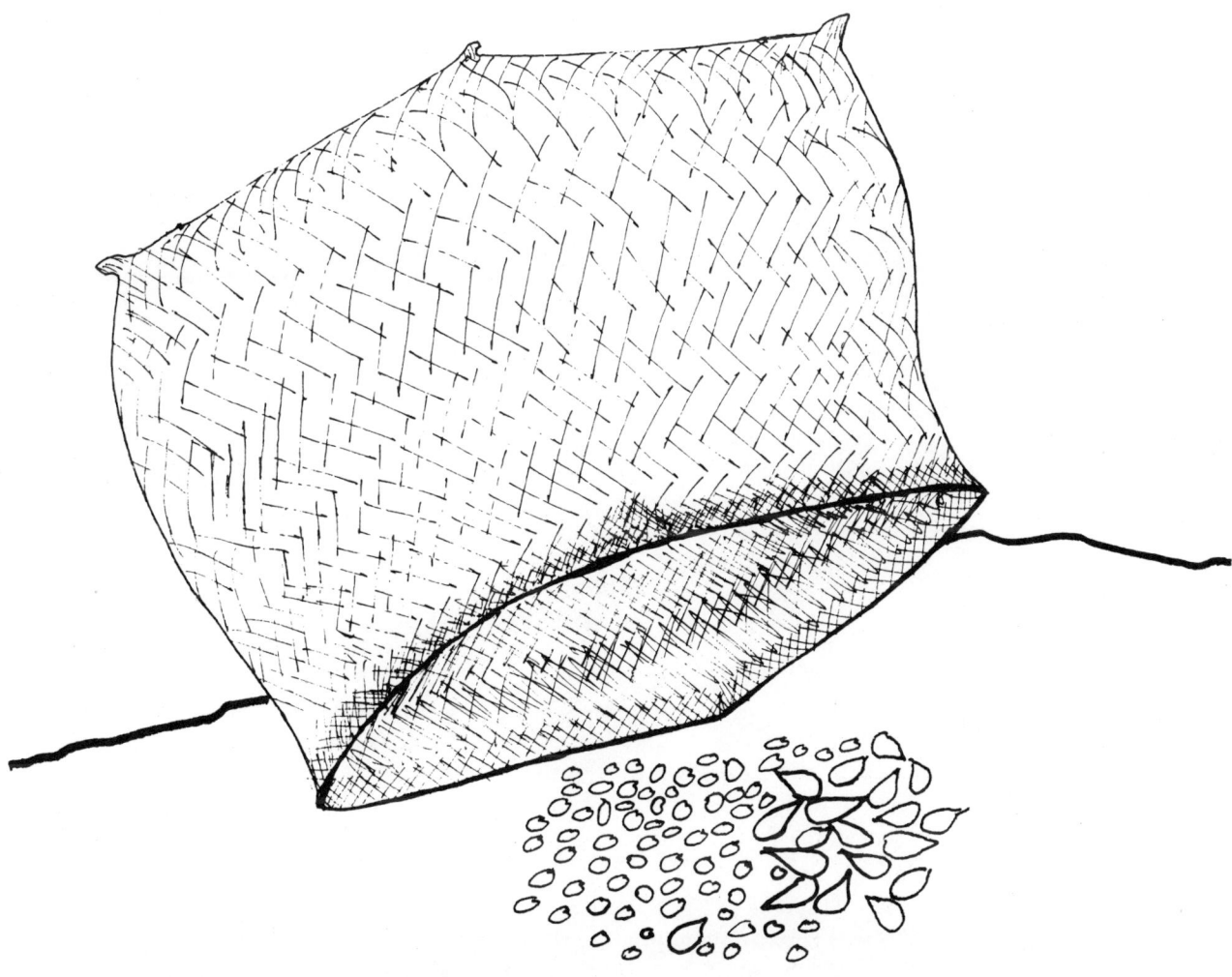

Figure 14.9 An Upside-Down Basket can be Set on Top of Drying Seeds to Protect Them from Birds, Insects, and Wind

cloth will separate seeds that may be stuck together. This is easy if the seeds are clean and dry. Only gentle rubbing is needed; hard pressure can scratch the seed coat and result in disease problems or drying of the embryo. Seeds from moist fruits such as squash, melons, tomatoes, chilis, and eggplants are brittle when dry, and will break if bent in half.

Seeds with some sort of pod or casing such as pulses are easier to store and plant if separated from this casing. There may be exceptions to this when pods offer some protection in storage. To remove pods after drying, any mature pods that have not opened can be threshed or rubbed to break them open, then winnowed in a light breeze as is done with grain. When dropped from shoulder or waist height the heavier seeds will fall straight down into an awaiting container while the pods and other debris will be blown away. Before storing, large pod-borne seeds should be dried for several days after removing their pods.

After drying, it is a good idea to keep seeds for several days at the same temperature at which they will be stored. If the seeds do not feel damp and do not stick to each other during this time they are probably dry enough for storage. The length of time to dry seeds varies greatly depending on the air humidity, drying conditions, seed size, and how clean the seeds are.

14.4 Saving Seed from Trees

Although many trees are propagated vegetatively (Chapter 7), some are grown from seeds with good results. Seedlings are also grown as root stock for later grafting. Seedlings have a stronger root structure than plants started from cuttings, especially during the first few years, and therefore are hardier under stress. However, if the tree is dioecious the sex of the seedling will not be known until it flowers, which could take many years. This is an important reason why some trees are not propagated from seeds.

To ensure that the seed collected is fully developed, only healthy, mature fruits, including those that have just fallen from the tree, should be gathered. The seed of some dryland garden trees such as olive, date, cashew, carob, and baobab can be easily stored for later planting. First, any fleshy part of the fruit or sweet pod which could attract pests or host seed-damaging bacteria and fungi should be removed. The seed is then dried in the shade for a few days, depending upon its size. Finally, seeds are stored in a cool, dry, ventilated place, in a breathable container such as a basket or bag.

Cashew seed can be stored for 7 to 12 months.[45] Olive, date, carob, and baobob remain viable for several years, although germination percentages may drop (section 6.6.1). We have planted baobob seed stored for over 15 years with excellent germination.

The hard outer coat of olive and stone fruit seed provides good protection during storage and can be cracked when it is time to plant. Seeds for planting pistachios are picked when the hard outer hulls turn blue green. These hulls are also useful for storage, but are thought to inhibit germination, and so they should be carefully cracked or removed before planting.[46]

Seeds of mangoes, avocados, and citrus, should be planted when fresh, although limited storage may be possible. In eastern Senegal, cleaned mango seed is briefly air dried in the shade and then stored for up to 100 days packed in ceramic containers with moist charcoal.[47] The charcoal acts as an evaporative cooling system keeping the seeds cool and moist but not wet. Similarly, clean, partially dried citrus seed can also be stored for short periods in ground charcoal.[48]

14.4.1 Cold Stratification

Cold stratification is the process of chilling seeds, which is required for good germination of some seeds and the production of healthy seedlings. Even if these seeds do germinate without stratification, the seedling is frequently dwarfed and growth is abnormal. Dryland tree seeds that require cold stratification are those of species or varieties that grow in high altitude or latitude drylands with a marked cold season, such as the stone fruits, olive, jujube, and pistachio. The seeds of these and other trees whose fruits ripen in these areas during the late summer and fall often need cold stratification.

Good temperatures for cold stratification are 2°C to 7°C (36°F to 45°F), but they can be lower.[49] The best way for dryland gardeners to stratify their tree seeds is to leave them exposed to the cold winter weather of their area. Selected seeds should not be stored in the house or other places where they would be protected from low winter temperatures. They can be buried in the ground, or in a container filled with moist sand or soil and left outdoors for the cold season, and then removed and planted when the cold weather has passed. Seeds can also be planted directly in the ground approximately 15-20 cm (6-8 in) deep at the beginning of the cold season.

The Navajo Native Americans living in the southwestern United States grow peaches, a crop introduced to the area by the Spanish several centuries ago. At the end of the hot season, fruits are cut up and dried for storage, and their seeds discarded nearby.[50] The seeds are stratified by exposure to the winter temperatures. The following spring the viable seeds produce seedlings, the best of which are selected and transplanted to permanent growing sites.

14.5 Seed Storage

Good seed storage conditions include low moisture and temperature, and protection against rodents and insects. High temperatures and moisture encourage seed-damaging fungi and bacteria and increase respiration, shortening the seed's life. Extremely high temperatures can kill the seed. Locally available storage containers and additives can prevent or minimize pest damage to stored seeds.

A sealed, airtight container can keep out moisture, rodents, and insects. A calabash or clay pot, for example, can be plugged and sealed with clay or wax. A piece of cloth dipped in hot wax can be draped over the container opening to seal it. The Tohono O'Odham of the Sonoran Desert traditionally stored seeds in small ceramic vessels called *hahawa*.[51] A piece of broken pottery was trimmed to fit the vessel's mouth where it was sealed in place using the heated sap of a local tree. Lidded jars and wooden or metal boxes also work well. On the other hand, a container that is closed but not sealed allows the gardener to check periodically on the condition of the seeds. However, this may lead to problems with moisture if the outside air is humid. Leather pouches, pots, cans, boxes, or jars can all be used. Table 14.3 summarizes dryland seed storage problems and responses.

14.5.1 Moisture and Temperature

In some dryland areas moisture may not be a problem and periodically opening containers or storing seed in closed but breathable containers like unglazed clay pots, baskets, cloth, or leather pouches is fine. However, even in very arid areas the increased humidity during a short rainy season can lead to damage of the stored seeds, greatly reducing their viability. Sealed containers of glass, metal, or glazed clay are nonbreathable or airtight and will keep out moisture (Figure 14.10).

Toasted grains or pulses can be added to absorb excess moisture in an airtight storage container.[52] The grains or pulses are toasted by slow heating without burning, which dries them out completely so they readily absorb water. As soon as the toasted materials have cooled to room temperature, they are mixed with the seeds and put into an airtight container which is then sealed. The stored mixture should contain twice as much toasted material as seed. Each time the container is opened, the old toasted grain or pulses must be replaced with some that are freshly toasted. Fresh ashes absorb moisture given off by stored seeds and are a good additive for this purpose, as well as for pest control as discussed in section 14.5.2. If available and inexpensive, corn starch, salt, and baking powder can also be used to absorb any moisture seeds give off. However, unlike ashes or toasted grains, these additives should not be mixed in directly with the seeds but kept in a cloth or paper bag inside the seed storage container. These additives, especially salt, draw so much moisture out of the seeds that seeds can become desicated and die if they are surrounded, for example, with salt. When using any additive it is a good idea to check the condition of the seeds regularly.

Table 14.3 Summary of Dryland Seed Storage Problems and Responses

Storage problem	Response
High temperatures	Shade, insulate, keep away from source of heat, take precautions for moisture
Moisture (encouraging fungi and bacteria)	Dry seeds thoroughly, add toasted grain and/or ashes to container, avoid heat
Rodents	Keep storage area clean, use sealed containers, "guards" on container legs, or stone-based container
Insects	Keep storage area clean, use sealed container, use additives; e.g., sand, ashes, smoke, herbs, oil

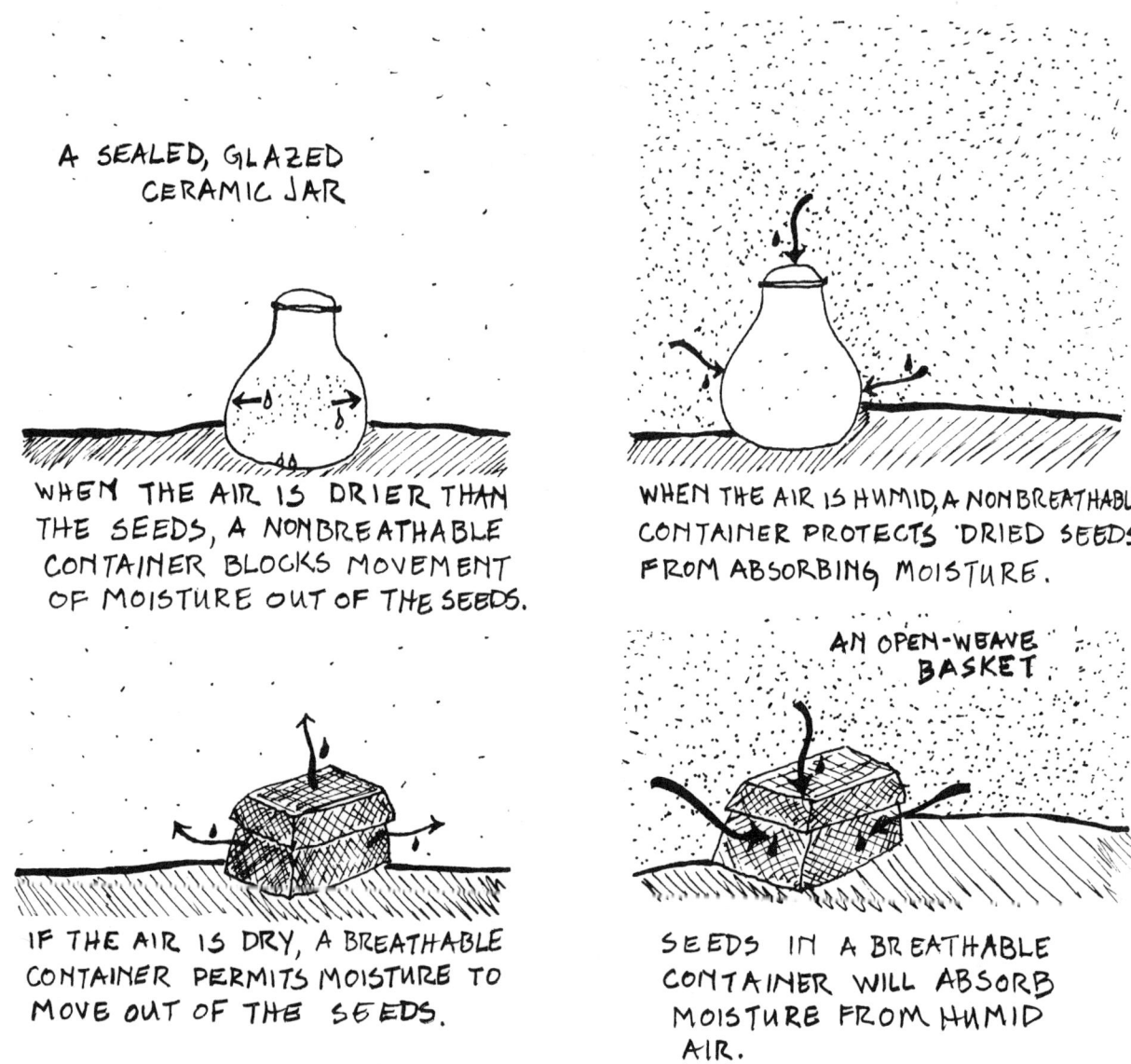

Figure 14.10 Breathable and Nonbreathable Storage Containers

Closed seed storage containers should be shaded from the sun and kept away from cooking fires or walls heated by the sun or fire. Even well-dried seeds contain some moisture, and will produce water vapor if they get hot. Containers with thick walls offer some insulation from rapid temperature changes.

14.5.2 Pest Control

There are two ways to deal with pest problems in stored seed. One is to repel pests by making the seeds or the storage environment undesirable. The second method is to kill the pests. Indigenous methods of pest control in stored seed and foods operate by repelling or killing pests while manufactured pesticides act by killing them.

As discussed in Chapter 13, manufactured synthetic pesticides pose a number of problems: they must be purchased; their availability may not be reliable; directions for correct and safe use may not be available or accessible; and they are poisonous not only to pests but to humans and other animals as well (section 13.2.4). Pesticides can not only poison those handling them or seeds treated with them, but can affect others now and in the future through contamination of rooms, containers, soil, and water.

Some indigenous, as well as recently developed methods for controlling pests in stored seed using inexpensive, locally available materials have been found to be just as effective as manufactured pesticides.[53] These methods are practical and much safer, and there is growing interest in them worldwide.

Just like people, insects and rodents living in a dry environment are attracted to moisture, so storage containers and areas should be kept dry. Keeping the storage area clean is also important because an unclean storage area attracts pests, offering them not only food but also places to hide or nest (Figure 14.11).

Some insects like weevils (Apionidae family) and bruchids (Bruchidae family) can lay their eggs on seeds such as beans while they are still in the garden. If conditions are right, the eggs will hatch later and the larvae will eat the seeds. Holes in the seeds and powder at the bottom of the container from the damaged seeds are signs of these pests and their small white larvae can probably be found inside or around the seeds. These and other pests in stored seeds will die if they cannot obtain enough oxygen. Therefore, storing seeds in sealed, airtight containers with as little extra air space as possible will help to reduce pest damage. This does not limit the amount of oxygen enough to harm the seeds themselves.[54]

Rats, mice, or other rodents can also be kept out by sealing the containers, or by storing seeds in containers with leg guards that prevent rodents from reaching the seeds. Indigenous storage bins for grains frequently include these leg guards (Figure 14.12). Another method used in some parts of Sahelian West Africa is to build the storage bin on a stone base which rodents cannot climb up or chew through.[55] Similar smaller containers may be appropriate for seeds, or a container of garden seeds can be placed inside such a bin.

In addition to the condition and care of the storage area, another way to reduce pests is by using a repelling substance, such as a local plant. For example, the compound azadirachtin contained in the leaves and seeds of neem trees (*Azadirachta indica*), which grow in the drylands of Africa and Asia, has been found to be an effective insect repellent.[56] One method used by farmers in western India and Pakistan is to grind the leaves into a paste which is mixed with clay and formed into seed storage containers.[57] In West Africa dried, powdered neem leaves are also used as an insect repellent.[58] Spicey, strong-smelling dried chili peppers and onion leaves, used in southern Nigeria in stored cowpea seed, may also repel weevils to some extent.[59] Some Native Americans used wild tobacco leaves

Figure 14.11 A Clean Storage Area with Sealed Ceramic Seed Jars in Burkina Faso
(After Dupriez and De Leener 1983:157)

(*Nicotiana rustica*) to repel insects from stored seeds.

Certain additives operate primarily by killing pests but may also repel them. Fine sand can be added to seeds in a ratio of one or more volumes of sand to one volume of seed.[60] Because it fills the spaces between seeds and because of its weight the sand prevents insects from moving around easily. Adult bruchid beetles, common pests in stored beans, are unable to move around enough to mate and reproduce in beans mixed with sand, and so the population dies out. The sand also scratches the thin wax coating of the insect's outer cuticle and its delicate limb joints, causing it to dry out and eventually die.

Dust additives operate the same way as sand although some may scratch while others absorb the insect's protective waxy layer, exposing it to dehydration. Ashes and finely ground limestone are other common additives that fill spaces between seeds and absorb insect's waxy protective layer.[61]

In the West African country of Togo, one volume of bean seeds is thoroughly mixed with between one and two volumes of cooking fire ashes.[62] The ashes should also completely cover the surface of the stored seeds. Some people feel that the ashes from burnt goat and cattle dung or from burning certain local trees are particularly effective.[63] Sand mixed with the seed coats of *Polygala butracea* or with the leaves and husks of *Cassia nigricans* provided good protection of cowpeas stored in locally made clay jars in northern Togo.[64]

Figure 14.12 A Traditional Storage Bin in Iran with Wooden Disks as Leg Guards Against Rodents
(After NAS 1978:73)

Another additive used to protect stored seeds from insect damage is oil. In Nigeria, weevil infestations in dried cowpeas are minimized by coating the seeds with a layer of groundnut or palm oil, (approximately 5-8 ml of oil/kg of seed, 0.1 fl oz of oil/lb of seed).[65] The oil is believed to act both chemically as an insect repellent, and mechanically by sealing air out of the peas and preventing growth of any weevil larvae they contain. In rural India, castor bean (*Ricinus communis*) oil is used to protect stored seed.[66] Similarly, neem oil was found to be very effective for controlling pests in stored cowpeas in northern Togo.[67] Hulled, ground neem seed powder was kneaded by hand to squeeze out the oil. These cowpeas had a germination rate of 27% after 8 months of storage, compared with only 2% for untreated seeds. Those stored in sand, however, had a better germination rate of 47%.

In many areas smoke is traditionally used to protect stored food, seed, and even houses from pests.[68] Seed or food can be stored over the cooking fire or in a raised container under which a fire is periodically built (Figure 14.13). The important points to remember about this method of pest control are not to overheat the seeds or food by placing them too close to the fire, and to only use containers such as loosely woven baskets that allow ventilation so that any moisture released due to increased temperatures will not be trapped inside. Bundles of unthreshed grains or pulses can be treated with smoke, but care must be taken that they do not dehisce while in storage.

14.6 Resources

Control of crop genetic resources has become an important issue of international debate, in part because it reflects larger questions about the relationship between the industrial and Third Worlds, the "north" and the "south." This debate is represented by a rapidly growing number of publications on genetic resources including *Shattering* (Fowler and Mooney 1990), *The Gene Hunters* (Juma 1989), *First the Seed* (Kloppenburg 1988) and *Altered Harvest* (Doyle 1985). For Spanish speakers and readers there is Daniel Querol's *Recursos Genéticos, Nuestro Tesoro Olvidado* (1988). In addition, this issue is being discussed now in both popular and academic periodicals all over the world.

Botany books can be good sources of information on plant genetics. For example, *The Biology of Plants* (Raven, et al. 1981:115-166) has a section on genetics and evolution. Cox and Atkins (1979:513-536) look at plant and animal genetics in agriculture.

Resources with simple practical information about seed saving and storage for the low-resource gardener are GTZ (1980) and Stoll (1987). The new On Farm Seed Project was started by the private US organization Winrock International Institute for Agricultural Development (see Chapter 19 for the address). The proj-

Figure 14.13 Using Smoke to Protect Stored Seeds and Grain from Pests (After FAO 1970:161)

ect is conducting experiments and workshops on on-farm seed storage methods in Senegal and The Gambia. They also publish a bilingual (French and English) newsletter *Seed Sowers/Les Semeurs*.

References

1. African Diversity 1990.
2. Cleveland and Soleri 1989.
3. Cleveland and Soleri 1987.
4. Brush 1986; Brush, et al. 1988.
5. Soleri and Cleveland 1989.
6. Oldfield 1984:27-32.
7. Richards 1986:131-146.
8. Plucknett, et al. 1987:3.
9. Atlin and Frey 1989.
10. IITA 1986:52-55; ICRISAT 1986:15, 33, 54.
11. Andrews 1989.
12. Cox and Atkins 1979:553.
13. Cleveland and Soleri 1989.
14. Gould 1988.
15. Buttel, et al. 1985; Doyle 1985:94-113,167-172.
16. Goldburg, et al. 1990.
17. Cox and Atkins 1979:520-521; Doyle 1985:1-17; NAS 1972; Oldfield 1984:22-23.
18. Cleveland and Soleri n.d.b.
19. Cleveland 1986.
20. Grivetti 1978.
21. Clawson 1985.
22. Richards 1985:98-99; Richards 1986:133-138,142,149.
23. Ferguson and Sprecher 1987.
24. Dupriez and De Leener 1987:199-200.
25. Dewey 1981:35.
26. Marten and Abdoellah 1988.
27. Harlan 1976; Plucknett, et al. 1987.
28. Crisp and Astley 1985; Plucknett et al 1987:3-4, 17-26.
29. Oldfield 1984:27-32.
30. Crisp and Astley 1985.
31. Oldfield and Alcorn 1987.
32. Fowler and Mooney 1990:206-207.
33. Kloppenburg and Kleinman 1987.
34. Kloppenburg and Kleinman 1987.
35. Kloppenburg and Kleinman 1987.
36. IBPGR 1987.
37. Kloppenburg and Kleinman 1987.
38. African Diversity 1990; Keystone Center 1990.
39. Zehrer 1980:52.
40. Nabhan and Rea 1988.
41. Leon and Withers 1986:70.
42. Leon and Withers 1986:72.
43. NAS 1978:50-53.
44. Nabhan, et al. 1981.
45. Garner, et al. 1976:193.
46. Hartmann and Kester 1983:628.
47. Bittenbender 1984:81; Tuck 1985.
48. Purseglove 1974:512.
49. Hartmann and Kester 1983:135.
50. Jett 1979.
51. Nabhan, et al. 1981:6.
52. DCFRN #8-1.
53. Zehrer, et al. 1980.
54. Zehrer 1980:115.
55. Zehrer 1980:48.
56. Ahmed and Grainge 1986; NAS 1980:114.
57. Ahmed and Grainge 1986:204.
58. Zehrer 1980:120.
59. Otuya 1986.
60. Zehrer 1980:110.
61. NAS 1978:55.
62. Zehrer 1980:114.
63. DCFRN #4-2.
64. Zehrer 1984:453-460.
65. King, et al. 1985:218.
66. Zehrer 1980:121.
67. Zehrer 1984:459.
68. FAO 1970:158-160; Dupriez and De Leener 1983:156; NAS 1978:55; Zehrer 1980.

15
Processing, Storing, and Marketing Food from the Garden

The majority of fruits and vegetables are most nutritious when eaten fresh from the garden without processing or storage of any kind. Yet year-round harvests may be limited by the climate, garden space, or lack of time. Also, some garden crops like mangoes or peaches can produce a larger harvest at one time than can be used. For example, market gardeners in southern Senegal say that at least 50% of their abundant, peak-season mango harvest rots each year.[1] In these situations processing and storing some of the produce is a way to save it for use later in the year.

In many areas there is a long tradition of food processing and storage which is important economically, nutritionally, and socially (Figure 15.1). However, in some cases the time and resources required for processing and storing garden produce just do not make sense under local conditions. This will be decided by the gardener or whoever else is doing the work.

The goals of processing garden produce are to:
- Maintain the best possible nutritional value.
- Require as little time and as few resources as possible.
- Use inexpensive, locally available inputs.
- Produce foods that appeal to local tastes and do not cause illness.

Marketing is another way gardeners or their households can use the garden harvest, and income is a strong incentive, especially for poor households. Marketing a small amount from a household garden is often an easy way to earn a little money, but marketing a large amount of produce, is far more complicated and risky. Reducing the gardener's dependence on factors beyond his control is the best way to cope with this.

15.1 Summary

Fresh garden crops harvested when ripe should be eaten or processed quickly for the greatest nutritional benefit. Most fruits are best when harvested ripe and handled gently to avoid bruising, although some may be harvested while still unripe.

Preservation and processing of garden foods affects different nutrients in different ways. Vitamins are the most sensitive: the longer the time from harvest to consumption, and the longer and hotter they are cooked, the greater the loss.

Drying is one of the easiest and most effective ways to preserve dryland garden produce if a few simple guidelines are followed. Sprouting, malting, drying, and fermenting of some foods increases the nutritional value per unit weight. Many popular indigenous foods are prepared these ways. Some foods, like olives, African locust beans, and cassava, can only be eaten after processing to remove toxins and anti-nutrients. Processing also produces foods that taste good.

Garden produce can be stored fresh or processed. Fresh produce has a relatively short storage life. Dried garden foods can be stored much longer using simple techniques to eliminate pests and microorganisms that cause spoilage.

Special considerations when harvesting garden produce for market include finding reliable and affordable transportation, picking fruits before they are fully ripe, and packing produce to protect it during transportation. Once at the market, simple measures such as shading and sprinkling fresh garden produce with water protects its quality and appearance.

15.2 Harvesting Garden Foods

Most garden foods have the highest nutrient content and best flavor when they are harvested as close as possible to the time they will be eaten or processed. The exceptions are pulses, cereals, sunflower and ses-

Figure 15.1 In Many Areas there is a Successful Tradition of Food Drying

ame seeds, and other crops that are allowed to ripen and dry on the plant before harvesting.

Harvesting fruit or leaves from perennials or young annuals can affect the plant's future production. For example, when harvesting apricots, care should be taken not to damage the fruiting spurs which will continue to produce fruit for 3 to 4 years. When harvesting leaves, enough must be left on the plant for adequate photosynthesis for future growth.

In some parts of Nigeria both the leaves and pods of okra are eaten. Studies there found that if only leaves on the lower half of the plant are harvested, pod production will not be reduced.[2] In Zambia, researchers working with Ethiopian mustard greens found that harvesting up to half of the total leaf area every 1 or 2 weeks increases leaf production.[3] However, harvesting three-quarters of the leaf area causes yields to drop. Research on the effects of harvesting cowpea leaves concluded that moderate leaf harvesting reduced pod production but the combined edible yield of pods and leaves (in dry weight) was greater for most varieties than the yield when only pods were harvested.[4]

As fruits ripen, there is an increase in the amount of vitamins they contain, especially vitamin C. Sugar content also increases, making them more flavorful. A color change from green to yellow, orange, or red is a sign of ripening in many fruits like tomatoes, peppers, stone fruits, papayas, mangoes, and bananas. Other fruits are green when ripe, but become softer, like avocados, or softer and sweet smelling, like some melons and guavas.

Many fruits bruise and spoil easily and should be carefully harvested and handled. Including a little of the stem, or *pedicel*, directly above the fruit when harvesting it helps reduce spoiling (Figure 15.2). Removing the place at the top of the fruit where it is attached to the plant often exposes some of the moist flesh under the fruit's skin, which easily rots. With dates, removing the attachment, or *cap*, allows dirt and insects to get deep inside the fruit. However, no matter how carefully harvested, the stem may drop off of very ripe fruit and this fruit should be eaten or processed as soon as possible.

Some fruits can be eaten before they are fully ripe. Unripe fruits add new flavors and textures to the diet and can be used differently than ripe ones. For example, both sweet peppers and chilis can be used when green. Green papayas and tomatoes can be cooked to reduce their bitterness. Green mangoes are used to make relishes, and in some areas such as southern Sudan we have seen them eaten fresh, sprinkled with

Figure 15.2 *Harvesting Garden Fruits with the Stem still Attached*

lime juice, salt, and chili powder. Bananas and plantains are higher in starch and lower in sugar before they ripen, and unripe ones are cooked and eaten.

Unripe fruits may also be eaten when other food is scarce. In southern Pakistan, for example, people whose food supplies are running low before the dates ripen, process and eat unripe dates.[5] The bright green, unripe dates called *kimri* are mashed in a basket, releasing some of their tannin-loaded juices. These *kimri* are then put in a ceramic jar which is wrapped in a blanket and left overnight. By the next day the *kimri* have turned a mud color and lost much of their bitterness.

15.3 Cooking and Using Garden Foods

Cooking can improve the nutritional value of food and helps soften or break down the cellulose in plant cell walls, making the nutrients inside available for digestion. It destroys toxins present in some uncooked pulses such as lima beans, cowpeas, and lentils. Cooking can also destroy some nutrients. However, great efforts need not be made to retain high levels of a nutrient in one food when the nutrient is abundant and readily available in the diet. For example, if a child is eating fresh guavas every day, the loss of vitamin C from cooking malted beans should not be a concern. Table 15.1 summarizes suggestions for minimizing nutrient losses in cooking.

15.3.1 Fresh Foods

The greatest concentration of nutrients is usually in the outer layers of fruits and vegetables, so trimming and peeling should be kept to a minimum. Gardening without toxic chemicals means that lots of peeling and trimming is not necessary. In urban areas, garden produce grown above ground, like leaves and fruits, often has toxic lead residues from the lead in gasoline (petrol) used in vehicles. This is especially true of gardens grown near busy city streets. Washing this produce in a mixture of vinegar and water can remove most of these residues.[6] If this is not possible it is a good idea to peel city-grown fruits, especially for children.

The hotter and longer most fresh fruits and vegetables are cooked, the more vitamins will be destroyed. Even though boiling and steaming does not raise the temperature much above 100°C (212°F), the boiling point of water, it can still destroy vitamins. Vitamin C is the most sensitive to heat, although vitamin A is also affected. A study of eight vegetable leaves commonly eaten in Ghana showed vitamin C losses of 44-78% when boiled for 10 minutes in a covered container with just enough water to cover the leaves.[7] Frying in oil can be much hotter than boiling or steaming, and therefore destroys more of these heat-sensitive vitamins.

Water-soluble vitamins, like vitamin C, niacin, riboflavin, and thiamin, are dissolved out of foods by water. To minimize these losses:

- Avoid cutting food into small pieces, exposing more surface area to air and water.
- Avoid soaking fresh fruits and vegetables before cooking (exceptions include some root crops like cassava).
- Cook for the shortest possible time with a minimal amount of water.
- Drink vitamin-rich cooking water, or use it in other dishes.

Table 15.1 Cooking and Nutrient Content of Foods

Nutrient sensitivity	Garden examples	Suggested cooking method
Heat sensitive (vitamins A, C, and thiamin)	Many fruits and vegetables	Where appropriate, eat produce fresh, soon after harvest; keep cooking time brief and temperatures low; avoid frying
Water soluble (vitamin C, niacin, riboflavin, thiamin)	Tomatoes, peppers, squash, greens	If cooking with water, try steaming with minimal amount of water; use this cooking water in other foods
Alkaline sensitive (vitamin C, thiamin)	Greens, pulses	Do not add bicarbonate of soda or ashes while cooking

- Do not salt raw fruits and vegetables because this draws out water containing dissolved nutrients.

Both cutting and cooking will cause losses in nutrients due to *oxidation*. Vitamins A, C, E, and folacin oxidize, meaning that their chemical structure is altered, and their nutritional value reduced. Exposure to the air (which contains 20% oxygen) and to heat both increase oxidation. For example, an orange cut open long before it is eaten loses vitamin C by oxidation.

Raising the pH of cooking solutions by adding bicarbonate of soda, ashes, or other alkaline substances can shorten the cooking time and improve the color of some vegetables. However, it destroys vitamin C and thiamin in foods. Similarly, acids contribute to the destruction of carotenoids, an important source of vitamin A in fruits and vegetables.[8]

15.3.2 Dried Foods

Dried fruits are a flavorful treat and can be eaten with or without first cooking them in water. Other dried foods such as green leaves, onions, okra, or tomatoes, are cooked before being eaten and are often used for making sauces or soups. Unlike fresh fruits and vegetables, presoaking most dried foods is recommended because it shortens cooking time and saves precious fuel, and the soaking water can be used for cooking. Dried leaves and small fruits only need presoaking for an hour or less. Pulses and tubers, which are bigger than leaves, have a greater volume and need to be soaked longer, often overnight. One volume of dried food can absorb two volumes or more of water (Figure 15.3).

Soaking dried pulses such as pigeon peas or chickpeas dissolves anti-nutrients like tannins and phytates into the water (section 2.10).[9] In these cases the soaking water should be poured on the garden, and fresh water used for cooking.

The seed coats of pulses are high in fiber which is fine for most adults but should be removed when making weaning foods or food for someone with a stomach or intestinal infection or diarrhea. Seed coats can be removed by soaking beans or seeds and rubbing the coats off while they are still wet, or by parching.[10] *Parching* is soaking pulses in oil or water and then drying them so that their seed coats crack and can be easily rubbed off. Lightly roasting groundnuts (which are not soaked) makes it easy to remove their papery coats.

15.4 Food Drying

Drying is one of the oldest and most widely used methods of processing food for storage. In Egypt, for example, a popular green called *mulukhiyah* (jute) is grown or purchased in large quantities during the warm season. The washed leaves are stripped from the stalks and dried on palm-fiber mats. Some people partially dry the *mulukhiyah* in the sun, then move it indoors to complete the drying, while others dry it entirely in the shade.

In central Mali the Dogon grow bunching onions in dry-season gardens.[11] The onions are harvested in two stages so that they may be dried for sale or household use. First, the green onion tops or leaves are removed. These are then pounded into a pulp which is formed into balls, with any extra liquid being squeezed out. These balls ferment and are left to dry in the sun for about 10 days. The day after removing their tops the onion bulbs are dug up. If possible they are eaten or sold immediately. Otherwise they are also pounded and formed into balls which are fermented and sun-dried for later sale or home use.

Drying preserves foods by removing the water which spoiling microorganisms need to grow. Since fresh fruits and vegetables contain about 80% water,[12]

Figure 15.3 One Volume of Dried Garden Produce can Absorb Two or More Volumes of Water

drying reduces their volume and weight substantially. It also concentrates their nutrients and preserves them for times when these nutrients may not be available in fresh foods. For example, an experiment in Senegal found that on average, after drying and storage for 6 months a 100-gm (3.5-oz) piece of mango contained 100% of the RDAs for vitamins A and C for children.[13]

Drying is a quick and easy method for preserving many garden foods, including leafy greens, okra, onions, tomatoes, eggplants, squash, roots, and tubers. Sweet fruits such as dates, cashew apples, figs, peaches, apricots, mangoes, bananas, papayas, and loquats do not need to be dried as completely as vegetables because their high sugar content also acts as a preservative (section 15.4.3).

High temperatures make drying go faster by increasing the rate of evaporation. However, sunlight destroys vitamins A and riboflavin, and high temperatures destroy vitamins A, C, folacin and thiamin[14] by increasing oxidation. One study showed that cowpea leaves dried in the open sun kept only 11% of their vitamin C and 42% of their vitamin A compared with 24% and 57% for leaves dried in the cooler shade.[15] Flavor and color are also lost by exposure to heat and light.

The best way to dry food is in the shade, using warm air to evaporate the moisture out of the food. Good air circulation is essential because high humidity encourages bacteria and fungi, which spoil the foods, ruin their taste, and cause illness. Many of the same principles and procedures discussed in Chapter 14 for drying and storing seeds are also appropriate when drying and storing food.

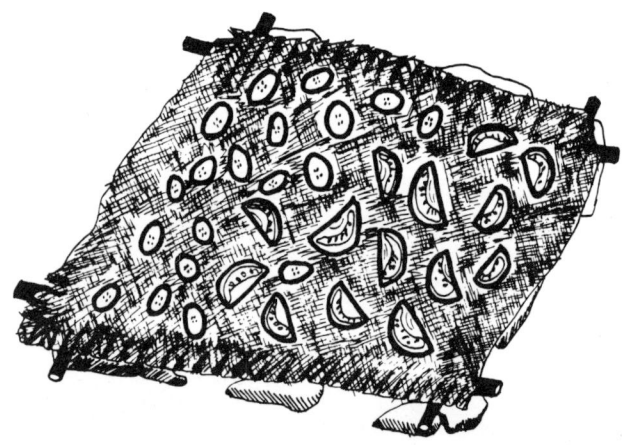

Figure 15.4 Food Drying on a Woven Mat

15.4.1 Materials for Drying

Woven mats or baskets are good drying surfaces and can also be used to cover and shade drying food because they are rigid and well ventilated. In addition, mats or baskets are often locally made. Cloth, netting, or wood can also be used. Painted surfaces or metals other than stainless steel should not be used because food can absorb bad flavors or poisons from them. Soft, wet fruits like tomatoes, bananas, and peaches must be laid out on a clean surface such as a woven mat, shallow basket, or wooden tray (Figure 15.4). Bunches of leaves or chilis and slices of firm-fleshed vegetables like eggplant, squash, and tubers such as sweet potato can be tied together and hung up to dry.

15.4.2 Preventing Contamination

Many foods, especially fruits with high sugar content, will darken as they dry. Drying foods should be examined for signs of spoiling if there is any other dramatic change in their color or surface texture. The drying area should be kept clean and free of dirt, sand, and rocks which can stick to the food and make it unpleasant and dangerous to eat. Protecting food from wind, which could coat it with sand and dirt, is also a good idea.

There are simple ways to protect drying food from insects such as ants, weevils, or flies which may eat it or lay eggs on it (Figure 15.5). A basket or other covering which permits air circulation across the foods protects them from flying insects and birds while providing shade (Figure 14.9 in section 14.3.2). Ants and other crawling insects can be discouraged by placing the dryer up on legs or on a pedestal standing in containers of water. Another method is to ring the legs with tree sap or other locally available sticky materials (Figure 15.6) so that anything crawling into it becomes stuck and will not reach the food.

15.4.3 Selecting and Preparing Produce for Drying

Most fruits, tubers, roots, and leaves for drying should be picked at the same stage of growth as for eating fresh. For example, fruits should be picked when ripe. The best time to pick herbs such as mint or oregano for drying is just before their flowers open. This is when they are most flavorful because they contain the highest quantity of aromatic oils. Any produce that is overripe or bruised will spoil, and contaminate other

15 Processing, Storing, and Marketing Food from the Garden

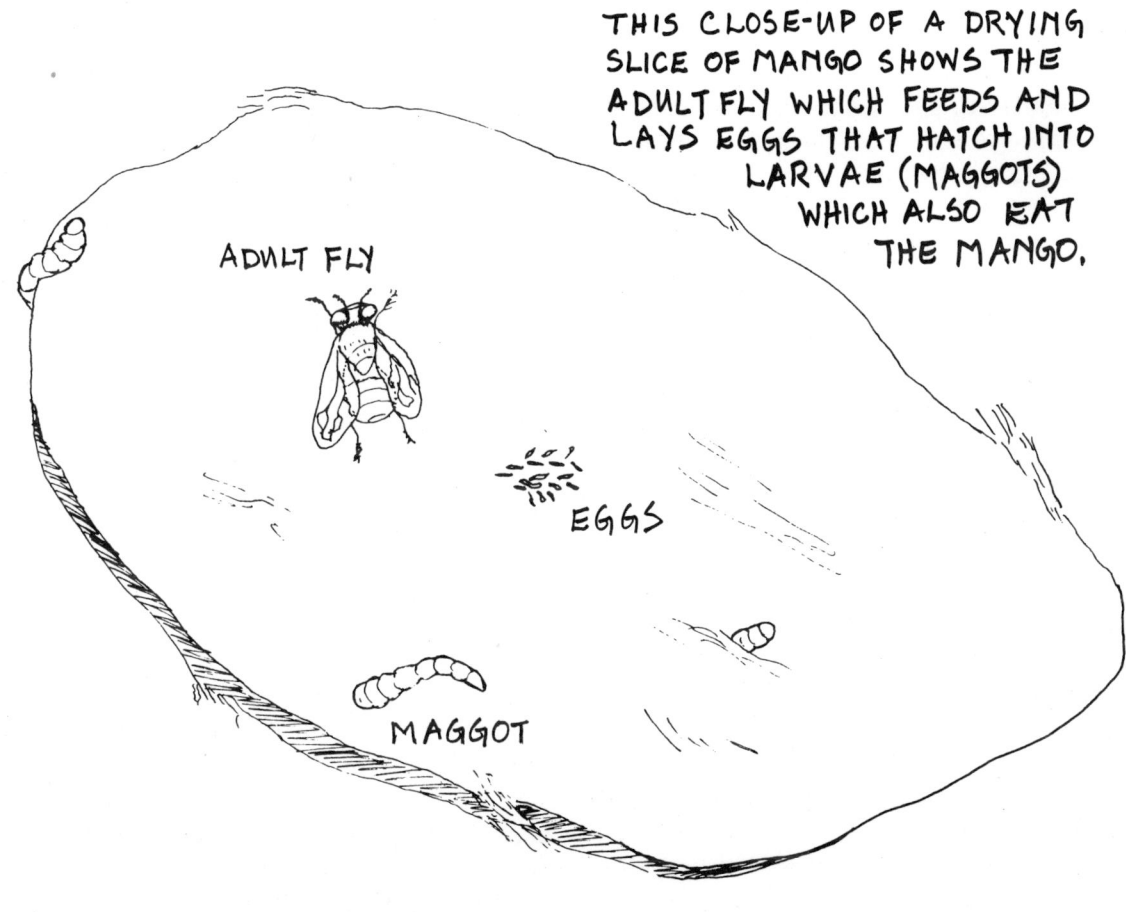

Figure 15.5 Some Insects will Lay Eggs on Drying Food

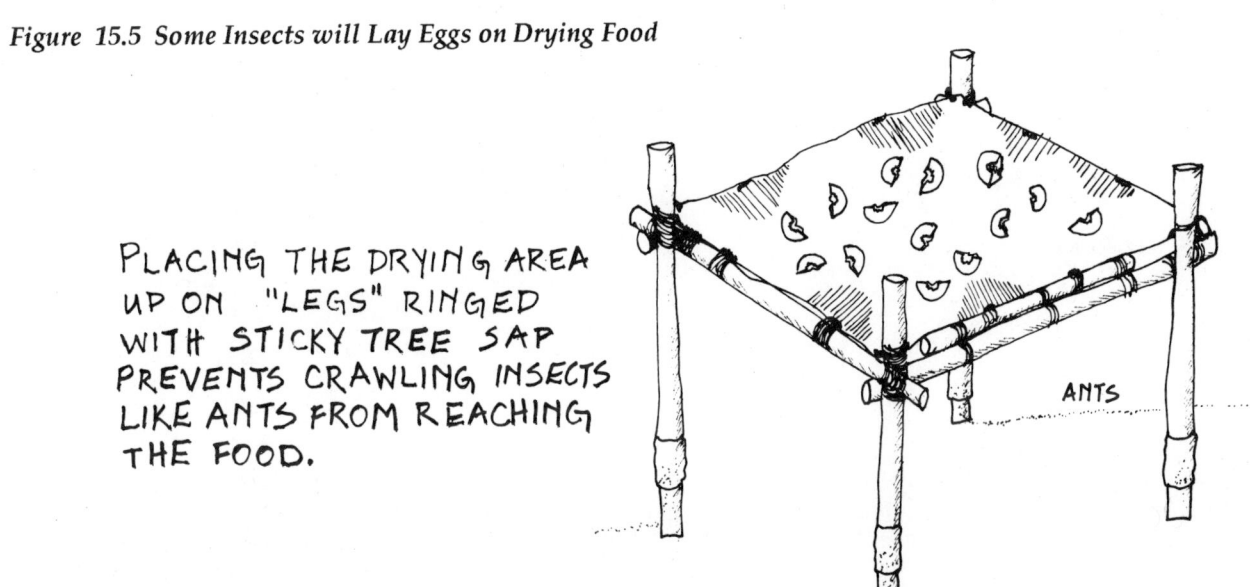

Figure 15.6 Protecting Drying Food from Crawling Insects

food stored with it. Washing the produce after harvesting removes dirt, insects, and their eggs. If the rind of the fruit or vegetable, such as some squash, is not to be eaten, it is often easier to remove it before drying.

Drying methods for different types of dryland garden produce are summarized in Table 15.2. Blanching is an optional process discussed in Box 15.1. Whether fruits and vegetables are sliced for drying depends on their size and moisture content. The thicker the slice the longer its drying time and the greater the chances of case hardening if the weather is hot and dry. Case hardening results when the exposed surface of the food dries faster than the inside, forming a barrier and sealing in the remaining moisture. Fungi and bacteria can grow easily in this moisture, spoiling the food. Cutting the food up into many thin slices, however, exposes more surface area to oxidation and results in a greater loss of vitamins. The best approach is to keep these considerations in mind and experiment.

Leaves such as amaranth are dried whole because they are so thin. Small okra may also be dried whole. Chilis, onions, and garlic are often left whole and hung in strands for drying and later storage. Bunches of leaves and herbs may also be dried this way. Tomatoes, eggplant, and okra, and root crops such as yams, sweet potatoes, and carrots, are often sliced into pieces approximately 1 cm (0.4 in) thick so that they dry thoroughly and quickly. Native Americans of southwestern North America sometimes dry cut up squashes and peaches on their rooftops (Figure 15.7), or hang long strips of squash from wooden poles to dry for later use.

Dates and grapes are dried whole, but larger, juicier fruits should be sliced. Peaches, loquats, and apricots can be cut in halves or quarters. Cutting mangoes, bananas, and papayas into slices about 1-2 cm (0.4-0.8 in) thick helps them dry faster.

Drying time will depend upon the climate, the food being dried, and how it will be stored. Turning the food over occasionally helps it dry more quickly and evenly. The longer the dried food is to be stored, or the hotter the storage conditions, the drier it should be. Leaves, chilis, roots, and tomatoes are usually dried until brittle. The high sugar content of fruits such as mangoes, bananas, stone fruits, figs, and dates acts as a preservative, therefore, they need be dried only until pliable and leathery. For example, the moisture content of leaves dried for storage is approximately 6-10%,[16] while dried sweet fruits containing up to 25% moisture can be stored without spoilage.[17] In some drylands such as southern Italy, tomatoes, sweet peppers, and eggplants are dried until leathery then layered in containers with herbs and garlic and covered with oil. The oil preserves and protects the vegetables from contamination so they need not be dried as thoroughly as they would if stored without the oil.

Table 15.2 Summary of Drying Methods

Type of produce	Preparation and drying method
Leaves: e.g., amaranth, other greens, cabbage leaves, herbs	Whole, on a flat, ventilated surface or hung in bunches (blanch 1 minute or less)
Onions, chilis, garlic	Whole, in hanging strands (no blanching)
Onions, peppers, tomatoes, okra, eggplants	Sliced, on a flat, ventilated surface (no blanching)
Root crops	Sliced, on flat, ventilated surface (blanch 3 minutes)
Cucurbits	Sliced, on a flat, ventilated surface; firm fleshed ones hung in strips (blanch 2 minutes)
Dates, grapes, figs	Figs and dates whole or split, on a flat, ventilated surface or strung; grapes in bunches or singly (no blanching)
Moist fruits: e.g., stone fruits, mangoes, bananas, papayas, loquats	Sliced, on a flat, well ventilated surface, turn regularly (no blanching)

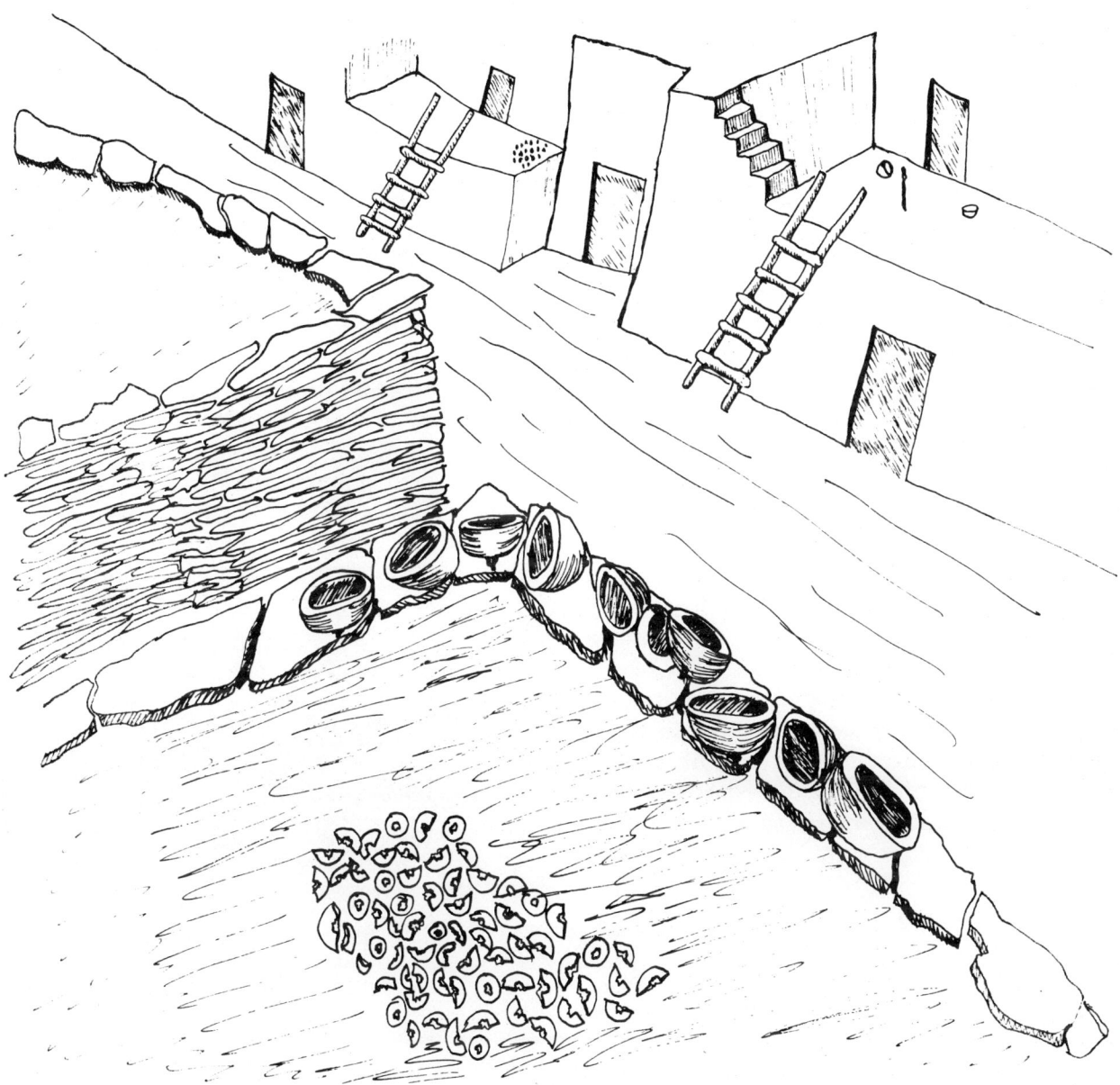

Figure 15.7 Squash and Peaches Cut and Drying on Rooftops in a Hopi Village
(After Kennard 1979:555)

15.5 Sprouting and Malting

Sprouting seeds improves their nutritional value and digestibility. Dried or toasted (malted) sprouts can be ground into flour and used to make flavorful, nutrient-dense foods (section 16.3.4).

15.5.1 Sprouting

Sprouting is the process of germinating seeds before eating or further processing. It has several benefits:[19]

- The amount and/or availability of vitamin C, iron, riboflavin, niacin, and phosphorus in some foods is increased.
- The amount of anti-nutrients such as phytates, tannins, and oligosaccharides are decreased (section 2.10).
- Starches are converted into simple sugars such as glucose, fructose, and maltose, improving taste and digestibility. This does not occur as quickly in pulses as it does in cereals.
- Since cooking is eliminated or reduced, fuel is saved.

> **Box 15.1**
> *Blanching*
>
> Nutritionists sometimes recommend *blanching*, a brief boiling or steaming, as a pretreatment before drying some foods. Tree fruits are not usually blanched, nor are chilis, onions, garlic, or herbs.
>
> Blanching does several things:
> - Kills enzymes in the food which might otherwise continue the ripening process after harvesting.
> - Softens and breaks down the cell walls making both drying and rehydration go faster.
> - Fixes the color of the food so it will not fade or turn brown.
>
> Blanching is not necessary, although it may improve the storage life, flavor, and appearance of some dried garden produce. In the dry season, getting food quickly and sufficiently dried to prevent contamination and continued ripening is usually easy. In the humid rainy season it can be more difficult, and blanching may be helpful. Otherwise, we do not recommend blanching because it uses cooking fuel and destroys nutrients, especially vitamins A and C. For example, fresh okra and amaranth lose 30% and 35% of vitamin A as well as 41% and 51% of their vitamin C, respectively, after blanching for 3 minutes in water.[18]
>
> If garden produce is blanched before drying, the following suggestions can minimize nutrient losses:
> - Instead of immersing it in water, blanch the produce with steam to reduce loss of water-soluble nutrients.
> - Steam only long enough to heat the produce thoroughly and soften it slightly. In many cases like leaf amaranth and carrots, the color will appear brighter when sufficiently steamed.
> - Change the water after each batch is steamed because sugars and enzymes from the food accumulate in the water, making the blanching less effective. This flavorful water can be drunk or used in cooking.
>
> Steaming can be done using a covered pot containing about 3 cm (1 in) of water. Clean rocks placed in the pot will support a clean piece of matting, sticks, or rigid stainless steel screen, keeping it and the food on it above the water (Figure 15.8). Wrapping the food in a cloth bundle and hanging this from a stick above the boiling water is another way to steam produce.
>
>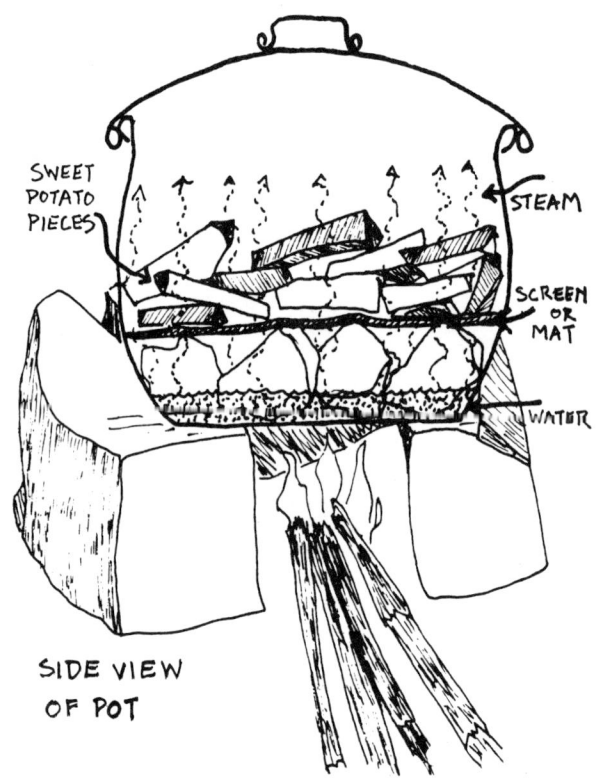
>
> *Figure 15.8 Blanching*

Mustards, radishes, sunflowers, sesame, pulses like the moth bean, cowpea, and pigeon pea, and some cereals such as wheat are among the many plants whose seeds can be sprouted. However, not all seeds produce sprouts good for eating. Plants with poisonous leaves such as tomatoes should not be used for sprouting. Sorghum seed sprouts form the poison hydrogen cyanide (HCN) (section 2.10) during digestion and should not be eaten.[20]

Sprouted seed can be eaten fresh, cooked, or malted (section 15.5.2). Sprouting takes about 2 to 4 days depending on the seed and the stage at which it is to be used (Figure 15.9). First, the seed is soaked until softened, usually overnight, depending on the seed coat and size. This water should not be drunk or used for cooking, but poured on the garden. Second, the softened seed is kept in a moist (but not wet), dark place, such as a covered bowl, jug, tray, or mat. Seeds are often spread between woven mats or leaves and sprinkled occasionally with water to keep them moist. Gently rinsing seeds, at least once a day in hot weather, helps prevent molds from growing.

When germination has occurred the sprouts are ready to be cooked or malted. Sprouts for eating raw or for cooking as a fresh vegetable can be left long enough to develop leaves if they do not get too tough. Before eating raw, sprouts should be exposed to sunlight for about 1 day, while keeping them moist and protected from high temperatures. Unlike harvested produce, the sprouts are living plants and their leaves turn green when exposed to sunlight, giving them a sweeter taste and increasing their vitamin A and C content. Before eating, the seed coat or hull can be removed by rinsing the sprouts.

15.5.2 Malting

Malting is drying or toasting sprouted seeds which are then ground into a meal or flour that can be cooked into a porridge or used for making beverages. The sprouted seeds are dried or toasted before they develop any leaves (Figure 15.10). The seed coats can be removed before the seeds are ground.

Because malting breaks starches down into simple sugars, malted flour absorbs less water than unmalted flour and so it makes a more nutrient-rich food (section 16.3). Malted porridge has a slightly sweet taste which children like. Although some heat-sensitive nutrients such as vitamin C will be destroyed during cooking, malting increases the iron, riboflavin, and niacin content of foods.[21]

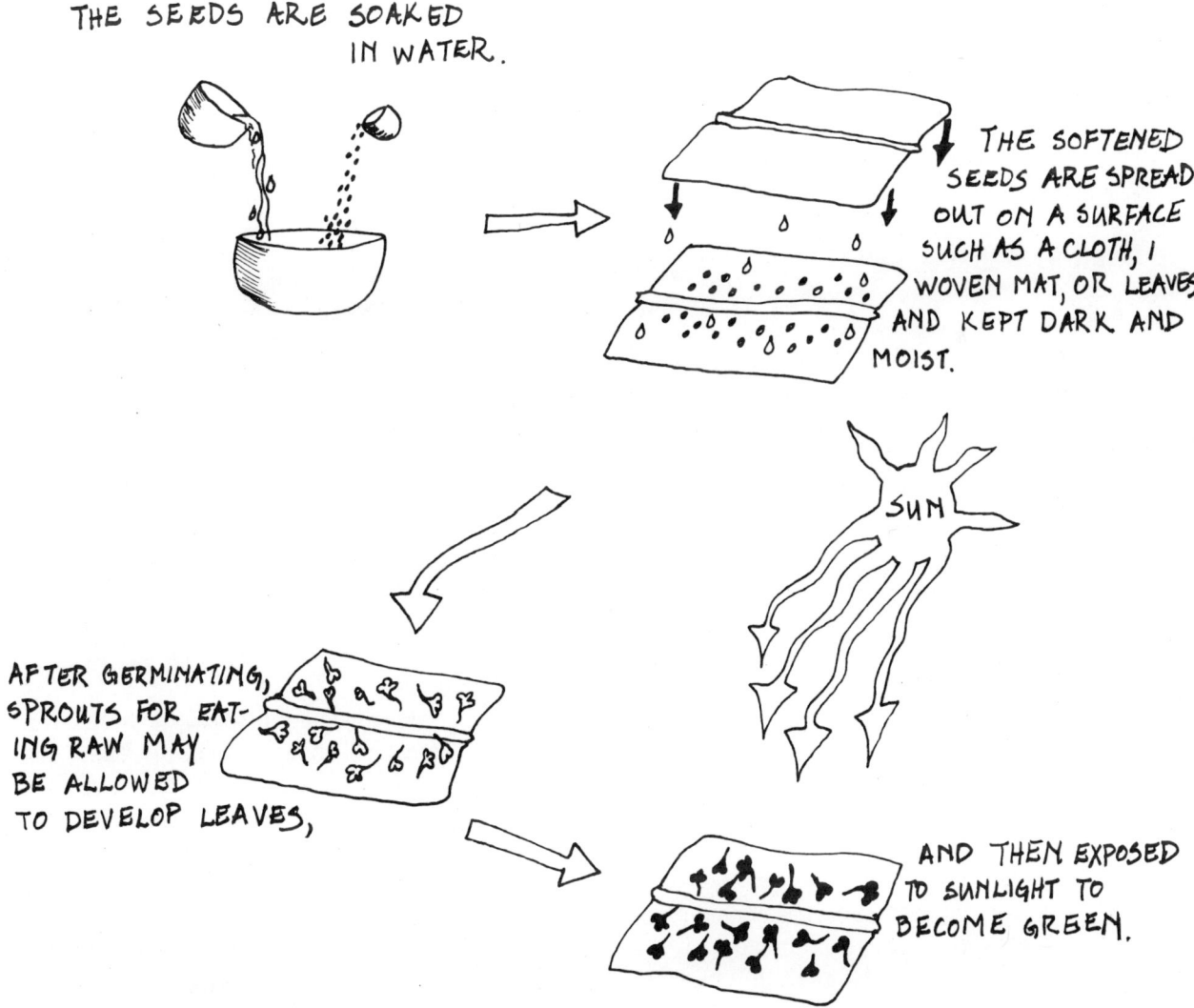

Figure 15.9 Sprouting

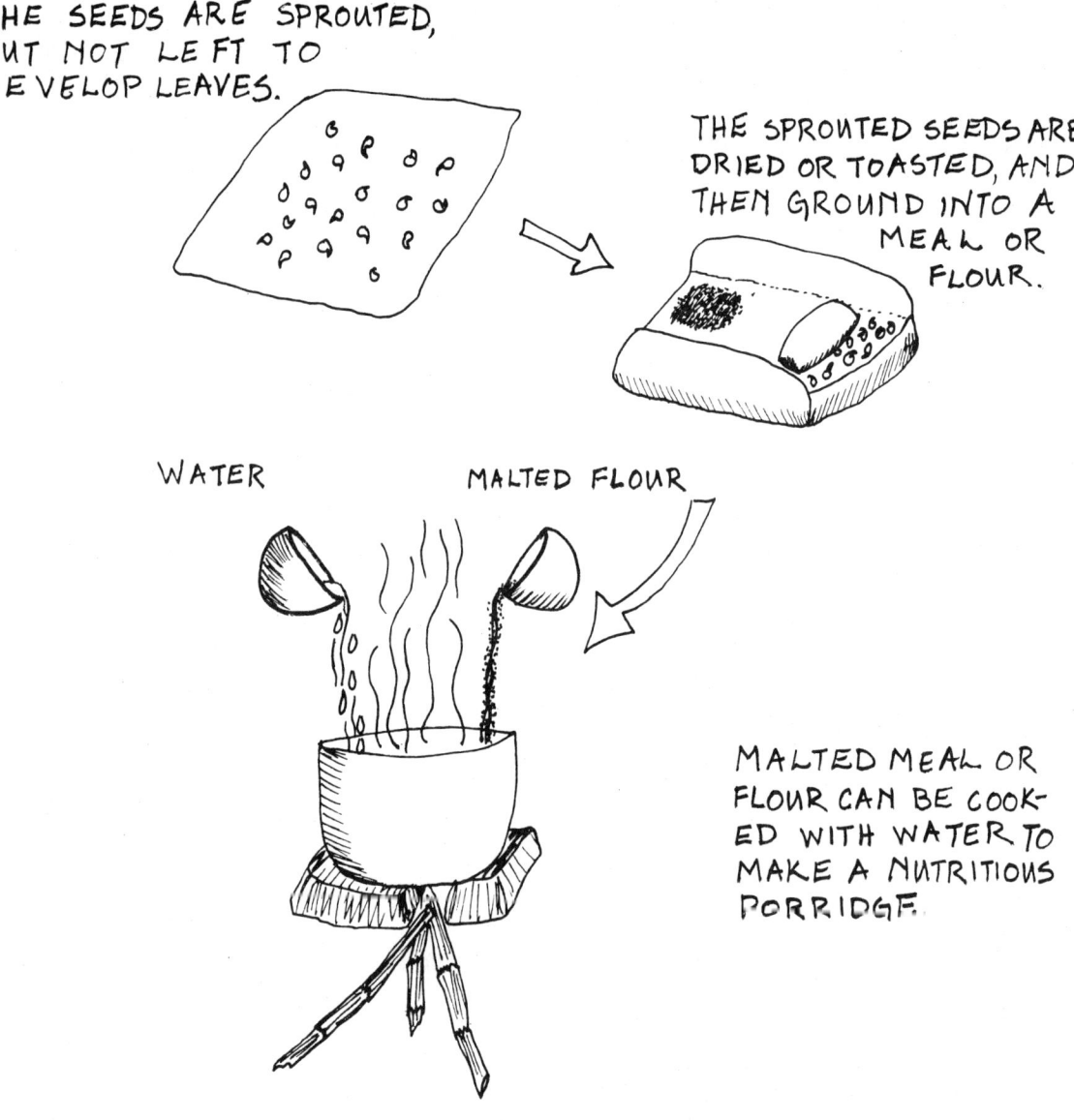

Figure 15.10 Malting

15.6 Fermentation

Fermentation is a process in which certain bacteria or fungi are allowed to grow in a food to improve its flavor and/or digestibility. As these microorganisms absorb and process nutrients and excrete waste products they affect the flavor, nutritional value, texture, and storage life of the food. The organisms break down proteins, convert starches to simple sugars, and can increase or make various vitamins and minerals more available. In other words, these organisms are feeding on and digesting the food.

Because the process of breaking foods down into their constituent parts has already begun, fermented foods can be easier for people to digest than the same food before fermentation. For example, the seeds of the African locust bean are inedible before fermentation and processing to make *dawadawa* or *iru*, a flavorful condiment used for sauces in West Africa. Fermentation can also reduce or eliminate some anti-nutrients such as phytates and oligosaccharides (section 2.10). In this book we do not have space to cover in detail how to develop fermentation techniques, and this should not be done unless expert advice is available. It is best to work with existing fermentation techniques. The following brief discussion is given to help readers understand the fermentation process and enable them to support improvements in indigenous food fermentation where appropriate.

Fermentation lengthens the storage life of foods because the presence of the fermenting organisms prevents other, spoiling organisms from becoming established. Often this is because fermentation changes the pH of the food, for example making it so acid that only *Lactobacilli* spp. bacteria will survive. This is a kind of fermentation called pickling which is discussed in section 15.6.1.

Care and experience are important in food fermentation, otherwise the wrong organisms may start growing and will spoil the food and make people sick. Traditional fermentation processes often include ways to select desirable organisms. For example, the Sudanese beverage *hulu mur* is made from sorghum, dates, tamarind, and ground spices. The spices are an important ingredient not only because they add flavor, but because they determine which organisms grow, thus influencing the kind of fermentation.[22] The spices also decrease oxidation rates, aiding preservation.

The method of fermentation varies according to the specific food and local processing traditions. The essential step is a period of "rest" during which the food is undisturbed while the fermenting organisms grow and multiply. In the case of *dawadawa*, the bean seeds are cooked several times and the seed coats removed before being packed in moist leaves for about 36 hours for fermentation.[23]

The balls of onion greens made in savanna West Africa (section 15.4), rice and pulse breads such as *idli* and *dhosai* from India, and *taamiah*, the fried bean and vegetable balls popular in Egypt,[24] are some fermented foods made with ingredients from dryland gardens. Sometimes fermented products such as *dawadawa* and onion balls are dried or smoked to halt the fermentation process at the point when the food is most flavorful and easy to preserve.

In Nepal, mustard, radish, and cauliflower leaves are fermented to make *gundruk*, a traditional condiment.[25] The leaves are dried, crushed, soaked in water, and then allowed to ferment in a warm, dark place for several weeks. Following fermentation the *gundruk* is sun dried, after which it can be stored for about 1 year without spoiling.

15.6.1 Pickling

Pickling is fermentation that results in an acidic (low pH) food. Bacteria (including many that produce lactic acid) and some yeasts, are the microorganisms responsible for this type of fermentation. In many cases the fruits or vegetables are soaked in a mixture of salt and water and/or vinegar, creating a medium that encourages the growth of the pickling microorganisms. As they multiply, the pH of the solution drops even further (it becomes more acid). High salt content in the pickling solution draws water out of the foods due to osmosis (section 5.2), also contributing to their keeping qualities.

The nutritional value of pickled foods is much lower than that of the same foods when fresh. However, they keep well, and are important as flavorful appetizers, which make the staple food more appealing.

Pickled garden foods are eaten in many parts of the world. In Egypt, carrots, turnips, onions, tomatoes, chilis, and sweet peppers are among the vegetables commonly pickled in the home using salt solutions. Vegetables are also pickled this way in Mexico and are often sold from small street stands where they are displayed in large glass jars (Figure 15.11). In Asia, leaf vegetables, root crops, citrus fruits, and mangoes are also pickled.

Olives are popular throughout the Mediterranean, Middle East, and southwestern North America. There are many ways to process olives, and fermentation is a common method. A salt brine is the medium and *Lactobacillus* spp. bacteria are responsible for the fermentation. In this case the brine should contain enough salt to keep an uncooked egg (still in its shell) afloat. The olives are left in this solution for 4 weeks to 3 months, depending on the type of olives and how ripe they are. When the olives no longer taste bitter they are placed in fresh, unsalted water for about 1 week, and the water is changed every other day. Finally the olives are marinated with an oil, vinegar, and salt solution, adding herbs and other condiments to taste.

15.7 Storing Garden Foods

Storing food can reduce a household's vulnerability to unforseen food shortages and saves produce for marketing later in the year. However, no matter what methods are used, there will almost always be a loss of nutrients during storage. Except for roots, tubers, and some nuts and seeds, most garden produce deteriorates rapidly in storage unless processed in some way. Processing and storing garden foods takes time and resources. For some poor households or those in remote locations, storing garden produce is well worth-

Figure 15.11 Selling Pickled Vegetables in Mexico

while. For other households it may not be worthwhile, and alternatives to storage should be considered.

15.7.1 Preharvest Storage

Some roots, bulbs, and tubers can be stored in the ground for several months or more after the crop has matured. This method will not work well if the soil is very wet, and so is best used to store crops maturing at the end of the rainy season. Cassava, garlic, Jerusalem artichoke, onion, potato, and sweet potato can be stored this way, but yams and tiger nuts cannot. It is a good idea to select some produce from different areas of the garden at each harvest to make sure there are no problems with pests or disease.

15.7.2 Postharvest Storage of Fresh Produce

Only produce in the best condition with no insect damage, disease, or bruising should be stored. Bruises are places where plant cells have been crushed and are good sites for the growth of microorganisms which spoil food. Damaged fruit or tubers should be stored separately after harvest and used or processed as soon as possible. Emerging shoots on onions, garlic, and root crops make them vulnerable to contamination and rot. Before storing freshly harvested bulbs, tubers, and roots, they should be allowed to dry until a skin has formed over any surface cuts, otherwise they will rot. Water is rapidly lost from cracks or insect holes in fruits like tomatoes, melons, loquats, or pomegranates, and these are signs that insects may already be inside the fruit. Keeping the storage area clean helps minimize pest problems (section 14.5.2).

Garden produce continues to transpire after harvesting, and to release carbon dioxide and heat from respiration. Water losses due to transpiration should be minimized by storing produce out of the sun and protecting it from heat or wind. The cool, shady conditions described in section 5.4 as lowering transpiration rates in living plants will do the same for harvested garden produce. Gentle air circulation helps remove the heat being given off by respiration. If heat or humidity is high it is better to spread produce out to minimize points of contact where heat and moisture from respiration and storage conditions become concentrated and lead to spoilage. Leaving the stem attached to fruits as discussed in section 15.2 also improves storage life.

If the weather is very hot and dry wilting can be reduced by sprinkling water on the produce or gently covering it with moist leaves, cloth, or mats, as long as there is still some air circulation. Some leaf vegetables, like arugola, leaf amaranth, or spinach, can be made into bundles and the base of the stems kept in a container of water for a day or so (Figure 15.12).

Roots and tubers can be stored loose in piles or in baskets or other well-ventilated containers, in a cool, dry, dark place. By checking them regularly, sprouts can be rubbed off and any other problems caught in the early stages.

Pumpkins, melons, and squash with hard rinds can also be stored whole in a cool, dark place. Any stored squash should be checked regularly for black spots on the surface, signs of spoilage due to bacteria inside the fruit. The Hopi Native Americans grow varieties of squash (*Cucurbita argyrosperma*) and watermelons that they have selected for their good storage qualities, among other things. For example, the watermelons have thick rinds and firm, sweet flesh. If these fruit are harvested in September and kept in a cool, dry corner of the house they will be good for eating throughout the winter and even as late as the following May.

The Pima Bajo Native Americans gather watermelons when not quite ripe, take them to a sandy place, and bury them to protect them from birds and other animals. They say that they choose open places where the heat of the sun will help ripen the buried watermelons.[26]

Nuts and seeds contain a large proportion of oils that are unsaturated fats which are liquid at room temperature, and an important nutrient. When exposed to

Figure 15.12 A Bundle of Arugola with its Stems in Water to Keep it Fresh

light and air for a long time oils become *rancid*. This means that they oxidize (section 15.3.1), acquiring an unpleasant, bitter taste and losing nutrients; rancid oils can make people very sick. Storing nuts and seeds in their shells and protecting them from heat, moisture, light, and pest damage reduces the chances of oils becoming rancid and prevents the growth of poisonous molds such as aflatoxin (section 2.10).

15.7.3 Storing Dried Produce

Before being stored in containers, dried foods should be given a "cooling off" period of 2 or 3 days, when they are kept at the temperature and humidity at which they will be stored and are occasionally turned or mixed. This final drying period prevents later moisture condensation in containers due to temperature change.

Dried fruits and vegetables continue to lose nutritional quality if exposed to light or heat during storage. Keeping them in a cool, dark part of the house or in containers that light cannot pass through, will improve their retention of vitamins A and C. Dried foods absorb moisture readily from the air, and should be checked occasionally during the rainy season for the molds and bacteria that tend to grow when the humidity is high.

Dried garden produce can be stored in covered containers such as calabashes, baskets, pots, cans, or jars. These containers can be sealed to keep insects and other pests out more effectively, but should be opened regularly to check for spoilage. The contents of an unsealed container, although more accessible for pests, can be constantly and easily surveyed. The discussion of airtight and breathable containers in section 14.5.1 is relevant for storing dried foods as well as seeds.

15.7.4 Storing Other Processed Garden Foods

Most of the storage guidelines for dried foods apply to the storage of other kinds of processed garden produce as well:

- Keep foods out of the direct sunlight and in opaque containers if possible.
- Avoid storing foods in warm places, such as near the cooking fire or a wall that receives much exposure to sun and heat.
- Unlike many dried and malted foods, some fermented and sprouted foods are eaten without cooking. To prevent spoiling or contamination it is especially important to keep the storage area clean and the container covered.
- When storing some fermented foods, containers made of certain materials should not be used. Pickled foods should never be stored in aluminum or iron containers, or in containers with lids made from those materials. Wood, ceramic, glass, and stainless steel containers are fine.

15.8 Marketing Garden Produce

Many of the same guidelines for harvesting, processing, and storing garden produce already discussed are appropriate for produce that will be marketed. However, there are some special considerations for handling produce for market because of the time between harvest and consumption and the transportation and sale of the produce in the marketplace.

15.8.1 Harvesting for Market

Harvesting should be done as close as possible to market time (section 15.2). The longer the produce is stored, the poorer its quality and the more likely it is to be damaged during transport. Some fruits, like tomatoes, avocados, mangoes, bananas, and melons should be harvested before they are fully ripe and soft, because they are less likely to be damaged if they are transported while still firm.

Ethylene is a hormone that promotes ripening and is given off as a gas by many fruits including bananas, guavas, melons, papayas, peaches, and avocados. For example, ripe banana peels give off ethylene and can be used to speed ripening of harvested produce at home or for market. Ethylene-producing fruits or their peels can be placed in a covered basket, bag, or other closed, breathable container with the fruits that are being ripened.

It is best to pick leaf crops in the cool, early morning while they are still full of water. The heat and sun during the day increase transpiration and can make even healthy, living plants wilt. Individual leaves of crops like amaranth and basil wilt quickly. They will last much longer if a stem or branch is harvested instead of single leaves. Bunches of these stems can be bundled together and placed in water to keep them fresh as described in section 15.7.2.

15.8.2 Transport from Garden to Market

How garden produce will get to market is an important consideration that should be planned for before market gardens are planted. The closer the market is the better, because produce is fresher, cost of transport and seller's time is less, and the local economy is supported.

However, markets are often far away and gardeners must pay to have their produce transported. If the cost and time involved in transporting produce to market is too great for individual gardeners, they can join together to form a cooperative. For example, in southern Senegal women formed gardening cooperatives to work together on problems of marketing garden produce. Asking members to contribute very small annual fees, the cooperatives pooled these funds and were able to maintain and operate a truck donated to the group by UNICEF.[27]

Once an affordable means of transportation has been found, garden produce must be packed and loaded so that it will not be damaged during the trip. Bruising or exposure to the sun during transport makes produce unmarketable or greatly diminishes its value. Coverings made of a variety of materials can be used to shade produce. Packing produce into containers padded by leaves, straw, or cloth helps protect it. In Egypt fresh produce is often packed into hand-built crates (Figure 15.13). The crates are made in village and town markets from the central ribs of date palm fronds. These readily available, rigid containers prevent soft fruits and vegetables from being crushed during transport to market.

Fresh garden produce kept in an airtight container in the sun will spoil rapidly. As the air inside the container gets hot, the produce wilts, giving off water vapor and creating a hot, wet environment perfect for the growth of spoiling microorganisms.

Many of the common causes of spoiling in fresh fruits and vegetables can be avoided by marketing dried garden produce. But dried foods can also spoil, and in the hot season care should be taken to keep them out of the sun and heat as much as possible. Dried foods should also be protected from contamination by dust, insects, and rodents. This can be done by wrapping them in cloth or leaves or storing them in containers during transportation to the market and even while in the marketplace.

If well cleaned and dried, seeds that are to be marketed are easy to package and transport. They should be kept out of the sun, especially if inside an airtight container.

Figure 15.13 Making Packing Crates from Palm Frond Ribs in Egypt

Figure 15.14 Protecting Garden Produce at the Market

15.8.3 Protecting Produce Quality at the Market

Exposure to heat and sun causes big losses of fruits and vegetables at many markets, especially in the hot season. In addition to the foods that are spoiled, the nutritional quality of any remaining food is signifiantly reduced. For example, if fresh okra and fluted pumpkin leaves are exposed to the sun for several hours, as often happens in the market, they lose approximately 57% and 67% of vitamin A-containing carotenoids and 63% and 74% of vitamin C, respectively.[28]

There are some simple ways to protect the nutritional quality and appearance of garden produce at the market. All goods including fresh and dried produce, processed foods, and seeds should be shaded. Covering them with a cloth, basket, or leaves shades them and gives some protection from insects and blowing dirt. However, this may be a problem because it prevents customers from seeing the goods and comparing them with others. Therefore finding a shady spot to sell from or setting up an umbrella or other kind of shade may be the best solution (Figure 15.14). Occasionally sprinkling fresh produce with water helps cool it and reduces wilting. Even if produce is to be marketed immediately, the concepts and techniques discussed in section 15.7.2 about storing fresh produce are useful.

15.9 Resources

The best ideas for locally appropriate food processing are often experienced local people. Project workers can find information elsewhere as to how to support and improve existing methods. High school or college texts on food science or nutrition can be helpful for explaining the basic principles of processing or preserving garden produce. Another resource may be extension service materials on simple food processing methods.

References

[1] Rankins, et al. 1989.
[2] Iremiren 1987.
[3] Mnzava 1986.
[4] Barrett 1987.
[5] FAO 1982b:164.
[6] Bassuk 1986.
[7] Watson 1976.
[8] Simpson 1983.
[9] Akpapunam and Achinewu 1985; Khokhar and Chauhaw 1986; Rao and Deosthale 1982.
[10] Aykroyd and Doughty 1982:45.
[11] Eskelinen 1977.
[12] Oomen and Grubben 1978:29.
[13] Rankins, et al. 1989.
[14] Cameron and Hofvander 1983:45-48.
[15] Maeda and Salunkhe 1981.
[16] Ali and Sakr 1981; Chen and Saad 1981.
[17] Ayres, et al. 1980:76.
[18] Akpapunam 1984.
[19] Akpapunam and Achinewhu 1985; Cameron and Hofvander 1983:66; Aykroyd and Doughty 1982:44,59-60.
[20] Panasiuk and Bills 1984.
[21] Cameron and Hofvander 1983:66.
[22] Agab 1985.
[23] Odunfa 1985:181.
[24] Aykroyd and Doughty 1982:48-49.
[25] Tamang, et al. 1988.
[26] Pennington 1979:160.
[27] Yoon 1983.
[28] Akpapunam 1984.

16
Weaning Foods from the Garden

Breast milk should be the child's primary food for the first year of its life, with breast-feeding continuing until 2 years of age or more. In addition to nourishment, breast-feeding is important for forming a loving bond between a mother and her baby (Figure 16.1). *Weaning* is the process of supplementing the child's diet of breast milk with weaning foods, starting a transition to an adult diet. It is a time of change for the child and its mother and other caretakers. Successful weaning must fit the schedules and needs of all involved.

Weaning is a critical period for children in drylands. Without nutritious and appetizing weaning foods malnutrition may easily result, leading, in turn, to underweight children who grow slowly and get sick easily. Malnutrition in childhood can cause major health problems later in life.

Weaning food ingredients from the garden can play an important role in the growth of healthy, well-nourished children. Although many commercially produced breast milk substitutes and weaning foods are advertised and promoted by manufacturers, breast milk and homemade weaning foods are better because they are more nutritious, widely available, much less expensive, and more easily prepared.

16.1 Summary

Nutritious weaning foods are essential for the healthy growth of children. Minor changes in adult foods and in the proportions of different kinds of these foods in the diet, can easily produce good, inexpensive weaning foods.

Many garden foods make excellent weaning foods, contributing nutrients essential for rapidly growing young children. Weaning foods should have a consistency and flavor that children like, and because of their small stomachs, must be nutrient-dense.

Because they are so rich in nutrients, weaning foods can spoil easily. Care should be taken in preparing and serving them to reduce chances of the child becoming sick since this can quickly undo the benefits of good care and nutritious foods.

The preparation of weaning foods must fit into people's daily schedule. The mother or others responsible for the child are often very busy and weaning foods requiring lengthy preparation are not appropriate. This is especially critical during the peak labor periods such as planting, weeding, and harvest time for mothers who are farmers, or year-round for mothers who are wage laborers working away from the home.

16.2 The Role of Weaning Foods

Weaning foods are first given as a complement to breast milk at approximately 4 to 6 months of age. At this time they are important as an introduction to solid foods, even though breast milk is still providing most of the infant's nutritional needs. One or two mouthfuls of a bland, peeled and mashed garden fruit or vegetable such as banana, papaya, avocado, steamed or boiled squash, or sweet potato after breast-feeding is a good start. If there is no interest in the food, it can be offered before breast-feeding, when the child is more eager to eat.

Each new food should be added to the weaning child's diet one at a time, in small quantities, with several days wait until another new food is introduced. This makes it easy to identify foods to which the child may be allergic. An *allergy* is a sensitivity to a specific substance that stimulates an immune response. The infant's undeveloped intestine allows some large molecules from foods to pass through into the circulatory system. There the immature immune sys-

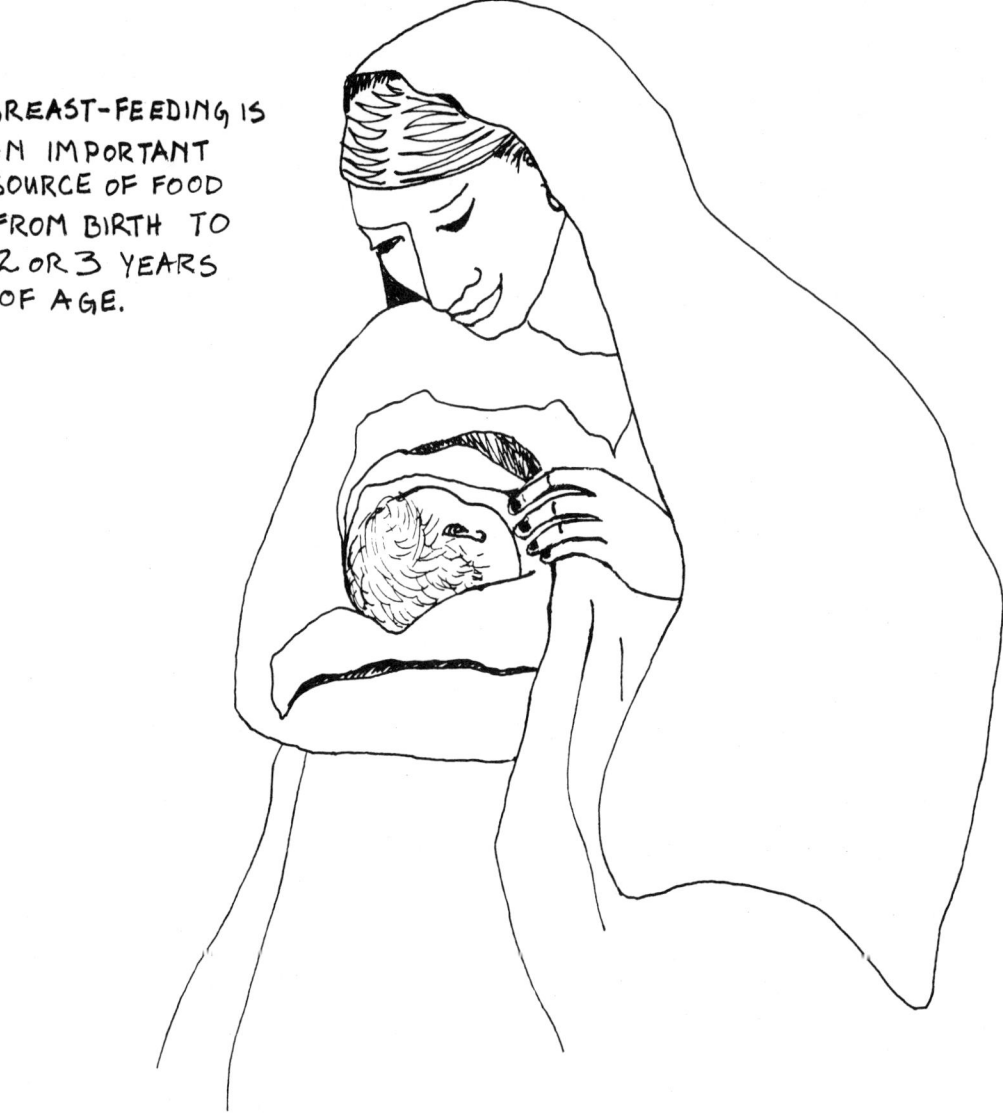

Figure 16.1 Breast-Feeding

BREAST-FEEDING IS AN IMPORTANT SOURCE OF FOOD FROM BIRTH TO 2 OR 3 YEARS OF AGE.

tem may respond with an allergic reaction like a rash or diarrhea. Breast milk contains immunoglobins which help protect the child from some allergies and illnesses. Citrus fruits, cow's milk, egg whites, and wheat are some foods commonly cited as sources of food-induced allergies in infants and young children. Not all children will be allergic to these foods and there are other foods that can cause allergies as well. If a child does have a negative reaction the food should be avoided, at least for a while.

As the child becomes familiar with solid foods, the frequency, quantity, and variety of feedings can be increased. Breast milk continues to be an important part of the child's diet at least until it is 2 years old. However, by 6 to 7 months of age, breast milk alone cannot sustain the child's healthy growth, and small quantities of nutritious weaning foods, given four or more times a day, become more and more essential.

Homemade weaning foods should be prepared from a combination of ingredients that supply energy, protein, vitamins, and minerals, all of which are needed by the rapidly growing child. Some nutritionists categorize weaning food ingredients into four groups, which are then combined into nutritionally complete weaning foods referred to as *multi-mixes*:[1]

- Main energy source (usually the staple crop, e.g., sorghum or maize).
- Protein supplement (e.g., beans, nuts, dried dark green leafy vegetables, eggs or meat).
- Vitamin and mineral supplement (e.g., fruits and vegetables).
- Energy-dense supplement (e.g., oil).

Local adult diets usually consist of a staple food flavored with a small amount of vegetable sauce or legumes. Weaning foods are created from adult foods by increasing the proportion of the most nutrient-rich ingredients (such as sauces), improving the consistency by thoroughly mashing the foods, and eliminating hot spices. A study in Nigeria of 228 weaning-age children found that those who were fed a modified adult diet of food that was well cooked and mashed were better nourished than those children fed a weaning diet of only unnutritious weaning foods such as *pap*, a watery maize gruel, or commercial weaning foods and formulas.[2]

It is important to distinguish between good weaning foods and a good weaning diet. Porridges made from local staple crops are excellent weaning foods. For example, in many areas a fermented grain porridge is an important part of the weaning child's diet, frequently supplying the main source of energy. When diluted with water these porridges become drinks, such as the *pap* described above, that are useful for preventing dehydration caused by diarrhea and vomiting (Box 16.1 in section 16.4) but are not adequate as weaning foods. The weaning diet must include other foods that will provide the additional nutrients needed by the growing child. Household gardens are an excellent source of ingredients that can be easily made into nutritious, good-tasting weaning foods, which are especially important as supplements to the staple food porridges traditionally used for weaning.

16.3 Nutrient Density

A ***nutrient-dense*** food is one that contains a relatively high proportion of a beneficial nutrient or nutrients per unit volume. Energy (calories), protein, vitamins A and C, and iron are the main nutrients that become deficient during weaning (sections 2.3.1 and 2.11). Children need more nutrients in proportion to their size than adults because they are growing so fast. For example, for each kilogram of body weight an infant needs more than three times more energy and protein than a moderately active adult woman.[3] This is why nutrient-dense weaning foods are so important, especially when a spoonful of food may be all the child can eat. A weaning diet of only the adult staple food or a thin, watery weaning gruel prepared from it simply does not contain a high enough concentration of nutrients. This causes children to become malnourished because their small stomachs do not allow them to eat enough of the food to obtain the nutrients they need.

16.3.1 Energy

Oil-rich garden fruits like olives and avocados are energy dense and make excellent additions to weaning foods. (Olives that are very salty or prepared with very spicy or acid flavorings are not appropriate.) This is also true of pastes made from garden nuts and seeds like sunflower, sesame, and melon seeds, groundnuts, cashews, pistachios, pine nuts, and almonds. In addition to calories, these pastes add protein, minerals, flavor, and a creamy consistency. Vegetable oils made from garden nuts or seeds, and fats like butter or *ghee* are other good, energy-dense additions to weaning foods. For example, 1 teaspoon (5 ml) of fat contains about 44 calories,[4] but a teaspoon of cooked sorghum flour contains only about 10 calories.[5] Thus, over 4 teaspoons of cooked sorghum porridge are needed to provide the same amount of energy as 1 teaspoon of fat (Figure 16.2). However, no more than 20-30% of the child's calories should come from fats.[6] A greater proportion of fats results in a diet

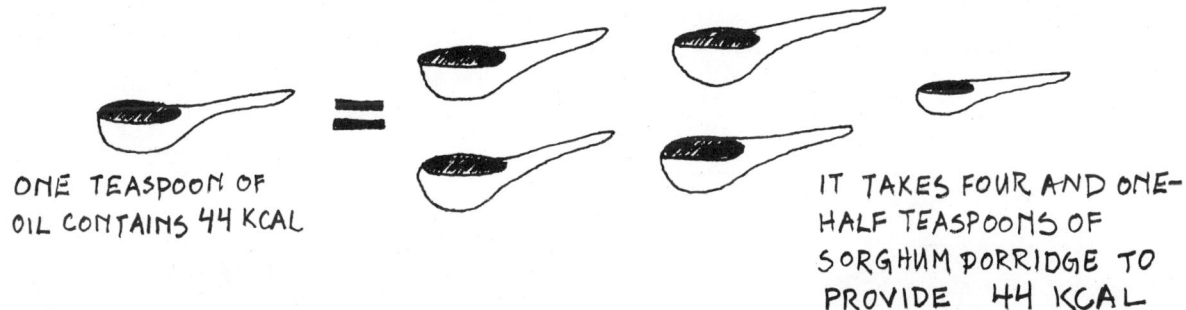

Figure 16.2 Comparing the Caloric Density of Oil and Cooked Sorghum Porridge

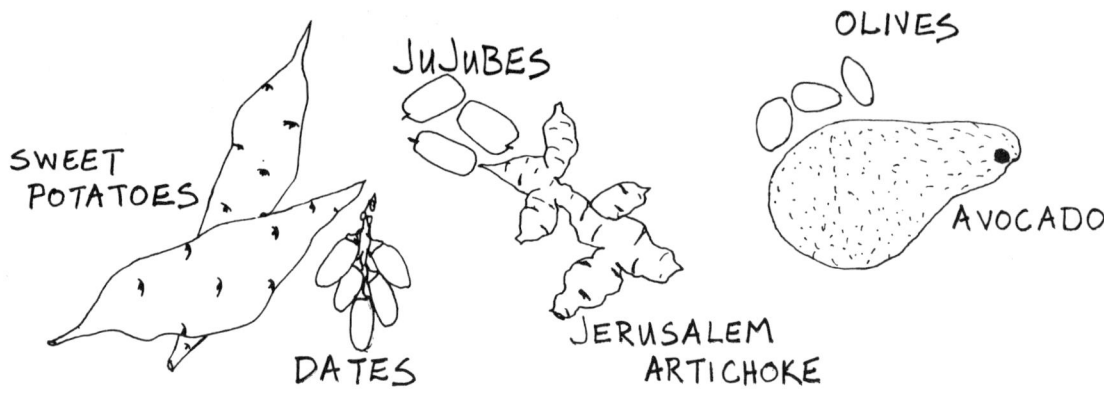

Figure 16.3 Some Garden Foods that Add Energy to the Weaning Diet

with inadequate quantities of other kinds of foods and the nutrients they provide.

Other garden foods contain concentrated energy in the form of carbohydrates. They include vegetables such as tiger nut and dried sweet potato, and yam, and many dried fruits such as dates, figs, and peaches (Figure 16.3). The caloric value of these foods is as high as, or higher than, that of dry cereals, which is about 350 kcal/100 gm (100 kcal/oz). All of these foods must be prepared so that weaning children can eat them.

16.3.2 Protein

Many high-energy garden foods are also good sources of protein, including groundnuts, cashews, sesame, squash, melon, and sunflower seeds, and pulses like cowpeas, beans, and lentils (Figure 16.4). These should all be thoroughly ground or mashed so they will not choke the child. Drying fresh leaves that contain 80% or more water significantly reduces their bulk and concentrates nutrients. The leaves of many dryland garden plants such as amaranth, jute, pumpkin, cowpea, and cassava contain 20-30% protein when dried.[7] For example, 100 gm (3.5 oz) of dried amaranth leaves can provide up to 28% of the protein requirements for a 1-year-old child.[8] Dried leaves can be easily added to a weaning porridge for the last 5 minutes that it is cooking.

Drying increases the protein density of insects and their larvae, which are an excellent source of calories and calcium. Some, like dried caterpillars, are rich in niacin, vitamin A, and riboflavin as well.[9] Foods like

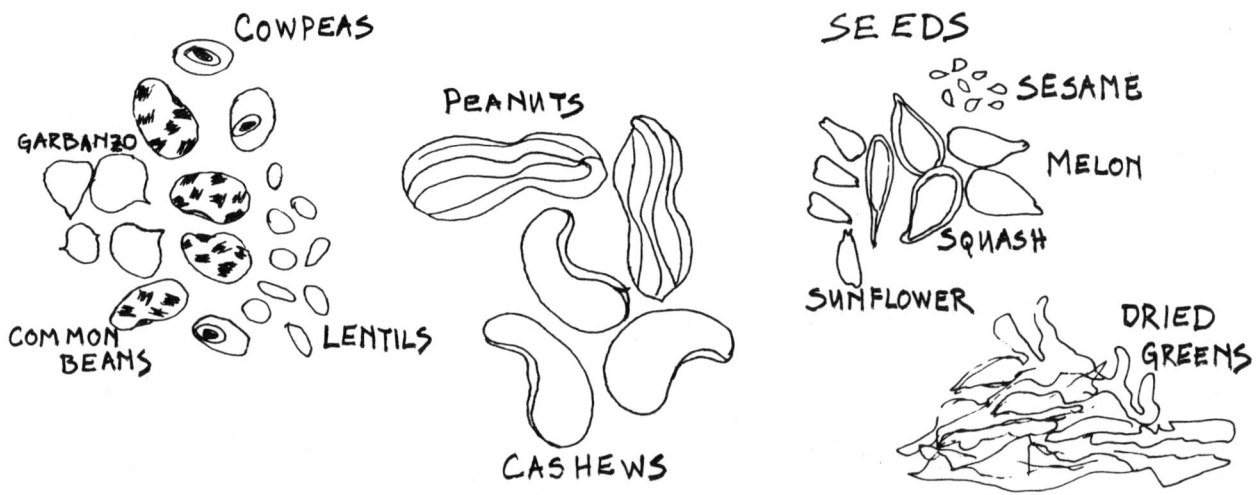

Figure 16.4 Some Garden Foods that Add Protein and Energy to the Weaning Diet

yogurt or cheese are good sources of proteins, fats, and other nutrients for weaning-age children. Meat from both wild or domesticated animals is another good source of protein although it is often very expensive or difficult to obtain.

The yeast used to brew local grain beers in many dryland areas is a nutrient-rich flavoring commonly added to soups, and makes an excellent weaning food. It is high in protein, iron, niacin, thiamin, and riboflavin. In northern Ghana, soups made with the yeast used to brew sorghum beer are traditionally used as weaning foods.[10] However, use of beer yeast may be inappropriate for some households because of religious dietary restrictions on alcohol and foods associated with it.

16.3.3 Vitamins and Minerals

Gardens are especially important for their contribution of vitamins and minerals (Figure 16.5) (sections

Figure 16.6 Many Children Enjoy Carrots which are Good Sources of Vitamins A and C

2.6 and 2.7). Vitamins can, however, be lost through cooking (section 15.3). Many of the orange and yellow tree fruits such as mangoes, oranges, apricots, peaches, and papayas are rich in vitamins A and C, and their natural sweetness appeals to children. Ripe cactus fruits contain much vitamin C and some calcium, and can be eaten fresh, although it may be better to strain out their large seeds and make a flavorful juice. Carrots and vegetables like yellow and orange squash, tomatoes, and sweet red and yellow peppers are good sources of vitamins A and C—they are flavorful weaning-food ingredients and a good source of those vitamins for older children as well (Figure 16.6).

Dark green leafy vegetables (DGLVs) are high in vitamin A and often easier to grow or less costly than many fruits. Fresh or dried leaves can be chopped and

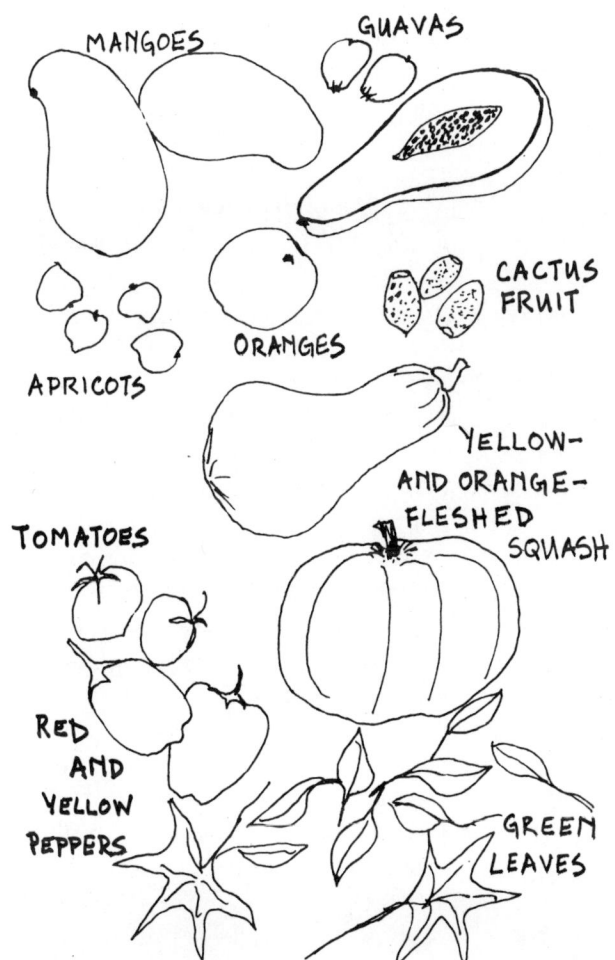

Figure 16.5 Some Garden Foods that Add Vitamins A and C to the Weaning Diet

their nutrient density and the availability of nutrients to the child. The seed coats of pulses such as dried beans, groundnuts, and sesame seeds should be removed because they are high in fiber. High-fiber weaning foods can inhibit absorption of nutrients and cause gas and discomfort. In addition, the coat or hull may contain anti-nutrients. For example, sesame seed hulls contain oxalates which inhibit absorption of the calcium contained in the seeds (section 2.10). Dark-colored hulls contain tannins that decrease protein and carbohydrate digestibility.[11]

16.3.4 Weaning Food Consistency

Smooth, semisolid weaning foods are easier to eat and more readily accepted than foods that are thick and coarse. For these reasons, and to avoid the danger of the child choking, all food fed to the child during the first few months of weaning should be thoroughly mashed. Mashing the food through a wire sieve makes it softer and smoother, and also removes the tough skins or seed coats of beans.

While water or other liquids can improve the consistency of weaning foods, they also dilute the foods' nutritional value. Adding a little fat such as oil, *ghee* (clarified butter), or butter to the weaning food is one way to give it an acceptable consistency without diluting its energy density. The fat helps moisten the food, adding calories without noticeably increasing the quantity of food the child must eat. Some fats provide other nutrients as well, such as the vitamin A in red palm oil and *ghee*.

By using malted flours (section 15.5.2) for making weaning porridge, nutrient density for an acceptable consistency will be greater than if the porridge were made with unmalted flours. Malted flour does not absorb as much water as unmalted flour because malting breaks starches down into sugars. Therefore, malted porridge can be made much thinner and easier for the child to eat, while still having a high energy density. For example, 100 gm (3.5 oz) of millet porridge made with unmalted flour contains 25 kcal and 0.4 gm (0.01 oz) of protein. One hundred gm (3.5 oz) of porridge of the same consistency made with malted millet flour has 83 kcal and 1.3 gm (0.05 oz) of protein.[12] To prepare 230 gm (8 oz) of semisolid porridge from unmalted flour requires 14 gm (0.5 oz) of flour and 213 gm (7.5 oz) of water. Two hundred and thirty gm (8 oz) of a porridge of the same consistency using malted flour takes 57 gm (2 oz) of flour and only 170 gm (6 oz) of water (Figure 16.8).

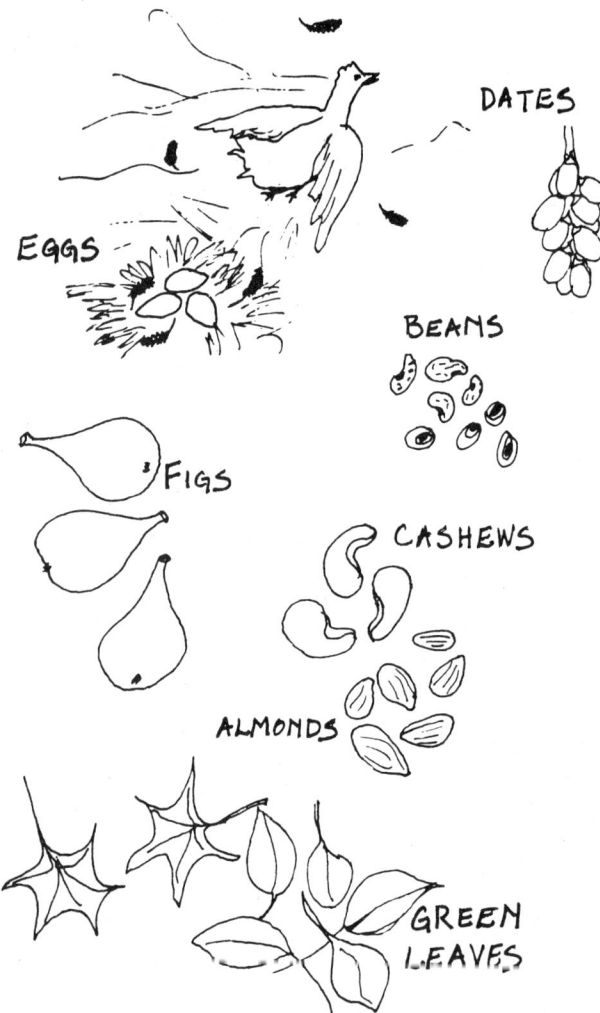

Figure 16.7 Some Garden Foods that Add Iron to the Weaning Diet

added to porridge at least 5 minutes before it has finished cooking. In some areas of Asia DGLVs may not be given to children because of dietary restrictions or because they may not taste good to children. Cooking greens and then rinsing them before eating improves their taste by removing some of the bitter-tasting compounds such as tannins or oxalic acid.

Iron is an important mineral for the healthy development of the young child (section 2.7.1). Dates, DGLVs, figs, nuts, and beans are all sources of iron from the garden (Figure 16.7). Eggs are another good source of iron which can be used in the weaning diet as the child gets older and if she has shown no allergic reaction to them. Lack of vitamin C can be a limiting factor in iron utilization and so the child must receive an adequate supply of this vitamin to avoid iron deficiency.

Processing and preparing weaning foods affects

TO MAKE 8 OZ OF PORRIDGE

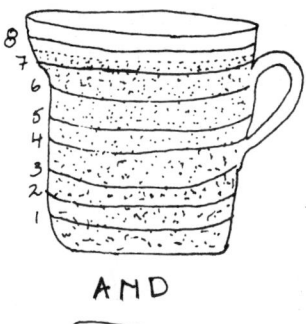

7.5 OZ OF WATER

AND

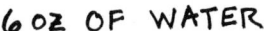

0.5 OZ OF UNMALTED MILLET FLOUR

6 OZ OF WATER

AND

2 OZ OF MALTED MILLET FLOUR

BOTH OF THESE MIXTURES YIELD A PORRIDGE WITH THE SAME SEMI-SOLID CONSISTENCY, BUT WITH DIFFERENT ENERGY VALUES.

60 KCAL/8 OZ

199 KCAL/8 OZ

Figure 16.8 Comparing the Caloric Density of Malted and Unmalted Millet Flour

16.4 Hygiene

One of the greatest health hazards for the weaning child is eating unclean weaning foods that cause sickness. When the child's body is experiencing the nutritional demands of rapid growth, the stress of illness makes good nutrition even more difficult to achieve. (Box 16.1 gives some suggestions about what to do should the weaning-age child become sick.)

There are several ways to minimize contamination problems in the preparation and serving of weaning foods. First, foods and any utensils used must be kept as clean as possible and away from disease vectors such as flies. Whether weaning foods are fed to infants with a spoon or directly by hand is not important (Figure 16.9). Whatever is used should be carefully cleaned and washed. A nutritionist working in northern Ghana noted that a nutrition education program which advised, among other things, using spoons for feeding weaning foods, was actually encouraging practices that were less hygienic than the traditional method of feeding directly by hand.[13] This was because the spoons often fell on the floor and were left uncleaned between feedings, while there was a strong traditional practice of washing the right hand before eating a meal or feeding a small child.

Second, any liquids added to weaning foods to produce the right consistency should be added during cooking. They should never be added to weaning foods that have already been cooked and removed from the heat, as this may introduce disease-causing microorganisms. These microorganisms will not be

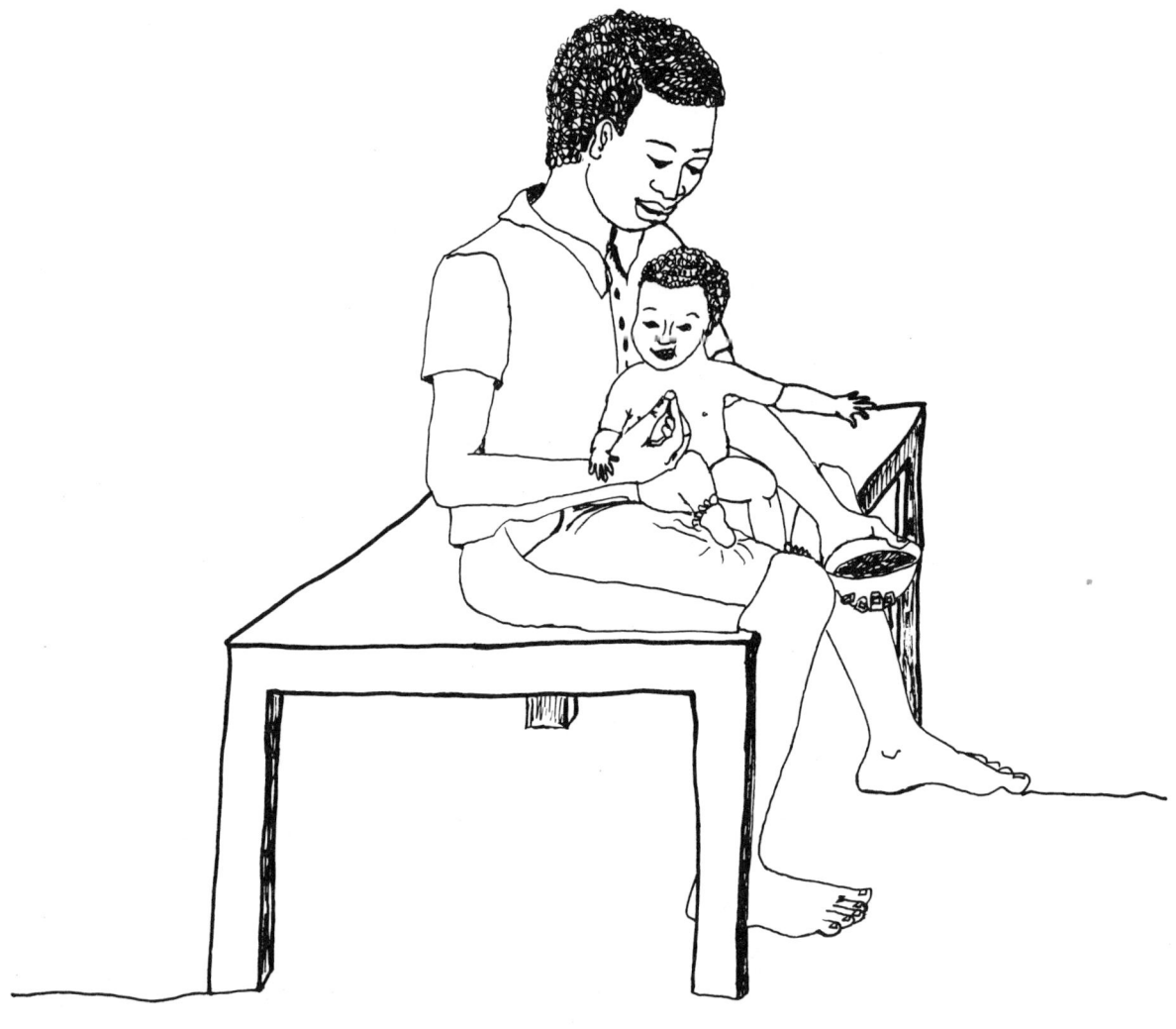

Figure 16.9 Weaning Foods can be Served by Hand

> **Box 16.1**
> **When the Weaning Child Becomes Sick**
>
> When weaning-age children do become sick, the loss of liquids due to diarrhea, vomiting, or sweating from fever can be serious. If these fluids are not replaced the child can easily become dehydrated and eventually may die from lack of liquids and nourishment. This is why it is essential to give food and liquids to a child with diarrhea. If the sickness continues and/or the child does not want to eat, she must keep drinking liquids.
>
> In addition to lack of fluids, dehydration means losses of the carbohydrates and minerals necessary for the body to absorb the liquids and nutrients it needs from the digestive system into the bloodstream. For this reason, although plain water is better than nothing, a drink that replaces some of those carbohydrates and minerals is the best.
>
> In the 1970s and early 1980s a mixture of water, salts, and sugar, known as **ORS** (*oral rehydration solution*), for treating diarrhea and dehydration was strongly promoted by international organizations like WHO and UNICEF (Chapter 19 lists these resource organizations). This campaign included the production of packets of premeasured salts and sugars, and recipes for homemade salt and sugar mixes. The campaign saved millions of lives but current research shows that cereal- or potato-based gruels are just as effective if not better, and for many households are easier, and less expensive to obtain.[16]
>
> A drinkable gruel made from the local staple grain (or root such as potato) is an available food that most households can easily make and give to their children when diarrhea or vomiting is a problem (this is not adequate as a weaning food, however). This drink may be simply a diluted version of the homemade weaning foods discussed earlier, although it should only contain a cereal and water, no oil or spices. In Mozambique mothers have a tradition of preparing such a gruel, called *papinha*, from wheat, rice, or various roots.[17] Spoonfuls or small sips of the gruel should be given to the child constantly, even if she has been vomiting. The goal is to get her to swallow as much or more liquid than she has lost through sickness. The child should be watched carefully and if problems persist, taken to a health clinic.

killed unless the food is boiled for a minimum of 5 minutes.[14] The same is true of any fats added to weaning foods. For this reason it is best to cook all ingredients together. Making the porridge a little thin allows for the tendency of most weaning porridges to thicken slightly as they cool.

Third, prepared weaning foods should not be stored for later use unless they are thoroughly recooked because their high protein and carbohydrate content make them ideal sites for bacterial growth. In warm weather, bacteria can begin growing within 2 hours even if the food is kept in a covered container in a relatively cool place.[15] Traditional, fermented weaning porridges that contain only the staple crop may be exceptions. The acidity of these fermented foods helps preserve them, preventing the growth of spoiling microorganisms. Leftover weaning food can be eaten by other members of the household while it is still fresh, but if given to the child it should be boiled again for at least 5 minutes. Again, the advantages of easy-to-make weaning foods are obvious, especially for households with limited time and/or cooking fuel.

16.5 Weaning as a Part of Daily Life

For weaning foods to fufill their important role in child nutrition, the preparation and use of these foods must fit into people's daily schedules, household economy, and their beliefs about the kinds of foods appropriate for weaning.

LOCAL PRACTICES In most areas there are strong beliefs and feelings about food habits. Weaning foods are often a special category of foods and there may be a number of rules and practices for what and how weaning-age children are to eat. Understanding what these practices are and why they are followed is the first priority for someone working with community members on improving the nutrition of weaning-age children.

ACCESSIBILITY AND COST Even the most nutritious weaning food will do little good if it is difficult to obtain and prepare. Commercial weaning foods are relatively expensive, and most Third World women have no experience in preparing them. Often, written directions are inadequate because even those women who are literate may not know the language in which directions are written. In contrast, home-prepared weaning foods can be made from ingredients that are easily available, inexpensive, and familiar to the mother, or whoever is responsible for preparing them. The household garden is an excellent source of such ingredients.

PREPARATION TIME   Child care is an important activity, but it must be integrated into a schedule of other work. Women's workloads affect the time available for weaning food preparation, and thus food quality and cleanliness. During periods of high labor demand, food for both children and adults is often prepared well in advance. During the period between when the food is prepared and when it is eaten conditions for contamination are perfect. It is likely that this is a major cause of increased child illness at these times.[18] Modifications of the normal household diet or simple, quick recipes based on the local staple food plus some garden produce are best. Other solutions to the problem could vary from decreasing women's workloads to exploring cooperative cooking arrangements.

WHO IS CARING FOR THE CHILD?   Although the child's mother is often the one responsible for providing weaning foods, this responsibility may be shared among other members of the household or even community. Older brothers and sisters frequently care for infants and small children so that their mothers can work (Figure 16.10). In recognition of this UNICEF began a "Child to Child program" in 1979, the International Year of the Child. Child to Child programs all over the world help older children take better care of younger ones.[19] An example of this is a community health program in Brazil which trains adolescents how to prepare weaning foods for their younger sisters and brothers, and neighbors.[20]

16.6 Resources

The *Manual on Feeding Infants and Young Children* (Cameron and Hofvander 1983) is a good resource for anyone involved with projects concerning the health of infants and young children. While there is much practical advice, some of the nutritional formulas and tables for making weaning foods will only be useful to someone in a clinical or academic setting.

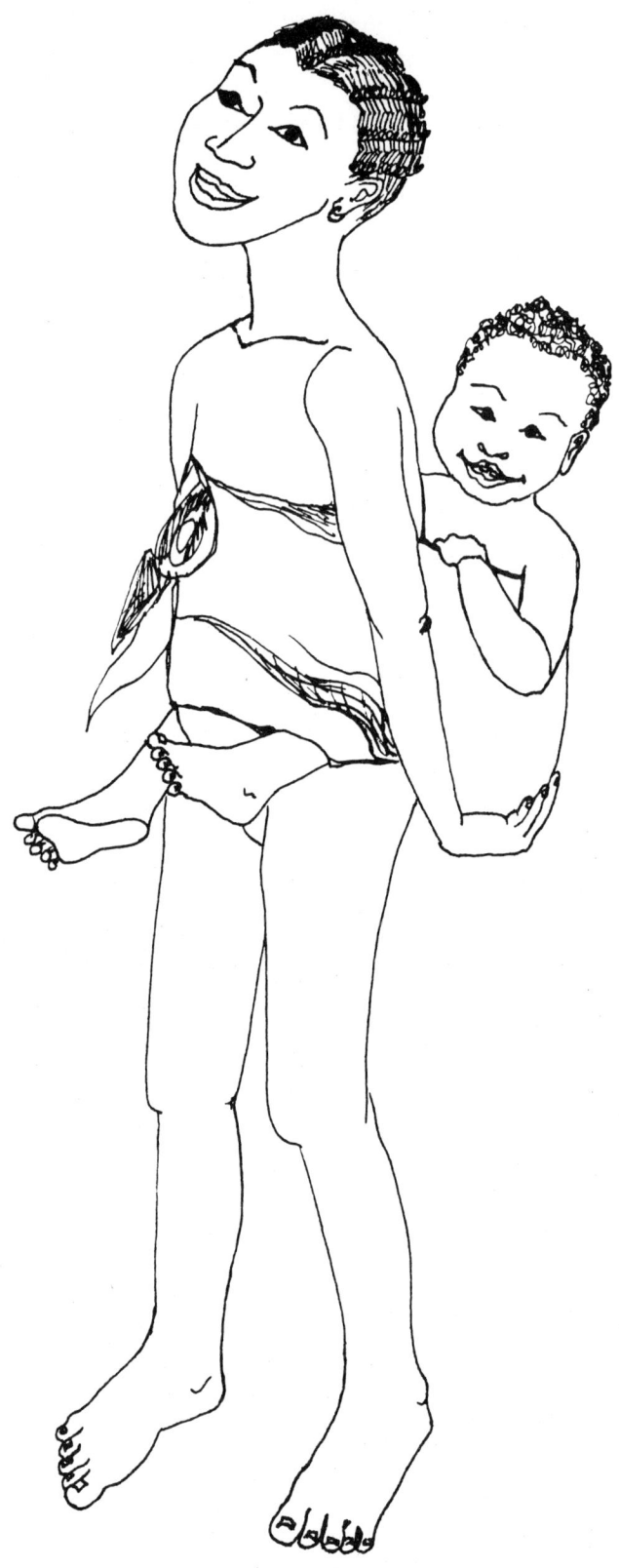

Figure 16.10 Children are often Responsible for the Care of Their Younger Brothers and Sisters

Chapter 7 ("Feeding the Family") in *Nutrition for Developing Countries* (King, et al. 1972) includes a discussion of weaning foods and their preparation. *Human Nutrition in Tropical Africa* (Latham 1979), which emphasizes infant and child health, is a very useful reference and textbook for health care workers. It includes a section (number 35) on recipes for infants and young children.

Guidelines for Training Community Health Workers in Nutrition (WHO 1986) is devoted primarily to maternal and child nutrition, and Module Four discusses weaning foods. Chapter 7 "The Malnourished Child" in *Primary Child Care: Book One* (King, et al. 1980) briefly discusses weaning foods while covering clinical diagnosis and treatment of malnutrition in infants and small children.

"The Best Diet for Small Children" in *Where There is No Doctor* (Werner 1977:121-124) gives a brief overview of important ideas for weaning and its effects on child health.

Jellife and Jellife (1978) is an academic review of both biological and sociocultural aspects of breastfeeding.

References

[1] Cameron and Hofvander 1983:117-118.
[2] Isenalumhe 1986.
[3] Cameron and Hofvander 1983:38.
[4] Cameron and Hofvander 1983:126.
[5] Calculated from Cameron and Hofvander 1983:189 and Leung, et al. 1968.
[6] Cameron and Hofvander 1983:61.
[7] Calculated from Leung, et al.1968.
[8] Oomen and Grubben 1978:31.
[9] Latham 1979:272.
[10] Gordon 1973:42-43, 434.
[11] Aykroyd and Doughty 1982:57.
[12] Cameron and Hofvander 1983:67.
[13] Gordon 1969:5.
[14] Cameron and Hofvander 1983:70.
[15] Cameron and Hofvander 1983:78.
[16] Greenough 1987:29-32.
[17] Werner 1986:14-15.
[18] Longhurst 1983:4.
[19] Werner and Bower 1982:24-1 to 24-30.
[20] Gibbons and Griffiths 1984:30.

17
Glossary

Following are abbreviations and equivalencies for metric and English units of measurement, atomic symbols and molecular formulas, and other abbreviations or acronyms that are used in this book.

17.1 Abbreviations Used in Measurements

a = acre
AF = acre foot, the amount of water needed to cover 1 acre 1 foot deep
AI = acre inch, the amount of water needed to cover 1 acre 1 inch deep
°C = degrees centigrade
cal = calorie
Cal = Calorie
cc = cubic centimeter = cm^3
cm = centimeter
dS = deciSiemens
°F = degrees farenheight
fl oz = fluid ounce
ft = foot
gm = gram
gal = gallon (US)
ha = hectare
hr = hour
in = inch
kcal = kilocalorie
kg = kilogram
km = kilometer
μ = micron
m = meter
mcg = microgram
mg = milligram
mi = mile
min = minute
mm = millimeter
mmhos = millimhos
MT = metric ton
oz = ounce
π = 3.14 (the ratio of a circle's circumference to its diameter)
pt = pint
S = Siemens
sec = second
T = ton (US)

17.2 Equivalencies in Units of Measurement

Area
1 km^2 = 100 ha
1 ha = 10,000 m^2 = 0.01 km^2 = 2.47 a = 107,639 ft^2
1 m^2 = 1 × 10^6 mm^2 = 10.76 ft^2
1 mi^2 = 640 a = 2.59 km^2
1 a = 43,560 ft^2 = 0.404 ha = 4,047 m^2
1 ft^2 = 0.0929 m^2

Distance
1 km = 1,000 m = 0.621 mi
1 m = 100 cm = 1,000 mm = 3.28 ft
1 cm = 10 mm = 0.394 in
1 mi = 1.609 km
1 ft = 0.3048 m
1 in = 2.54 cm = 25.4 mm
1 μ = 0.0001 mm

Volume
1 m^3 = 1,000 liters = 26.42 gal = 35.31 ft^3
1 liter = 1,000 ml = 1,000 cm^3 = 0.264 gal = 34 fl oz = 1 mm depth of water from rain or irrigation on 1 m^2 = 61.02 in^3 = 0.035 ft^3
1 ft^3 = 7.48 gal = 28.31 liters
1 gal = 8 pt = 3.785 liters = 0.1337 ft^3

1 pt = 16 fl oz
1 fl oz = 1.804 in³ = 29.59 cm³
1 AF = 3.26 x 10⁵ gal = 1.23 x 10⁶ liters = 12 AI
1 AI = 2.72 x 10⁴ gal = 1.03 x 10⁵ liters

Weight
1 MT = 1,000 kg = 1.103 T
1 kg = 1,000 gm = 2.205 lb
1 gm = 1,000 mg = 1.0 x 10⁶ mcg
1 mg = 1,000 mcg
1 T = 2,000 lb = 0.906 MT
1 lb = 16 oz = 0.453 kg
1 oz = 28.4 gm

Energy
1 Cal (with upper case "C") = 1,000 cal (with lower case "c") = 1 kcal
Temperature in °C = (Temperature in °F - 32) x 5/9
Temperature in °F = (Temperature in °C x 9/5) + 32
Freezing point of water = 0°C = 32°F
Boiling point of water = 100°C = 212°F

17.3 Atomic Symbols and Molecular Formulas

Al = aluminum
B = boron
C = carbon
Ca = calcium
$CaCO_3$ = calcium carbonate
Ch = chromium
Cl = chlorine
Co = cobalt
CO_2 = carbon dioxide
Cu = copper
$C_6H_{12}O_6$ = simple sugar
Fl = florine
Fe = iron
H = hydrogen
H_2CO_3 = carbonic acid
H_2O = water
$H_2PO_4^-$ = soluble phosphate
I = iodine
K = potassium
K_2O = potassium oxide
Li = lithium
Mb = molybdenum
Mg = magnesium
$MgCO_3$ = magnesium carbonate
Mn = mangenese
Mo = molybdenum
N = nitrogen
Na = sodium
NH_3 = ammonia gas
NH_4^+ = ammonium
NH_2CONH_2 = urea
NO_3^- = nitrate
O = oxygen
P = phosphorus
S = sulfur
Se = selenium
Si = silica
SO_4^- = sulfate
Zn = zinc

17.4 Other Abbreviations and Acronyms

AW = plant available soil water
Bt = *Bacillus thuringiensis*
CAM = crassulacean acid metabolism
CBH = carbohydrate
CEC = cation exchange capacity
C:N = carbon to nitrogen ratio
d = crop root depth
DGLV = dark green leafy vegetable
DDT = dichlorodiphenyltrichloroethane
DNA = deoxyribonucleic acid
Ea = water application efficiency
EAA = essential amino acid
EC_c = salinity of soil allowing acceptable yield
EC_w = salinity of irrigation water
ET = evapotranspiration
ETa = actual evapotranspiration
ETm = maximum evapotranspiration
ETo = theoretical evapotranspiration
FC = field capacity
i = irrigation interval (in days)
Id = irrigation water applied
IPM = integrated pest management
kc = crop coefficient
LR = leaching requirement
NGO = nongovernmental organization
ORS = oral rehydration solution
p = probability
P = precipitation
PCM = protein calorie malnutrition
PEM = protein energy malnutrition
ppm = parts per million
pr = proportion of plant available soil water which a crop can use without ETa becoming less than ETm

PVO = private voluntary organization
PWP = permanent wilting point
R = runoff
R% = runoff percent
RDA = recommended dietary allowance

RE = retinol equivalents (vitamin A)
W = amount of water needed to bring soil in root zone to field capacity
WCE = watery compost extract

18
Some Crops for Dryland Gardens

In the text we use common English names for crops and plant families, and in this Chapter we provide their scientific names. This is not meant to be a list for putting together demonstration gardens; it is not an exhaustive list and does not include the many important local crops and crop varieties.

The best sources for information about dryland garden crops are local gardeners and farmers. Researchers at university botany departments can also be helpful. Useful books include Acland 1971; Dupriez and De Leener 1987; FAO 1983, 1988; Irvine 1969; Kassam 1976; NAS 1975, 1979, 1989b; Purseglove 1974, 1983; Purseglove, et al. 1981; van Epenhuijsen 1978.

18.1 Common English and Scientific Names for Some Crops and Crop Groups

In the text we use the English common names of dryland garden crops. Here we list them alphabetically (with alternative English names in parentheses), and provide the genus and species name for each. Also listed are names for groups of crops.

agave, *Agave* spp.
African fan palm, *Borassus aethiopum*
African locust bean, *Parkia biglobosa*
almond, *Prunus dulcis*
amaranth (some species and varieties grown primarily for leaves, others for grain), *Amaranthus* spp.
apple, *Malus* spp.
apricot, *Prunus armeniaca*
artichoke, globe, *Cynara scolymus*
artichoke, Jerusalem, *Helianthus tuberosus*
arugola (rocket salad), *Eruca sativa*
asparagus, *Asparagus officinalis*
asparagus bean (black-eyed pea, cowpea, yard-long bean), *Vigna unguiculata*
artichoke, globe, *Cynara scolymus*
artichoke, Jerusalem, *Helianthus tuberosus*
aubergine (brinjal, eggplant, garden egg), *Solanum* spp.
avocado, *Persea americana*

Bambara groundnut, *Voandzeia subterranea*
banana, *Musa* spp.
baobab, *Adansonia digitata*
basil, *Ocimum basilicum*
bean, African locust, *Parkia biglobosa*
bean, asparagus (black-eyed pea, cowpea, yard-long bean), *Vigna unguiculata*
bean, broad (fava bean), *Vicia faba*
bean, common, *Phaseolus vulgaris*
bean, Egyptian (hyacinth bean), *Lablab niger*
bean, fava (broad bean), *Vicia faba*
bean, jack (horse bean), *Canavalia ensformis*
bean, hyacinth (Egyptian bean), *Lablab niger*
bean, horse (jack bean), *Canavalia ensformis*
bean, lima, *Phaseolus lunatus*
bean, mat (moth bean), *Vigna aconitifolia*
bean, mung (golden or green gram), *Phaseolus aureus*
bean, tepary, *Phaseolus acutifolius*
bean, yard-long (asparagus bean, black-eyed pea, cowpea), *Vigna unguiculata*
beet, *Beta vulgaris*
bird pepper, *Capsicum frutescens*
black-eyed pea (asparagus bean, cowpea, yard-long bean), *Vigna unguiculata*
black gram, *Phaseolus mungo*
bottle gourd, *Lagenaria siceraria*
brinjal (aubergine, eggplant, garden egg), *Solanum* spp.
broad bean (fava bean), *Vicia faba*
broccoli, *Brassica oleracea*

cabbage, *Brassica oleracea*
cactus, Indian fig, *Opuntia ficus-indica*

343

cactus, prickly pear, *Opuntia* spp.
carob, *Ceratonia siliqua*
carrot, *Daucus carota*
cashew, *Anacardium occidentale*
cassava (manioc, tapioca, yuca), *Manihot esculenta*
cauliflower, *Brassica oleracea*
cereals, grain crops, e.g., sorghum, millet, maize, wheat, rice, barley
chayote (choyote), *Sechium edule*
chard, *Beta vulgaris*
chenopods, *Chenopodiaceae* family
chickpea (garbanzo), *Cicer arietinum*
chili, *Capsicum annuum*, *Capsicum frutescens*
Chinese cabbage, *Brassica* spp.
choyote (chayote), *Sechium edule*
cilantro (coriander), *Coriandrum sativum*
citrus, *Citrus* spp.
coffee, *Coffea* spp.
collard, *Brassica* spp.
common bean, *Phaseolus vulgaris*
corn (maize), *Zea mays*
coriander (cilantro), *Coriandrum sativum*
cowpea (asparagus bean, black-eyed pea, yard-long bean), *Vigna unguiculata*
cucumber, *Cucumis sativus*
cucurbits, *Cucurbitaceae* family
cumin, *Cuminum cyminum*
crucifers, *Cruciferae* family

date palm, *Phoenix datylifera*
doum palm, *Hyphaene thebaica*

eggplant (aubergine, brinjal, garden egg), *Solanum* spp.
egusi melon, *Cucumeropsis edulis*
Egyptian bean (hyacinth bean), *Lablab niger*
epazote, *Chenopodium ambrosioides*
Ethiopian mustard, *Brassica carinata*

fava bean (broad bean), *Vicia faba*
fennel, *Foeniculum vulgare*
fig, *Ficus* spp.
fluted pumpkin, *Telferia occidentalis*

garbanzo (chick pea), *Cicer arietinum*
garden egg (aubergine, brinjal, eggplant), *Solanum* spp.
garlic, *Allium sativum*
garlic, great-headed, *Allium ampeloprasum*
globe artichoke, *Cynara scolymus*
golden gram (green gram, mung), *Phaseolus aureus*
goosefoot (lambsquarter), *Chenopodium* spp.
gourd, bottle, *Lagenaria siceraria*

gourd, wax (white gourd), *Benincasa hispida*
gram, black, *Phaseolus mungo*
gram, golden (green gram, mung), *Phaseolus aureus*
grape, *Vitis* spp.
grapefruit, *Citrus paradisi*
grass pea, *Lathyrus sativus*
green gram, *Phaseolus aureus*
groundnut (peanut), *Arachis hypogaea*
groundnut, Bambara, *Voandzeia subterranea*
groundnut, Hausa, *Kerstingiella geocarpa*
guava, *Psidium guajava*
gumbo (lady's finger, okra, okro), *Abelmoschus esculentus*

Hausa potato, *Solenostemon rotundifolius*
Hausa groundnut, *Kerstingiella geocarpa*
hibiscus (roselle), *Hibiscus sabdariffa*
horse bean, *Canavalia ensiformis*
hot pepper (chili), *Capsicum frutescens*
hyacinth bean (Egyptian bean), *Lablab niger*

Indian fig cactus, *Opuntia ficus-indica*
Indian spinach, *Basella alba*

jack bean, *Canavalia ensiformis*
jack fruit (jak fruit), *Artocarpus heterophyllus*
jaltomata, *Jaltomata procumbens*
Jerusalem artichoke, *Helianthus tuberosus*
jujube, *Ziziphus jujuba*, *Z. mauritiana*
jute (mallow), *Corchorus olitorius*

kale, *Brassica* spp.
kenaf, *Hibiscus cannabinus*

lady's finger (gumbo, okra, okro), *Abelmoschus esculentus*
lambsquarter (goosefoot), *Chenopodium* spp.
legumes, *Leguminosae* family
lemon, *Citrus limon*
lemon grass, *Cymbopogon citratus*
lemon verbena, *Aloysia triphylla*
lentil, *Lens esculenta*
lettuce, *Latuca sativa*
lima bean, *Phaseolus lunatus*
lime, *Citrus aurantifolia*
loquat, *Eriobutrya japonica*
luffa, *Luffa* spp.

maize (corn), *Zea mays*
mallow (jute), *Corchorus olitorius*
mango, *Mangifera indica*

manioc (cassava, tapioca, yuca) *Manihot esculenta*
marjoram, *Oreganum vulgare*
marrow, *Cucurbita pepo*
mat bean (moth bean), *Phaseolus aconitifolia*
melon, *Cucumis melo*
mesquite, *Prosopis* spp.
millet, common, *Panicum* spp.
millet, bulrush, *Pennisetum typhoideum*
millet, finger, *Eleusine coracana*
millet, foxtail, *Setaria italica*
mint, *Mentha* spp.
moth bean (mat bean), *Phaseolus aconitifolia*
mulberry, *Mora* spp.
mung bean (golden or green gram), *Phaseolus aureus*
mustards, *Brassica* spp.

Natal plum, *Carissa grandiflora*
nectarine, *Prunus dulcis*
Niger seed, *Guizota abyssinica*

okra (gumbo, lady's finger, okro), *Abelmoschus esculentus*
olive, *Olea europaea*
onion, *Allium* spp.
orange, *Citrus sinensis*
oregano, *Oreganum vulgare*

palm, African fan, *Borassus aethiopum*
palm, date, *Phoenix datylifera*
palm, doum, *Hyphaene thebaica*
palm, palmyra, *Borassus flabellifer*
papaya (pawpaw), *Carica papaya*
parsley, *Petroselinum crispum*
passion fruit, *Passiflora edulis*
pawpaw (papaya), *Carica papaya*
pea, *Pisium sativum*
pea, black-eyed (asparagus bean, cowpea, yard-long bean), *Vigna unguiculata*
pea, grass, *Lathyrus sativus*
pea, pigeon, *Cajanus cajan*
peach, *Prunus persia*
peanut (groundnut), *Arachis hypogaea*
pear, *Pyrus communis*
pepper, sweet, *Capsicum annum*
pepper, bird, *Capsicum frutescens*
pepper, hot (bird, chili), *Capsicum frutescens, Capsicum annuum*
pigeon pea, *Cajanus cajan*
pineapple, *Ananas comosus*
pistachio, *Pistacia vera*
plantain, *Musa* spp.

plum, Natal, *Carissa grandiflora*
pomegranate, *Punica granatum*
potato (white potato), *Solanum tuberosum*
potato, Hausa, *Solenostemon rotundifolius*
potato, sweet, *Ipomoea batatas*
prickly pear cactus, *Opuntia* spp.
pulses, legumes grown for their seeds which are often dried before cooking and eating
pumpkin, *Cucurbita pepo*
purslane, *Portulaca oleracea*
pyrethrum, *Chrysanthemum cinerariaefolium*

quince, *Cydonia oblonga*
quinoa, *Chenopodium quinoa*

radish, *Raphanus sativus*
rocket salad (arugola), *Eruca sativa*
Rose family, Rosaceae
roselle (hibiscus), *Hibiscus sabdariffa*
rosemary, *Rosmarinus officinalis*

sage, *Salvia officinalis*
sapote, *Casimiroa edulis*
sesame, *Sesamum indicum*
shea butter tree, *Butryospermum paradoxum*
sorghum, *Sorghum* spp.
sour orange, *Citrus* spp.
spinach, *Spinacia oleracea*
spinach, Indian, *Basella alba*
squash, *Cucurbita* spp.
stone fruit, *Prunus* spp.
sugarcane, *Saccharum* spp.
sunflower, *Helianthus annus*
sweet pepper, *Capsicum annuum*
sweet potato, *Ipomoea batatas*

tamarind, *Tamarindus indica*
tangerine, *Citrus reticulata*
tapioca (cassava, manioc, yuca), *Manihot esculenta*
teosinte, *Zea* spp.
tepary bean, *Phaseolus acutifolius*
thyme, *Thymus vulgaris*
tiger nut, *Cyperus esculentus*
tomatillo, *Physalis* spp.
tomato, *Lycopersicon esculentum*
tree tomato, *Cyphomandra betaceae*
turnip, *Brassica rapa*

umbellifers, Umbelliferae family
urd, *Phaseolus mungo*
wax gourd (white gourd), *Benincasa hispida*

watermelon, *Citrullis lanatus*
white gourd (wax gourd), *Benincasa hispida*
white potato, *Solanum tuberosum*
wheat, *Triticum* spp.
winter squash, *Cucurbita* spp.

yam, *Dioscorea* spp.
yard-long bean (asparagus bean, black-eyed pea, cowpea), *Vigna unguiculata* yuca (cassava, manioc, tapioca), *Manihot esculenta*

18.2 Important Dryland Garden Plant Families

Because of characteristics common to their members, it is sometimes more useful to refer to plant families rather than individual plants.

Onion family (Alliaceae)
  Allium ampeloprasum, great-headed garlic
  Allium sativum, garlic
  Allium spp., onion

Chenopod family (Chenopodiaceae)
  Beta vulgaris, beet root, chard
  Chenopodium quinoa, quinoa
  Chenopodium spp., goosefoot (lambsquarter), and many wild greens
  Spinacia oleracea, spinach

Citrus family (Rutacea)
  Citrus aurantifolia, lime
  Citrus limon, lemon
  Citrus paradisi, grapefruit
  Citrus reticulata, tangerine
  Citrus sinensis, orange

Composite family (Compositae)
  Chrysanthemum cinerariaefolium, pyrethrum
  Cynara scolymus, globe artichoke
  Guizotia abyssinia, Niger seed
  Helianthus annus, sunflower
  Helianthus tuberosus, Jerusalem artichoke
  Latuca sativa, lettuce

Crucifer family (Cruciferae)
  Brassica oleracea, broccoli, cabbage, cauliflower, collard, kale
  Brassica rapa, turnip
  Brassica spp., Chinese cabbage, Indian mustard, rape
  Eruca sativa, rocket salad
  Raphanus sativus, radish

Cucurbit family (Cucurbitaceae)
  Benincasa hispida, wax gourd
  Citrullus lanatus, watermelon
  Cucumeropsis edulis, egusi melon
  Cucumis melo, melon
  Cucumis sativus, cucumber
  Cucurbita spp., pumpkin, squash, marrow
  Lagenaria siceraria, bottle gourd
  Luffa spp., luffa
  Sechium edule, chayote (choyote)

Mint family (Labiatae)
  Mentha spp., mint
  Ocimum basillicum, basil
  Origanum vulgare, oregano and marjoram
  Rosmarinus officinalis, rosemary
  Salvia officinalis, sage
  Thymus vulgaris, thyme

Legume family (Leguminosae)
  Arachis hypogaea, peanut or groundnut
  Cajanus cajan, pigeon pea
  Canavalia ensiformis, horse or jack bean
  Ceratona siliqua, carob
  Cicer avietinum, chickpea or garbanzo
  Kerstingiella geocarpa, Hausa groundnut
  Lablab niger, hyacinth or Egyptian bean
  Lathyrus sativus, grass pea
  Lens esculenta, lentils
  Parkia biglobosa, African locust bean
  Phaseolus acutifolius, tepary beans
  Phaseolus aureus, mung bean or green or golden gram
  Phaseolus lunatus, lima bean
  Phaseolus mungo, black gram
  Phaseolus vulgaris, common bean
  Pisum sativum, pea
  Vicia faba, fava bean or broad bean
  Vigna aconitifolia, mat or moth bean
  Vigna unguiculata, asparagus bean, black-eyed pea, cowpea, yard-long bean
  Voandzeia subterannea, Bambara groundnut

Malva family (Malvacea)
  Abelmoschus esculentus, okra, lady's finger, gumbo
  Hibiscus cannabinus, kenaf
  Hibiscus sabdariffa, roselle or hibiscus

Palm family (Palmae)
 Borassus aethiopicum, African fan palm
 Borassus flabellifer, palmyra palm
 Hyphaene thebaica, doum palm
 Phoenix datylifera, date palm

Rose family (Rosaceae)
 Eriobutrya japonica, loquat
 Malus spp., apple
 Prunus americana, apricot
 Prunus dulcis, almond
 Prunus persica, peach, nectarine
 Prunus spp., stone fruits
 Pyrus communis, pear

Solanum family (Solanaceae)
 Capsicum annuum, sweet pepper
 Capsicum annuum, chili, hot pepper
 Capsicum frutescens, bird pepper, hot pepper, chili
 Cyphomandra betaceae, tree tomato
 Lycopersicon esculentum, tomato
 Physalis spp., tomatillo
 Solanum spp., eggplant, aubergine, brinjal, garden egg
 Solanum tuberosum, potato

Umbellifer family (Umbelliferae)
 Coriandrum sativum, coriander, cilantro
 Cuminum cyminum, cumin
 Daucus carota, carrot
 Foeniculum vulgare, fennel
 Petroselinum crispum, parsley

19
Resource Organizations

The following is a list of resource organizations that may be of use to those working with gardens in drylands. This is not a complete list, there are many more groups, both local and international, whose work is relevant. We include here brief descriptions; some written by us, some from the organizations themselves, and some taken from the list of organizations in AGRECOL and ILEIA's *Towards Sustainable Agriculture* (AGRECOL/ILEIA 1988).

AGRECOL DEVELOPMENT INFORMATION c/o Oekozentrum, CH-4438 Langenbruck, Switzerland, telephone 062-60 14 20 or 062-60 14 60

AGRECOL Development Information was founded in 1983 and is "an information center within the network for sustainable agriculture in Third World countries." It promotes balanced ecological systems, independence from external inputs, local self-reliance, and socially and economically sound agriculture through a documentation center providing answers to questions and contacts with consultants. A good source for unusual documents such as reports on small PVO projects. They encourage contributions to the center.

AMERICAN FRIENDS SERVICE COMMITTEE (AFSC) 1501 Cherry Street, Philadelphia, Pennsylvania 19102-1479, USA, telephone (215)241-7000, telex 247559 AFSC UR

AFSC has numerous small projects in the Third World, and in the United States, aimed at helping improve food production, empower the poor and reduce military tension.

APPROPRIATE TECHNOLOGY INTERNATIONAL (ATI) c/o Volunteers in Asia, PO Box 4543, Stanford, California 94305, USA, telephone (415)326-8581

ATI produces the *Appropriate Technology Sourcebook* (Darrow and Saxenian 1986), and a microfiche library of AT publications for Third World development projects.

ASIAN VEGETABLE RESEARCH AND DEVELOPMENT CENTER (AVRDC) PO Box 42, Shanhua, Tainan, 741, Taiwan, telephone 06-5837801, fax 06-5830009

Most of AVRDC's work appears to be devoted to large-scale commercial aspects of vegetable production and industrial-style gardens. A source for technical publications on some Asian crops. AVRDC has supported model garden research and garden extension in Asia (Llemit and Pura 1988).

BIO-INTEGRAL RESOURCE CENTER (BIRC) PO Box 7414, Berkeley, California 94707, USA, telephone (415)524-2567

The BIRC is a source for books and other educational materials on integrated pest management and biological control of agricultural and household pests.

BOARD ON SCIENCE AND TECHNOLOGY FOR INTERNATIONAL DEVELOPMENT (BOSTID) National Research Council 2101 Constitution Ave., NW, Washington, District of Columbia 20418, USA

"BOSTID examines ways to apply science and technology to problems of economic and social development through overseas programs, research grants, studies, advisory committees, workshops, and other mechanisms." It is the source for several of the NAS publications listed in Chapter 20; many of these publications are available free of charge to Third World organizations.

CENTER FOR INDIGENOUS KNOWLEDGE FOR AGRICULTURAL AND RURAL DEVELOPMENT (CIKARD), Technology and Social Change Program, Iowa State University, 324 Curtis, Ames, Iowa 50011, USA, telephone (515)294-0938

CIKARD "focuses its activities on preserving and using the local knowledge of farmers and other people around the world." CIKARD has a library and publishes a quarterly newsletter.

CENTRO INTERNATIONAL DE AGRICULTURA TROPICAL (CIAT) Apartado Aereo 6713, Cali, Colombia, telephone 57 3-675050

CIAT is a member of the CGIAR. It conducts research on *Phaseolus* spp. beans, cassava, rice, and tropical pastures.

CONSULTATIVE GROUP ON INTERNATIONAL RESEARCH (CGIAR) 1818 H St., NW, Washington, DC 20433 USA, telephone (202)334-8028, telex 440098

CGIAR supports a system of international agricultural research centers (IARCs). It is dominated by Western countries, is housed in the World Bank, and focuses on industrial-style agriculture and development. The IARCs all distribute publications on their research which are listed, along with those of a number of other international agricultural research organizations, in IRRI (1989).

CULTURAL SURVIVAL 53A Church Street, Cambridge, Massachusetts 02138, USA, telephone (617)495-2562

Cultural Survival was founded in 1972 to help small societies survive the rapid changes that are destroying them. It provides funds and expertise for projects around the world. Publishes papers, books, special reports and the *Cultural Survival Quarterly*.

DEVELOPING COUNTRIES FARM RADIO NETWORK (DCFRN) 595 Bay Street, Toronto, Ontario M5G2C3 Canada, telephone (416)593-3752, fax (416)593-3820

DCFRN produces well-written radio scripts on a wide range of topics oriented toward small-scale producers and promoting the use of low-cost local resources. The scripts contain many ideas for extension and production techniques, and are backed up with references. Radio scripts available in English, French and Spanish. Member radio stations worldwide receive resource materials.

ENVIRONMENT LIASON CENTER INTERNATIONAL (ELCI) PO Box 72461, Nairobi, Kenya, telephone (254-2)562015, 562022, 562172, fax 562175

ELCI is an NGO established in 1974 in Kenya to promote communication and cooperation among NGOs worldwide and to serve as a link between them and the United Nations Environment Program (UNEP). It promotes environmentally sustainable development and agriculture, organizes workshops and conferences, and publishes the bimonthly *Ecoforum*.

ENVIRONNEMENT ET DEVELOPPEMENT DU TIERS MONDE (ENDA) BP 3370, Dakar, Sénégal, telephone 221-21 60 27

ENDA's activities are "research, workshops, publications, and inquiry service on environment, development, and sustainable food production."

FOOD AND AGRICULTURE ORGANIZATION OF THE UNITED NATIONS (FAO) Via delle Terme di Caracalla, 00100 Rome, Italy, telephone 57971, fax 5782610, 57973152

The Distribution and Sales Section at this address will send free copies of *FAO Books in Print* and *List of Documents*. These publications contain lists of FAO documents available to the general public and local addresses in about 80 countries where they can be purchased. Technical publications on agriculture tend to be oriented toward large-scale, capital-intensive production, but provide good data. FAO publications on "traditional crops," agroforestry, and nutrition are more appropriate for gardens, and contain good references.

FOOD FIRST (see Institute for Food and Development Policy)

DEUTSCHE GESELLSCHAFT FUR TECHNISCHE ZUSAMMENARBEIT (GTZ) GmbH Postfach 5180 Dag-Hammarskjold-Weg 1+2, D 6236 Eschborn/Ts. 1, Germany, telephone (06196)79-0, fax (06196)79-1115

GTZ is the overseas development organization of the German government. GTZ has published workshop proceedings and research reports on postharvest storage and pest control for Third World agriculture, and on a number of other topics relevant for gardens.

HESPERIAN FOUNDATION PO Box 1692, Palo Alto, California 94302, USA, telephone (415) 325-9017

Hesperian Foundation promotes grassroots community health care through consulting, projects in Mexico, and excellent training and reference materials including *Where There is No Doctor* (Werner 1977) and *Helping Health Workers Learn* (Werner and Bower 1982).

INFORMATION CENTRE FOR LOW-EXTERNAL INPUT AGRICULTURE (ILEIA) PO Box 64,

Kastanjelaan 5, 3830 AB Leusden, the Netherlands, telephone 033-94-30-86, fax 033-94-07-91

ILEIA "collects and disseminates information on low-external input agriculture for Third World countries, publishes the *ILEIA Newsletter* and bibliographies on sustainable agriculture (AGRECOL/ILEIA 1988; Carlier 1987), organizes workshops, and has an inquiry service."

INSTITUTE FOR FOOD AND DEVELOPMENT POLICY/FOOD FIRST (IFDP) 1885 Mission Street, San Francisco, California 94103, USA, telephone (415)864-8555

Through public education programs, research, and publications Food First emphasizes world hunger as a result of economic and political systems, not food scarcity or natural disasters. A source of publications and other educational materials. *Diet for a Small Planet* (Lappé 1982) and *Food First* (Lappé and Collins 1978) are standard references.

INTERMEDIATE TECHNOLOGY DEVELOPMENT GROUP (ITDG) Myson House, Railway Terrace, Rugby CV21 3HT, United Kingdom

"ITDG helps people in rural areas of the Third World to aquire the appropriate tools and technologies to work themselves out of poverty." ITDG publishes a newsletter, and provides consultants.

INTERMEDIATE TECHNOLOGY PUBLICATIONS 9 King St., London WC2E 8HW, United Kingdom

The outlet for many appropriate technology publications.

INTERNATIONAL CENTER FOR AGRICULTURAL RESEARCH IN THE DRY AREAS (ICARDA) PO Box 5466, Aleppo, Syria, telephone 55 0465, telex 331206 SY

A member of CGIAR, ICARDA conducts "research on farming systems (dry areas of north Africa, Middle East), barley, lentils, fava beans, wheat, and chick pea."

INTERNATIONAL COUNCIL FOR RESEARCH IN AGROFORESTRY (ICRAF)PO Box 30677, Nairobi, Kenya, telephone 29867, telex 22048

ICRAF is an independent organization supporting research and publications on agroforestry, including a quarterly magazine, *Agroforestry Today*. Some of their publications, as well as the magazine, are available free to Third World organizations.

INTERNATIONAL CROPS RESEARCH INSTITUTE FOR THE SEMI-ARID TROPICS (ICRISAT) Patancheru PO, Andrha Pradesh 502 324, India, telephone 262251, telex 42203 ICRI IN

ICRISAT is a member of CGIAR. Though ICRISAT is mostly oriented toward sorghum and millet, it also conducts research on a variety of production topics of interest to gardeners.

INTERNATIONAL DEVELOPMENT RESEARCH CENTRE (IDRC) Communications Division PO Box 8500 Ottawa, Ontario K1G 3H9, Canada

Financed by the government of Canada, IDRC's publications including those on agriculture, food, nutrition, and health, are sometimes available free of charge to Third World organizations.

INTERNATIONAL FEDERATION OF ORGANIC AGRICULTURE MOVEMENTS (IFOAM) IFOAM General Secretariat, Okozentrum Imsbach, D-6695 Tholey-Theley, Germany, telephone 49(0)6853-5190

IFOAM publishes a quarterly bulletin and sponsors regular international scientific conferences supporting organic, sustainable agriculture.

INTERNATIONAL FOUNDATION FOR DEVELOPMENT ALTERNATIVES (IFDA) Case Postale 1260 Nyon, Switzerland, telephone 41(22) 618282, telex 28840 IFDA CH

IFDA publishes a biannual journal, the *IFDA Dossier* with contributions on diverse topics from many people and regions. IFDA promotes dialogue and discussion between individuals and organizations interested in alternative approaches to development.

INTERNATIONAL INSTITUTE FOR TROPICAL AGRICULTURE (IITA)PO Box 5320, Ibadan, Nigeria, telephone 413440, telex 31417 TROPIB NG

IITA is a member of CGIAR; it conducts "research on farming systems (Africa), rice, maize, cassava, cocoyam, cowpea, yam, sweet potato, and soybean."

INTERNATIONAL ORGANIZATION OF CONSUMERS UNIONS (IOCU) PO Box 1045, Penang, Malaysia, telephone (04)20391, telex MA40164 APIOUC

"IOUC promotes international cooperation in the testing of goods and services and in all other aspects of consumer information, education, and protection." It has a membership of over 100 consumer's groups from 50 countries and has been an active participant in efforts to protect consumers rights and health on issues concerning the use of infant formulas and pesticides. IOCU published the *Pesticide Handbook* (1984), and is the coordinator for the international Pesticides Action Network (PAN).

NATIVE SEEDS/SEARCH (NS/S) 2509 N. Campbell Ave., #325, Tucson, Arizona 85719, USA, telephone (602)327-9123

An example of a regional seed conservation organization, NS/S focuses on folk varieties of the southwestern United States and northern Mexico.

OXFAM UK 274 Banbury Rd, Oxford OX2 7DZ, United Kingdom, telephone (0865)56777, fax (0865)57612

OXFAM UK has projects in the Third World, and produces publications useful for project workers, as well as educational materials.

POSTHARVEST INSTITUTE FOR PERISHABLES (PIP) 314 University of Idaho Library, Moscow, Idaho 83843, USA

PIP is a documentation center for publications and unpublished documents on food storage and processing. Copies of documents are available free of charge on request.

RURAL ADVANCEMENT FUND INTERNATIONAL (RAFI) PO Box 655, Pittsboro, North Carolina 27312 USA, telephone (919)542 1396

"RAFI works for the conservation of genetic resources within the broader framework of rural development, and does research, public education, lobbying, and consulting on both policy and technical matters."

ROYAL BOTANIC GARDENS, KEW ("KEW GARDENS") Kew, Richmond, Surrey TW9 3AE, United Kingdom

KEW has a Survey of Economic Plants for Arid and Semi-Arid Tropics (SEPASAT) project that collates "information about plants from the dry tropics that have been reported to be useful (in any way) but that have not been commercially exploited on a large scale." Write for more information on inquiries that can be made of the data base; presently supplied free of charge for nonprofit institutions.

TERRES ET VIE Rue Laurent Delvaux, 13, 1400 Nivelles, Belgique

Terres et Vie publishes and distributes the books by Chleq and Dupriez (1984), Dupriez (1982), and Dupriez and De Leener (1983, 1987). These handbooks for extension workers and secondary school students have lots of excellent photographs, and frequently include descriptions of indigenous methods.

UNITED NATIONS CHILDREN'S FUND (UNICEF) 3 UN Plaza, New York, New York 10017 USA, telephone (212)326-7000, telex 175989 TRT

UNICEF works on international policy as well as having field programs that promote improved child survival and well-being. They have published several booklets on household gardens for nutrition; Sommers (1984), UNICEF (1985).

WINROCK INTERNATIONAL INSTITUTE FOR AGRICULTURAL DEVELOPMENT Petit Jean Mountain, Route 3, Morrilton, Arkansas 72110, USA, telephone (501)727-5435, telex 910-720-6616 WI HQ UD

A Winrock project of interest to gardeners is the On Farm Seed Project which publishes the newsletter *Seed Sowers/Les Semeurs* (section 14.6).

WORLD HEALTH ORGANIZATION (WHO) 1211 Geneva 27, Switzerland

WHO publishes many documents on nutrition and public health. Contact the Nutrition Unit or Distribution and Sales Service.

20
References

All references cited in the text are included in this chapter. Following the publication information in each citation, the chapter in the text that cites the reference appears in brackets. We have annotated some of the references that we feel would be of most use to readers. Those publishers marked with an asterisk (*) are listed in Chapter 19, Resource Organizations.

ABRAHAMSE, Tanya and Angela M. BRUNT (1984) An investigation into pesticide imports, distribution and use in Zambia with special emphasis on the role of multinational companies. Insect Science Application 5(3):157-173. [Ch.13]

ACLAND, J.D. (1971) East African Crops. Longman, London, UK (by arrangement with FAO). 252pp. [Ch.7,18]

ACHINEWHU, S.C. (1986) Some biochemical and nutritional changes during the fermentation of fluted pumpkin (*Telferia occidentalis*). Plant Foods and Human Nutrition 36:97-106. [Ch.2]

ACPP (Alternative Crop Protection Project) (n.d.) Watery compost extract as fungicide. ACPP, c/o McKean Rehabilitation Centre, PO. Box 53, Chiang Mai 50000, Thailand. mss., 3pp. [Ch.13]

ADAMS, William M. (1986) Traditional agriculture and water use in the Sokoto Valley, Nigeria. The Geographical Journal 152:30-43. [Ch.11,12]

ADENIJI, M.O. (1977) Studies on some aspects of control of the yam nematode, *Scutellonema bradys*. Acta Horticulturae, No. 53:249-255. [Ch.13]

AFRICAN DIVERSITY (1990) Diversity News. African Diversity 2-3:1-4. [Ch. 14]

AGAB, Muna Ahmed (1985) Fermented food products 'hulu mur' drink made from *Sorghum bicolor*. Food Microbiology 2:147-155. [Ch.15]

AGRECOL/ILEIA (1988) Towards Sustainable Agriculture. AGRECOL*, Langenbruk, Switzerland, and ILEIA*, Leusden, The Netherlands. Part One (24pp) lists documents, periodicals, and organizations. Part Two (24pp) is a bibliography. [Part II, Ch.19].

AGRIOS, George N. (1988) Plant Pathology. 3rd ed. Academic Press, New York, USA. xvi+803pp. [Ch.9,13] A standard textbook with good descriptions of life cycles and mostly high-tech identification techniques; emphasizes chemicals, very little on nonchemical control or ecology.

AHMED, Saleem, and Michael GRAINGE (1986) Potential of the neem tree (*Azadirachta indica*) for pest control and rural development. Economic Botany 40:201-209. [Ch.14]

AHN, Peter M. (1970) West African Soils. Oxford University Press, London, UK. xii+332pp. [Ch.9,10] Semipopular. An excellent introduction to practical soil science, with basics applicable anywhere. Emphasizes the changes that occur moving north and south, between arid and humid regions of West Africa. Discusses soil changes and plant nutrition in traditional agriculture.

AKPAPUNAM, Maurice A. (1984) Effects of wilting, blanching and storage temperatures on ascorbic acid and total carotenoids content of some Nigerian fresh vegetables. Plant Foods for Human Nutrition 34:177-180. [Ch.15]

AKPAPUNAM, Maurice A., and S.C. ACHINEWU (1985) Effects of cooking, germination and fermentation on the chemical compostion of Nigerian cowpea (*Vigna unguiculata*). Plant Foods for Human Nutrition 35:353-358. [Ch.2,15]

ALI, H.M., and I.A. SAKR (1981) Drying of vegetables in Egypt. *In* Food Drying. Gordon Yaciuk, ed. Pp. 15-19. IDRC*, Ottawa, Canada. [Ch.15]

ALTIERI, Miguel A., and Matt LIEBMAN (1986) Insect, weed and plant disease management in multiple cropping systems. *In* Francis 1986:183-218. [Ch.8]

AMES, Bruce N. (1983) Dietary carcinogens and anticarcinogens. Science 221:1256-1264. [Ch.2]

ANDREWS, David J. (1989) Cereal breeding in Africa. Paper based on a seminar given at Summer Institute for African Agricultural Research, University of Wisconsin-Madison, USA. 21pp. [Ch.14]

ARLOSOROFF, S., G. TSHANNELRL, D. GREY, W. JOURNEY, A. KARP, O. LANGENEFFER, and R. ROCHE (1987) Community Water Supply: The Handpump Option. World Bank. [Ch.12] Summarizes findings of a 5-year project on the testing and technical and managerial development of handpump-based water supply systems. [Reviewed by F. Carroll]

ARNON, I. (1975) Physiological Principles of Dryland Crop Production. *In* Gupta 1975:3-145. [Ch.8]

ATLIN, Gary N., and Kenneth J. FREY (1989) Breeding crop varieties for low-input agriculture. American Journal of Alternative Agriculture 4(2):53-58. [Ch.14]

ATKINSON, D., D. NAYLOR, and G.A. COLDRICKA (1976) The effect of tree spacing on the apple root system. Horticultural Research 16:89-105. [Ch.6]

ATTEH, Oluwayomi D. (1987) Pesticide flow and government attitude to pests and pesticides in Kwara State, Nigeria. *In* Tait and Napompeth 1987:86-91. [Ch.13]

AUMEERUDDY, Y., and F. PINGLO (1989) Phytopractices in Tropical Regions. UNESCO/Laboratoire de Botanique Tropicale, Montpellier, France. 71pp. [Ch.7] Many extremely interesting but very brief descriptions about traditional manipulation of individual plants.

AYERS, R.S., and D.W. WESCOT (1985) Water Quality for Agriculture. Revision 1. FAO Irrigation and Drainage Paper, 29. (First published, 1976) FAO*, Rome. xii+174pp. [Ch.5,11,12] Covers salinity, infiltration, toxicity and other problems with examples of using water of various qualities from around the world.

AYKROYD, W.R., and Joyce DOUGHTY (1982) Legumes in Human Nutrition. (Revised by Joyce Doughty and Ann Walker, first published 1964) FAO*, Rome. viii+152pp [Ch.15,16]

AYRES, John C., J. Orvin MUNDT, and William E. SANDINE (1980) Microbiology of Foods. W.H. Freeman, San Francisco, California, USA. [Ch.15]

BAKER, Kenneth F. (1987) Evolving concepts of biological control of plant pathogens. Annual Review of Phytopathology 25:67-85. [Ch.13]

BARKOW, J.H. (1972) Hausa women and Islam. Canadian Journal of African Studies 6:317-328. [Ch.3]

BARRETT, Robert P. (1987) Integrating leaf and seed production strategies for cowpea. Pulse Beat [Bean/Cowpea CRSP, Michigan State University, East Lansing, USA] [Ch.15]

BASSUK, Nina L. (1986) Reducing lead intake in lettuce. Hortscience 21:993-995. [Ch.9,15]

BELL, Morag, Richard FAULKNER, Patricia HOTCHKISS, Robert LAMBERT, Neil ROBERTS, and Alan WINDRAM (1987) The Use of Dambos in Rural Development, with Reference to Zimbabwe. Loughborough University, UK and University of Zimbabwe. Final Report of ODA Project R3869. xii+151pp. + appendices. [Ch.9,10,11,12,13] A survey and case study of gardens in valley areas usually watered by shallow wells. Focuses on soils and water, and emphasizes the importance of dambo gardens to households.

BENESON, Abram S. (1985) Control of Communicable Diseases in Man. 14th ed. The American Public Health Association, Washington DC, USA. [Ch.11]

BENZ, B.F., L.R. SANCHEZ-VELASQUES, and F.J. SANTANA MICHEL (1990) Ecology and ethnobotany of *Zea Diploperennis*: Preliminary investigations. Maydica 35:85-98. [Ch.8]

BERKES, F., D. FEENY, B.J. MCCAY, and J.M. ACHESON (1989) The Benefit of the Commons. Nature 340:91-93. [Ch.3]

BERNARD, H. Russel (1988) Research Methods in Cultural Anthropology. Sage Publications, Newbury Park, California, USA. 520pp. [Ch.4] A handbook of field methods emphasizing quantitative measurements.

BETTOLO, G.B. Marini, ed. (1988) Towards a Second Green Revolution: From Chemical to New Biological Technologies in Agriculture in the Tropics. Elsevier, Amsterdam, The Netherlands. xi+530pp. (Source of Pimentel 1988)

BHOWMIK, P.C., and J.C. DOLL (1984) Allelopathic effects of annual weed residues on growth and nutrient uptake of corn and soybeans. Agronomy Journal 76(3):383-388. [Ch.8]

BIRKELAND, Peter W. (1984) Soils and Geomorphology. Oxford University Press, London. xiv+372pp. [Ch.9]

BITTENBENDER, H. C. (1984) Handbook of Tropical Fruits and Spices. Department of Horticulture, Michigan State University, East Lansing, Michigan, USA. v+127pp. [Ch.14]

BITTENBENDER, H. C. (1985) Home gardens in less

developed countries. HortScience 20:645-649. [Part I]

BLAKIE, Piers (1985) The Political Economy of Soil Erosion in Developing Countries. Longman, London, UK. [Ch.9]

BLALOCK, Hubert M., Jr. (1972) Social Statistics. 2nd ed. McGraw-Hill, New York, USA. xiv+583pp. [Ch.4]

BLAYLOCK, J.R., and A.E. GALLO (1983) Modeling the decision to produce vegetables at home. American Journal of Agricultural Economics 65:722-729. [Ch. 3]

BLEIBERG, Fanny M., Thierry A. BRUN, Samuel GOIHMAN, and Emile GOUBA (1980) Duration of activities and energy expenditure of female farmers in dry and rainy seasons in Upper-Volta. British Journal of Nutrition 43:71-82. [Ch.2]

BODLEY, John H. (1990) Victims of Progress. 3rd ed. Mayfield Publishing Company, Mountain View, California, USA. ix,261pp. [Part I] A well-written history of the physical and cultural destruction of indigenous peoples around the world as a result of colonialism and the industrial revolution. Current movements for self-determination are also covered.

BOULOS, Loutfy and M. Nabil EL-HADIDI (1984) The Weed Flora of Egypt. The American University in Cairo Press, Cairo, Egypt. iv,178pp. [Ch.8]

BRADFIELD, Maitland (1971) The Changing Pattern of Hopi Agriculture. Royal Anthropological Institute Occasional Paper No. 30. Royal Anthropological Institute, London, UK. 65pp. [Ch.6]

BRIDGE, John (1987) Control strategies in subsistence agriculture. In Brown and Kerry 1987:389-420. [Ch.13]

BRIERLEY, J.S. (1976) Kitchen gardens in the West Indies, with a contemporary study from Grenada. The Journal of Tropical Geography 43:30-40. [Ch. 3]

BROWN, Lester R., Alan DURNING, Christopher FLAVIN, Hilary FRENCH, Jodi JACOBSON, Marcia LOWE, Sandra POSTEL, Michael RENNER, Linda STARK, and John YOUNG (1990) State of the World 1990: A Worldwatch Institute Report on Progress Toward a Sustainable Society. W.W. Norton and Company, New York and London. xii+253pp. [Part I] (Source of Durning 1990) This report is published yearly and is a widely cited source of information on the destruction of natural resources and efforts to halt it.

BROWN, R.H. and B.R. KERRY, eds. (1987) Principles and Practice of Nematode Control in Crops. Academic Press, Sydney, Australia. (Source of Bridge 1987; Maas 1987)

BROWNRIGG, Leslie (1985) Home Gardening in International Development: What the Literature Shows. League for International Food Education, Washington, DC, USA. ca.341pp. [Part I,Ch.2,3] This is the most comprehensive review of household garden projects to date and has much valuable information for anyone involved in garden programs. It is out of print and the publisher is defunct, but it was widely distributed around the world, for example, to USAID missions.

BRUN, Thierry, Fanny BLEIBERG, and Samuel GOIHMAN (1981) Energy expenditure of male farmers in dry and rainy seasons in Upper-Volta. British Journal of Nutrition 45:67-75. [Ch. 2]

BRUN, Thierry, Jaqueline REYNAUD, and Simon CHEVASSUS-AGNES (1989) Food and nutritional impact of one home garden project in Senegal. Ecology of Food and Nutrition 23:91-108. [Ch.2,3] A rare attempt at garden project evaluation. Findings included: gardens had no direct nutritional impact after nearly 20 years; wild foods are important nutritionally; garden income is socially important.

BRUN, Thierry, and Michael C. LATHAM, eds. (1990) Maldevelopment and Malnutrition. World Food Issues, Volume 2. Center for the Analysis of World Food Issues, Cornell University, Ithaca, New York, USA. (Source of Campbell, et al. 1990; Latham 1990)

BRUSH, Stephen B. (1986) Genetic diversity and conservation in traditional farming systems. Journal of Ethnobiology 6(1):151-167. [Ch.14]

BRUSH, Stephen B., Mauricio BELLON CORRALES, and Ella SCHMIDT (1988) Agricultural development and maize diversity in Mexico. Human Ecology 16(3):307-328. [Ch.13]

BRYSON, Reid A., and F. Kenneth HARE (1974) Climates of North America. Elsevier Publishing Co., Amsterdam, The Netherlands. [Ch.11] Includes Mexico.

BULL, David (1982) A Growing Problem: Pesticides and the Third World Poor. OXFAM*, Oxford, UK [Ch.13] A general overview promoting IPM. Documents problems in late 1970s and early 1980s.

BUNCH, Roland (1982) Two Ears of Corn: A Guide to People-Centered Agricultural Improvement. World Neighbors, 5116 N. Portland Ave., Oklahoma City, Oklahoma, 73112 USA. vii+251pp. [Ch.4] Based on World Neighbors' many years of experience, the book gives some practical pointers useful for anyone involved in development and is valuable because of its emphasis on villager participation. The chapter on assessment, however, does not give much in the way of specific ideas about assessment

techniques.
BURN, A.J., T.H. COAKER, and P.C. JEPSON (1987) Integrated Pest Management. Academic Press, London, UK. xi,474pp. (Source of Cammel and Way 1987)
BUTTEL, Fredderick, Martin KENNEY, and Jack KLOPPENBURG, Jr. (1985) From green revolution to biorevolution: Some observations on the changing technological bases of economic transformation in the Third World. Economic Development and Cultural Change 33:31-55. [Ch.14]
BYE, Robert A. (1981) Quelites: Ethnoecology of edible greens—past, present and future. Journal of Ethnobiology 1(1):109-123. [Ch.8]

CAMERON, Margaret and Yngve HOFVANDER (1983) Manual on Feeding Infants and Young Children. 3rd ed. Oxford University Press, Oxford, UK. [Ch.2,15,16] A good resource with much practical advice including recipes for weaning foods.
CAMMELL, M.E., and M.J. WAY (1987) Forecasting and monitoring. In Burn et al. 1987:1-26. [Ch.13]
CAMPBELL, T. Colin, Chen JUNSHI, Thierry BRUN, Banoo PARPIA, Qu YINSHENG, Chen CHUNMING, and Catherine GEISSLER (1990) Can developing nations avoid the diseases of affluence? The case of China. In Brun and Latham 1990:56-63. [Ch.2]
CAMPBELL, T. Colin, Thierry BRUN, Chen JUNSHI, Feng ZULIN, and Banoo PARPIA (n.d.) Erythrocyte glutathione reductase and riboflavin intakes in China. mss. [Ch.2]
CARLIER, Hans (1987) Understanding Traditional Agriculture: Bibliography for Development Workers. ILEIA*, Leusden, The Netherlands. 114pp. [Ch.19]
CARLONI, Alice S. (1981) Sex disparities in the distribution of food within rural households. Food and Nutrition 7(1):3-12. [Ch.2]
CARROLL, C. Ronald, John H. VANDERMEER, and Peter ROSSET (1990) Agroecology. McGraw-Hill, New York, USA. 641pp. [Part II]
CFA (California Fertilizer Association, Soil Improvement Committee) (1980) Western Fertilizer Handbook. 6th ed. Interstate Printers and Publishers, Inc., Danville, Illinois, USA. [Ch.9,11]
CHACON, J.C., and S.R. GLIESSMAN (1982) The use of the "non-weed" in traditional agroecosystems of southeastern Mexico. Agro-Ecosystems 8:1-11. [Ch.8]
CHAMBERS, Robert (1983) Rural Development: Putting the Last First. Longman, London, UK. [Ch.4,9] A good discussion of why and how most development projects overlook those most in need, and suggestions for change.
CHAMBERS, Robert (1988) Managing Canal Irrigation: Practical Analysis from South Asia. Cambridge University Press, London, UK. [Ch.12]
CHAROENKIATKUL, Somsri, Aree VALYASEVI, and Kraisid TONTISIRIN (1985) Dietary approaches to the prevention of vitamin A deficiency. Food and Nutrition Bulletin 7(3):72-76. [Ch. 2]
CHATELIN, Yvon (1979) Une Epistemologie des Sciences Du Sol. Office de la Recherche Scientifique et Technique Outre-Mer, Memoire No. 88. [Ch.9] [Reviewed by L. Busch, Agriculture and Human Values 2(4):66-67.]
CHEN, T.S., and S. SAAD (1981) Folic acid in Egyptian vegetables: The effect of drying method and storage on the folicin content of *mulukhiyah* (*Corchorus olitorius*). Ecology of Food and Nutrition 10:249-255. [Ch.15]
CHLEQ, Jean-Louis, and Hugues DUPRIEZ (1984) Eau et Terres en Fuite, Métier de l'Eau du Sahel. Terres et Vie*, Nivelles, Begium. (In French) [Ch.9,11,12]
CLAWSON, David L. (1985) Harvest security and intraspecific diversity in traditional tropical agriculture. Economic Botany 39(1):56-67. [Ch.14]
CLEVELAND, David A. (1980) The Population Dynamics of Subsistence Agriculture in the West African Savanna: A Village in Northeast Ghana. Ph.D. Dissertation, Department of Anthropology, University of Arizona, Tucson, Arizona, USA. University Microfilms International, Ann Arbor, Michigan, USA. xviii+363pp. [Ch.11]
CLEVELAND, David A. (1982) Economic and dietary contributions of urban gardening in Tucson. Paper presented at the Annual Meeting of the American Anthropological Association, Washington, DC, USA. [Ch.2]
CLEVELAND, David A. (1986) Culture and horticulture in Mexico. Culture and Agriculture No. 29:1-5. [Part I, Ch.4,14]
CLEVELAND, David A. (1990) Development alternatives and the African food crisis. In Confronting Change. Stress and Coping in African Food Systems, Vol. 2. R. Huss-Ashmore and S. Katz, eds. Pp. 181-206. Gordon and Breach, New York, USA. [Ch.3]
CLEVELAND, David A. (1991) New crop varieties in a green revolution for Africa: Implications for sustainability and equity. In The Political Economy of Famine. The Class and Gender Basis of Hunger.

Stress and Coping in African Food Systems, Vol. 3. R.E. Downs, D.O. Kerner and S.P. Reyna eds. Gordon and Breach, New York, USA. [Part II]

CLEVELAND, David A., Thomas V. ORUM, and Nancy FERGUSON (1985) Economic value of home vegetable gardens in an urban desert environment. HortScience 20(4):694-696. [Ch.3,10]

CLEVELAND, David A., and Daniela SOLERI (1987) Household gardens as a development strategy. Human Organization 46(3):259-270 [Part I, Ch.3,9,14]

CLEVELAND, David A., and Daniela SOLERI (1989) Diversity and the new green revolution. Diversity 5(2&3):24-25. [Ch.14]

CLEVELAND, David A., and Daniela SOLERI (n.d.a) Unpublished data on seasonal water use in mixed crop desert gardens. [Ch.10]

CLEVELAND, David A., and Daniela SOLERI (n.d.b) Household gardens in an irrigated district of northern Pakistan. mss. [Ch.14]

CLEVELAND, David A., and Daniela SOLERI (n.d.c) The quest for environmentally and socially sustainable agriculture: Production, diversity, and stability. mss. [Part II, Ch.14]

COLLINS, G.N. (1914) A drought-resisting adaptation in seedlings of Hopi maize. Journal of Agricultural Research 1(4):293-392. [Ch.6]

CORNIA, Giovanni A. (1985) Farm size, land yields and the agricultural production function: An analysis of fifteen developing countries. World Development 13:513-534. [Ch.3]

COX, George W., and Michael D. ATKINS (1979) Agricultural Ecology: An Analysis of World Food Production Systems. W.H. Freeman, San Francisco, California, USA. 721pp. [Part II, Ch.5,11,14] An excellent introduction to the field.

CRISP, P., and D. ASTLEY (1985) Genetic resources in vegetables. Progress in Plant Breeding 1:281-310. Butterworths & Co., Borough Green, Sevenoaks, Kent, UK [Ch.14] A good review of the topic, categorizing vegetables into four groups based on the nature of their genetic diversity.

CROUCH, D., and C. WARD (1988) The Allotment: Its Landscape and Culture. Faber and Faber, London. [Part I, Ch.3] Historical and social analysis of allotment (community) gardens in Great Britain.

DADA, L.O., and D.A.V. DENDY (1987) Preliminary study of the effect of various processing techniques on the cyanide content of germinated sorghum. Tropical Science 27:101-104. [Ch.2]

DALY, Herman (1989) Sustainable development: Some basic principles. Manuscript of keynote address to the Hoover Institution's conference on Population and Resources, February 1, 1989, Stanford University, Palo Alto, California. 15pp. [Ch.3]

DALY, Herman, and John COBB (1989) For the Common Good. Beacon Press, Boston, Massachusetts, USA. vii+482pp. [Ch.3] Convincingly demonstrates that the dominant world economic policy of economic growth destroys society and the environment. Gives suggestions for alternatives.

DANCETTE, C., and A.E. HALL (1979) Agroclimatology applied to water management in Sudanian and Sahelian zones of Africa. In Hall, Cannell, and Lawton 1979:98-118. [Ch.11]

DAVIS, Litton, and Robert A. BYE (1982) Ethnobotany and progressive domestication of *jaltomata* (*Solanaceae*) in Mexico and Central America. Economic Botany 36(2):225-241. [Ch.8]

DARROW, Ken, and Mike SAXENIAN (1986) Appropriate Technology Sourcebook. Volunteers in Asia, Stanford, California, USA. (Available from Appropriate Technology International* [ATI]). ca 800pp. [Ch.19]. "A review of 1,150 of the most useful appropriate technology books from around the world." A complete "library" of all the books reviewed in this sourcebook is available on microfiche, also from ATI.

DCFRN (Developing Countries Farm Radio Network) (1979-1987) Developing Countries Farm Radio Packages, Numbers 1 to 12. DCFRN*, Toronto, Canada. [Ch.6,8,9,13,14] An excellent series of extension packets written for radio programs in the Third World.

DELGADO, Christopher L. (1979) The Southern Fulani Farming System in Upper Volta: A Model for the Integration of Crop and Livestock Production in the West African Savannah. African Rural Economy Paper No. 20. Department of Agricultural Economics, Michigan State University, East Lansing, Michigan, USA. [Ch.9]

DELGADO DURAN, Francisco Oscar (1984) Mulching as a Means of Producing Vegetable Crops Under a Limited Water Supply. M.S. Thesis, Department of Plant Sciences, University of Arizona, Tucson, Arizona, USA. [Ch.10,14]

DENEVEN, William M. (1980) Latin America. In Klee 1980:217-244. [Ch.9,12]

DeWALT, Billie (1985) Mexico's second green revolution: Food for feed. Mexican Studies/Estudios Mexicanos 1(1):29-60. [Part I, Ch.2,3,14]

DEWEY, Kathryn G.(1981) Nutritional consequences of the transformation from subsistence to commercial agriculture in Tabasco, Mexico. Human Ecology 9(2):151-187. [Ch.2,3,14]

DIXON-MUELLER, Ruth (1985) Women's Work in Third World Agriculture. International Labor Office, Geneva. xi+151pp. [Ch. 3]

DONAHUE, Roy L., Raymond W. MILLER, and John C. SHICKLUNA (1983) Soils: An Introduction to Soils and Plant Growth. Prentice Hall, Inc., Englewood Cliffs, New Jersey, USA. xv+667pp. [Ch.9,10,11,12] Beginning college text with generally good explanations of soil science basics, though sometimes unclear. Methodology and field application are based on expensive equipment and large-scale commercial production largely inappropriate for the Third World and for household gardens.

DONEEN, L.D., and D.W. WESCOT (1984) Irrigation Practice and Water Management. Revision 1. FAO Irrigation and Drainage Paper 1. (First published 1971). FAO*, Rome. [Ch.10,12]

DOORENBOS, J., and A.H. KASSAM, with others (1979) Yield Response to Water. FAO Irrigation and Drainage Paper 33. FAO*, Rome. ix+193pp. [Ch.10] Contains brief guidelines for calculating maximum yield and crop water requirements with application to optimizing yields and total production under conditions of limited water supply; detailed data on crop water requirements and yield for 26 major commercial crops.

DOORENBOS, J., and W.O. PRUITT, in consultation with others (1977) Guidelines for Predicting Crop Water Requirements. FAO Irrigation and Drainage Paper 24. (Revised ed.) FAO*, Rome. 144pp. [Ch.10,12] Gives detailed calculation methods for estimating water requirements, pan evaporation, crop coefficients, crop evapotranspiration and their application to design of irrigation systems. Oriented toward large-scale monoculture.

DOYLE, Jack (1985) Altered Harvest: Agriculture, Genetics, and the Fate of the World's Food Supply. Penguin Books, New York. xix+502pp. [Ch.14]

DREGNE, H. E. (1976) Soils of Arid Regions. Elsevier Scientific Publishing Co., Amsterdam, The Netherlands. [Ch.9]

DUFFIELD, Mary Rose, and Warren D. JONES (1981) Plants for Dry Climates: How to Select, Grow and Enjoy. H.P.Books, Tucson, Arizona, USA. 176pp. [Ch.8] Mostly ornamentals, but a number of fruit and nut trees. Focuses on southwestern United States, but applicable to northern Mexico.

DUGGAN, William (1985) Irrigated gardens, Molepolole, Botswana. In ILO 1985:7-20. [Ch.3]

DUPRIEZ, Hugues (1982) Paysan d'Afrique noire. (In French) Terres et Vie*, Nivelles, Belgium. 256pp. [Ch.19]

DUPRIEZ, Hugues and Philippe DE LEENER (1983) Agriculture Tropicale en Milieu Paysan African. (In French) Terres et Vie*, Nivelles, Belgium. 280pp. [Ch.5,9,10,11,13,14] Clearly written with numerous illustrative photographs and drawings. Emphasizes wisdom and adaptability of traditional systems. Covers semiarid to humid Africa with most examples from West Africa, including many from drylands.

DUPRIEZ, Hugues, and Philippe DE LEENER (1987) Jardins et Vergers D'Afrique. (In French) Terres et Vie*, Nivelles, Belgium. 354pp. [Part I, Ch.8,14,18] The best extension handbook we have seen for people working with household gardens in sub-Saharan Africa. Incorporates some traditional gardening techniques; discussions are illustrated with numerous excellent photographs of gardens and plants. The first section of the book discusses techniques and the second is a listing of 86 domesticated and nondomesticated plants used in the region, including their names in a number of local languages, and a brief description of how the plants are used and how they are propagated.

DURNING, Alan B. (1990) Ending poverty. In Brown, et al. 1990:135-153. [Part I]

DUTT, Gordon R. (1981) Establishment of NaCl-treated catchments. In Dutt, et al. 1981:17-21. [Ch.11]

DUTT, Gordon R., C.F. HUTCHINSON, and M. ANAYA Garduño (1981) Rainfall Collection for Agriculture in Arid and Semiarid Regions. Proceedings of a Workshop, 1980. Commonwealth Agricultural Bureaux, Slough, UK. (Source of Dutt 1981)

EARLY, Daniel K. (1977) Cultivation and uses of amaranth in contemporary Mexico. In Rodale Press 1977:39-60. [Ch.8]

EASTMAN, Susan J. (1988) Vitamin A Deficiency and Xerophthalmia. Reprint of Assignment Children 1987-3. UNICEF*, New York and Geneva. 84pp. [Ch.2]

EL AMAMI, Sleheddine (1979) Utilization of runoff waters: The "meskats" and other techniques in Tunisia. African Environment 3(3-4):107-120 (Available from ENDA*). Poorly translated from French, but some idea of the techniques do come through. [Ch.8,11]

ESKELINEN, Riitta K. (1977) Dogon Agricultural Systems: Sociological Aspects Relating to Development Interventions. Report to The Research Foundation of State University of New York, Binghamton, New York, USA. mss. iii+93pp. [Ch.3,15]

EVENARI, Michael, Leslie SHANAN, and Naphtali TADMOR (1982) The Negev: The Challenge of a Desert. 2nd ed. Harvard University Press, Cambridge, Massachusetts, USA. ix+437pp. [Ch.6,7,11]

EVETT, Steven R. (1983) Erosion and Runoff from Sodium Dispersed, Compacted Earth Water Harvesting Catchments. M.S. Thesis, University of Arizona, Tucson, Arizona. University Microfilms International, Ann Arbor, Michigan, USA. [Ch.11]

EVETT, Steven R. (1985a) Advisory report on small-scale irrigation and the African Food Systems Initiative in Lesotho. Prepared for Peace Corps/OTAPS. Unpublished report. [Ch.12]

EVETT, Steven R. (1985b) Personal communication. [Ch.11]

FAIRBOURN, Merle L., and H.R. GARDNER (1972) Vertical mulch effects on soil water storage. Soil Science Society of America Proceedings 36:823-827. [Ch.10]

FAIRBOURN, Merle L., and H.R. GARDNER (1974) Field use of microwatersheds with vertical mulch. Agronomy Journal 66(6):740-744. [Ch.10,11]

FAO (Food and Agriculture Organization, United Nations) (1961) Agricultural and Horticultural Seeds. FAO*, Rome. 531pp. [Ch.6]

FAO (1970) Handling and Storage of Food Grains. FAO*, Rome. 350pp [Ch.14]

FAO (1976-77) Better Farming Series. FAO Economic and Social Development Series No. 3. FAO*, Rome. [Ch.5]

FAO (1982a) Food Composition Tables for the Near East. FAO*, Rome. x+265pp. [Ch.2] Has three tables with: 1) energy, protein, vitamin and mineral, 2) amino acid, and 3) fatty acid content. Also very brief descriptions of about 110 prepared foods listed in the tables.

FAO (1982b) Date Production and Protection. With Special Reference to North Africa and the Near East. FAO Plant Production and Protection Paper 35. FAO*, Rome. [Ch.7,8,13,15]

FAO (1983) Food and Fruit-Bearing Forest Species 1: Examples from Eastern Africa. FAO Foresty Paper 44/1. FAO*, Rome. xiii+172pp. [Ch.8] This is a good example of the wealth of local tree resources that can be used instead of imported exotics for dryland gardens. Oriented toward commercial applications but gives basic data helpful for garden planting.

FAO (1988) Traditional Food Plants. A Resource Book for Promoting the Exploitation and Consumption of Food Plants in Arid, Semi-Arid and Sub-Humid Lands of Eastern Africa. FAO*, Rome. x+593 pp. [Ch.8,18]

FAO (1989) Forestry and Food Security. FAO Forestry Paper 90. FAO*, Rome. viii+128. [Ch.3]

FAO/WHO (1973) Energy and Protein Requirements. FAO Nutrition Meeting Report Series No. 52. FAO*, Rome. [Ch.2]

FELDMAN, Lewis J. (1988) The habits of roots. Bioscience 38(9):612-618. [Ch.5]

FERGUSON, Anne E., and Susan SPRECHER (1987) Women and plant genetic diversity: The case of beans in the Central Region of Malawi. Paper presented at the Annual Meeting of the American Anthropological Association, Chicago, Illinois, USA. [Ch.14]

FERRANDO, R. (1981) Traditional and Non-Traditional Foods. FAO*, Rome. [Ch.2]

FISCHER, R.A. and Neil C. TURNER (1978) Plant productivity in the arid and semiarid zones. Annual Review of Plant Physiology 29:277-317. [Ch.10]

FLEURET, Patrick (1985) The social organization of water control in the Taita Hills, Kenya. American Ethnologist 12:103-118. [Ch.12] Brief nonquantitative description of a small-scale irrigation system contructed and maintained by hand labor, and its social organization.

FLINN, J.C., and D.P. GARRITY (1986) Yield Stability and Modern Rice Technology. IRRI Research Paper Series No. 122. International Rice Research Institute (IRRI), Manila, Philippines. [Ch.10,14]

FOLBRE, Nancy (1984) Household production in the Philippines: A non-neoclassical approach. Economic Development and Cultural Change 32:303-330. [Ch.2,3]

FOWLER, Cary, and Pat MOONEY (1990) Shattering: Food, Politics and the Loss of Genetic Diversity. University of Arizona Press, Tucson, Arizona, USA. xvi+278pp. [Ch. 14]

FRAENKEL, Peter (1987) Water-Pumping Devices: A Handbook for Users and Choosers. IT Publications*. 196pp. [Ch.12] A detailed, comprehensive, and practical review of the options available for pumping and lifting water on a small scale, especially for irrigation. Demonstrates costs and general suitability of different technical options including human, animal, renewable, and fossil fuel powered devices. [Reviewed by F. Carroll]

FRANCIS, Charles, ed. (1986) Multiple Cropping Systems. Macmillan Publishing Company, New York, USA. (Source of Altieri and Liebman 1986; Gliessman 1986)

FRAISER, Gary W., and Lloyd E. MEYERS (1983) Handbook of Water Harvesting. Handbook No. 600. Agricultural Research Service, USDA, Washington, DC, USA. 45pp. [Ch.11]

GALT, A.H., and J.W. GALT (1978) Peasant use of some wild plants on the island of Pantelleria, Sicily. Economic Botany 32:20-26. [Ch.10]

GARCIA, R., L.E. CALTAGIRONE, and A.P. GUTIERREZ (1988) Comments on a redefinition of biological control. Bioscience 38:692-694. [Ch.13]

GARDNER, Walter H. (1979) How water moves in the soil. Crops and Soils Magazine, November 1979:13-18. [Reprints available from American Society of Agronomy, Inc., 677 South Segoe Road, Madison, Wisconsin 53711, USA] [Ch.10] Clear, simple explanation with many photographs from the laboratory demonstrating the principles.

GARNER, R.J., Saeed Ahmed CHAUDHRI, and the staff of Commonwealth Bureaux of Horticulture and Plantation Crops (1976) The Propagation of Tropical Fruit Trees. Commonwealth Agricultural Bureaux, Farnham Royal, Slough, UK. xv+556pp. [Ch.7,8,14]

GERSHON, Jack, Yen-ching CHEN, and Jen-fong KUO (1985) The AVRDC Garden Program 1983-84. AVRDC*, Shanhua, Tainan, Taiwan. [Ch.2]

GIBBON, David, and Adam PAIN (1985) Crops of the Drier Regions of the Tropics. Longman, London. x+157pp. [Ch.5]

GIBBONS, Gayle, and Marcia GRIFFITHS (1984) Program Activities for Improving Weaning Practice. World Association of Public Health Associates (for UNICEF*), Geneva, Switzerland. 54pp. [Ch.16]

GLADWIN, Christina H., and John BUTLER (1984) Is gardening an adaptive strategy for Florida family farmers? Human Organization 43:208-215. [Part I, Ch.3]

GLIESSMAN, Stephen R. (1986) Plant interactions in multiple cropping systems. In Francis 1986:83-95. [Ch.8]

GLIESSMAN, Stephen R., ed. (1990) Agroecology: Researching the Ecological Basis for Sustainable Agriculture. Springer-Verlag, New York. xiv+380pp. [Part II] (Source of Lumsden 1990)

GOLDBURG, Rebecca, Jane RISSLER, Hope SHAND, and Chuck HASSEBROOK (1990) Biotechnology's Bitter Harvest: Herbicide Tolerant Crops and the Threat to Sustainable Agriculture. A Report of the Biotechnology Working Group (available from RAFI*). 73pp. [Ch.14]

GORDON, Gillian (1969) The evaluation of nutrition programs. Paper read at the Institute of Social and Statistical Research, Legon, Ghana. mss. [Ch.16]

GORDON, Gillian (1973) An Evaluation of Nutrition Education Given to Pregnant and Nursing Women in Six Communities in the Upper Region of Ghana. M.Sc. Thesis, Department of Nutrition and Food Science, University of Ghana, Legon, Ghana. 435pp. [Ch.16]

GOULD, Fred (1988) Evolutionary biology and genetically engineered crops. BioScience 38(1):26-33. [Ch.13,14]

GRAY, Robert F. (1963) The Sonjo of Tanganyika: An Anthropological Study of an Irrigation Based Society. Oxford University Press for the International African Institute, London. xii+181pp. [Ch.12]

GREENOUGH, William B. (1987) Status of cereal-based oral rehydration therapy. In Symposium Proceedings. Cereal-Based Oral Rehydration Therapy: Theory and Practice. Pp.29-32. International Child Health Foundation, PO Box 1205, Columbia, Maryland 21044, USA. [Ch.16]

GRIFFITHS, J.F., ed. (1972) Climates of Africa. Elsevier Publishing Co., Amsterdam, The Netherlands. xv+604pp. [Ch.11] Contains chapters explaining climates of different regions of the continent. Many tables including one giving monthly and yearly mean maximum, and mean minimum temperature, and mean, mean maximum and mean minimum precipitation for many locations in Africa.

GRIVETTI, Louis E. (1978) Nutritional success in a semi-arid land: Examination of Tswana agro-pastoralists of the eastern Kalahari, Botswana. The American Journal of Clinical Nutrition 31:1204-1220. [Ch.2,14]

GRÜN, Ingolf, Michel BECK, John S. CALDWELL, and Marilyn S. PREHM (1989) Development and testing of integrative methods to assess relationships between garden production and nutrient consumption by low-income families. Virginia Polytechnic and State University, Blacksburg, Virginia, USA. mss. 26pp. [Ch.4]

GTZ (Deutsche Gesellschaft für Technische Zusammernarbeit) (1980) Post Harvest Problems. GTZ*, Germany. 258pp+33pp annex. [Ch.14] (Source of Zehrer 1980; Zehrer, et al. 1980)

GUPTA, J.P., and G.N. GUPTA (1983) Effects of grass mulching on growth and yields of legumes. Agricultural Water Management 6:375-383. [Ch.10] Reports results of 2 seasons' experiments in semiarid India on mulching of legume grains. Does not state whether grass mulch applied was fresh or dry.

GUPTA, U.S., ed. (1975) Physiological Aspects of Dryland Farming. Allanheld, Osmun & Co., Montclair, New Jersey, USA. xv+391pp. (Source of Arnon 1975; Larson 1975)

GUTMAN, Pablo (1987) Urban agriculture: The potential and limitations of an urban self-reliance strategy. Food and Nutrition Bulletin 9:37-42. [Ch.3]

HALDERMAN, Alan D. (1977) Irrigation: When? How Much? How? (Revised) Bulletin A20. College of Agriculture, University of Arizona, Tucson, Arizona, USA. 10pp. [Ch.10]

HALL, A.E., G.H. CANNELL, and H.W. LAWTON, eds. (1979) Agriculture in Semi-Arid Environments. Springer-Verlag, Berlin, Germany. xvi+340pp. (Source of Dancette and Hall 1979; Hall, Foster, and Waines 1979; Henderson 1979; Lawton and Wilke 1979)

HALL, A.E., K.W. FOSTER, and J.G. WAINES (1979) Crop adaptation to semi-arid environments. In Hall, Cannell, and Lawton 1979:148-179. [Ch.5]

HAMMOND, Peter (1966) Yatenga: Technology in the Culture of a West African Kingdom. The Free Press, New York. xi+231pp. [Ch.3,9] Offers a brief description of wet- and dry-season gardens, though confusing in places.

HARLAN, Jack R. (1976) Genetic resources in wild relatives of crops. Crop Science 16:329-333. [Ch.14]

HARTMANN, Hudson T., and Dale E. KESTER (1983) Plant Propagation: Principles and Practices. 4th ed. Prentice-Hall, Inc. Englewood Cliffs, New Jersey 07632, USA. 727pp. [Ch.5,6,7,8,14] A standard reference in the field.

HASWELL, M.J. (1975) The Nature of Poverty. Macmillan, London. [Ch.3]

HENDERSON, D.W. (1979) Soil management in semi-arid environments. In Hall, Cannell, and Lawton 1979:224-237. [Ch.10]

HILL, Dennis S. (1983) Agricultural Insect Pests of the Tropics and Their Control. 2nd ed. Cambridge University Press, Cambridge, UK. xii+746pp. [Ch.13] Extensive coverage of insect pests with good line drawings. Emphasizes synthetic chemical pesticides and "improved" crop varieties as the best approach.

HILL, Dennis S. and J.M. WALLER (1982) Pests and Diseases of Tropical Crops. Volume 1. Principles and Methods of Control. Longman, London. xvi+175pp. [Ch.13] An easy-to-read handbook with same emphases as Hill (1983).

HOFKES, E.H. (1983) Water Pumping for Rural Water Supply. ENDA Third World Documents Series No. 21-81. ENDA*, Dakar, Senegal. 52pp. [Ch.12]

HORTICULTURAL ABSTRACTS (1984) 55(5):283. [Ch.8]

IBPGR (International Board for Plant Genetic Resources) (1987) IBPGR Annual Report 1986. IBPGR Headquarters, Crop Genetic Resources Center, Plant Production and Protection Division, FAO*, Rome. vi+89pp. [Ch.14]

ICRISAT (International Crops Research Institute for the Semi-Arid Tropics) (1986) ICRISAT in Africa. ICRISAT*, Andhra Pradesh, India. 60pp. [Ch.14]

IDRC (International Development Research Centre) (1980) Nutritional Status of the Rural Population of the Sahel. Report of a working group, Paris, France, April 28-29 1980. IDRC*, Ottawa, Canada. [Ch.2]

IGBOANUGO, A.B.I. (1986) Phytotoxic effects of some Eucalypts on food crops, particularly on germination and radicle extension. Tropical Science 26:19-24. [Ch.8]

IITA (International Institute of Tropical Agriculture) (1986) Annual Report and Research Highlights 1985. IITA*, Ibadan, Nigeria. 145pp. [Ch.14]

ILO (International Labor Organization) (1985) Rural Development and Women: Lessons From the Field. Vol. 1: Women in Production and Marketing and Their Access to Credit. ILO/DANIDA/80/INT/35. International Labor Office, Geneva, Switzerland. (Source of Duggan 1985; Milimo 1985)

IMMINK, M.D.C., D. SANJUR and M. COLON (1981) Home gardens and the energy and nutrient intakes of women and preschoolers in rural Puerto Rico. Ecology of Food and Nutrition 11:191-199. [Ch.2]

IOCU (International Organization of Consumers Unions) (1984) The Pesticide Handbook. Profiles for Action. IOCU*, Penang, Malaysia. 165pp. [Ch.13] Lists major pesticides and their hazards. Some background papers on global issues and brief case studies of 3 Third World countries.

IREMIREN, G.O. (1987) Effects of artificial defoliation

on the growth and yield of okra (*A. esculentus*) Experimental Agriculture 23(1):1-7. [Ch.15]

IRRI (International Rice Research Institute) (1989) Publications of the International Agricultural Research and Development Centers. IRRI*, Los Baños, Philippines. 547pp. [Ch.19]

IRVINE, F.R. (1969) West African Crops. Oxford University Press, London. ix+272pp. [Ch.2,7,18] General introduction to specific crops in humid to semi-arid areas with emphasis on commercial crops. Appears to be based mostly on author's experience in Nigeria and Ghana.

ISENALUMHE, Anthony E. (1986) Modified adult meals: A plausible alternative to orthodox weaning foods in a Nigerian community. Hygie 5(4):14-19. [Ch.16]

JELLIFE, D.B. (1972) Commerciogenic malnutrition? Nutrition Reviews 30(9):199-205. [Ch.3]

JELLIFE, D.B. and E.F.P. JELLIFE (1978) Human Milk in the Modern World. Oxford University Press, London. [Ch.16]

JENSEN, M.E., ed. (1980) Design and Operation of Farm Irrigation Systems. The American Society of Agricultural Engineers, St. Joseph, Michigan, USA. xi+829pp. [Ch.12] (Source of Merriam, et al. 1980; Stegman, et al. 1980)

JETT, Stephen C. (1979) Peach cultivation and use among the Canyon de Chelly Navajo. Economic Botany 33(3):298-310. [Ch.14]

JOHNSON, Allen (1974) Ethnoecology and planting practices in a swidden agricultural system. American Ethnologist 1:87-101. [Ch.9]

KASSAM, A.H. (1976) Crops of the West African Semi-Arid Tropics. ICRISAT*, Andrha Pradesh, India. vii+154pp. [Ch.5,7,18] Covers 23 important crops in depth: ecology, cultivation, diseases, and pests. Garden crops include cowpea, groundnut, soybean, cassava, yam, sweet potato, cocoyam, potato, tomato, onion, pepper, okra, roselle, sesame, and sugarcane.

KENNEDY, E. (1983) Determinants of family and preschool food consumption. Food and Nutrition Bulletin 5(4):22-29. [Ch.3]

KENNEDY, W.K., and T.A. ROGERS (1985) Human and Animal Powered Water-Lifting Devices. IT Publications*, London. 111pp. [Ch.12]

KEYSTONE CENTER (1990) Madras Plenary Session: Final Consensus Report of the Keystone International Dialogue Series on Plant Genetic Resources. Genetic Resources Communication Systems, Washington, DC, USA. 38pp. [Ch.14]

KHOKHAR, Santosh, and B.M. CHAUHAW (1986) Antinutritional factors in moth bean (*Vigna aconitifolia*): Varietal differences and effects of methods of domestic processing and cooking. Journal of Food Science 51(3):591-594. [Ch.2,15]

KING, J., D.O. NNANYELUGO, H. ENE-OBONG, and P.O. NGODDY (1985) Household consumption profile of cowpea (*Vigna unguiculata*) among low-income families in Nigeria. Ecology of Food and Nutrition 16:209-221. [Ch.14]

KING, Maurice H., Felicity KING, and Soebagyo MARTODIPOERO (1980) Primary Child Care. Book One. A Manual for Health Workers. Corrected ed. Oxford University Press. (available from IT Publications*) xi+315pp. [Ch.16]

KING, Maurice H., Felicity M.A. KING, David C. MORLEY, H.J. Leslie BURGESS, and Ann P. BURGESS (1972) Nutrition for Developing Countries. With Special Reference to the Maize, Cassava and Millet areas of Africa. Oxford University Press, Nairobi. vars. pp. [Ch.16] A book for schoolteachers and literate field workers involved with family health and nutrition in East Africa. The simple style, East African examples, and numerous illustrations make it an accessible book for those with no previous formal training in nutrition. Each chapter concludes with a section of "Things to do," activities to demonstrate or reinforce the topics covered in the chapter.

KIRKBY, Anne V.T. (1973) The Uses of Land and Water Resources in the Past and Present Valley of Oaxaca, Mexico. Memoirs of the Museum of Anthropology, University of Michigan, No. 5. Prehistory and Human Ecology of the Valley of Oaxaca, Vol. 1. Museum of Anthropology, University of Michigan, Ann Arbor, Michigan, USA. [Ch.11]

KLEE, Gary A., ed. (1980) World Systems of Traditional Resource Management. John Wiley and Son, New York, USA. (Source of Denevan 1980; Manners 1980)

KLEER, Jerzy, and Augustyn WOS, eds. (1988) Small-Scale Food Production in Polish Urban Agglomerations. Food-Energy Nexus Report No. 26, Food-Energy Nexus Program, United Nations University, Paris. 63pp. [Part I, Ch.3]

KLOPPENBURG, Jack Jr. (1988) First the Seed: The Political Economy of Plant Biotechnology, 1492-

2000. Cambridge University Press, New York. [Ch.14] An account of how seeds have come increasingly under the control of private enterprise in the USA. Chap. 8, "Heterosis and the Social Division of Labor" is a good social history of hybrid maize seed.

KLOPPENBURG, Jack Jr., and Daniel Lee KLEINMAN (1987) The plant germplasm controversy. Bioscience 37(3):190-198. [Ch.14]

KOEGEL, R.G. (1977) (reprinted 1985) Self-Help Wells. FAO Irrigation and Drainage Paper 30. FAO*, Rome. 77pp. [Ch.11] A brief review of several techniques for both small- and large-diameter wells. Not a practical manual, but a guide to "provide ideas", with many pictures and diagrams.

KOGAN, Marcos (1986) Ecological Theory and Integrated Pest Management. John Wiley and Sons, New York, USA. (Source of Levins 1986)

KOLARKAR, A.S., K.N.K. MURTHY, and N. SINGH (1983) Khadin: A method of harvesting water for agriculture in the Thar Desert. Journal of Arid Environments 6:59-66. [Ch.11]

KOLARKAR, A.S., Y.V. SINGH, and A.N. LAHIRI (1983) Use of discarded plastic infusion sets from hospitals in irrigation on small farms in arid regions. Journal of Arid Environments 6:385-389. [Ch.12]

KOURIK, Robert (1986) Designing and Maintaining Your Edible Landscape Naturally. The Edible Landscape Book Project, Santa Rosa, California, USA. xxi+370pp. [Ch.8]

KUMAR, S.K. (1978) Role of the Household Economy in Child Nutrition at Low-Incomes: Case Study in Kerala. Occasional Paper No. 95. Department of Agricultural Economics, Cornell University, Ithaca, New York, USA. 78pp. [Ch.2,3]

LADD, Edmund J. (1979) Zuni Economy. *In* Ortiz:1979:492-497. Includes a brief description of Zuni sunken bed gardens of southwestern North America. [Ch.9,11]

LAGEMANN, Johannes (1977) Traditional African Farming Systems in Eastern Nigeria: An Analysis of Reaction to Increasing Population Pressure. Weltform-Verlag, Munich, Germany. [Part II, Ch.3,9]

LAL, Rattan (1987) Managing the soils of sub-Saharan Africa. Science 236:1069-1076. [Ch.9].

LAPPÉ, Frances Moore (1982) Diet for a Small Planet. revised ed. (First published 1971) Ballantine Books, New York, USA. 432pp. From Food First*. [Ch.2,19] An excellent discussion of how the food system in the developed world is related to hunger in the rest of the world. Good explanation of protein complementarity for meatless diets.

LAPPÉ, Frances Moore, and Joseph COLLINS, with Cary FOWLER (1978) Food First: Beyond the Myth of Scarcity. Revised ed. Ballantine Books, New York, USA. From Food First*. xvii+619pp. [Ch.19] A powerfully-written argument that it is not scarcity or over-population that causes hunger, but inequality of control over productive resources. Concludes with suggestions for change and personal involvement.

LARSON, K.L. (1975) Drought injury and resistance of crop plants. *In* Gupta 1975:147-165. [Ch.8]

LA ROVERE, Emilio Lebre (1985) Food and Energy in Rio de Janeiro: Provisioning the Poor. Food-Energy Nexus Research Report No. 13, Food-Energy Nexus Program, United Nations University, Paris. 59pp. [Ch.3]

LATHAM, Michael C. (1979) Human Nutrition in Tropical Africa. 2nd ed. FAO Food and Nutrition Series No. 11. FAO*, Rome. xi+286pp. [Ch.2,16] Written "with special reference to community health problems in East Africa." An extremely useful reference and textbook for health care workers, especially in east Africa. It includes many photographs and thorough explanations of basic nutrition and nutrition related diseases, recommendations for policy planners and recipe suggestions for household and institutional preparation of foods. Infant and child health is discussed throughout with section 35 giving "recipes for infants and young children." Overall it emphasizes practical, locally appropriate responses to nutritional need, although only mentions gardens briefly (pp.183,184).

LATHAM, Michael C. (1984) Strategies for the control of malnutrition and the influence of the nutritional sciences. Food and Nutrition 10(1):5-35. [Ch.2]

LATHAM, Michael C. (1990) Innapropriate modernization and Westernization as causes of malnutrition and health disorders in non-industrialized countries. *In* Brun and Latham 1990:86-93. [Part I, Ch.2]

LAWTON, H.W. and P.J. WILKE (1979) Ancient agricultural systems in dry regions. *In* Hall, Cannell, and Lawton 1979:1-44. [Ch.12]

LEACH, Edmund (1961) Pul Eliya: A Village in Ceylon. A Study of Land Tenure and Kinship. Cambridge University Press, London, UK. [Ch.12]

LEON, Jorge, and Lyndsey A. WITHERS, eds. (1986) Guidelines for Seed Exchange and Plant Introduction in the Tropics. FAO*, Rome. [Ch.14]

LEONARD, David (1980) Soils, Crops and Fertilizer Use: A What, How and Why Guide. 3rd ed. (First published 1967) Reprint R-8. Peace Corps, Washington, DC. iii+162pp. [Ch.9,11] Includes some useful information on soils and practical tests for gardeners, though one-half of this book is on commercial fertilizers. Written for Peace Corps Volunteers, with much United States English slang.

LEUNG, Woot-Tsuen Wu, with Marina FLORES (1961) Food Composition Table for use in Latin America. Institute of Nutrition of Central America and Panama, Guatemala City, Guatemala, and National Institutes of Health, Bethesda, Maryland, USA. xi+145pp. [Ch.2]

LEUNG, Woot-Tsuen Wu, with Felix BUSSON and Claude JARDIN (1968) Food Composition Table for use in Africa. FAO*, Rome; Public Health Service, U.S. Department of Health, Education and Welfare, Bethesda, Maryland, USA. ix+306pp. [Ch.2,16]

LEUNG, Woot-Tsuen Wu, Ritva Rauaheimo BUTRUM, Flora Huang CHANG, M. Narayana RAO, and W. POLACCHI (1972) Food Composition Table for use in East Asia. FAO*, Rome; National Institutes of Health, U.S. Department of Health, Education and Welfare, Bethesda, Maryland, USA. xiii+334pp. [Ch.2]

LEVI, John, and Michael HAVINDEN (1982) Economics of African Agriculture. Longman*, London. vii+175pp. [Ch.2,3] A good introduction to economics from the viewpoint of the small-scale farmer. Does not require any previous economics or math.

LEVINS, Richard (1986) Perspectives in integrated pest management: From an industrial to ecological model of pest management. In Kogan 1986:1-18. [Ch.13]

LONGHURST, Richard (1983) Agricultural production and food consumption: Some neglected linkages. Food and Nutrition 9(2):2-6. [Ch.2,16]

LUMSDEN, R.D., R. GARCIA-E., J.A. LEWIS, and G.A. FRIAS-T. (1990) Reduction of damping-off disease in soils from indigenous Mexican agroecosytems. In Gliessman 1990:83-103. [Ch.13]

MAAS, P.W. (1987) Physical methods and quarantine. In Brown and Kerry 1987:265-291. [Ch.13]

MACNAB, A.A., A.F. SHERF, and J.K. SPRINGER (1983) Identifying Diseases of Vegetables. Pennsylvania State University, College of Agriculture, University Park, Pennsylvania, USA. 62pp. [Ch.13]

MAEDA, E.E., and D.K. SALUNKHE (1981) Retention of ascorbic acid and total carotene in solar dried vegetables. Journal of Food Science 46:1288-1290. [Ch.15]

MANNERS, Ian R. (1980) The Middle East. In Klee 1980:39-65. [Ch.12]

MARTEN, Gerald G., and Oekan S. ABDOELLAH (1988) Crop diversity and nutrition in West Java. Ecology of Food and Nutrition 21:17-43. [Ch.14]

MAYER, A.M. and A. POLJAKOFF-MAYBER (1975) The Germination of Seeds. 2nd ed. Pergamon Press, Oxford, UK. [Ch.6]

MERRIAM, J.L., M.N. SHEARER, and C.M. BURT (1980) Evaluating irrigation systems and practices. In Jensen 1980:721-760. [Ch.10]

MILIMO, Mabel C. (1985) Chikuni fruit and vegetable producer's co-operative society, Zambia—A case study. In ILO 1985:21-35. [Ch.3]

MING, Wang and Sun YUN-WEI (1986) Fruit trees and vegetables for arid and semi-arid areas in northwest China. Journal of Arid Environments 11:3-16. [Ch.10]

MNZAVA, Namens (1986) Compensatory leaf and seed yield increase in vegetable rape (*Brassica carinata*). [Abstract] Hortscience 21(3) [Ch.15]

MONDAL, R.C. (1974) Farming with a pitcher: A technique of water conservation. World Crops March/April:94-97. [Ch.12]

MORGAN, W.T.W. (1974) The South Turkana expedition. Scientific Papers X. Sorghum gardens in south Turkana. The Geographical Journal 140:80-93. [Ch.9,11] Description of sorghum gardens cultivated by Turkana pastoralists in arid northern Kenya. Details of hydrology and soils but comparatively little on cultivation or socioeconomic aspects.

NABHAN, Gary P. (1979) The ecology of floodwater farming in arid southwestern North America. Agro-Ecosystems 5:245-255. [Ch.11] Brief, general description of some of the techniques used traditionally by the Tohono O'Odham and other groups.

NABHAN, Gary P. (1983) Papago Fields: Arid Lands Ethnobotany and Agricultural Ecology. Ph.D. dissertation, University of Arizona, Tucson, Arizona, USA. [Ch.8]

NABHAN, Gary P., Cynthia ANSON, and Mahina DREES (1981) Kaicka: Seed Saving the Papago-Pima Way. Meals for Millions/Freedom from Hunger Foundation, Tucson, Arizona, USA. [Ch.14]

NABHAN, Gary P., and Amadeo REA (1988) Plant

domestication and folk-biological change: The upper Piman/devil's claw example. American Anthropologist 89:57-73. [Ch.14]

NABHAN, Gary P., and Thomas SHERIDAN (1977) Living fence rows of the Rio San Miguel, Sonora, Mexico: Traditional technology for floodplain management. Human Ecology 5:97-111. [Ch.9,13]

NAS (National Academy of Sciences) (1972) Genetic Vulnerability in Major Crops. NAS, Washington, DC, USA. [Ch.14]

NAS (1973) Toxicants Occurring Naturally in Foods. 2nd ed. NAS, Washington, DC, USA. [Ch.2] (Source of Oberlas 1973; Singleton and Kratzer 1973)

NAS (1975) Underexploited Plants with Promising Economic Value. NAS, Washington, DC, USA. ix+188pp. (Available from BOSTID*) [Ch.18]

NAS (1978) Post Harvest Food Losses in Developing Countries. NAS, Washington, DC, USA. 200pp. (Available from BOSTID*) [Ch.14]

NAS (1979) Tropical Legumes: Resources for the Future. NAS, Washington, DC, USA. x+331pp. (Available from BOSTID*) [Ch.18]

NAS (1980) Firewood Crops. NAS, Washington, DC. xi+233pp. (Available from BOSTID*) [Ch.13,14]

NAS (1986) Common Property Resource Management. NAS, Washington, DC, USA. xi+631pp. (Available from BOSTID*) [Ch.3]

NAS (1989a) Alternative Agriculture. NAS, Washington, DC, USA. xiv+448pp. [Ch.3]

NAS (1989b) Lost Crops of the Incas: Little-known Plants of the Andes with Promise for Worldwide Cultivation. National Academy Press, Washington, DC, USA. xii+407pp. (Available from BOSTID*) [Ch.18]

NAS (1990) Saline Agriculture: Salt Tolerant Plants for Developing Countries. NAS, Washington, DC, USA. viii+133pp. (Available from BOSTID*) [Ch.5] Part one, "Food" (pp. 17-49), lists salt tolerant conventional crops as well as many lesser known grains and tree fruits and seeds, along with references and research contacts.

NIÑEZ, Vera (1987) Household Gardens: Theoretical and Policy Considerations. Agricultural Systems 23(1987):167-186. [Part I]

NISSEN-PETERSEN, Erik (1982) Rain Catchment and Water Supply in Rural Africa: A Manual. Hodder & Stoughton, London, UK. Available from IT Publications*, London. x+83pp. [Ch.11] Based on author's 4 year's experience in Makindu, Kenya, annual average rainfall = 1,000 mm in 2 seasons. Concentrates almost entirely on ferocement storage sytems, probably most appropriate for school or community gardens because of cost in time, skill, and resources. No data on cost or availability of materials.

NOAA (National Oceanic and Atmospheric Administration) (1987) Local Climatological Data. 1986 Annual summary of comparative data. Tucson, Arizona.NOAA, Washington, DC, USA. 8pp. [Ch.11]

NOKES, Jill (1986) How to Grow Native Plants of Texas and the Southwest. Texas Monthly Press, Austin, Texas, USA. [Ch.7]

NRC (National Research Council) (1989) Recommended Dietary Allowances. 9th ed. Office of Publications, NAS, Washington, DC. x+284pp. + foldout table of RDAs. [Ch.2] This is revised about every 5 years. However, due to disagreement, there were 9 years between the 9th and 10th editions. The RDAs are "designed for the maintenance of good nutrition of practically all healthy people in the USA," and in many ways are not entirely appropriate for most developing countries where there is a much lower intake of protein and fat, higher activity levels, much higher levels of infectious diseases, and greater heat stress. They do, however, provide a widely recognized reference point for those areas where no local RDAs have been developed. See FAO/WHO (1973) and UNU (1979) for more appropriate standards for energy and protein.

OBERLEAS, Donald (1973) Phytates. In NAS 1973:363-371. [Ch.2]

O'BRIEN-PLACE, Patricia M. (1987) Evaluating Home Garden Projects. U.S. Department of Agriculture, Office of International Cooperation and Development, Nutrition Economics Group, in cooperation with U.S. Agency for International Development, Bureau for Science and Technology, Office of Nutrition, Washington, DC, USA. 31pp. [Ch.4] One of the only efforts to systematically address the problem of evaluating garden projects. It is preliminary and meant to stimulate further revision in the field, but this has not yet occurred. It has a rather narrow economic focus and requires a large number of quantitative measurements.

ODUNFA, S.A. (1985) African fermented foods. In Microbiology of Fermented Foods, Vol. 2. Brian J.B. Wood, ed. Elsevier Applied Science Publisher, London, UK. Pp.155-191. [Ch.2,15]

OFUYA, T.I. (1986) Use of wood ash, dry chili pepper fruits and onion scale leaves for reducing *Callosobruchus maculatus* (Fabricius) damage in cow-pea seeds during storage. Journal of Agricultural Science 107:467-468. [Ch.14]

OGUNTOYINBO, Julius and Paul RICHARDS (1978) Drought and the Nigerian farmer. Journal of Arid Environments 1:165-194. [Ch.11]

OLDFIELD, Margery L. (1984) The Value of Conserving Genetic Resources. National Park Service, U.S. Department of the Interior, Washington, DC, USA. (Reprinted in 1989 by Sinauer Associates, Sunderland, Massachusetts, USA.) xxii+360pp. [Ch.14]

OLDFIELD, Margery L., and Janice B. ALCORN (1987) Conservation of traditional agroecosystems. Bioscience 37(3):199-208. [Ch.14]

OMOHUNDRO, John T. (1985) Efficiency, sufficiency, and recent change in Newfoundland subsistence horticulture. Human Ecology 13(3):291-308. [Part I]

ONWUEME, I.C. (1978) The Tropical Tuber Crops: Yams, Cassava, Sweet Potato, and Cocoyams. John Wiley and Sons, New York, USA. [Ch.13] Detailed information on botany, life cycle, common pests and diseases, propagation, and cultivation of each plant; methods of harvest, storage, and food preparation are also discussed. Traditional techniques suitable for household gardens are discussed throughout, in addition to other techniques. Well documented with reference to research findings.

OOMEN, H.A.P.C., and G.J.H. GRUBBEN (1978) Tropical Leaf Vegetables in Human Nutrition. Communication 69. Department of Agricultural Resarch, Koninklijk Institut voor de Tropen, Amsterdam, The Netherlands. 140pp. [Ch.2,15,16]

ORR, David (1988) Food alchemy and sustainable agriculture. Bioscience 38:801-802. [Part II]

ORTIZ, Alfonso, ed. (1979) Handbook of North American Indians. Volume 9. Southwest. The Smithsonian Institution, Washington, DC, USA. xvi+701pp. A survey of native peoples in this dryland region including some information on gardens. (Source of Ladd 1979; Kennard 1979)

OUANGRAOUA, Hamado (1988) Protecting the garden. The IDRC* Reports 17(4):18-19. [Ch.13]

PACEY, Arnold (1978) Gardening for Better Nutrition. IT Publications*, London, UK. 64pp. [Part I] This booklet is still in print. It is based on a review of PVO garden projects, and shows that many fail because of not understanding the local situation. The discussion of production techniques in the second half of the book is not very useful.

PACEY, Arnold, with Adrian CULLIS (1986) Rain Water Harvesting: The Collection of Rainfall and Runoff in Rural Areas. IT Publications*, London, UK. viii+216pp. [Ch.9,11] This is a good general introduction to the topic, primarily at a scale appropriate for household gardens. However, the organization and descriptions are sometimes a bit confusing. It is not a field guide or technical manual, although much factual information is presented. The authors emphasize throughout that spending a great deal of time and resources gathering data and making detailed design calculations, although theoretically desirable, may in practice be inappropriate to the needs of the poor. Rather, they emphasize the value of traditional water harvesting techniques, the importance of the socioeconomic setting, and the need for experimentation involving both development professionals and local people.

PACEY, Arnold, and Philip PAYNE, eds. (1985) Agricultural Development and Nutrition. Hutchinson and Company, London, and Westview Press, Boulder, Colorado, USA. 255pp. [Ch.4]

PAGE, W. W., and Paul RICHARDS (1977) Agricultural pest control by community action: The case of the variegated grasshopper in southern Nigeria. African Environment 2(4) and 3(1):127-141. (Available from ENDA*) [Ch.13] Reports investigations near Ibadan of indigenous knowledge of this serious pest and how to control it.

PANASIUK, Oksana, and Donald D. BILLS (1984) Cyanide content of sorghum sprouts. Journal of Food Science 49:791-793. [Ch.15]

PASSMORE, R., B.M. NICOL, and M. Narayana RAO (1974) Handbook on Human Nutritional Requirements. World Health Organization, Geneva, Switzerland. 66pp. (Reprinted 1981) [Ch.2]

PENNINGTON, Campbell W. (1979) The Pima Bajo of Central Sonora, Mexico. Vol. I. The Material Culture. University of Utah Press, Salt Lake City, Utah, USA. [Ch.15]

PIMENTEL, David (1988) Pesticides: Energy use in chemical agriculture. In Bettolo 1988:157-175. [Ch.13]

PIMENTEL, David, and Marcia PIMENTEL (1979) Food, Energy and Society. Edward Arnold, London, UK. viii+165pp. [Ch.3] Discusses energy efficiencies for different types of agriculture with many tables showing data for specific case studies. Makes the point that industrial food production is much less energy efficient than small-scale, low-input production.

PIWOZ Ellen G., and Fernando E. VITERI (1985) Studying health and nutrition behavior by examining household decision-making, intra-household resource distribution, and the role of women in these processes. Food and Nutrition Bulletin 7(4):1-31. [Ch.3]

PLUCKNETT, Donald L., Nigel J.H. SMITH, J.T. WILLIAMS, and N. Murthi ANISHETTY (1987) Gene Banks and the World's Food. Princeton University Press, Princeton, New Jersey, USA. xv+247pp. [Ch.14] Enthusiastic promotion of gene banks as the only way of conserving crop genetic diversity.

POPENOE, Paul (1973) The Date Palm. Field Research Projects, Coconut Grove, Miami, Florida, USA. [Ch.7]

PRATT, Brian, and Jo BOYDEN, eds. (1985) The Field Directors' Handbook: An OXFAM Manual for Development Workers. Oxford University Press, Oxford, UK. (Available from OXFAM UK*) 512pp. [Ch.4]

PURSEGLOVE, J.W. (1974) Tropical Crops: Dicotyledons. Corrected ed. Longman*, London, UK. 719pp. (First published in 1968 in 2 volumes) [Ch.6,9,14,18] A standard reference. Covers origin and distribution, cultivars, ecology, plant structure, pollination and fruit set, germination, chemical composition, propagation, husbandry, major diseases, major pests, improvement, and production; gives major references for each plant considered. Excellent line drawings.

PURSEGLOVE, J.W. (1983) Tropical Crops: Monocotyledons. Revised ed. Longman, London, UK. 607pp. (first published in 1972 in 2 volumes). [Ch.6,18] Same format as Purseglove 1974.

PURSEGOLVE, J.W., E.G. BROWN, C.L. GREEN and S.R.J. ROBBINS (1981) Spices. vols. 1 and 2. Longman, London, UK. xi+813pp. [Ch.18] A standard reference. Covers pepper, cinnamon, cassia, nutmeg, mace, clove, pimento, and chilis in vol. 1, and ginger, tumeric, cardamon, vanilla, and coriander in vol. 2. Also discusses related crops in less detail.

QUEROL, Daniel (1988) Recursos Genéticos, Nuestro Tesoro Olivadado: Aproximación Técnica y Socioeconómica. Industrial Grafica S.A., Lima, Peru. xviii+218pp. [Ch.14]

RADEWALD, John D. (1977) Nematode Diseases of Food and Fiber Crops of the Southwestern United States. Priced Publication 4083. Division of Agricultural Sciences, University of California, Berkeley, California, USA. 64pp. [Ch.13] Oriented to large-scale growers. Simply written introduction and very good drawings and colored photographs for identifying nematode damage. Most of the discussion of control, however, advocates dangerous synthetic chemicals.

RADWANSKI, S.A., and G.E. WICKENS (1981) Vegetative fallows and potential value of the neem tree (*Azadirachta indica*) in the tropics. Economic Botany 35:398-414. [Ch.13]

RAMA MOHAN RAO, M.S., V. RANGA RAO, M. RAMA CHANDRAM, and R.C. AGNIHOTRI (1977) Effect of vertical mulch on moisture conservation and yield of sorghum in Vertisols. Agricultural Water Management 1:333-342. [Ch.10] Report of experiments carried out between 1971-76 in semiarid India showing positive effects of vertical mulch on production.

RANKINS, Jenice, Sampson HOPKINSON, and Mouhamadou DIOP (1989) Palatability and nutritional significance of solar dried mangoes for Senegal. Ecology of Food and Nutrition 23:131-140. [Ch.15]

RAO, P. Udayasekhara, and Yeshwant G. DEOSTHALE (1982) Tannin content of pulses: Varietal differences and effects of germination and cooking. Journal of the Science of Food and Agriculture 33:1013-1016. [Ch.2,15]

RAVEN, Peter H., Ray F. EVERT, and Helena CURTIS (1981) Biology of Plants. 3rd ed. Worth Publishers, New York, USA. [Ch.9,14]

RICE, Elroy L. (1983) Pest Control with Nature's Chemicals: Allelochemics and Pheromones in Gardening and Agriculture. University of Oklahoma Press, Norman, Oklahoma, USA. xiii+224pp. [Ch.8,13]

RICE, Elroy L. (1984) Allelopathy. 2nd edition. Academic Press, Orlando, Florida, USA. xi+422pp. [Ch.8]

RICHARDS, Paul (1985) Indigenous Agricultural Revolution: Ecology and Food Production in West Africa. Hutchinson & Co., London, UK. [Ch.14]

RICHARDS, Paul (1986) Coping with Hunger: Hazard and Experiment in an African Rice-Farming System. Allen and Unwin, London, UK. [Ch.14] A case study documenting indigenous agricultural techniques for managing ecological variables and crop genetic resources a stable production.

ROBERTS, Daniel A. and Carl W. BOOTHROYD (1984) Fundamentals of Plant Pathology. 2nd ed. W.H. Freemman, New York, USA. xvi+432pp. [Ch.13]

RODALE PRESS, ed. (1977) Proceedings of the First Annual Amaranth Seminar, July 29, 1977. Rodale Press, Inc., Emmaus, Pennsylvania, USA. 132pp. (Source of Early 1977)

RUTHENBERG, Hans (1980) Farming Systems in the Tropics. 3rd ed. Oxford University Press, Oxford, UK. xxii+424pp. [Ch.8] An often-cited reference on agricultural systems in the tropics, including drylands. Very little on gardens per se, see pp. 73,76-77,127.

SAMSON, Jules A. (1986) Tropical Fruits. 2nd ed. Tropical Agriculture Series, Longman, New York, USA. ix+335pp. [Ch.6]

SANYAL, Bishwapriya (1986) Urban Cultivation in East Africa. Food-Energy Nexus Research Report No. 14. Food-Energy Nexus Program, United Nations University, Paris. 75pp. [Ch.3]

SAUL, Mahir (1981) Beer, sorghum and women: Production for the market in rural Upper Volta. Africa 51(3):746-764. [Ch.4]

SCHMUTTERER, Heinz and K.R.S. ASCHER, eds. (1984) Natural Pesticides from the Neem Tree and other Tropical Plants. GTZ*, Eschborn, Germany. 587pp. (Source of Zehrer 1984)

SCHUMACHER, E.F. (1973) Small is Beautiful: Economics as if People Mattered. Harper and Row, New York. viii+305pp. [Ch.3]

SCHWERDTFEGER, Werner, ed. (1976) Climates of Central and South America. xii+532pp. Elsevier Publishing Co., Amsterdam, The Netherlands. [Ch.11]

SCOTT, Earl Price (1976) Indigenous Systems of Exchange and Decision Making Among Smallholders in Rural Hausaland. Michigan Geographical Publication No. 16, Department of Geography, University of Michigan, Ann Arbor, Michigan, USA. xxiv+303pp. [Ch.3]

SCUDDER, Thayer (1962) The Ecology of the Gwembe Tonga. Manchester University Press, Manchester, UK. [Ch.11]

SCUDDER, Thayer (1982) Regional Planning for People, Parks and Wildlife in the Northern Portion of the Sebungwe Region. Working Paper 3/1982. Department of Land Management, Faculty of Agriculture, University of Zimbabwe, Harare, Zimbabwe. [Ch.11]

SHANAN, L., and N.H. TADMOR (1979) Micro-Catchment Systems for Arid Zone Development: A Handbook for Design and Construction. 2nd ed. Centre of International Agricultural Cooperation, Hebrew University, and Ministry of Agriculture, Jerusalem and Rehovot, Israel. vii+73pp. [Ch.11] Summarizes existing practices and proposes design criteria with figures and illustrations. Based on projects in Israel, but takes into account work done elsewhere.

SHARMA, K.D., O.P. PAREEK, and H.P. SINGH (1982) Effect of runoff concentration on growth and yield of jujube. Agricultural Water Management 5:73-84. [Ch.11] Reports 5 years of experiments in semiarid India. Effect of catchment slope, length, and CGAR on fruit production. No statistical analysis, no data on catchment treatment or erosion.

SHRIMPTON, Roger (1989) Vitamin A deficiency in Brazil: Perspectives for food production oriented interventions. Ecology of Food and Nutrition 23:261-271. [Ch.2]

SIMPSON, Kenneth L. (1983) Relative value of carotenoids as precursors of vitamin A. Proceedings of the Nutrition Society 42:7-17. [Ch.15]

SINGLETON, V.L., and F.H. KRATZER (1973) Plant phenolics. In NAS 1973:309-345. [Ch.2]

SMALE, M. (1980) Women in Mauritania: The effect of drought and migration on their economic status and implications for development problems. Office of Women in Development, U.S. Agency for International Development, Washington, DC, USA. [Ch.3]

SMITH, P.D., and W.R.S. CRITCHLEY (1985) The potential of runoff harvesting for crop production and range rehabilitation in semiarid Baringo. In Soil and Water Conservation in Kenya. Proceedings of a Second National Workshop. Pp. 305-322. Institute for Development Studies and Faculty of Agriculture, University of Nairobi, Nairobi, Kenya. [Ch.11]

SMITH, R.J., and N.H. HANCOCK (1986) Leaching requirements of irrigated soils. Agricultural Water Management 11:13-22. [Ch.12] An alternative method of calculating LR is developed based only on salinity of irrigation water and acceptable salinity of soil. This method is compared with other widely used methods, from which it differs significantly.

SOLERI, Daniela (1989) Hopi gardens. Arid Lands Newsletter 29:11-14. [Ch.6,9]

SOLERI, Daniela and David A. CLEVELAND (1989) Hopi Crop Diversity and Change: A Report on a Preliminary Survey of Hopi Crop Genetic Resources. Project Report of Sources of Seed in a Native American Farming Tradition: Hopi Crops After A Half Century of Culture Change. 59pp. Native Seeds/SEARCH*, Tucson, Arizona, USA. [Ch.12,14]

SOLON, Florentino, Tomas L. FERNANDEZ, Michael C. LATHAM, and Barry M. POPKIN. (1979) An Evaluation of strategies to control vitamin A deficiency in the Philippines. American Journal of Clinical Nutrition 32:1445-1453. [Ch.2]

SOMMER, Alfred, Ignatius TARWOTJO, Gusti HUS-

SAINI, and Djoko SUSANTO (1983) Increased mortality in children with mild vitamin A deficiency. The Lancet 1983 2:585-588. [Ch.2]

SOMMER, Alfred, Ignatius TARWOTJO, Edi DJUNAEDI, Keith P. WEST, A.A. LOEDEN, and Robert TILDEN (1986) Impact of vitamin A supplementation on childhood mortality. The Lancet 1986 1:1169-1173. [Ch.2]

SOMMER, Alfred, Ignatius TARWOTJO, and Joanne KATZ (1987) Increased risk of xerophthalmia following diarrhea and respiratory disease. American Journal of Clinical Nutrition 45:977-980. [Ch.2]

SOMMERS, Paul (1984) Dry Season Gardening for Improving Child Nutrition. UNICEF*, New York, New York, USA. 48pp. [Part I]

SPRADLEY, James P. (1979) The Ethnographic Interview. Holt, Rinehart and Winston, New York, USA. vi+247pp. [Ch.4,9]

SPRADLEY, James P. (1980) Participant Observation. Holt, Rinehart and Winston, New York, USA. xi+195 pp. [Ch.4]

STEGMAN, E.C., J.T. MUSICK, and J.I. STEWART (1980) Irrigation water management. *In* Jensen 1980:763-816. [Ch.10,12]

STERN, Peter (1979) Small Scale Irrigation: A Manual of Low-Cost Water Technology. Intermediate Technology and International Irrigation Information Center, London, UK, and Bet Dagan, Israel. (Available from IT Publications*) 152pp. [Ch.10,11,12] A brief general introduction covering some basic information, but mostly for applications larger than household gardens; not enough detail for field use; few references given. Little consideration of traditional systems of irrigation or of how Western techniques presented in more detail could be integrated with existing techniques, e.g. planting in rows is usually assumed (pp.43-44).

STOLER, Ann (1979) Garden use and household economy in Java. *In* Agriculture and Rural Development in Indonesia. G.E. Hansen, ed. Pp. 242-254. Westview Press, Boulder, Colorado, USA. [Ch.2,3]

STOLL, Gaby (1987) Natural Crop Protection: Based on Local Farm Resources in the Tropics and Subtropics. 2nd ed. Verlag Josef Margraf. 187pp. [Ch.13] Order from TRIOPS, Tropical Scientific Books, Raiffeisenstr. 24, D-6070 Langen, FR Germany. Available in English, German and Spanish. An annotated bibliography emphasizing the use of local resources. Has sections for specific crops and for field and storage methods.

STONE, M. Priscilla, Barbara PERQUIN and Sarr HAMIDOU (1987) Vegetable Production Along the Senegal River. A Reconnaissance Survey of Gardens in the Brakna and Gorgol Regions. Mauritania Agricultural Research Project II, College of Agriculture, University of Arizona, Tucson, Arizona, USA. [Ch.3]

TAIT, Joyce and Banpot NAPOMPETH, eds. (1987) Management of Pests and Pesticides: Farmers' Perceptions and Practice. Westview Press, Boulder, Colorado, USA. (Source of Atteh 1987)

TAKAHASHI, K., and H. ARAKAWA (1981) Climates of Southern and Western Asia. Elsevier Publishing Co., Amsterdam, The Netherlands. xiii+333pp. [Ch.11]

TAMANG, Jyoti P., Prabirk SARKAR, and Clifford W. HESSELTINE (1988) Traditional fermented foods and beverages of Darjeeling and Sikkim: A review. Journal of the Science of Food and Agriculture 44(4):375-385. [Ch.15]

TARWOTJO, Ignatius, Alfred SOMMER, Tito SOEGIHARTO, Djoko SUSANTO, and MUHILAL (1982) Dietary practices and xerophthalmia among Indonesian children. American Journal of Clinical Nutrition 35:574-581. [Ch.2]

THOMSON, James T. (1980) Preliminary evaluation: OXFAM Micro-Catchment Project, Ouahigouya, Upper Volta. mss. submitted to OXFAM UK*. [Ch.11]

TITILOYE, E.O., E.O. LUCAS and A.A. AGBOOLA (1985) Evaluation of fertilizer value of organic waste materials in south western Nigeria. Biological Agriculture and Horticulture 3:25-37. [Ch.9]

TODARO, Michael (1985) Economic Development in the Third World. 3rd ed. Longman, New York, USA. [Part I, Ch.3] A popular textbook; makes the standard assumptions about the necessity of economic growth and industrial agriculture.

TRICAUD, Pierre-Marie (1987) Urban Agriculture in Ibadan and Freetown. Food Energy Nexus Research Report No. 23, Food Energy Nexus Program, United Nations University, Paris, France. 45pp. [Ch.3]

TROEH, Frederick R., J. Arthur HOBBS and Roy L. DONAHUE (1980) Soil and Water Conservation for Productivity and Environmental Protection. Prentice Hall, Englewood Cliffs, New Jersey, USA. xv+718pp. [Ch.9,10] Beginning textbook, with basics in easy-to-read style, although oriented toward large-scale production in the United States. Chapter 14, "Water Conservation" (pp. 454-489), is on drylands.

TUCK, Brian (1985) personal communication and draft mss., Le Sahel: An Agricultural Production Manual. [Ch.14]

TULLY, Dennis (1988) Culture and Context in Sudan: The Process of Market Incorporation in Dar Masalit. State University of New York Press, Albany, New York, USA. xiii+306pp. [Ch.8,12]

UDS (Faculty of Agriculture, Forestry and Veterinary Science, University of Dar es Salaam) (1983) Proceedings of a Workshop on Resource-Efficient Farming Methods for Tanzania, May 16-20, 1983. Rodale Press, Emmaus, Pennsylvania, USA. 128pp. [Ch.9]

UNEP (United Nations Environment Program) (1983) Rain and Stormwater Harvesting in Rural Areas. Published for UNEP by Tycooly International, Dublin, Ireland. vii+238pp. [Ch.11] Based on limited review of the literature and personal observations, from a discussion by UNEP consultants in 1979. Provides a general review of a great variety of water harvesting and floodwater farming methods with many photographs and drawings but few technical details.

UNESCO (United Nations Educational, Scientific and Cultural Organization) (1977) Map of the World Distribution of Arid Regions. MAB Technical Notes 7. UNESCO, Paris. 54pp+world map 1:25,000,000 scale, 100 x 65 cm, in color. [Ch.1] This is the best map of world drylands we have seen. Accompanying booklet gives brief summaries of climate, vegetation, and land use for dryland countries or regions and diagrams showing the relationship between rainfall and potential ETm for 93 representative sites.

UNICEF (1985) Gardening for Food in the Semi-Arid Tropics. A Handbook for Programme Planners. A WHO/UNICEF Joint Nutrition Support Programme Publication. UNICEF*, New York, New York, USA. [Part I]

USDA (United States Department of Agriculture) (1975) Soil Taxonomy: A Basic System of Soil Classification for Mapping and Interpreting Soil Surveys. Agriculture Handbook No. 436. Soil Conservation Service, USDA, Washington, DC, USA. [Ch.9]

USDA (1982) Composition of Foods: Fruits and Fruit Juices. Agricultural Handbook No. 8-9. USDA, Washington, DC, USA. vi+283pp. [Ch.2]

USDA (1984a) Composition of Foods: Vegetables and Vegetable Products. Agricultural Handbook No. 8-11. USDA, Washington, DC, USA. vi+502pp. [Ch.2]

USDA (1984b) Composition of Foods: Nut and Seed Products. Agricultural Handbook No. 8-12. USDA, Washington, DC, USA. v+137pp. [Ch.2]

USDA (1989) Composition of Foods: Cereal Grains and Pasta. Agricultural Handbook No. 8-20. USDA, Washington, DC, USA. iv+137pp. [Ch.2]

VAN DEN BOSCH, Robert, P.S. MESSENGER, and A.P. GUTIERREZ (1982) An Introduction to Biological Control. Plenum Press, New York, USA. xiv+247pp. [Ch.13] Written as a beginning college text. Clear explanations with many examples.

VAN DOORNE, J.H. (1985) A review of small-scale irrigation schemes in Kenya. AGL/MISC/2/85. FAO*, Rome. 90pp. [Ch.11]

VAN EPENHUIJSEN, C.W. (1978) La Culture des Légumes Indigènes au Nigéria. FAO*, Rome. (Also in English) xiv+108pp. [Ch.8,18] Promotes the production of local vegetables, but industrial growing methods. Includes descriptions of local Nigerian vegetables. See Brownrigg (1985) for description of the FAO project out of which this publication came.

VARISCO, Daniel Martin (1983) Irrigation in an Arabian valley. Expedition 25(2):26-34 [Ch.10]

WALLERSTEIN, Immanuel (1974) The Modern World-System I: Capitalist Agriculture and the Origins of the European World-Economy in the Sixteenth Century. Academic Press, New York. xiv+410pp. [Ch.3]

WARREN, D.M., L.J. SLIKKERVEER and S.O. TITILOLA, eds. (1989) Indigenous Knowledge Systems: Implications for Agriculture and International Development. CIKARD*, Ames, Iowa, USA. 186pp. [Part I]

WATERLOW, J.C. (1982) Nutrient needs for man in different environments. In Food, Nutrition and Climate. Kenneth Blaxter and Leslie Fowder, eds. Pp. 271-283. Applied Science Publishers, London. [Ch.2]

WATSON, J.D. (1976) Ascorbic acid content of plant foods in Ghana and the effects of cooking and storage on vitamin content. Ecology of Food and Nutrition 4:207-213. [Ch.15]

WATT, S.B. (1978) Ferrocement Water Tanks and Their Construction. IT Publications*, London, UK. [Ch.11]

WATT, S.B. and W.E. WOOD (1979) Hand Dug Wells and Their Construction. 2nd ed. IT Publications*, London, UK. 253pp. (First published 1977) [Ch.11,12]

Based on field experience and clearly written with many useful diagrams and photographs. The 8 middle chapters describe in detail the construction of one type of well: "a reinforced concrete lined circular shaft well of 1.3 m internal diameter, excavated through sedimentary soils to an open aquifer having a water table some 20-30 m below ground surface. The wellhead is open, extraction of water will be by buckets and ropes, and construction is by 'self-help' methods employing local labour" (53). Details for making tools and equipment and for estimating labor, costs, and materials needed are given. The orientation is toward programs that will be building a number of wells and can therefore justify some fairly costly equipment. The authors give some alternative methods appropriate for building one or two wells using more local and less expensive materials; the principals remain the same.

WELTZIEN, H.C. and N. KETTERER (1986) Control of downy mildew, *Plasmopara viticola* (de Bary) Berlese et de Toni, on grapevine leaves through water extracts from composted organic wastes. Journal of Phytopathology 116:186-188. [Ch.13]

WERNER, David (1977) Where There is No Doctor. The Hesperian Foundation*, Palo Alto, California, USA. [Ch.2,11,13,16] An excellent manual for field use emphasizing prevention and inexpensive medical care. Available in Arabic, French, Spanish, and many other languages.

WERNER, David (1986) Report Concerning Diarrhea Control in Mozambique. Based on a 1986 visit by David Werner to Mozambique as a consultant to the Ministry of Health. The Hesperian Foundation*, Palo Alto, California, USA. 32pp. [Ch.16]

WERNER, David, and Bill BOWER (1982) Helping Health Workers Learn: A Book of Methods, Aids and Ideas for Instructors at the Village Level. The Hesperian Foundation*, Palo Alto, California, USA. [Ch.4,16] See the Resources section of Chapter 4 for a description of this valuable book.

WESTPHAL, E., et al. (1981) L'Agriculture Autochtone au Cameroun. Miscellaneous Papers 20 (1981), Lanbouwhogeschool, Wageningen, The Netherlands. H. Veenman & Zonen B.V., Wageningen, The Netherlands. (In French) 175pp. [Part II]

WESTPHAL, E., et al. (1985) Cultures Vivrières Tropicales avec Référence Spéciale au Cameroun. Pudoc, P.O. Box 4, 6700 AA Wageningen, The Netherlands. (In French) 514pp. [Part II]

WHO (World Health Organization) (1982) Control of Vitamin A Deficiency and Xerophthalmia. WHO Technical Report Series 672. WHO*, Geneva, Switzerland. 70pp. [Ch.2] Popular to semipopular review of current knowledge on vitamin A deficiency, its assesment, distribution, treatment and control. Advocates gardens as a good source of vitamin A while warning that vitamin A content of foods given in food composition tables is unreliable and that more research is needed on how to improve consumption of indigenous DGLVs.

WHO (1986) Guidelines for Training Community Health Workers in Nutrition. (Revised ed.; First ed., 1981). Prepared by K. Bagchi. WHO*, Geneva, Switzerland. [Ch.16] A good resource for training in maternal and child health and nutrition. A series of brief, accessible training modules provide a framework for addressing nutritional needs of mothers and children. These modules could easily be integrated with garden activities.

WILKEN, Gene C. (1977) Manual irrigation in Middle America. Agricultural Water Management 1:155-165. [Ch.12] Describes manual irrigation from shallow wells; common for high-value vegetable crops in semiarid highlands.

WILKEN, Gene C. (1987) Good Farmers: Traditional Agricultural Resource Management in Mexico and Central America. University of California Press, Berkeley, California, USA. [Ch.8,9]

WMO (World Meteorological Organization) (1983) Guide to Climatological Practices. 2nd ed. Secretariat of the WMO, Geneva, Switzerland. [Ch.11]

WOLF, Eric R. (1982) Europe and the People Without History. University of California Press, Berkeley and Los Angeles, California, USA. xi+503pp. [Ch.3]

WORLD NEIGHBORS (1985) Introduction to Soil and Water Conservation Practices. World Neighbors, 5116 N. Portland Ave., Oklahoma City, Oklahoma 73112, USA. 33pp. [Ch.9]

WORSLEY (1984) Three Worlds: Culture and World Development. University of Chicago Press, Chicago, Illinois, USA. xiv+409pp. [Ch.3]

WRIGHT, Peter (1984) Report on runoff farming and soil conservation in Yatenga, Upper Volta. Report to OXFAM, Oxford, UK. (Cited in Pacey and Cullis 1986) [Ch.9]

YOON, Soon Young (1983) Women's garden groups in Casamance, Senegal. Assignment Children 63/64:133-153. [Ch.3,15]

YOUTOPOULOS, Pan A. (1985) Middle-income classes and food crises: The "new" food-feed competition.

Economic Development and Cultural Change 33:463-483. [Ch.2]

ZEHRER, W. (1980) Traditional methods of insect pest control in stored grain. *In* GTZ 1980:98-129. [Ch.14]

ZEHRER, W., E. WEGMANN and D. AKOU-EDI (1980) The effect of traditional pest control substances on the development of *Callosobruchus maculatus* in stored beans. *In* GTZ 1980:148-157. [Ch.14]

ZEHRER, W. (1984) The effect of the traditional preservatives used in northern Togo and of neem oil for control of storage pests. *In* Schmutterer and Ascher 1984:453-460. [Ch.14]

ZIMMERMAN, Sonia D. (1982) The Women of Kafr al Bahr: A Research Into the Working Conditions of Women in an Egyptian Village. (English translation by Rosemary Risseeuw) Research Centre, Women and Develpment, State University of Leiden, Institute for Social and Cultural Studies, Stationsplein 10, 2312 AK Leiden, The Netherlands. [Ch.3]

Index

Page numbers in *bold italics* indicate the place in the text where the word or phrase is defined.

adhesion *71*, 192
adventitious roots; *see* roots, adventitious
advertising 36, 209, 248, 327
A-frame level 178-180
aggregates; *see* soil, aggregates
agrochemicals 63
    synthetic fertilizers 166-167
    synthetic pesticides 246-250, 302
    harming plants 267
air layering (marcottage); *see* layering, air
Alfisols 154-155; *see also* soil orders
alleles *288*, 289
allelopathy *138*
allergy (to food) *327*-328
alluvium *174*, 186, 223
    used in floodwater gardening 219-221
altitude 22
amino acids *19*, 20
anaerobic decay
    in compost *171*, 174
    in soil 234
anemia (iron deficiency) 23, 25, 28, 209
animal manure; *see* manure, animal
anions *161*
annuals *79*-80
    transplanting 130, 133
anthers *82*
anti-nutrients *26*-28
    aflatoxin *27*-28, 321
    ascorbase *27*
    effect of sprouting on 315-316
    hydrogen cyanide (HCN) *27*, 316
    in weaning foods 332
    mycotoxin *27*
    oligosaccharides *27*
    oxalates *26*, 332
    phytates 25, *26*, 311
    tannins *27*, 310, 311, 332
ants 244, 251
aphids 243-244, 253, 255
apical
    dominance *113*
    grafting *114*
application efficiency (Ea) 198-199, 217-218, *228*, 229
approach (attached) scion grafting *111*
aquifer *221*
    artesian *221*
    open *221*
Aridisols 154-155, 159; *see also* soil orders
arthropods *250*
ascorbic acid; *see* vitamin C
asexual reproduction; *see* vegetative propagation
ashes
    as a soil amendment 163
    in compost 171, 174
    for seed storage 301, 303
assessment, project *47*-62; *see also* interview; survey
    biases in 49, 50-52
    deciding information needed 53
    types of 48-49
assumptions 44, 294
    effect on assessment 50-52
    development approaches and 7-9
    of conventional economic theory 32-36
awns *292*
Azadirachta indica; *see* neem

Bacillus thuringiensis* (Bt) 245-246
back-translating *54*
bacteria *262*; *see also* disease in plants, bacterial
    in food fermentation 319, 335
    in the soil 70
bars (of soil water potential) *192*

bark *68*
barriers against pests
   for protecting the garden 258-261
   for protecting seeds and food 303-304, 312-313
basin beds; *see* garden beds, sunken
baseline survey *48*
bedrock *158*
beetles 243, 245, 247, 248
benefit/cost ratios; *see* evaluation of projects
beriberi 15, 23
beta carotene *22*; *see also* vitamin A
biennials *79*-80, 104-105
bilharzia; *see* schistosomiasis
biological control (of pests and pathogens) *244*-246, 253; *see also* pests
biotechnology 290
blanching *316*
branch
   angle *144*-146
   collar *147*-148
breast-feeding (lactation) 17, 327-328
   nutritional requirements of 13, 17, 19, 22, 23, 25
breast milk 327-328
   pesticide contamination of 248
budding 111
   chip *113*
   T-bud (shield) *112*-113
buffering agent *170*
bulbs *104*
burning
   and nematode control 257
   and soil pH 163
   as weed control 141
   diseased plants 263-264

C_4 72, 74
calcium (Ca)
   garden yield of 28
   in the diet 26
   in the soil 154, 160, 168
calendars for assessment 56-58
caliche 158, *160*, 168
cambial tissue (cambium) *68*
   and grafting 108-109, 112, 113, 114, 116, 117
   and notching 149
canal
   flow, measuring 200-201
   systems 228-229
capillary action *192*, 193
capsaicin *247*
carbohydrates (CBHs) *19*, 70, 330

carbon:nitrogen ratio (C:N) 164, 171-173, 203
carotenoids 311; *see also* vitamin A
catchment *212*
catchment to garden area ratio (CGAR) *217*-218, 220
caterpillars 246, *251*-253; *see also* insects, as food
cation *161*, 170
cation exchange capacity (CEC) *161*
cellulose *26*, 171, 310
cereals *19*
chelates *167*, 170
chemicals 245-250 ; *see also* agrochemicals; pesticides
   botanical *246*-247
   in plants and pest resistance 242
child growth charts 15-17
children; *see also* weaning; weaning foods
   caring for other children 336
   special nutritional needs of 14-15, 20, 21-22, 28, 327-329
   what to do when they are sick 335
chlorophyll 22, *70*, 161
chloroplasts *70*, 71
chlorosis *160*
   and disease 262
   and nutrient deficiencies 164, 168
   asymmetrical 273
   patterns and diagnosis 273, 276-277
   symmetrical 273
cholesterol *26*
chromosomes 288-289
clay *155*, 156
   bentonite 155
   kaolinite 155, 214
   montmorillonite 155, 214
   sealing catchments with 214
   sesquioxide 155, 214
   water-holding capacity of 190
clone *99*
cold stratification of seeds *300*-301
cohesion
   in the plant *67*, 71
   in the soil 192
commerciogenic malnutrition *36*
common property resources 35
commons 35
community groups 43, 49, 322
community participation 7-8, 48-49
compost *171*-172
   amount in transplant holes 130-131
   avoiding salt in 235
   C:N ratio in 171-173
   pH in 171, 174

composting
    allelopathic plants 139
    fast 174
    trench bed *172*
containers; *see* food storage, containers; planting, in containers; seed storage, containers; water, storage
contour bunds/berms *176*-180
    and water harvesting 214
conveyance (of irrigation water) *227*-228
    efficiency *228*
cooperation
    and marketing 42-43, 322
    and resource control 35
cork *68*
corm *105*
cortex *68*, 265-266
cortical sloughing *265*-266
cotyledons *68*, 86, 91
crassulacean acid metabolism (CAM) *72*
crop
    adaptations to pests and pathogens 242-243
    breeding, indigenous 286, 287, 292
    breeding, industrial 288, 290-292
    coefficient (kc) 202
    diversity in 287-292
    introducing new 285-287
    rotation 139, 244
    trap *243*-244
    varieties and drought 34, 195-196
    varieties, local; *see* folk crop varieties
cross-pollination; *see* pollination, cross-
crown *101*
cultivation
    and erosion 193
    and physical properties of soil 156
    and water in the soil 192,193
    for weed control 141
cuttings *99*, 101-103
    from trees, hardwood *99*, 101-102
    from trees, softwood *99*, 101-102
    heel of *101*
    storage of 102

damping-off 89, 93, 96, 269
daylength; *see* photoperiod
deciduous *80*
    trees, as seasonal source of shade 129-130, 138, 206
    trees, cuttings from 101-102
    trees, notching 149
    trees, transplanting 129
dehisce *295*
dehydration, emergency care for in children 335
demineralization *17*
development, approaches to 2-4, 7-9, 50-51, 63-64
diagnosing garden problems 265-281
    abnormal growth 273, 278-279
    failure to thrive 271
    of established plants 270-271
    of fruit 273, 280-281
    of leaves 273, 276-277
    of seeds, seedlings, and recent transplants 95-96, 268-271
    wilts 273, 274-275
diarrhea, emergency care for in children 335
dicots *68*, 108, 109
diet
    factors affecting 59
    surveying for assessment 58-59
    value of diversity in 292-293
    weaning 329-333
diminishing marginal returns *33*; *see also* efficiency; gardens, investments in
dioecious; *see* flowers, dioecious
direct-lift *237*; *see also* water-lifting
disease in people; *see also* malnutrition
    and iron 25
    and spoiled weaning foods 334-335
    and vitamin A 15, 21
    and water quality 208-209
    effect on nutritional needs 18-19
    in children, emergency care of 335
    malaria 19, 56, *209*
    seasonality of 19, 56
disease in plants *261*-265; *see also* biological control; pathogens (plant disease)
    bacterial 262, 264, 265
    control with crop rotation 139
    control with fallow 244
    controls in plants 243
    controls in the environment 243, 263
    diagnosing 265-281
    fungal 262-265
    localized *262*
    mechanical control of vectors 243-244, 263
    mycorrhizae and resistance to 161
    spread of 101, 148, 255, 263, 264
    systemic *262*-263
    viral 262-264
disproportional stratified sample; *see* sample, stratified, disproportional

diversity 63
    and returns to labor 36-37
    and the diet 292-293
    center of *293*-294
    genetic 287-292, 293
    in agriculture 63, 286-287, 292-293
    in mixed planting 138
dormancy; *see* seed, dormancy
drainage; *see also* waterlogging
    in containers 127-129
    in planting sites 131-132, 158, 160
    problems due to erosion 174
    testing 128, 158
drip line *267*
drought *71*
    -adapted 72, 189, 195-196
    -adapted, distinguished from heat-tolerant 74
    avoidance *72*
    deciduous 74, *80*
    effect on yield 74, 195-196
    -escaping *72*
    resistance *72*
    tolerance *74*
drying (food) 311-315
    and marketing 41-42
    blanching for 316
    case hardening 314
    controlling pests 312-313
    effect on nutrient content of food 19, 20, 312, 330
    guidelines for 314
    indigenous examples of 311
    seeds; *see* seed, drying
drylands 2
    formation of soils in 154-155
    most important nutritional problems in 28

economic development 31-36
economy *36*
efficiency *32*-33
    application of water (Ea) 198-199, 217-218, *228*, 229
    conveyance *227*-228
    irrigation *227*-228, 229
    used in evaluation 48-49
    water-lifting 236-237
elephantiasis (filariasis) *209*; *see also* disease in people, and water quality
embryo; *see* seed, embryo
emergence (seedling) *86*
    failure 95-96

emitters (for trickle irrigation) *231*
energy
    content and drying food 19, 312, 330
    content and malting food 332-333
    fats as a source of 26, 329
    garden sources for weaning foods 330-331
    people's requirements for 13, 19
    seasonality of needs or sources 17, 19
    sources of 14, 19, 26, 329
Entisols 154-155; *see also* soil orders
environmental
    contamination by pesticides 248-249
    degradation 33, 60, 153; *see also* erosion
    sustainability 3, 63-64
enzymes *19*, 167, 316
epidermis *68*, 71, 161
erosion *172*-183
    and cultivation 156
    and slope 175-180
    by raindrops 181
    by wind 182
    gully *175*
    on rainwater catchments 217
    rill *175*
    sheet *175*
essential amino acids (EAAs) *19*
ethylene *322*
evaluation of projects *48*-49
evaporation *193*; *see also* evapotranspiration
    and salt buildup 235-236
    from plants 71; *see also* transpiration
    from raised beds 183-185
    from soil 189
    reduced by mulches 203
evapotranspiration (ET) *193*; *see also* evaporation; transpiration
    actual (ETa) *195*
    and water stress 195
    management to reduce excess 193, 203-206, 138
    maximum (ETm) *193*, 202, 217-218, 221
    theoretical (ETo) *202*
evergreen *80*
    trees, cuttings from 101
    trees, transplanting 129
ex situ conservation *293*-294
exoskeleton *250*
extracted (water) *227*-228
eyes (on a tuber) *104*

F_1 and F_2 hybrids 290

fallow, to control pests and pathogens 244, 257
fats 26, 329; *see also* vitamin, fat-soluble
feasibilty study *48*
feces; *see* manure
fermentation
    for cleaning seeds 296, 297
    for foods *see* food processing, fermentation
fertilization (of flowers) 85
fertilizers; *see* agrochemicals, synthetic fertilizers; soil, fertilizers
fiber in the diet 25, 26
    processing to reduce 310, 311
field capacity (FC) 161, *190*, 193, 198-199, 228
filariasis (elephantiasis) *209*; *see also* disease in people, and water quality
float method (of water measurement) *200*-201
flood recession gardening *220*-221
floodwater gardening *219*-221; *see also* irrigation, root zone
flowers *79*, 80-81
    anthers *82*
    as clues to type of pollination 82
    controlling pollination of 83
    dioecious *80*-81
    life span of 80
    monoecious *80*-81, 83
    perfect *80*-81, 83
    silks *83*
    tassels *83*, 290
flumes *228*
folacin (folate, folic acid) 23
    effect of food processing on 310, 312
folate, folic acid; *see* folacin
folk crop varieties (or indigenous crop varieties) 34, *285*, 286-295
    and disease resistance 263
    and nitrogen fixation 164
    and pest and pathogen management 243
    control of and compensation for 293-295
food distribution (household) 28-29, 59
food, drying; *see* drying (food)
food frequency survey 59
food processing
    drying; *see* drying (food)
    effect on nutrient content 310-311, 315-318
    fermentation 27, *318*-320, 329, 335
    for market 40-42
    malting 27, *317*-318, 332-333
    oxidation during *311*, 314, 321
    parching *311*

food storage 319-322
    containers 321-322
frass *255*, 273
freeboard *220*
fruit *85*
    cap *309*
    harvesting 309-310
    how borne on trees 146-147
    pedicel *309*
    problems, diagnosing 273, 280-281
    processing and nutrient content 310, 312
    pruning to improve production of 142, 144
    using unripe 309-310
fungi 70, 172, *262*; *see also* disease in plants, fungal

galls *133*, 255
gametes *79*, 80, 85, 288
garden beds; *see also* nursery beds
    raised 183-185, 235
    sunken (basin) 183-184, 229-230
gardens
    diversity and nutrient production 292-293
    economic contribution to household 37-39
    effect on nutritional status 28-29
    floodwater *219*-221
    indigenous *8*, 9, 50, 52
    industrial 2, *8*, 51
    investments in 32-34, 40-42; *see also* efficiency
    location and soil temperature 158
    management and pest and pathogen control 243
    management and production 241-242
    market 38, 42, 322-324
    mixed 137-138, 292-293
    model *8*
    problems, diagnosing; *see* diagnosing garden problems
    started by projects 7, 8, 9, 50
    urban 38, 170, 267, 310
    yields 28, 36-37, 195-196
genes *79*, 288-289, 290
genotype *288*
germination *86*
    epigeal *86*, 91
    hypogeal *86*
    percentage *96*-97
    temperature requirements for 86
    test 96-98
girdling *101*, 148
glume *292*
goiter *27*

grafting 68, *108*-116; *see also* budding
   apical *114*
   approach (attached scion) *111*
   guidelines for 116
   incompatability in 109-110
   reasons to use 109
   scion *108*-116
   sliced-approach *111*, 112
   stock *108*-116
   topworking *114*-115
grasshoppers 251-252
gravel *155*
green manure; *see* manure, green
green revolution 9, 35
gross national product (GNP) 35
groundwater *221*
grub *251*

halophobic plants *75*
halophytic plants *75*; *see also* salt-tolerant plants
hand-pollination; *see* pollination, hand-
hardening off *133*
harvesting 307-309
   and nutrient content of foods 27, 309
   and returns to labor 37
   and the nutrient cycle 168
   and yields 309
   for market 322
   rainwater; *see* indigenous examples of, rainwater harvesting; rainwater harvesting
heat tolerance *74*, 189
heel (of a cutting) *101*
hemoglobin *25*
herbaceous *80*
heterogeneous (crop varieties) 287, *288*
heterozygous (plants) 287, *288*, 289
hilum *87*
homogeneous (crop varieties) 287, *288*
homozygous (plants) 287, *288*, 289
honeydew (from aphids) *255*
hookworm 18, 25, *209*; *see also* disease in people, and water quality
host *241*
household *2*
   control of income within 43
   decision making 35
   food distribution within 28-29, 59
   garden *2*
   representation of members in assessment 49
humus 156, *169*
hungry season *17*

husk *298*
hybrid 290; *see also* seed, hybrid

illness; *see* disease in people
inbred *290*
inbred lines *290*, 291
income; *see* gardens, economic contribution to household
indigenous *3*
   agriculture 9, 153, 286-287, 292-293
   crop varieties; *see* folk crop varieties
   gardens; *see* gardens, indigenous
   knowledge *8*, cataloging 50, 154
   soil classification systems 154
indigenous examples of
   cistern wells 222
   cold stratification 300-301
   controlling erosion 182
   crop selection and testing 287, 292, 295
   determining planting time 210
   flood recession gardening 221
   food drying 311
   food fermentation 318-319
   food storage 321
   gardens; *see* gardens, indigenous
   irrigation systems 228-230
   management and use of wild plants 139, 140
   maintaining varietal diversity 292
   measuring irrigation water 199
   mulch 203
   nursery beds 123
   pest control 244, 251, 258, 259, 260, 261
   rainwater harvesting 213
   seed planting 91
   seed storage 301, 303-304
   shades and windbreaks 183, 205, 206
   soil classification 154
   sunken garden beds 183
   transplanting 133
   water-lifting 237-238
   water spreading 220
   weaning foods 331
   weed classification and management 139, 140
induced deficiency (in plants) *166*
inductive (function of statistics) *54*
industrial
   agriculture 63, 153, 286
   gardens *8*
   seeds 286-287, 290-292
infiltration rate (of water) *175*, 192-193
   and irrigation 229-232

and runoff 214-215
and soil characteristics 156
improving with organic matter 169
improving with vertical mulch 203-204
insect-pollination; *see* pollination, insect-
insects *250*-255
    as food 23, 25, 251, 253, 330
    as pollinators; *see* pollination, insect-
    beneficial, and pesticides 247-248, 250-251
    life cycles; *see* life cycle (of insects)
    pests, boring 255
    pests, chewing 251-253
    pests, sucking 253-255
    pupating 255
    resistance to controls 246
in situ conservation *293*
integrated pest management (IPM) 247
intercropping; *see* mixed planting
interplanting; *see* mixed planting
interview; *see also* assessment, project; survey
    formal 52-56
    informal 52; *see also* assessment, project; participant observation
    tabulating responses of *56*
investment and returns as evaluation criteria 48
ion (soil nutrients) *161*
    toxicity in plants 208
iron (Fe) in the diet 25
    absorption and phytates 25, 26
    disease increasing the need for 18
    effect of malting on 317
    garden sources for weaning foods 332
    garden yield of 28
    in the soil 156, 160, 168
ironstone (laterite) 158, *160*
irrigation *227*-236
    and human disease 208-209
    and root growth 68-70
    and type of garden bed 183
    efficiency *227*-228, 235
    frequency 201-202
    furrow *230*-232
    measuring depth of 197, 202
    pitcher *233*-234
    problems 234-236
    rate of application 229-231
    requirements 196-200
    root zone *233*
    scoop *237*
    spreading nematodes *256*
    sprinkler 93, *233*-234, 273
    trickle *231*

junk food 36

kwashiorkor 15, *19*; *see also* malnutrition

labor; *see* work
lactation; *see* breast-feeding
land-use classification *154*-155
larvae *251*; *see also* insects, as food
laterite; *see* ironstone
latitude
    and plant daylength requirements 75
    and vitamin D requirements 22
layering 68, *114*-120
    air (marcottage) *117*-118, 120
    pruning for 142-143
    simple 117, 118, 119
leach (salts) *208*, 236
leaching requirement (LR) *208*, 236
lead, reducing amount in garden crops 170, 310
leaf
    color 166, 200; *see also* chlorosis
    cuticle *74*
    orientation and drought avoidance 72, 74
    problems, diagnosing 208, 253, 273, 276-277
    signs of ion toxicity 208
    surface, characteristics effecting transpiration 74
    temperature 71, 74, 166, 200
leaves
    anti-nutrients in 26, 332
    as weaning foods 330
    harvesting 309, 321, 322
    processing and nutrient content 310, 311, 312, 314
    value in the diet 20, 21-22, 23, 25, 26, 28, 330, 331-332
leeward *183*
legumes, and nitrogen fixation 161, 164-165
leptospirosis *209*; *see also* disease in people, and water quality
life cycle (of insects); *see also* pupate
    fruitflies 255
    grasshoppers 251, 252
    moths 251, 252
    of pests and pathogens, and their control 244, 263
life cycle (of plants) *79*-80
    and coping with drought 72, 74
    and effect of drought on yield 195-196
    and escaping pests and pathogens 242
    use in controlling weeds 141
life span (of plants) *79*-80

loam; *see* soil, loam
log scale (pH) *163*
long-term trends, and assessment 60

maggot *255*, 313
malaria 19, 56, *209*; *see also* disease in people
malnutrition *11*; *see also* nutritional deficiencies (in people)
    and disease 11-12, 18-19, 21
    kwashiorkor 15, *19*
    marasmus 15, *19*
    pellagra 15, *25*
    protein-calorie malnutrition (PCM) *19*-20, 28
    protein-energy malnutrition (PEM) *19*-20, 28
    risk of during weaning 327, 329
    scurvy 15, *23*
    xerophthalmia 15, *21*
manure
    animal 161, 163, 169-171
    ashes for seed storage 303
    as a pest repellent (feces) 259
    green *164*, 168
    human 161, 171
maps, as an assessment tool 60
marasmus 15, *19*; *see also* malnutrition
marcottage; *see* layering, air
market surveys 58
marketing
    and cooperation 42-43
    and distribution of resources 34-35
    and local control 39-40
    and timing 42
    and transport of produce 322
    and women; *see* women, and marketing
    control of income from 43
    harvesting for 322
    investments in 40-42
    protecting produce for 322-324
measuring
    canal flow 200-201
    CGAR 217-218
    leaching requirement (LR) 236
    rainfall 210-212
    runoff percent (R%) 214-216
    slope 217-219
    water requirements 197-202
    watering depth 197, 202
microcatchments *214*-217, 220
microenvironments
    and mixed planting 138
    and trellises 150
    in nursery beds 123-124
microorganisms
    and availability of phosphorus in soil 166-167
    and nitrogen fixation 164
    controlling in weaning foods 334-334
    controlling soil-borne plant pathogens 265
    in food 19, 311, 312, 319
    in soil 164, 168, 169
    role in composting 171-172
    soil pH and availability of nutrients to 161-163
middleperson *42*
minerals
    for human nutrition 25-26, 331-332
    in soil 155, 159, 160, 161, 166-168
mixed planting (intercropping, interplanting) 89-90, 137-138
    and risk spreading 286, 292-293
    and water conservation 189, 204-206
monitoring, project *48*
monocots *68*
monoecious flowers; *see* flowers, monoecious
mulch *202*-204
    encouraging pests 243
    surface 170, *203*
    vertical *203*-204
mulching 71
    and water conservation 189, 193
    seeds and seedlings 94
    to control soil temperature 158, 203
    to control weeds 141
mycorrhizae 70, *161*, 166, 262

neem (*Azadirachta indica*)
    for controlling nematodes 257
    in homemade pesticides 247
    to protect stored seeds 303
nematodes 139, *209*, 255-259; *see also* neem, for controlling nematodes
    and human disease 209
    in nursery beds 124
net protein utilization (NPU) *20*
niacin 24-25, 330
    effect of food processing on 310-311, 312, 315, 317
nitrogen (N) 163-165
    deficiency in plants due to waterlogging 234
    fixation 70, 164
    in commercial fertilizers 166-167
    lost through burning 163
nodules (root)
    and nitrogen fixation *164*
    distinguishing from nematode knots 255

notching *148*-149
nursery beds *123*-124, 251
nutrients *11*
    availability to plants 161-163
    cycles in the garden 161-162
    deficiency in plants 164, 168, 273, 276, 277
    density in weaning foods *329*-333
    garden sources of 14, 19-26
    garden yields of 19, 20
    needed by plants 159-168
nutrition (human) *11*
    education 29, 36
nutritional deficiencies (in people) 14-15, 28; *see also* malnutrition

oedema (in people) 20; *see also* kwashiorkor
offsets *105*-107
oral rehydration solution (ORS) 335
organic matter *168*-172
    advantages of 156, 166, 170
    and control of nematodes 257
    and water infiltration rate 192, 193
osmosis *67*, 68, 71, 75
output/input ratio; *see* evaluation of projects
ovary *82*, 85
ovule *80*, 85, 288
Oxisols 154-155, 214; *see also* soil orders

parasite 164, *241*, 262
parenchyma cells *109*
participant observation *50*
participation; *see* community participation; representativeness (of samples)
patent *294*
pathogens (plant disease) *241*, 246
    cleaning tools and controlling 101, 148, 256, 264
    control of 263-265
    soil-borne 244
pellagra 15, *25*; *see also* malnutrition
perfect flowers; *see* flowers, perfect
permanent wilting point (PWP) *192*
    and water movement in plants 67
permeability (of aquifer) *221*, 223; for soil *see* soil, permeability
perennials *79*-80
    and returns to labor 36-37
    transplanting 133
pests *241*; *see also* biological control (of pests and pathogens); insects
    and mixed planting 138
    and stored food 322
    and weeds 140
    barriers/fences for 258-261, 312, 313
    cleaning tools and controlling 101, 148, 256, 264
    controlling in stored seeds 302-305
    controlling with fallow 244
    controls in plants 242
    controls in the environment 243
    frightening away 261
    mechanical control of 243-244
    problems, diagnosing 265-281
    repellents for 247, 259
    resistance to controls 246
    traps 254-255, 264
pesticides *241*
    acaricides *249*
    and beneficial insects 247-248
    fungicides *248*
    herbicides *248*
    nematicides *249*
    poisoning and emergency care 248-250
    safe homemade 247
    synthetic 246-250, 302
pH
    and food processing 310, 311, 319
    of compost 171, 174
    of soil 155, *161*-163, 169, 170
phloem 68, 101, 108, 115, 148
phosphorus (in the soil) 70, 159, 166-167, 170
photoperiod *75*-76
photosynthesis 67, 68, *70*-71
    effect of salinity on 75
plant available soil water (AW) *192*, 198-199
plant pathologists *261*
planting
    density 89-90, 98, 124, 181
    depressions 87-88, 128
    depth 91
    direct 129
    in containers 125-129
    sites 129-133
    timing and controlling pollination 83
    timing, drought, and yields 196
plants; *see also* life cycle (of plants)
    nutrients needed by 159-168
    soil pH and availability of nutrients to 163
plant spacing; *see* planting, density
plinthite *160*
pollen grains *80*, 82, 83, 85
pollination *82*
    controlling 83, 290
    cross- *82*, 83, 85, 290

hand- 80, 82, 83, 84, 85
insect- *82*, 83, 250-251
self- *82*, 83, 290
wind- *82*, 83
population 54
porosity (of an aquifer) *221*; for soil *see* soil, porosity
potassium (in the soil) 159, 163, 167
preformed vitamin A *22*; *see also* vitamin A
pregnancy, special nutritional needs of 13, 17, 19, 22, 23, 25
probability *54*, 210-211
production; *see also* gardens, investments in; gardens, yield
   and industrial crop breeding 287
   and stability and diversity 292
   of nutrients and garden diversity 292-293
   possibility curve 33
propagation, vegetative 99-120; *see also* cuttings; grafting; layering
   disinfecting tubers for 256
proportional stratified sample; see sample, stratified, proportional
protein
   complementarity *20*, 29
   consumption and calcium utilization 26
   garden sources of 14, 20
   garden yield of 20, 28
   implications of increasing consumption of 20-21
   synthesis 20
protein-calorie malnutrition (PCM) *19*-20, 28
protein-energy malnutrition (PEM) *19*-20, 28
provitamin A *22*; *see also* vitamin A
pruning *142*-149
   by insects 242
   to control disease 262, 264
   to encourage shoot production 80, 105, 115-117
   transplants 135
pulses 20, *23*, 25, 26, 27
   saving seeds of 295, 300
pumps; *see also* water-lifting
   Archimedean screw *238*
   chain and washer *238*
   displacement *238*
   mechanical 239
   power for 238-239
pupate *255*; *see also* life cycle (of insects); insects

rain 210-212
   calculating probability 210-211
   infiltration rate 175, 182
   intensity 175
   rainwater harvesting 212-219
raised garden beds 183-185
   and waterlogging 235
   for shallow topsoils 158
rancid (oils) 321
recommended dietary allowances (RDAs) *12*-13
   garden yields of 28
repellents
   for garden pests 247, 259
   for seed storage pests 303-306
representativeness (of samples) *49*, 54, 97
reproductive growth *79*
   and drought 74, 159
respiration *70*
   effect of salinity on 75
   in harvested foods 312
   in stored seeds 301
resources
   control of 34-35
   mapping for assessment 60
retinol *21*, 22; *see also* vitamin A
Rhizobium bacteria 70
   nitrogen fixation by 164
riboflavin (B_2) 23, 330
   effect of food processing on 310, 312, 315, 317
rickets 15, 22
risk *34*, 229
   and crop varieties 34, 196, 286
   and diversity in agriculture 286, 287, 292
   and marketing 40, 42
   and probability of rainfall 210, 211
   spreading and dormancy in wild plants 86
root-bound 131, 133-134, 271
roots 68-70
   adventitious *99*, 114, 135
   and irrigation or watering patterns 68-70
   and nutrient uptake 161; *see also* nitrogen, fixation
   and soil texture and structure 70
   and transplanting 131, 134
   and waterlogging 70
   compensatory growth of *68*
   depth and irrigation 229
   depth and water requirements 198-199
   fibrous *68*
   growth and planting density 89-90
   hairs *68*
   lateral *68*
   nodules on; *see* nodules (root)
   pruning 143
   secondary *68*
   suckers *108*

tap *68*
tertiary 68
testing the health of 265-267
root zone irrigation *233*
   with vertical mulch 203-204
runoff 175
   and erosion 217
   calculating for water harvesting 214-217
   percent (R%) *214*-217
   ratio *214*
   storing 223-224

Saline *208*
salinity
   and irrigation 228, 235-236
   and soil structure 156
   effect on transpiration, photosynthesis, and
      respiration 75
   of water, measuring 208
   symptoms in plants 75, 208
salt buildup 208, 235-236
   in pitcher irrigation 233
   on containers 127
   on raised beds 184-185
salt-tolerant (plants) *75*, 171, 208, 236
sample
   for an assessment interview *54*
   of root for problem diagnosis 267
   random *54*
   selection for seed germination tests 97
   stratified, disproportional *55*
   stratified, proportional *55*
sand *155*, 156
   added to soil covering seeds 91
   for protecting stored seeds 303, 304
   water-holding capacity of 190
sap *68*, 161
savings; *see* gardens, economic contribution to
      household
schistosomiasis (bilharzia) 18, *209*; *see also* disease in
      people, and water quality
scion *108*-116
scurvy 15, *23*; *see also* malnutrition
seasonality
   and marketing 41, 42
   and patterns of child growth 16, 17
   in assessment 56-58, 59
   of food 17, 19, 58
   of work 17-18, 19
seed
   after-ripening of 296

   banks 293-294
   -borne disease 263-264
   case hardening of 299
   control of and compensation for 293-295
   dehiscing *295*
   diversity in 287-292
   dormancy 76, *86*
   drying 299-300
   embryo 86, 96, 288, 295
   germination *86*
   hybrid 287, *290*-292
   mulching and shading 94
   planting depth 91
   planting site 87-89
   preparation for planting 87
   viability *85*, 97
seed coat
   preparing for planting 86-87, 295
   removing from food 311
seedlings
   and water stress 93
   failure to emerge 95-96
   mulching and shading 94
   problems, diagnosing 268-269
   watering 93-94
seed saving
   cucurbits 296
   maize 298
   peppers 297
   pod-bearing crops 295
   seed-bearing flower heads 297-298
   soft small-seeded fruit 297
   trees 300-301
seed storage
   containers 301, 302
   controlling moisture and temperature 301-302
   controlling pests 302-305
   problems and responses 301
self-pollination; *see* pollination, self-
setts *104*
Seventh Approximation (soil classification) 154-155
sexual reproduction (in plants) *79*-86; see also
      fertilization (of flowers); pollination
shading 71, 204-206
   and water conservation 189
   seeds and seedlings 94
shoot system *68*
side dressing (fertilizer) *167*
significance (in statistics) *54*
silks 83
silt *155*, 156

simple layering; *see* layering, simple
slope 175-176, *217*
    and irrigation 229-231
    and rainwater catchments 217-219
    and sprinkler irrigation 234
    effect on erosion 176
    effect on soil pH 163
    measuring 217-219
smoke, to protect stored seeds 304; *see also* seed storage
social sustainability 3-4, 64
soil
    acid, acidic *163*; *see also* pH, of compost; pH, of soil
    aggregates *156*, 172, 181
    basic, alkaline *163*; *see also* pH, of compost; pH, of soil
    -borne pathogens 244, 265
    characteristics for catchments 214-215
    classification systems *154*-155
    clayey 155, 182, 190, 192, 214
    cobbles *155*
    color 156
    cubes 123, 125-126
    depth for nursery beds 124
    erosion; *see* erosion
    fertilizers *164*, 166-167, 168-172
    formation in drylands 154-155
    horizon *158*; *see also* soil profile
    improvement 89, 19, 139; *see also* organic matter
    loam *155*-156, 182, 214
    microorganisms 70
    nutrients 159-168
    permeability *156*, 181
    pH; *see* pH, of soil
    pores *156*, 190, 192
    porosity *156*, 181
    ring test 157
    sandy 155, 182, 190, 192, 214
    saturated *190*
    settling out test 157
    sterilization *128*
    suppressive *265*
    tests 157, 161
    type and planting depth 91
    water-holding capacity of; *see* water-holding capacity
    water potential *192*
soil orders
    Alfisols 154-155
    Aridisols 154-155, 159
    Entisols 154-155
    Oxisols 154-155, 214
        Vertisols 154-155, 204, 214
soil profile *158*
    horizons *158*, 160
    impermeable layers in 131, 160, 234-235
soil structure *156*
    and erosion 172
    and evaporation 193
    and humus 169
    and infiltration of irrigation water 229, 231
    and root growth 70
soil temperature 76, 126, 158
    and disease 263, 269
    controlled by mulch 158, 203
soil texture *155*-156
    and indicators of water deficit 200
    and infiltration of irrigation water 229, 231
    and root growth 70
    different layers and transplant growth 130-131
    different layers and water movement 193, 194
    tests for 157
springs *221*
sprinklers, for seeds and seedlings 93-94; *see also* irrigation, sprinkler
spurs; *see* trees, spurs
stigma *82*
stock *108*-116
stoma, stomata *71*, 72, 74, 166, 167
strongloidiasis *209*; *see also* disease in people, and water quality
subsoil *158*, 172
sulfur (in the soil) 163, 168, 170
surface mulch; *see* mulch, surface
survey (assessment) *52*; *see also* assessment, project; interview
sustainability 3-4, 63-64
symbiosis *70*, 161, 164
systemic pesticides/poisons 248, 249; for diseases *see* disease in plants, systemic

tassels; *see* flowers, tassels
temperature
    and composting 174
    and plant growth 76, 243
    for cold stratification 300
    inhibiting flower fertilization 83, 85
    of food processing and nutrient content 310, 312
    of leaves 71, 74, 166
    of soil; *see* soil, temperature
    regulation in plants 71

requirements for germination 86
terminal
    bud *113*
    fruit-bearing (in trees) 146-147
terraces 180-181
testing
    depth of water infiltration 197, 202
    root health 265-267
    seed germination 96-98
    soil for water deficit 200
    soil texture 157
    thiamine (B_1) 23
    effect of food processing on 310, 311, 312, 317
thinning 98
Third World *2*
    gardens in 1-2, 8-9; *see also* gardens, indigenous
threshold rainfall *214*
timing
    and controlling pollination 83
    and controlling spread of disease 263
    and crop health 243
    and pest control 251, 255
    and weed management 140-141
    as a crop selection criterion 292
    of drought and yield 196
    of transplanting 129
topsoil *158*
total dissolved solids (in water) *208*
trace elements *25*
tragedy of the commons 35
translation, and assessment 52, 53-54
transpiration 67, *71*-74; *see also* evapotranspiration
    effect of salinity on 75
    excess 189, 193
    limited by plants' physical characteristics 74
    of harvested foods 321, 322
    pruning to reduce 135, 143
    reduced by drought deciduousness 80
transplanting 98, 129-136
    bare-root *129*
traps (for pests) 247, 254, 255, 264
trees
    branch angles 144-146
    central leader form 143, 144, 145
    how fruit is borne 144, 146-147
    nursery beds for 124
    open center form 143, 144, 145
    planting sites for 130-132
    pruning 143-149
    saving seeds from 300-301
    spurs *146*, 309
    vegetative propagation of 99-102, 105-120
    trellising 149-151, 264
    trench bed gardening (trench composting) *172*
    tube level
    to measure contours 178
    to measure slope 217, 219
tuberous roots *104*
tubers *104*
    disinfecting for propagation 256

urban gardens; *see* gardens, urban
urine
    as a source of nitrogen 171
    as an ingredient in homemade pesticides 247
    spreading disease 209

values, and their effect on policy 9, 294
varieties; *see* crop varieties
vascular system *67*-68, 161
vegetative growth *79*
    and drought 74, 195
vegetative propagation; *see* propagation, vegetative
vertical mulch; *see* mulch, vertical
Vertisols 155, 204, 214; *see also* soil orders
viability; *see* seed, viability
vigor
    and pest and pathogen resistance 243
    of seeds 96
virus *262*; *see also* disease in plants, viral
volume head product *236*; *see also* water-lifting
vitamin(s) 13, 14, 15, *21*-25, 29
    content of fruits 309
    effect of food processing on 310-318
    fat-soluble *21*
    for weaning foods 331-332
    oxidation of *311*, 314
    production and garden diversity 292-293
    water-soluble *21*, 310
vitamin A 21-22, 28, 29, 330
    effect of food processing and storage on 310, 311, 312, 316, 317, 321
    garden sources for weaning foods 331-332
vitamin C (ascorbic acid) 23
    and iron deficiency anemia 25
    effect of food processing and storage on 309, 310, 311, 312, 315, 316, 317, 321
    garden sources for weaning foods 331
vitamin D 22-23
wasps, for biological control of pests 245, 248, 253
water
    application efficiency (Ea) 198-199, 217-218, 228-

229
  conservation with mulches 202-204
  conveyance of *227*-228
  deficit and soil texture 200
  extracted *227*-228
  free (gravitational) *190*
  loss in the garden 189, 193
  management, goals of 189
  movement in plants 67-71
  movement in the soil 192-194
  quality and human disease 208-209
  requirements of the garden 196-202
  saline *208*
  salinity, measuring 208
  sources for gardens 207ff
  spreading *220*
  sprouts *101*
  storage 190-192, 223-224
water-holding capacity 154, 156, 182, 190-192
  and irrigation calculations 198-199, 202
  and organic matter 170
watering; *see also* irrigation
  containers 127
  depth 197
  seeds and seedlings 93-94
  transplants 132-133
water-lifting 236-239
  Archimedean screw *238*
  chain and washer pump *238*
  displacement pump *238*
  efficiency of 236-237
  Persian wheel *238*
  power for 236-239
waterlogging *192*, 234-235
  and damping-off 269
  and irrigation 228

  and root growth 70
  and soil horizons 158, 160
water stress (water deficit) *71*, 74
  and root and shoot growth 93
  and seedlings 93
  and yield 74, 195-196
water table *221*
  irrigation 233; *see also* flood recession gardening
  raised by irrigation 228
watery compost extract (WCE) 263
weaning 14, 17, *327*, 334-336
weaning foods 329-333
  commercial 327, 329, 335-336
  multi-mix *328*
  storing 335
weeds *139*-141
  control with mixed planting 138
  hosting pathogens 263
  hosting pests 244
weirs *200*
well-being, effect of income or savings on 29, 35-36, 39
wells *221*-223
  artesian *221*
  cistern *222*
  hand-dug 222
  large-diameter *222*
  small-diameter *223*
Western science 3, 9, 67, 78
wilting
  and water requirements 200
  as a result of disease 262
  diagnosing 273, 274-275
  in hardening off 133
  in harvested foods 321, 322, 324
  in transplants 134

wilting point; *see* permanent wilting point (PWP)
wind
    erosion by 182
    protection in sunken garden beds 183
    pruning to reduce damage by 142, 143
windbreaks *182*, 183, 189, 204-206
wind pollination; *see* pollination, wind-
windward *183*
women
    and control of gardens 29, 40
    and household food distribution 28-29
    and marketing 40, 41
    impact of environmental degradation on 60
    income and child nutrition 28-29, 39
special nutritional needs 17-18, 19, 23, 25, 26
    woody *80*
    work (labor) 32
    effect on people's nutritional needs 17-18
    seasonality of 17-18, 19

xerophthalmia 15, *21*; *see also* malnutrition
xylem *68*, 108, 115

yield *195*
    and agricultural intensity 36
    and diversity 286
    effects of drought and water deficit on 74, 195-196
    of gardens; *see* gardens, yields
    stability *286*

zinc
    and phytates in food 26
    in human nutrition 25
    plant use of in the soil 70